Technology Application in Tourism Fairs, Festivals and Events in Asia

Azizul Hassan
Editor

Technology Application in Tourism Fairs, Festivals and Events in Asia

Springer

Editor
Azizul Hassan
Tourism Consultants Network
The Tourism Society
London, UK

ISBN 978-981-16-8069-4 ISBN 978-981-16-8070-0 (eBook)
https://doi.org/10.1007/978-981-16-8070-0

This Springer imprint is published by the registered company Springer Nature Singapore Pte Ltd.
The registered company address is: 152 Beach Road, #21-01/04 Gateway East, Singapore 189721, Singapore

Introduction

This is an unconditional reality that the tourism industry in Asia is getting exposed to innovative technologies than ever before. This proposed book provides the latest research in the application of innovative technology to the tourism industry, covering the perspectives, innovativeness, theories, issues, complexities, opportunities, and challenges of tourism fairs, festivals, and events in Asia. This book, a blend of comprehensive and extensive effort by the contributors and editors, is designed to cover technology application in tourism fairs, festivals, and events in Asia. The application and practice of technologies in tourism, including the relevant niches of fairs, festivals, and events, are also covered. This book focuses on the importance of technologies in tourism. This also highlights technology in a comprehensive manner that is impacting the tourism industry in Asia as well as the constraints it is facing. Contents of this book deal with distinct topics such as tourism promotion, technology-driven sustainable tourism development, social media, accessibility, and so on to cover fairs, festivals, and events.

Several research studies are conducted outlining the contributions and importance of technologies in tourism, in general. However, researching the tourism industry of Asia so far has attracted very few researchers. Some contributions are made, but not sufficiently. Considering the ongoing trend of technology application in the tourism industry in Asia, very few research attempts have been made aiming towards exploring diverse aspects. This book will be a significant contribution towards the very limited knowledge in this identified research area with the example of selected Asian countries. The book is designed to accommodate both qualitative and quantitative research liking theory and practices. The book has clear focus on outlining the research issues. Each chapter of the book highlights the methodology that was used with rationale for the use. The book has addressed several revisions to integrate and unify the themes or frameworks with the chapters. The possible market of this book will be Asia. Tourism is expanding enormously across the world, creating more demands for effective technologies. This book will be a reading companion mainly for tourism students in higher academic institutions. It will also be read by relevant policy planners and industry professionals. Apart from them, it will be appreciated by researchers having keen interest in the Asian tourism industry.

Almost every aspect of tourism events, fairs, and festivals is influenced by technology. Most tourists, on the other hand, are intelligent enough to take advantage of technology's advantages, such as convenience, precision, and speed, while avoiding some of its drawbacks, such as user-friendliness, letter size, and lack of personal touch. Technology can save us time and money. The tourism industry emphasizes the importance of the human element and recognizes the role of tourism events, fairs, and festivals in the development of experiences. This enables visitors to take advantage of a variety of benefits and savings while booking hotels and other products and services online in order to attend and participate in events, fairs, and festivals. This book has 23 chapters in 8 parts (i.e., theoretical discussions, influential factors, perceptions and effects, social media, destination branding and national image, cases, challenges and solutions, and directions for future research), taking into account the fascinating qualities and importance of technology use in Asian tourism events, fairs, and festivals. The following are brief descriptions of each chapter.

In Chap. 1, Maidullah, Hossain, Akhter, and Hassan write on transmogrifying tourism events in Industry 4.0. Following the authors, festival tourism is currently a well-known type of tourist product all around the world. Many nations have various types of festivals in order to attract tourists from other countries, enabling cultural interaction between or among the participating countries. Although festivals are designed to provide tangible experiences, the revolutionary influence of technology on the tourist sector has forced the organizer to reconsider the planning process. The impact of technology on festival tourism is discussed in this chapter. Similarly, big data concepts may be used for Internet of Things (IoT) sensors, which aid festival management by allowing for easy tracking of data via real-time big data analytics. The outcomes of this study will add to the literature on festival tourism by addressing how tourism businesses are altering their business models and how festival visitors are interacting with new technology. The chapter asserts that industry 4.0 hyper technology enables event festival organizers to plan high-quality events and services while also adding value to the consumer experience in a variety of ways, including genuine information, ease of event booking, updated knowledge, and other aspects.

Sebata and Mollah in Chap. 2 make theoretical discussions on technology application in tourism festivals in Asia. Tourism festivals in Asia nowadays have adopted technology to improve the quality of events, the experience of attendees, and the marketing potential. Technology is enhancing tourist festival demand, inventiveness, and attendance as a result of this development. This increases market value by creating a modernized, industrialized, and commercialized industry that can compete effectively in the global tourism market. Although it is expected that the use of advanced technology in festivals will be accompanied by a number of technical challenges, as well as financial and human resource constraints, it is important to note that if appropriate and acceptable models are identified, as well as factors that influence technological application, a long-lasting impression on the minds of festival attendees can be created. This research was carried out to provide a theoretical overview of the use of technology in tourist festivals as well as to inform the creation of strategy in the Asian tourism festival business. Effective integration of modern digital technology in Asian tourism festivals is expected to offer a one-of-a-kind leisure experience, particularly in the form of "smart tourism festivals." Besides sustainability, these have a significant impact on important festival experience outcomes such as satisfaction and memorability. The COVID-19 pandemic has accelerated current technological application procedures in the tourism industry, particularly in key Asian economies such as China, South Korea, Japan, Malaysia, India, Singapore, Bangladesh, Hong Kong, and Thailand. This study draws on previous research to present a theoretical overview of tourism festivals in Asia, including themes of digital technology's impact on festivals, the creation of smart tourism festivals, and the underlying difficulties and prospects.

In Chap. 3, Kuri, Ananya, Islam, and Hassan theorize tourism philanthropy and the role in tourism fair, festival, and event stakeholders for global prosperity. Tourism fairs, festivals, and events can, of course, be a problem for nearby towns, causing traffic congestion and parking concerns, as well as possible increased noise—for every positive influence, there is sure to be something that damages local life! However, event organizers, particularly the larger ones, are aware of this and are striving to address it by donating to local charities and performing volunteer work. Tourism philanthropy is a compassionate act of assisting people and safeguarding the environment. Tourism festivals, fairs, and events stakeholders are essential in establishing a long-term link between society, culture, economy, and the natural environment. They play a critical role in establishing a long-term link between society, culture, economics, and the natural environment, which makes residents happy. This chapter uses the concept of responsible tourism to "do good" by philanthropically benefiting people and places. This exploratory study is based on interviews with chosen respondents to develop strategies for co-creating a tourism philanthropy program using their knowledge and experiences, according to the focus group research approach. There are several unique motivational methods that motivate people to participate in philanthropic programs as part of a positive global citizenship program, which may be initiated at any point in our lives through moral and environmental education. This research employs a multi-stakeholder interaction strategy. It explains how tourism fairs, festivals, and events may contribute to global prosperity by ensuring community and environmental well-being through contribution, commitment, and awareness.

Rahman, Akter, Hossain, and Hassan, in Chap. 4, outline the application of technology for pilgrimage and Halal tourism event. Religious tourism, pilgrimage/Hajj tourism, and Halal tourism are all growing in popularity in Asia and the Middle East. Hajj and Halal tourism events are a burgeoning industry that is gaining traction. Tourists visit various locations for a number of reasons. From the dawn of humanity, traveling for Halal or spiritual reasons appears to have been frequent. Halal tourism starts with a religious event, in which the Hajj is viewed as a pilgrim traveling for religious reasons. Religious, social, traditional, and landscape patterns all intertwine in the anticipation and selection of trip destinations, and travelers evaluate more than just Halal requirements. Despite the fact that some studies have focused on spiritual terminology and classification or tourist effect study, the majority of the studies are in the areas of Halal, spirituality, and pilgrimages. This chapter employs a research methodology to synthesize current literature and concepts in the realm of technological applications in pilgrimages and Halal tourist events. It underlines the viability of tourism as a viable idea with enormous potential in the realm of pilgrimage and Halal tourism. This chapter makes a number of suggestions for growing pilgrimage, religious value, and Halal tourist activities in Asian and Middle Eastern countries.

In Chap. 5, Johara, Jilani, and Uddin explain how social media literacy, psychological capital, and work engagement influence employees' morale in event and tourism industry. The latest outbreak of COVID-19 has wreaked havoc on the event and tourism industries. As a result, high turnover, disengagement, and low loyalty have emerged as issues affecting the mental health of personnel in this industry. To break free from the grip of these burgeoning challenges, a thorough examination of employees' social media literacy, mental outlook, optimism, and behavior is required, as these factors have a significant impact on employee performance, which, in turn, affects the achievement of the event and tourism industry's triple bottom line. This chapter will examine the event and tourism industry's employee morale by revealing the predictive influence of employees' social media literacy, psychological capital, and work engagement. The current investigation will unravel the moderating effects of age and tenure of serving employees, in addition to the mediated mechanism employing job engagement. Researchers prefer to use quantitative tools to implement a deductive reasoning strategy. With a quantitative framework of inquiry, partial least square-based structural equation modeling will be used. This study adds to our understanding while also bringing together new ideas. The current study's key contribution will be to provide highly important insights into how to improve employees' psychological health in order to engage them in more productive work and boost their morale. Furthermore, the goal of this chapter is to gain a better knowledge of practitioners' perspectives on social media literacy and psychological capital, both of which can affect employee engagement and morale at work. The importance of staff morale and engagement will be presented, followed by various practical solutions.

Chapter 6 by Salimullah and Nabi discusses an experience of cultural heritage–based tourism for the protection of sacred tourism sites, festivals, and events in Bangladesh over augmented reality. Bangladesh has several problems in maintaining and progressing its historical locations, resources, and heritages, as well as making them the center of attention in the actual world. One of the most important ways to safeguard the country's precious tourism destinations is to provide proper advice and assistance. The country has considered various possible solutions such as elevations of sites, structures with digitalized systems, wind and water-resistant materials, fire-safe design, and emergency back-ups in order to address the existing challenges to protect historical resources and buildings. The country's historic cities also improve public infrastructure by raising roads and bridges, reducing energy consumption, installing permeable pavements and green alleys, reconfiguring and narrowing sidewalks, modifying curbs and gutters, strengthening sewage systems, utilizing buffers and setbacks from seashores, and demolishing unstable structures.

The state must eventually implement steps to safeguard any monument, artifact, or location of great aesthetic and historical value or interest from distortion, destruction, or removal, according to Bangladesh's constitution (Article XXIV). We also have the 1968 Antiquities Act, which outlines the methods for protecting and preserving items related to our country's history and legacy. Cultural heritage (CH) preservation in situ might be challenging at times owing to its nature (e.g., beach nourishment and weak communication). Virtualization through a computer-stimulated environment of cultural property and digitalization through audio-visual components, sound, or other sensory stimuli provided via technology are frequently regarded the finest solutions for displaying and monitoring cultural resources. By utilizing CH assets in natural application contexts, digitalization attempts to increase the added value of CH assets (protection of CH, education, tourism industry, advertising, fashion, films, music, publishing, video games, and TV). As a result, augmented reality–enabled services will aid in data gathering, preservation, and the development of novel ways for visual effects representations of tangible or intangible assets and locations, as well as compatible metadata formats that facilitate data sharing and archiving.

In Chap. 7, Rahman, Hossain, Akter, and Hassan show that tourism events, technology innovation, and social media are global platforms. Web 2.0 and social media have had a huge impact on humans as a result of technological advancements and internet usage, and they have rapidly dominated over the previous few decades. Facebook, Instagram, LinkedIn, Twitter, WhatsApp, WeChat, blogs, and YouTube have all become vital tools and global platforms for promoting tourism events and festivals toward tourism destinations. A large number of active Facebook users write updates and share photos on a regular basis, which helps the tourist and commercial sectors build relationships with their customers. Customers use social media sites to research tours, make travel decisions, and share their knowledge of certain events and festivals. The tourism business is impacted by event organizers in particular; however, there is still a gap between event and festival groups since they are still inept to exploit their full potential. This study examines the issues and challenges as well as the multi-phase experience of using social media to promote tourism events and festivals, in order to understand the nature, extent of social media use in the festival, and how social media promote consumers in terms of providing better information on tourism events.

In Chap. 8, Rahman, Rana, Akter, and Hassan elaborate technological innovation as a destination branding tool for festivals and events tourism. Destination branding is a type of marketing that is used to promote a destination's image. Destination branding and its techniques based on people's or tourists' engagement, creativity, and invention, with branding strategies depending on progressive information and communication technologies, are used to focus a country's image. Because tourism destinations have been developed and progressed in events and festival portfolios as planned creativity to attract tourists and support their brand by using new technology's elements, the application of technology innovation becomes most significant in the tourism and hospitality industry. Tourists visiting an event or festival celebration will be able to learn about specified mobile applications, event management software, artificial intelligence programming, mixed reality station, and data gathering tools if they read this book chapter. As a result, events and festivals are becoming a more important part of destination branding for technological innovation since they encourage the use of tools like apps and internet direction, which boosts tourist growth. Despite this, the obstacles and gaps in the application of new tourism and hospitality technology are discovered in this study. This chapter examines the many modern technologies that are required while presenting an event, such as an artificial intelligence chat box and tailored suggestions to pique tourists' interest.

Chapter 9 by Alam, Milon, Rahman, and Hassan discusses the application of technology in tourism events, education, and training to enhance a nation's image. The use of technology in the tourism sector has a major impact on tourists' perceptions of event education, training, and tourism events, which may aid in the shaping of a country's image. The use of digital technologies that can be utilized to enhance general employability skills, problem-solving services, work attitudes, and occupation-specific abilities when an individual works in their companies can affect the condition of a nation's image. Asia, as the world's biggest continent, accommodates a wide range of tourism events. Many research studies are being carried out on visitors' perspectives on the use of technology in tourism events, education, and training programs. However, there is still need for improvement in this field of study. The review chapter focuses on the information obtained by visitors as a result of the use of technology in event education, training, and tourism events to improve a country's image. Tourism event education and training may assist the nation in forming a realistic economic shape by utilizing sophisticated technology. This book chapter highlights a gap in technology application in employing educational and training events to reconstruct the national image as a glaring problem. It also looks into the importance of using technology in tourist events for teaching and training, which might assist to improve the country's image. The study offers a successful methodology formulation and implementation in tourism events, education, and training for visitors to gain a better understanding of how to use technology.

Chapter 10 by Kakon features a revolutionary transformation of technological adaptation in tourism events, fairs, and festivals in Bangladesh. An event, fair, or festival is a public celebration arranged by a community to commemorate a special feature. Tourist interest in these festivals has grown to the point where they have become an exclusive tourist attraction and destination image generator. According to many surveys, today's tourists have a mixed interest in learning about diverse events while simultaneously enjoying increased comfort as a result of enhanced tourism arrangements. And it is a well-founded truth that, more than ever before, the quest for adaptation of any influencing aspect in the tourism business is centered on the use of technology for its unavoidable favorable consequences. Nowadays, people attend a variety of tourism events, fairs, and conferences. Attending various tourism events, fairs, and festivals has become a popular method to get away from the tension and monotony of modern complex life. At the same time, today's festival goers value smart tourism technology experiences for improved consumption, more appealing experiences, and a higher likelihood of returning. A number of studies have indicated that incorporating better technological applications into the tourism setting has streamlined the entire process for both the organizer and the guest. The purpose of this chapter is to have a more in-depth discussion on Bangladesh's technology embrace in the event, fair, and festival tourism sector, as well as its overall impact on the sector's evolution.

Chapter 11 by Nabi and Salimullah is about that the promotion of catamaran tourism and the challenges of the blue economy in Bangladesh. Catamaran tourism has the potential to be one of Bangladesh's most profitable civic enterprises. As one of the world's major resource-based industries, it contributes significantly to global GDP and creates a considerable number of employments in Bangladesh. Coastal and marine tourism, as well as related resources such as forestry and fishing, might develop as a vibrant and increasing sector in Cox's Bazar, Kutubdia, Patuakhali, and Moheshkhali. These locations, which are blessed with magnificent seascapes and beaches, have important and diverse natural, cultural, indigenous, and heritage effects that inhabitants and visitors may easily access. Recreational fishing is becoming much popular, and it may be expanded as part of catamaran tourism in those regions. Again, in places endowed with coastal and marine tourism, Bangladesh can create standard and sophisticated cruise shipping destinations, which is the major emphasis of this chapter. Activities such as sea kayaking, whale watching, SCUBA diving, coastal sightseeing, beach walking, surfing, and touring, as well as attractions, parks, accommodations, festivals, and special events with a marine theme or tourism experience, may be included in the emerging catamaran tourism in these areas. Fundamental statistics gaps outlining the size of coastal and marine tourism are significant concerns. Because catamaran tourism may confront well-known environmental difficulties, sustainable tourism should be promoted and protected in order to promote and maintain "sustaining local economies, local cultures, and local ecosystems." Bangladesh, as a host community, may use its physical environment and social fabric to promote natural resource regeneration and future production, as well as acknowledge the importance of people and communities, cultures, and lifestyles in the tourist experience.

In Chap. 12, Roy B. R. critically explains that in India, information and communication technology (ICT) is used in tourism events, fairs, and festivals. Tourism has established itself as one of the world's fastest growing industries. Tourists must be supplied with amenities, both in terms of infrastructure and service quality. In recent decades, the tourist sector has extensively relied on ICT to improve service quality. Almost all major sectors of the tourism industry, such as accommodation; transportation—airlines, cruise ships, and railways; restaurants; travel intermediaries; and destination management companies, have used various ICT applications such as augmented reality (AR) and virtual reality (VR), chatbots and facial recognition (AI), Internet of Things (IoT), livestreaming, and blockchain. For enhanced performance and promotion, there has been a global trend in employing these technologies in meetings, events, conference fairs, and festivals, which make up a substantial portion of the MICE segment of tourism. This chapter examines various ICT technologies utilized in the field of events, fairs, and festivals throughout the world, as well as reviews how well these technologies have been employed in the Indian setting and examining future prospects.

In the thirteenth chapter, Gupta, Meena, and Jingar show that digital tourism, as a new era of tourism, caters to the needs of modern-day travelers and has become a focused trend. Virtual tourism, as a new era of tourism, caters to the demands of modern-day tourists and has become a concentrated subject. In recent years, virtual reality (VR) has become increasingly popular in tourism events, and efforts are being made to integrate the two to give travelers a real-time experience. The implementation of virtual reality technology was highlighted as a new avenue in the digital growth of tourist events. The goal of this study is to illuminate the possible benefits of introducing virtual reality into tourist events in order to boost their image. There are two primary goals: first, to assess the factors that impact how visitors from various cultural backgrounds choose tourism events, and second, to investigate the function of virtual interaction in the marketing of tourism events. In this work, the impact of virtual input on the visit is investigated at two stages: before the visit, during the decision-making process, and at the time of the visit, during the experience evaluation process. The antecedents of flow and flow behaviors in 3D virtual tourism events were closely connected to visitors' plans to conduct a real trip, according to the study. By engaging in a VR tourism event and observing increased flow, tourists might develop consideration and knowledge of their future tourism event preferences. A more thorough investigation of the simulated environment and its influence on visitor decision-making is required. Technology plays a critical role in the success of the tourism business. The chapter's main goal is to help readers comprehend the notion of virtual reality in tourism and the advantages of using virtual reality platforms in the industry. The negative effects of incorporating virtual reality into the tourist sector have been proven. This study adds to our knowledge of how to create informative and entertaining tourism events in 3D virtual environments to attract future online and real-world visitors.

Chapter 14 by Kalsoom B. and Alam aims in comprehending digital technologies in tourism events in Pakistan and outlines methods to improve them. The practice of creating gatherings and festivals dates back thousands of years, to the prehistoric age. It includes non-tourism procedures as well as theoretical foundations for tourism event management. The current condition of tourist events, management, and understanding of digital applications encourages the inclusion of subcultures highlighted through social media platforms, electronic, and associated digital submissions methods. The digital improvement and comprehension of tourism activities were aided by knowledge development and cluster instigation of the events. By adding the survey approach, this study underlines the function of the digital phases in the exercise of upgrading the continuing tourism events in Pakistan. The participants are contacted via an online channel, and their responses are gathered and analyzed as needed. Through multidisciplinary and post-disciplinary technological techniques, tourist events are found to cross beyond disciplinary boundaries. The diversity and number of tourism-related products, promotion, and feedback have increased thanks to Internet channels. To safeguard tourism destinations, the report proposes filling gaps in existing research and overcoming non-tourism-related occurrences in the digital and technological era. The study recommends meeting the gaps in existing research and overcoming non-tourism-related events in the digital and technology era, to secure tourism destinations for extended benefits of tourism events in Pakistan.

Chapter 15 by Nawarathna, Arachchi, and Bandara discusses the viewpoint of event managers on information technology and its challenges in the context of Sri Lankan event tourism. In a global setting, the event sector is quickly growing and contributing to business and leisure tourism. Because of increased regulation and government and corporate involvement in events, the environment has become significantly more complicated, and many event companies have implemented information technology. As a result, the study's main goal was to identify the obstacles to information technology adoption in Sri Lankan event tourism. Primary data was collected from event managers who are members of the Sri Lanka Association of Professional Conference, Exhibition, and Event Organizers using the purposive sample methodology. The acquired data were transcribed and processed using content analysis as part of a qualitative data analysis method. The outcomes of the study show that the Sri Lankan event business does not make effective use of information technology. The cost of adopting information technology in the Sri Lankan event industry and further maximizing the awareness of attendees and organizers, investing in the latest event technologies, increasing government involvement in policy framework, and creating a private-public partnership for long-term sustainability resulted in cost, resistance to change, tech-savvy employees, lack of government support, and security issues.

Chapter 16 by Hossain, Mohiuddin, Ahsan, Mahmud, Amzad, and Irshad explains events, festivals, and fairs in China as a technology-based tourism development mechanism. China is a tourist-friendly country with many ancient landmarks such as the Great Wall and terracotta warriors. With the aid of tourism events, festivals, and fairs, this chapter intends to investigate the role of technology-based tourism development mechanisms in China. The chapter focuses on the creation of tourism apps that may have an impact on tourism events, festivals, and fairs, as well as improve tourism and quality of life (TQOL). Traditional tourism development tactics are extensively studied by academicians; however, the significance of technology-based apps in connecting social networks with events, festivals, and fairs receives little attention. The study examines how China draws tourists in the modern day through the use of technology. In this chapter, the topic of China's perception is discussed. The chapter indicates that with successful technology application in tourism festivals, events, and fairs in China, tourists' opinions could be positive, ensuring tourism and quality of life (TQOL).

Chapter 17 by Pitanatri and Wiarthi analyzes approaching the paradox of loving and hating of Indonesian cultural technology applications. Culture has evolved into a medium for self-expression and imagination as a term having major implications for human life. At the same time, cultural tourism and cultural heritage have become key factors in luring travelers to Indonesia, raising concerns about cultural continuity. The rising fusion of tourism and culture raises a slew of crucial considerations concerning the destination's overall viability. As the tourist discussion continues, academics are growing increasingly worried about issues of cultural authenticity. The migration of tourists between sites is becoming increasingly unstoppable as technology and modes of mobility progress. Though it has been criticized as not being "authentically Indonesian,', technology-infused events can provide space and possibilities for culture to thrive. This chapter intends to elaborate on the dynamic evolution of culture tourism, which is now intimately tied to technology, by looking at two cultural tourist centers in Indonesia, namely Bali and Yogyakarta. This chapter will also look at how cultural events with a technological twist create a love-hate relationship in Indonesia. To polish the prose, a variety of cultural events in Bali and Yogyakarta will be offered. Technology and culture tourism can be mutually helpful and supportive.

Chapter 18 by Osman outlines the application of technology in tourism fairs, festivals, and events in Malaysia. Technology has a significant impact on the tourism industry, especially the management of large events and festivals. It has dominated the road to event success, particularly in terms of how it is altering guest expectations and how it may be used to create a win-win situation for both the organizer and the attendee. As technology advancements exceed the purpose, mailing lists, pamphlets, and flyers are becoming obsolete. This chapter elucidates how technology advancement is being used to create a unique atmosphere and memorable experience for tourists, based on sharing from industry players involved in organizing and managing major events and festivals in Malaysia, such as Citrawarna, Floral festival, International Regatta, Langkawi International Maritime and Aerospace Exhibition (LIMA), and many more. It focuses on how technology has aided the events business, specifically in terms of outreach, promotion, and event management.

Chapter 19 by Hossain, Ahsan, Nadi, Ahamed, and Neyamah deliberates the post-COVID-19-pandemic strategy for exploring the role of technology in tourism events, festivals, and fairs in the United Arab Emirates (UAE). The purpose of this chapter is to investigate the role of technology in tourism events, festivals, and fairs in the UAE following the COVID-19 pandemic. The chapter focuses on a variety of technical apps based on cutting-edge technology that may have an impact on tourism events, festivals, and fairs. The pervasive role of technology apps in sustainable tourism development in combination with tourist festivals, events, and fairs is not well documented in the literature. Due to the advancement of the latest technology application in tourism, the study identifies how tourists are affected by technology application, revealing in particular an increased tourism development and how tourists are continually enthralled by and attracted to tourism festivals, events, and fairs. The viewpoint of the UAE is discussed in this chapter. The chapter demonstrates how incorporating technology into tourism festivals, events, and fairs can help the UAE achieve long-term tourism development, particularly in the post-COVID-19 pandemic period.

Chapter 20, written by Çiğdemli and Avcı, analyzes the Turkish examples of technology in tourism events, fairs, and festivals. Festivals, fairs, and events are major tourist attractions in terms of destination promotion, awareness, and formation of a favorable image, as well as the social and cultural development of local people, along with economic advantages to the national and regional economies. To enhance the number of people who attend festivals, fairs, and events, which have significant economic, social, and cultural benefits, as well as to boost reputation, visitor management methods should be organized in a predictable and acceptable manner for consumer preferences. In event marketing and management, the use of digital communication, technology, and media channels allows collaboration, making it easier for participants, managers, travel agencies, and other tourism-related support sectors, as well as local people, to reach a large number of visitors and establish an active connection between the organizer and the participant. Online ticket sales for event activities, ticket recording in a "cloud" system, websites and applications, website designs and linkages, mobile applications, social networks, forums and blogs, digital platforms, virtual and augmented reality applications, and gamification are examples of these applications. Tourism and trade fairs, music festivals, film festivals, traditional Turkish festivals, gastronomy festivals, science and informatics festivals, culture and arts events, and sport activities held in Turkey are featured in this chapter. The research covers the provinces that host large number of events in Turkey, including Istanbul, Ankara, Antalya, and Izmir.

Chapter 21 by Hassan and Quader outlines socio-cultural challenges and solutions of tourism festivals, events, and technology application in Asia. Tourism festivals, fairs, and events have benefited from the introduction of new digital technology applications, which have increased the number of visitors, introduced new services and goods, and strengthened local communities. However, in certain situations, the negative influence of digital technology application may make it difficult to successfully manage tourism festivals, fairs, and events, thereby affecting socioeconomic growth. The current socio-cultural disadvantages of digital technology application to tourist festivals, fairs, and events in Asia were addressed in this study, as well as suggested alternative measures to overcome them in order to assess the influence of digital technology applications in tourist festivals, fairs, and events. This study explores the literature review technique to solve socio-cultural issues of digital technology applications in tourist festivals, fairs, and events in order to evaluate the effect of digital technology applications in tourism festivals, fairs, and events. It highlights the major findings and their theoretical and practical consequences, as well as prospective solutions for overcoming socio-cultural disadvantages by creating tourism event and festival standards, therefore increasing visitor engagement and satisfaction.

Chapter 22 by Mohamad addresses new technology and tourism post COVID-19 pandemic. ICT has aided firms in the hotel and tourism industries in lowering costs, improving customer service and experience, increasing operational efficiency, and increasing customer retention. Tourism study has extensively explored and analyzed the impact of the Internet on tourism. Specific research on tourism fairs, festivals, and events, on the other hand, has received little attention. Internet and marketing are the key elements of ICT that have a powerful impact on tourism fairs, festivals, and events following the Covid-19 pandemic. Tourism fairs, festivals, and events bring people together, and social media serves as an excellent platform for participants to share their experiences. Tourism fairs, festivals, and events bring people together, and social media provides a great platform for participants to share their own experiences with other Internet users. Furthermore, mobile communication technology, such as mobile tablets and smartphones, has become increasingly important, allowing individuals to stay informed about changes to festivals and events, as well as special offers from businesses. As a result, it is critical for businesses in tourism fairs, festivals, and events to properly leverage the element of IT for business continuity. In other words, the use of ICT in tourism fairs, festivals, and events will help businesses operate more efficiently while also making festivals and events more fun. As a result, it is critical for businesses in tourism fairs, festivals, and events to properly leverage the element of IT for business continuity.

Chapter 23 by Ghosh and Anjum discusses future research directions for tourism events, festivals, and technology application in Asia. How much can technological advancements help Asia's festival tourist destinations? One of the oldest strategies for drawing travelers from all over the world was to commemorate certain festivals. Because of Asia's diverse religious, cultural, and gastronomic cultures, this geographical proximity has long been regarded as one of the world's most promising festival tourism regions. As a result, this research aims to investigate the benefits and drawbacks of festival tourism in contemporary Asia. Initially, it was created to reveal the scope and features of Asia's diverse tourism events and festivals, such as Japan's Gion Matsuri and Nozawa Onsen fire festival, Thailand's Songkran festival, China's Snow and Ice festival, India's Holi festival, Bangladesh's Bishwa Ijtema, and Thai-Chinese and Japanese culinary cuisine tours, which have all enticed global tourists since their inception. This study aimed to show how existing technical equipment such as digital marketing, social media-based promotion, digital opinion and public polls, e-ticketing services, and digital broadcastings have contributed to Asia's universal flourishing of wonderful events. Using qualitative research approach, this study has progressed to gather secondary data-based content analysis. It has also been encouraged to compare the extent and distillation of these technological equipment in world-renowned European and North American tourist festivals such as La Tomatina and Rio de Janeiro's Carnival with the existing aspects of Asian tourism events. Finally, this chapter vows to conclude by addressing the present limits in the application of technology in Asian tourist events in the hopes of securing its position as one of the top sustainable tourism festival locations in the days ahead.

London, UK Azizul Hassan

Contents

Part I Theoretical Discussions

1 **Transmogrifying Tourism Events in Industry 4.0: An Analysis from Industry and Tourist Perspectives** 3
Sana Maidullah, Mohammad Rokibul Hossain, Fahmida Akhter, and Azizul Hassan

2 **Technology Application in Tourism Festivals in Asia: Theoretical Discussions** 25
Emmanuel Sebata and Md. Ruhul Amin Mollah

3 **Tourism Philanthropy: The Role of Tourism Fair, Festival and Event Stakeholders** 41
Bapon Chandra Kuri, Sadia Afrin Ananya, Synthia Islam, and Azizul Hassan

4 **Pilgrimage and Halal Tourism Event: Application of Technology** 63
Muhammad Khalilur Rahman, Shaharin Akter, Mohammad Mainul Hossain, and Azizul Hassan

Part II Influential Factors

5 **How Do Social Media Literacy, Psychological Capital and Work Engagement Influence the Employee Morale of the Hospitality and Tourism Industry?** 79
Fatema Johara, Munshi Muhammad Abdul Kader Jilani, and Md. Aftab Uddin

Part III Perceptions and Effects

6 **Protection of Sacred Tourism Sites, Festivals and Events in Bangladesh over Augmented Reality: An Experience of Cultural Heritage Based Tourism** 99
A. H. M. Salimullah and Rakiba Nabi

Part IV Social Media

7 **Technology Innovation and Social Media as a Global Platform for Tourism Events** 121
Muhammad Khalilur Rahman, Mohammad Mainul Hossain, Shaharin Akter, and Azizul Hassan

Part V Destination Branding and National Image

8 **Technology Innovation as a Destination Branding Tool for Festivals and Events Tourism** 135
Muhammad Khalilur Rahman, Md. Sohel Rana, Shaharin Akter, and Azizul Hassan

9 **Technology Application in Tourism Event, Education and Training for Making a Nation's Image** 149
Md Rabiul Alam, Md Rashed Khan Milon, Muhammad Khalilur Rahman, and Azizul Hassan

Part VI Cases

10 **Technological Adaptation in Tourism Events, Fairs, and Festivals: Way to a Revolutionary Transformation in Bangladesh** 167
Kaniz Kakon

11 **Promoting Catamaran Tourism in Bangladesh: The Challenges of Blue Economy** 181
Rakiba Nabi and A. H. M. Salimullah

12 **Information and Communication Technology Application in Tourism Events, Fairs and Festivals in India** 209
Saroop Roy B. R.

13 **Digital Development of Tourism Events in India** 221
Sachin Gupta, Kiran Meena, and Priyanka Jingar

14 **Understanding and Improving Digital Tourism Events in Pakistan** 233
Kalsoom B. and Mehtab Alam

15 Information Technology and Its Challenges Towards Sri Lankan Event Tourism: Event Managers' Perspective 249
A. M. D. B. Nawarathna, R. S. S. W. Arachchi, and W. M. A. H. Bandara

16 Technology-Based Tourism Development Mechanism in China with Events, Festivals and Fairs 261
Syed Far Abid Hossain, Kazi Mohiuddin, Faiza Tanaz Ahsan, Hasan Mahmud, Arslan Amzad, and Shazia Irshad

17 Approaching the Paradox: Loving and Hating Technology Applications of Indonesia's Cultural Events 279
Putu Diah Sastri Pitanatri and Luh Yusni Wiarti

18 Technology Application in Tourism Fairs, Festivals and Events in Malaysia 301
Sharina Osman

19 Exploring the Role of Technology Application in Tourism Events, Festivals and Fairs in the United Arab Emirates: Strategies in the Post Pandemic Period 313
Syed Far Abid Hossain, Faiza Tanaz Ahsan, Armana Hakim Nadi, Musanna Ahmed, and Hafsa Neyamah

20 Technology Usage in Tourism Events, Fairs and Festivals: Samples from Turkey 331
Aslı Özge Özgen Çiğdemli and Ceren Avcı

Part VII Challenges and Solutions

21 Tourism Events, Festivals and Digital Technology Applications in Asia: Socio-Cultural Drawbacks and Ways to Overcome 345
H. M. Kamrul Hassan and Mohammed Shahedul Quader

22 Post COVID-19 Pandemic: The New Technology and Tourism Revolution for Tourism Fairs, Festivals and Events 363
Noraihan Mohamad

Part VIII Directions for Future Research

23 Tourism Events, Festivals and Technology Application in Asia: Directions for Future Research 391
Ratan Ghosh and Iffat Anjum

About the Authors 413

Part I
Theoretical Discussions

Chapter 1
Transmogrifying Tourism Events in Industry 4.0: An Analysis from Industry and Tourist Perspectives

Sana Maidullah, Mohammad Rokibul Hossain, Fahmida Akhter, and Azizul Hassan

Introduction

Being glorified as an "emerging giant", festival tourism acts as a critical mechanism in ensuring the sustainability of national economic growth (Doe et al., 2021). Festival Tourism stimulates the diffusion and commemoration of local culture, customs, and ceremonies, thus attracting domestic and international tourists, promoting tourism, successfully representing own culture and pride, and contributing to the development of the community. Many cities worldwide have shifted their focus to festival tourism and attained sustainable economic prosperity (Hernández-Mogollón et al., 2014). The running of the bulls in Pamplona, the tomato battle "Tomatina" in Bunyola, and the Beer Festival "Oktoberfest" in Munich are few examples that make these countries famous over the world. Festival tourism is a phenomenon that attracts non-residents to visit a place during festival seasons (Doe et al., 2021; Liu et al., 2019). Quinn (2006) opined that Tourism events and festivals foster authentic cultural values and enrich visitors' experiences. It also spreads the reputation of host countries, improves infrastructures for festival-related venues, promotes interactions of diverged cultures, and contributes to the national development of the economy. Ziyadin et al. (2019) highlighted technology as a significant factor in event

S. Maidullah
Indian Institute of Management Sirmaur (IIM Sirmaur), District Sirmaur, Himachal Pradesh, India

M. R. Hossain (✉)
Faculty of Business Studies, Premier University, Chattogram, Bangladesh
e-mail: rokibul.fbs@puc.ac.bd

F. Akhter
School of Business Administration, East Delta University, Chattogram, Bangladesh

A. Hassan
Tourism Consultants Network, The Tourism Society, London, UK

A. Hassan (ed.), *Technology Application in Tourism Fairs, Festivals and Events in Asia*, https://doi.org/10.1007/978-981-16-8070-0_1

growth and success, making events more interactive and famous and visible in every corner of the world for modern tourists to enjoy event tourism to its full potential. Mistilis and Dwyer (2000) also highlighted the importance of information technology in tourism and MICE tourism growth. The first evidence of technology adoption in the tourist event business began in 1988 when Switch Company (THISCO) introduced a 1D barcode to develop the lead generation method for trade shows.

The world experienced technological innovations to facilitate and standardize festival events and tourism within the next ten years. For instance, PC Nametag and Lasers Edge developed software for printing name badges for delegates. Similarly, in 1996–1997, McGettigan created the first online system for purchasing hotel rooms of Holiday Inn Corporation called Real-Planner. Layton founded Pass-Key, the first online group reservation system, developed APEX, XML, and a data map to support the meetings. Radio Frequency Identification (RFID) and Smartphone technology are beneficial to manage overcrowded gatherings like religious gatherings (Yamin et al., 2008). Further, in 1990, live music performance came into the scenario where individuals can enjoy music by a live performance with a digital media network (Holt, 2010). App technology emerged in 2010 at the festival scene (Luxford & Dickinson, 2015).

In modern days, a variety of technologies are used by the event organizer to promote and attract the participants in the events, such as the use of social media as a search engine tool (Ngernyuang & Wu, 2020), new and virtual landscapes, as well as the heightened sensory feelings and imaginations technology (Robertson et al., 2015), mobile technology (Luxford & Dickinson, 2015) and usage of biometric technology (Norfolk & O'Regan, 2020) to provide the better services in terms of engagement, safety, and accuracy of data. Event tourism should concentrate on science, technological innovation, and social connectivity (Cuffy & Nair, 2020). These technologies provide ease of access to information. New digital innovations (such as social networks, smartphones, analytics, or wearable systems) are used to implement critical business transformations (such as satisfying customers, reducing costs, or developing new business models (Liere-Netheler et al., 2018). Nowadays, social media is a widely used platform to connect customers and receive their feedback (Luxford & Dickinson, 2015).

Experts (Lagarense et al., 2018) on digital transformation stated that the websites and e-commerce platforms are powerful tools in the entire sports tourism activity to promote the sport tourism events to existing and potential customers. Technologies such as RFID and sensor participant monitoring, music festival sourcing platforms, actions predictive analytics, the Global Positioning System (GPS), and conversion tracking are used in event tourism. Likewise, Unurlu (2021) reported that social networks are integrated digital marketing in event tourism. Smartphone Apps technology can alter the user experience of music festivals and indicate that apps developers need to focus more on event anticipation and on-site experience (Luxford & Dickinson, 2015).

To keep pace with the travel lovers'-changing tastes and expectations and attain competitive advantages, there is the need to embrace the boons of Industry Revolution 4.0. But fortunately, there is an example of industry revolution 4.0

technology such as Internet of Things (IoT), Big Data, Virtual Reality (VR), Augmented Reality (AR), cloud technology, Artificial Intelligence (AI), high-performance computing, and Blockchain, etc., are few such blessings (Happ & Ivancsó-Horváth, 2018). Car et al. (2019) identified IoT applied in different tourism areas such as Geo-location technologies, Medical (wellness) tourism, Ecotourism, Retail tourism services, Virtual concierge, airport management. For example, through geolocation technologies, tourists have a good travel experience as it allows them to track different locations from route planning to accommodation arrangement and journey planning. Furthermore, VR tourism ensures a holistic picture of hotels and tourism destinations by reducing consumers' risk by familiarizing them with unknown goals (Kim et al., 2020). In a nutshell, the above researchers have mentioned that, digitalization will assist in implying a paradigm shift in event management in which key actors and users are in a win-win connection.

The chapter is divided into several sections: the first section introduces the topic; the following section illustrates classifications of tourism events and festivals. Further sections shed light on a wide array of technological tools used in tourism events and festivals and the impact of technology on business and tourism experiences. Finally, the chapter provides concluding remarks that illuminate how festival tourism has been transmogrified in Industry Revolution 4.0 and reshapes customers' desires and experiences.

Classification of Different Types of Tourism Events and Festivals

Tourist events are any social gathering with a tourism component, such as arranging stay and transportation. According to Getz (2008), all planned events can boost the country's tourism sector. Events are divided into different categories such as business events, sporting events, festivals and other cultural festivities, the Olympics, world fairs, and other mega-events. Similarly, Getz (2005) described various events based on their magnitude and importance to the community and economy, including local, regional, periodic landmarks, and infrequent mega-events. Festivals are events related to agriculture, art, culture, and ritual of people's moment of life in which some sports accompany people. Based on the liturgical calendar, a festival is held at a specified period and is repeated every year. Festivals all over the world are open to the general public, attracting a considerable number of tourists. Many studies have classified festivals as short-term events (Saleh & Ryan, 1993), formal events that appeal to tourists (Uysal & Gitelson, 1994), and events that reflect community values, ideology, identity, and continuity (Getz et al., 2010). Firstly, extensive religious festivals are held in various locations around the world, such as traditional festivals in Japan (Roemer, 2007, 2010), religious festivals associated with ethnic minorities in New Zealand (Tondo, 2010), puppet theatre festivals related to traditional beliefs in Indonesia (Osnes, 2010). Christian festivals in Poland are examples (Cudny et al.,

2011). Secondly, several art festivals play an important role in societal cohesion and cultural exchange among people worldwide and perform as an economic catalyst for the country's development. Art festivals are increasingly seen as unique factors to attract tourists and work as tools for destination image-makers. Art festivals are considered as a means of generating national revenue, such as the Macau Art Festival (McCartney & Ip Si Kei, 2018), Pink Night Festival (Mariani & Giorgio, 2017), Rio Lisboa Music Festival (Dinis et al., 2020), Chinese Music Festival (Chen & Lei, 2021), Jawa Jazz Festival (Saragih, 2021), Edinburgh Art Festival (Lafarge, 2017), Gertrude Street festival (Edensor & Sumartojo, 2018). To commemorate our rich past, culture, and customs, we organize festivals. They are designed to share the love with loved ones to celebrate significant moments and feelings of our life. In one way or another, all festivals are cultural in some form. Many different sorts of festivals are held each year, including national, religious, and seasonally based.

Different Types of Technological Tools Used in Tourism Events and Festivals

In the last seventy years, technologies have changed from analog to digital, collision to collaboration, content to context, an event to events without end, local to cloud, hardware to software, and human staff to technical staff (Lee et al., 2017). Mobile phones and the internet accelerate the adoption and use of Smart technology universally. Ericsson introduced the first Smartphone in 1997, and Apple launched the multi-touch iPhone in 2007, and Android launched Smartphone in 2008. In 2007, mobile and Bluetooth technology was used in festivals to deliver the details to the visitors (Luxford & Dickinson, 2015).

A multitude of research has examined various facets of online marketing communication (Koba, 2020), Facebook event sites (Lee et al., 2012), social media, and social media content (Bukhari et al., 2012; Mizzau et al., 2019; Marine-Roig et al., 2017; Süli & Martyin-Csamangó, 2020; Cuomo et al., 2021), and Twitter (Koba, 2020; Saragih, 2021). In addition, RFID technology (Yamin et al., 2008), Autonomic computing, IoT, and media technologies like Skype and webcams were also studied for event marketing and promotion (Marine-Roig et al., 2017; Sterritt, 2002). These technologies also implied smooth event operations and increased user engagement to attend the festival event (Saragih, 2021). Robertson et al. (2015) underlined the relevance of future Cerebral Systems. Emotional Capture (Feel the Beat in Your Pants), Semantic Web, Handpicked Content, Digital Experience (Dream Management), Thought Control (Multi-touch Application), and Contact Lens technologies in a festival event. Robertson et al. (2015) also discovered that working with these technology organizers can provide the most positive and valuable experience to the festival visitors, minimize negative experiences that occur by mishappening, insignificant events, and maximize psychological and social well-being.

Social Media Networking

Social media refers to a collection of platforms that connect people to people by sharing posts, pictures, and videos. Many types of social networking sites are accessible, such as Facebook, Twitter, LinkedIn, Instagram, and WhatsApp, allowing individuals to create friends and exchange information with them.

Blogs are personal websites written by people who are passionate about a subject and serve as a way to share that passion with the rest of the world while fostering an active community of readers who comment on the author's posts. YouTube, Flickr, and Vimeo are video and photo sharing sites that greatly simplify the process of sharing and commenting on photos and videos. Chat rooms and message boards are online gathering places where people can meet and discuss various topics, with the main feature being that anyone can start a discussion thread. Wikis are open-source websites that anyone can edit and maintain. Event organizers can use social media platforms effectively because it is one of the powerful tools for marketing and promotion. These are available at low cost and able to reach the maximum audience in less time.

Recently, the study conducted by MacKay et al. (2017) identifies Twitter as a social media for festival promotion, and according to these authors' observation, 1169 Tweets and 483 posts were collected by Twitter. Two-thirds of Twitter's posts were collected during the festival, one-third before the festival, and minimal activity after the festival. The above study recognizes that maximum people use social media differently at the various time frame in the event.

Further, Danielsen's (2017) study identifies microblogging strategies on Twitter increase local artist music festival exposure and improve online music streaming. Another empirical evidence by Arasli et al. (2021) shows that professional updates on social media sites such as Twitter, LinkedIn, Facebook, and WhatsApp contribute to festival satisfaction and the desire to attend again. Also, to make the event more engaging, event organizers spent a significant amount of time creating exciting content such as creating videos, posts, blog posts, personalized text messages, and animated posters. Event study identifies that Facebook likes, posts organizers strive to create relevant and personalized content to strengthen the brand's connection with social media users to promote the event (Chugh et al., 2019). And factor comments can also determine user engagement. Their research reveals that posted video, rather than posted images, has a significant impact on user engagement. Also, rather than advertisement posts, comments and shares by the developed post can be seen as the most authentic source of measuring engagement. The festival's organizers also employ many strategies to understand how their customers use social media to participate. Bukhari et al. (2012) also identify the implication of social media for extensive event management. They suggested that it help in analyzing public sentiment, public dialogue about remarkable citywide events, and trending topics.

Similarly, the study of Lee et al. (2012) supports the above study finding that event organizers can capture the participant's emotion with the help of Facebook comment. With the use of social media analytic, organizer maximizes their understanding to capture the different stages of shopping and recognize, in which stage customers could feel the difficulties in purchasing. Therefore, to improve customers' purchasing behavior through online media, Facebook and Instagram introduce the third-party vendors' direct sales, providing ease of purchase and increasing brand advocacy (Lindsey-Mullikin & Borin, 2017). Overall findings thus suggest that social media play a significant role in marketing festival events through word-of-mouth publicity and building relationships with the festival brand (Hudson et al., 2015). Furthermore, Ngernyuang and Wu (2020) determine that social media is used as a search engine tool for festival tourism promotion. As per the study by Lindsey-Mullikin and Borin (2017), social media marketing needs proper planning and strategies to make the event successful.

RFID Technology

RFID technology is available in a band, tag, and wristband; these devices use the radio signal and sensor technique to facilitate the event organizer and event goer. RFID technology enables the event organizer to experience valuable benefits such as less fraud, control access, traffic pattern analysis, increases on-site spending, and fun or novelty. Similarly, event attendees are also getting a unique experience by using RFID technology such as Less time waiting in lines, cashless payment, and social integration. Yamin et al. (2008) also identified the importance of RFID technology application in the event like the Hajj to manage the crowd. In addition, Mantoro et al. (2011) identified the application of RFID to control the navigation system, tracking people, and location-based service by integrated via mobile app. In the context of music events, the use of RFID technology was identified as improving the process of music events (Friedrich-baasner et al., 2017). Further, Mowafi et al. (2013) identified the Hajj locator and WISP (Wireless Identification and Sensing Platform) are used to track, monitor, and manage the movement of humans in mass event gatherings. Different RFID technologies in the form of electromagnetic waves are used as crowded management tools that determine the number of people in the event to prevent an incident like a stampede, which is commonly used to occur in the mega event and religious festival (Tofigh et al., 2020).

Mobile Application

Event planners create a custom mobile app for an event using mobile event app software without knowing how to code. These apps enable the creation of branded mobile applications that include schedules, maps, attendee information, exhibitor information, speaker bios, and more. These template-based app programs allow event planners to choose a design, special event features and customize the content. Attendees can then download these apps on their mobile devices to access vital event-related statistics. Other event management platforms or event planning software often use mobile event app creation software as a supplement. There are top ten mobile applications such as CrowdCompass, Whova, Socio, EventMobi, Hopin, Bizzabo, Attendify, SpotMe, PheedLoop, Airmeet. These mobile apps provided a quick and optimized service for registration, ticketing, scheduling, networking, streaming, video room, chats poll, live display, lead retrieval, analytical, email, and more to festival-goers and event organizers. The research by Ziyadin et al. (2019) identified the importance of mobile apps in event management in many ways, such as target audience massages, use of social networks, help in the navigation and information search. Mobile apps also use to send the personalized greeting and notifications on arrival at the event, measurement of timespan at the event, to understand the participant interest in the event. Similarly, Li et al. (2019), Luxford and Dickinson (2015), Lexhagen et al. (2005) also reported the importance of smartphones as experience changers, particularly for Smartphone users to attract attendees in the event like a music festival. In contrast, Fonseca et al.'s (2020) study identify less mobile phone use during Portugal music festivals due to a lack of knowledge about the application. One more study by Van Winkle et al. (2019) applied the Unified Theory of the Acceptance and Use of Technology for mobile adoption to a festival setting. Results reveal mobile technology enhances the festival experience. Finally, the study of Koukopoulos and Koukopoulos (2017) reports two essential factors of an event; the first is dissemination; and the second is exploitation, which required attention. This study thus stresses maintaining the factors mentioned above in Smartphone applications to support the carnival organizer.

Big Data

Big Data is the massive amount of data that is used in many ways for business at present. The term Big Data was introduced in 1990 by John R Mashey (Lohr, 2013). In industry 4.0, vast amounts of Data in both organized and unorganized forms are generated. However, there is no formal definition of Big Data; many researchers provide various reports for Big Data. For example, Laney (2001) defined Big Data with 3V, which denotes Volume, Variety, and Velocity that were extended to 4V with Value, which identifies the usefulness of Big Data with Volume, Variety, Value, and Velocity (Gantz & Reinsel, 2011). Different types of Big Data have pertained to tourism research, making incredible innovations in business. All such Big Data came from three sources: (1) users, who generated UGC data (such as online textual and photo data); (2) devices, which generated device data (such as GPS data, mobile roaming data, Bluetooth data, RFID data, WIFI data, and meteorological data); and (3) operations, which generated transaction data (such as web search data, online booking data, webpage visiting data, etc.). The following way can describe it as Cárdenas et al. (2013), and Cuomo et al. (2021) reported the use of Big Social Data (social media user-generated content) as a tool to understand the tourist experience for planning the cultural tourism or event tourism.

Groenli et al. (2018) informed that Big Data principles are applied to the IoT sensor, which helps in festival management by easily tracked information through real-time Big Data analytics. Therefore, it has the fact-full insights to decide for resolving the issue. Furthermore, Li et al. (2016) also considers the importance of Big Data analytic through the spatial pattern to validate the accuracy of travel databases. Similarly, the use of big data is also recognized in the study of the Spring Festival and the Mid-Autumn Music Festival of China (Pan & Lai, 2019), China Sport and Cultural Festival study (Hwang et al., 2020), China Spring Festival (Hu et al., 2017) and local flower festival to know tourist satisfaction (Lee & Lee, 2019).

Virtual Reality and Augmented Reality Tools

VR is a term that refers to interactive pictures or movies that allow the spectator to explore a scene in 360 degrees. Unlike a traditional video picture, which is recorded from a single point of view, VR production covers every aspect of a site. VR may be utilized in the event and tourism industries to capture events and locations in a new and engaging way (Thomson et al., 2021). VR Production has accomplished with the use of dedicated cameras, rigs, and software. After then, the material is watched via a VR headset, a standard PC, or a mobile device. VR can be used for marketing, for viewing 360-degree video in high resolution. VR tourism photography and videos provide travel opportunities to people who are unable to travel. Instead of presenting brochures and computer screens to tourists, travel agencies may offer them a virtual experience. This strategy may also be utilized to significantly affect trade fairs and events to capture the public's attention rapidly. Many researchers recognize virtual reality's importance in providing personalized service in event management (Ekici & Toksoz, 2021; Bustard et al., 2019). Tourism event and festival planners may utilize VR technology to create memorable and unique experiences for users; Virtuality enhances brand interaction and allows firms to stand out from the crowd. Also, VR will enable users to explore a scene on their own. All of these will maximize customer service and organization (Fig. 1.1).

Technological Impacts on Business

Significant events may be challenging to plan. It takes an event manager to ensure that every aspect of the event is appropriately handled so that everything runs smoothly. It entails the following steps: event planning, budgeting, venue selection, obtaining permits, transportation, entertainment, catering, and security. Therefore, event planning technology appears as an essential tool for providing ease to business and business organizations in managing workforce, machine, money, material, and methods. With that inner and outer environment of the organization also work by technology like product marketing, promotion, and dissemination of information all are managed by an organization through technology and Smart technology.

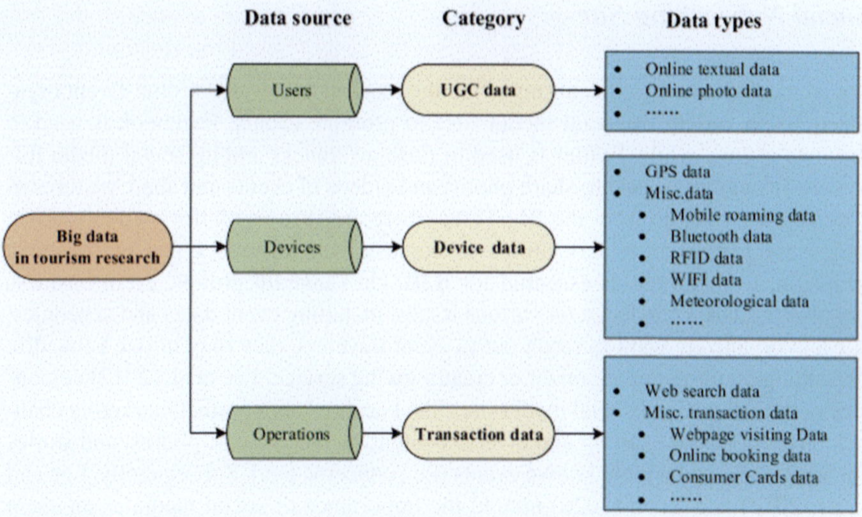

Fig. 1.1 Big data in tourism research. (Source: Li et al., 2018)

Furthermore, Koukopoulos and Koukopoulos (2017) stated that technology in outdoor events like sport, festivals, and cultural events succeeds in three areas: audience engagement, active engagement, and performing a functional role to support activities such as navigation or location services. Some devices are used to improve the benefit of the spectators, such as Download Festival from the UK: Rock Festival, Trotting Pal is a mobile factor insignificant. System for a trotting track, m-Group, and Boston Marathon is used to disseminate the annual event of Boston marathon information in the streets of Boston (Koukopoulos & Koukopoulos, 2017). The growing revolution in internet technologies provides a strategic direction for increasing the experiential value for tourism customers. Therefore, event service providers should use electronic commerce as a strategic tool when designing and implementing their event service packages. The modern marketplace has undergone a significant transformation from the industrial to the technological age. The above statement shows that technology-assisted customer service functions (e.g., paying bills, responding to inquiries, updating account records, tracking orders), retail and business-to-business purchase transactions, learning, and knowledge seeking. Tourism service providers can use digital technologies and innovative features to focus on customer engagement to increase customer satisfaction, retention, and organizational performance. To meet these goals, tourism and event service providers must interact with customers and use customer-generated data to fine-tune tourism service design and delivery to meet these goals. The tourism service episodes leave an indelible and memorable impression on the customers' minds. A study by Opute et al. (2020) showed that tourism service providers and related supporting service providers can benefit from the use of technological media not only to engage with customers but also to coordinate the network of relationships and interdependencies that exist between them (e.g., travel service, restaurants, etc.).

Social Networking Site

Social media had a significant impact on the tourism and events sector. Event organizers use a variety of social media sites to promote events. Facebook is used to promote events, while Twitter is used to raise awareness among social media followers. Instagram is used to share photos and videos of events and share pictures of special guests. Aside from the Big Three, there are various platforms that tourism and event service providers utilize to connect with diverse types of travelers. LinkedIn, for example, is essential for B2B. On LinkedIn groups, event business organizers share knowledge on various issues, including event dates and schedules. If event organizers want to reach out to event travelers, they may utilize LinkedIn, depending on the event organizer or creator giving service. Lee et al. (2012) demonstrates the influence of social media sites like Facebook as a perfect marketing channel. Users generate exciting and diverse content such as videos, photos, and stories for small local and widely known events (i.e., concerts, etc.). Subsequently, Lee and Paris (2013) and Koba (2020) uphold the importance of social media as an event marketer. Every day, an increasing number of active social media users provide the opportunity to business organizers to use SNS (social networking site) as an advertisement tool and decision-making tool where event-goers could engage themselves. In line with the above statement, the studies of Ngernyuang and Wu (2020) and Garay and Morales (2020) identify the potential of social media as the most critical advertisement tool in today's time for the music festival organizer to advertise events. Furthermore, the empirical study of Hudson et al. (2015) identifies social media's role to engage and develop the brand image within event-goers. In the later stage, it also influences the word-of-mouth publicity about the event in their network.

Similarly, Lee et al. (2012) identify the intention to go to a music festival influenced by Facebook posts and event pages. One more study in the same direction by Hoksbergen and Insch (2016) pinpoints the importance of Facebook to knowing tourist reasons and status to attend the festival. In addition, Kim et al. (2018) found that the intention to participate in the festival is influenced by one's attitude toward using social media sites, driven by perceived enjoyment and utility. They also recommended that the festival's management company ensure that the right social media content should be interactive and use the right content to find the direction about the festival easily. Kim et al. (2018) identified the social media influencer's vital role in influencing festival-goers to read the content of the festival's information. Influencers have determined the role of demographic management and how it works differently on social media. As per Lee et al. (2012), for different demographic users such as male users to access new, newest, and correct information while female users enjoy reading online reviews of the previous organization's demographics for sharing knowledge and relevant issues about the festival (Kim et al., 2018).

Moreover, social media user-generated content can retain both new and experienced customer experience through customer-to-customer communication at a meager cost and promote their event on a large scale (Marine-Roig et al., 2017). Event organizers use Instagram to set an end date and time, send the notification to the attendee, sell the ticket, send the reminder to the people about the deadline. The event organizer also used Facebook to generate creative stories about the festival or event to engage festival-goers. On the Facebook event page, organizers can create the tag feature to tag festival special guests and artists and discuss questions, answers, and important announcements. They can cause the pre-sale code for early bird purchases. In addition, the event organizer also uses the Facebook event page for ticket sales as well. Some studies identify (Bukhari et al., 2012; Lee et al., 2012) social media help organizers to capture the emotion of festival attendees. It helps the organizer plan festival operations as per the requirement of the festival-goer and raises social media sharing. Social media has also changed the loyalty program; acquiring a new client is more expensive than retaining the old one. Social media sites provide a romantic environment to maintain customer bonds and engagement. Customer posts their travel photos and videos on the brand page and can collect their loyalty points. Loyalty points can strengthen a bond with the brand.

RFID Impact on Business

RFID technology has changed the way of planning an event from manual to automatic. RFID is one of the most complex and emerging technologies of the tourism industry. It improved the tourism and events business process in many utilize days such as hotel process, cruise ship, resorts, and theme parks—RFID-enabled operations and marketing which can save costs, boost profits, and improve visitor experiences. A variety of examples are used to spark people's interest in how RFID may be implemented, raise understanding of practical difficulties with the technology, and drive managerial advice to help businesses better use it and avoid mistakes.

The research of Friedrich-baasner et al. (2017) shows the vast potential for RFID technology to improve the number of event activities. For example, RFID helps to Access control and ticket sales (accountability of event goer), purchase goods and services (cashless payment), personal logistic (record of a working hour of employees). They also identified that RFID bracelets and RFID readers handle access control systems, Point of Sale (POS) systems, and personal logistics in the event successfully. Furthermore, many types of research provided empirical evidence of RFID technology usability in Mega-events, festivals, and religious gatherings (Yamin et al., 2008; Mantoro et al., 2011; Mitchell et al., 2013; Yamin, 2019). However, RFID is also unable the organizer to manage attendees' data, customer support lead generation, seat planning, and competition management.

Big Data Impacts on Business

Variety, Volume, Velocity, and Value are the 4Vs of big data. From several sources, firms in the business world produce a sizable volume of data. As per the needs and demands of festival-goers, event organizers give facilities for festival-goers to use their data. The festival organizer configures and captures the value, which helps them redesign the tourist experience and add value to the existing product (Cuomo et al., 2021). A similar study conducted by Hwang et al. (2020) reports the problem of the overlapping festival, which cannot generate a reasonable profit. Thus, to solve this problem, they suggested the Ministry of Culture, Sports, and Tourism collect the SNS data to improve the visitor viewpoint.

Furthermore, Big Data technology helps to understand the attractiveness of the festival. During the Spring Festival travel rush in China, mass movement of people happens every year, leading to pressures on the transportation system. To solve this issue, Li et al. (2016) conducted a study. This study by Li et al. (2016) outlines that Big Data can help understand the travel trend and travel flow of the people in China and report that intense travel occurs five days before the Spring Festival and six days after the Spring Festival. Regarding China Sports and Cultural Festivals, the study of Hwang et al. (2020) identifies the crucial role of social network site data analysis through R to solve visitor issues. Likewise, Hu et al. (2017) recognize that Big Data is also applied to understand the urban mobility pattern of China's Spring Festival in terms of mass and high pressures on transportation. The IoT and Big Data analytics help (Groenli et al., 2018) to save the organizers from revenue loss, which is depicted in the following model (Fig. 1.2).

Virtual Reality and Augmented Reality Impacts on Business

Between 2020 and 2024, the global AR and VR market will reach USD 30.7 billion and grow at a CAGR of over 35%. However, these figures show a strong desire to use AR and VR in all major industries worldwide. VR is, of course, the most popular form of entertainment and gaming. A growing number of sectors, including tourism, real estate, and education, are also beginning to recognize the benefits of this

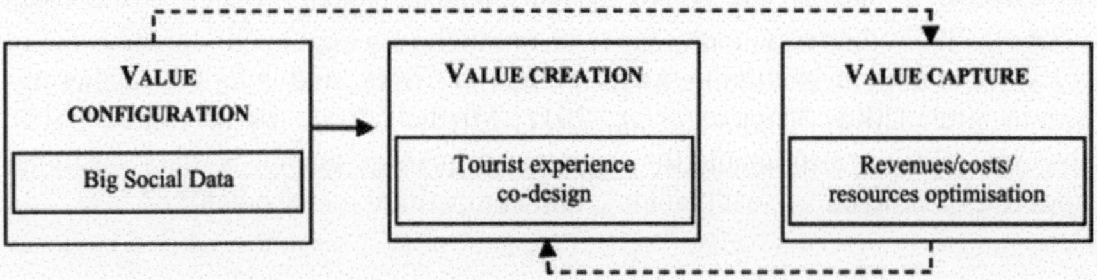

Fig. 1.2 The tourist value system creation. (Source: Cuomo et al., 2021)

new technology. Increasingly, businesses are recognizing the value of VR and AR in enhancing their customer experiences (CX). VR and AR are proving to be new opportunities for companies to develop unique experiences. The event industry can also benefit from virtual reality. As per Waida's (n.d.) report, 88% of event planners planned VR events. They also organized an AR event, with 87% of them doing so. Virtual event planning is clearly in demand at the moment. VR and AR can be utilized before, during, and after any event. To help the organizer organize the event and wow visitors' experience on the day of the event. Here are a few examples of how an organizer may use this technology in their circumstances: training event staff members and planning team, showcasing sponsorship packages, taking advantage of fully immersive trips, speeding up the RFP process, adopt an AR app that drives event traffic. Ekici and Toksoz (2021) recently highlighted the usage of AR and VR in event management as a way to give more tailored services.

Similarly, according to Bustard et al. (2019), users have a strong desire for real-time event integrated knowledge, followed by augmented and highly immersive experiences. If an event is held in an actual location, VR can help to design truly memorable experiences. VR and AR have provided several benefits to event organizers, such as Higher audience engagement, Viral promotion after the event, and Increased sales. Even (citation) recognizes that the framework of VR for Artists and content providers, especially musicians, would be an intriguing new medium. VR composers are specifically interested in VR because of its simplicity of content distribution and its ability to produce an immersive sound experience.

Technological Impact on Tourist Experience

It is considered that the technological revolution first facilitates the information dissemination process through some social networking sites such as Facebook, Twitter, and Instagram. Remarkably, the festival-goers or tourists benefit in many ways through today's AI base Smart technology, and tourists get to know about the exact date, time, and better navigation idea about the festival or special event. According to Law et al. (2009), computer technology improves the customer experience. Cuomo et al. (2021) also stated that the digital revolution alters the consumer journey pattern. Furthermore, Mizzau et al. (2019) discovered that a well-coordinated digital and social media presence enhanced the festival's social identification and participation environment. Industry 4.0 Smart technologies such as mobile applications (Van Winkle et al., 2017; Kim et al., 2018; Talantis et al., 2020), AR experience (Olya et al., 2020), and social media (Garay & Morales, 2020) contribute to the event's enjoyment, intrigue, and customer experience (Carlson et al., 2016).

Social Networking Site

Social media is the communication tool, which disseminates and circulates notice and news in a second to all connected followers and subscribers of Facebook, Instagram, YouTube, and twitters. Social media empower the tourist experience and decision-making by knowing what to do, where to go, and when to participate in any tourism-related event. The study by Hoksbergen and Insch (2016) reports that Facebook event pages help festival-goers in five ways as: first, operational (updates, the ability to obtain timely details, the ability to buy tickets); second, social (the ability to see and interact with other people who are attending, making it possible for others to know that they were present, being a member of the R&V Facebook group, promoting their interests and public image through exposure and promotion); third, emotional (activated memories and feelings of the festival's atmosphere Excitement that has been initiated); fourth, interactive (collaboration and engagement have increased, there is value in both functional features and involvement); and fifth, aesthetic value (the ability to view photos via social media, the design's aesthetic appeal, the ability to view videos via social media). Similarly study by Süli and Martyin-Csamangó (2020) reported that social media supports the young generation planning travel. Garay and Morales (2020) also identified the benefit of social media from the user's point of view. Furthermore, previous experiences of the festival-goer play a moderator role in the formation of attitude. To use the social media site, drive the intention to attend the festival (Kim et al., 2018).

Mobile Application Technology

Mobile applications have become an essential tool for festival-goers. These provide several facilities like festival participants for moving in and out of different social settings. These mediate the experience of the festival in diverse social contexts decoupled from temporal and spatial boundaries. As a result, the participants can control their social interactions with individuals in their local spatial presences, those at the festival with whom they desire to maintain contact, and larger groups beyond the festival site with whom they may be involved in other continuing stories (Luxford & Dickinson, 2015). Similarly, mobile application technology is also part of the Smart experience (Bustard et al., 2019). Again, Li et al. (2019) identified the importance of an official mobile app for a festival, directly impacting festival participant overall satisfaction. One more study identified the importance of mobile apps and reported that users found that mobile app superb usability on the festival, which influences the user overall event satisfaction (Ziyadin et al., 2019; Talantis et al., 2020). In addition, in the setting of a science festival, the AR experience of festival attendees was highly correlated with the demographic variable (Olya et al., 2020). According to Talantis et al. (2020), event-goers use app technology and perceived usefulness. It was a critical predictor in the context of conference satisfaction, as mobile apps manage various things such as the customer's profile, activity feed, event and meeting schedule, name of the performers, photo gallery, interactive map, and contact details. Thus, a mobile application is vital for collecting and co-creating the customer's experience (Van Winkle et al., 2017). In addition, the event goer can also use mobile technology to capture the unforgettable experience and publish it to their social media accounts to share it with others (Van Winkle et al., 2017).

Virtual Reality and Augmented Reality Tool

Recently conducted surveys by Wolhuter (2021) have found that customers are 86% more likely to pay a higher price or service if they have a positive customer experience. Businesses can create revolutionary customer experiences that have never been seen before, thanks to VR technology. A company will continuously add new elements to their environments as technology advances, and hardware (headsets and consoles) becomes more affordable. In the next five years, 84% of stakeholders predict that VR will surpass in-store CX initiatives. VR is an incredible feeling with 360-degree video and music that transports consumers to a virtual world. Participant may feel that they are in their room in one minute and on top of a mountain in the next, experiencing a rollercoaster with breathtaking views and a virtual adrenaline rush. Recently, Wreford et al. (2019) reported that Live events cannot be replaced by existing 360-degree VR technologies, based on data analysis. Those who responded indicated that VR might lead to good connections with event organizations and companies. That it stands, VR does not deliver the same social and sensory fulfilment as live events do at the live event time. Event organizers and current attendees alike can benefit from the usage of VR. This VR technology has provided several benefits to consumers, such as Try before you buy, Immersive can take the virtual tour, and participants can attend Multiple Events from a Single Location and enhance their experience.

Conclusion

Rapidly growing technologies change the shape of events and festivals to provide more immersive, efficient, safe, cost-effective, easy booking and entry into the event, online and cashless payment, and hassle-free service to festival-goers and organizers. Attendees' experiences like communication, planning, and networking can be improved with event technology. Technology makes the job of an event organizer easier and allows them to solve the earlier issue due to the lack of technology. Technology empowers them to concentrate on the critical aspects of the event. Indeed, social media, mobile event applications, RFID, IoT, and artificial intelligence embedded devices replace human staff with the technology to provide a valuable experience to the customer. Before technology, the event was organized through a manual process, and organizers made a list of all the participants, contact by contact. Marketing and promotion of the event were minimal. The promotion technique was also tiny, and it was available in the form of print media only a type of event or live event. Advancements in intelligent technology enable virtual events and hybrid events, increase the engagement with the targeted audience to promote the business internationally and nationally, increase brand awareness and advocacy, increase quality leads, educate the customer, retain the customer and increase sales. A variety of user-generated content on popular social media sites, Facebook, Twitter, and YouTube, was analyzed using Big Data analytics, social media marketing analytics, and machine learning to determine which consumers would most likely attend the event. These technologies enhance the opportunities to capture customer value and purpose or enable the business organization to understand customer needs, wants, and desires. As technology advances, new ways to enthrall event participants become available. Discover which recent event technology trends and technologies will benefit guests the most, and then put them in place to boost engagement. They will try to design interesting and impressive content rather than irritated and suffocated. Overall, organizers need to use the right technology with the right strategies to allow the technology benefits. These technologies, which we have discussed in the present book chapter, play an enormous role in business management and consumer decision-making.

References

Arasli, H., Abdullahi, M., & Gunay, T. (2021). Social media as a destination marketing tool for a sustainable heritage festival in Nigeria: A moderated mediation study. *Sustainability, 13*(11), 6191.

Bukhari, I., Wojtalewicz, C., Vorvoreanu, M., & Dietz, J. E. (2012, 13–15 November). Social media use for large event management: The application of social media analytic tools for the Super Bowl XLVI. In *2012 IEEE conference on technologies for Homeland Security (HST).* Waltham, MA, pp. 24–29.

Bustard, J. R. T., Bolan, P., Devine, A., & Hutchinson, K. (2019). The emerging smart event experience: An interpretative phenomenological analysis. *Tourism Review, 74*(1), 116–128.

Car, T., Pilepić Stifanich, L., & Šimunić, M. (2019). Internet of things (Iot). In *Tourism and hospitality: Opportunities and challenges. Tourism in southern and Eastern Europe: Creating innovative tourism experiences: The way to extend the tourist season* (pp. 163–173). Faculty of Tourism and Hospitality Management, University of Rijeka.

Cárdenas, A. A., Manadhata, P. K., & Rajan, S. P. (2013). Big data analytics for security. *IEEE Security & Privacy, 11*(6), 74–76.

Carlson, J., Rahman, M. M., Rosenberger, P. J., & Holzmüller, H. H. (2016). Understanding communal and individual customer experiences in group-oriented event tourism: An activity theory perspective. *Journal of Marketing Management, 32*(9–10), 900–925.

Chen, Y., & Lei, W. S. (Clara). (2021). Behavioral study of social media followers of a music event: A case study of a Chinese music festival. *Journal of Hospitality and Tourism Insights, 4*(2), 163–182.

Chugh, R., Patel, S. B., Patel, N., & Ruhi, U. (2019). Likes, comments and shares on social media: Exploring user engagement with a state tourism Facebook page. *International Journal of Web Based Communities, 15*(2), 104–122.

Cudny, W., Gosik, B., & Rouba, R. (2011). Festiwal kultury chrześcijańskiej w Łodzi jako przykład produktu turystycznego-wydarzenie. *Zeszyty Naukowe Uniwersytetu Szczecińskiego. Ekonomiczne Problemy Usług, 66*, 89–100.

Cuffy, V. V., & Nair, B. B. (2020). Events tourism: A critical debate for the 21st century. In V. V. Cuffy, F. Bakas, & W. J. L. Coetzee (Eds.), *Events tourism: Critical insights and contemporary perspectives* (pp. 251–256). Routledge.

Cuomo, M. T., Tortora, D., Foroudi, P., Giordano, A., Festa, G., & Metallo, G. (2021). Digital transformation and tourist experience co-design: Big social data for planning cultural tourism. *Technological Forecasting and Social Change, 162*, 120345.

Danielsen, A. (2017). The mediated festival: Live music as trigger of streaming and social media engagement. *Convergence, 25*(4), 714–734.

Dinis, M. G., Eusébio, C., & Breda, Z. (2020). Assessing social media accessibility: The case of the rock in Rio Lisboa music festival. *International Journal of Event and Festival Management, 11*(1), 26–46.

Doe, F., Preko, A., Akroful, H., & Okai-Anderson, E. K. (2021). Festival tourism and socioeconomic development: Case of Kwahu traditional areas of Ghana. *International Hospitality Review*. https://doi.org/10.1108/ihr-09-2020-0060

Edensor, T., & Sumartojo, S. (2018). Reconfiguring familiar worlds with light projection: The Gertrude street projection festival, 2017. *GeoHumanities, 4*(1), 112–131.

Ekici, R., & Toksoz, D. (2021). New trends in event marketing. In K. Birdir, S. Birdir, A. Dalgic, & D. Toksoz (Eds.), *Impact of ICTs on event management and marketing* (pp. 156–176). IGI Global.

Fonseca, D., Reis, J. L., Teixeira, S., & Peter, M. K. (2020). Mobile applications at music festivals in Portugal. In Á. Rocha, J. Reis, M. Peter, & Z. Bogdanović (Eds.), *Marketing and smart technologies* (pp. 190–201). Springer.

Friedrich-baasner, G., Heim, D., Hauser, M., Gramer, W., & Winkelmann, A. (2017, 20–22 September). Improving the processes of music events with radio frequency identification. In *2017 IEEE international conference on RFID technology & application (RFID-TA)*. Warsaw, pp. 119–124.

Gantz, J., & Reinsel, D. (2011). *Extracting value from chaos*. Retrieved from: http://www.emc.com/collateral/analyst-reports/idcextracting-value-from-chaos-ar.pdf. Accessed 11 Aug 2021.

Garay, L., & Morales, S. (2020). Decomposing and relating user engagement in festivals' virtual brand communities: An analysis of Sónar's Twitter and Facebook. *Tourist Studies, 20*(1), 96–119.

Getz, D. (2005). *Event management and event tourism* (2nd ed.). Cognizant.

Getz, D. (2008). Event tourism: Definition, evolution, and research. *Tourism Management, 29*(3), 403–428.

Getz, D., Andersson, T., & Carlsen, J. (2010). Festival management studies: Developing a framework and priorities for comparative and cross-cultural research. *International Journal of Event and Festival Management, 1*(1), 29–59.

Groenli, T., Flesch, B., Vatrapu, R., & Bergner, H. (2018, 10–13 December). Internet of things big data analytics: The case of noise level measurements at the Roskilde music festival. In *2018 IEEE international conference on big data (big data)*. Seattle, WA, pp. 5153–5158.

Happ, E., & Ivancsó-Horváth, Z. (2018). Digital tourism is the challenge of future. *Knowledge Horizons – Economics, 2*(2), 9–16.

Hernández-Mogollón, J. M., Folgado-Fernnández, J. A., & Duarte, P. (2014). Event tourism analysis and state of the art. *European Journal of Tourism, Hospitality and Recreation, 5*(2), 83–102.

Hoksbergen, E., & Insch, A. (2016). Facebook as a platform for co-creating music festival experiences: The case of New Zealand's Rhythm and Vines New Year's Eve festival: Facebook as a platform. *International Journal of Event and Festival Management, 7*(2), 84–99.

Holt, F. (2010). The economy of live music in the digital age. *European Journal of Cultural Studies, 13*(2), 243–261.

Hu, X., Li, H., & Bao, X. (2017, 16–18 June). Urban population mobility patterns in spring festival transportation: Insights from Weibo Data. In *2017 International conference on service systems and service management*. Dalian, pp. 1–6.

Hudson, S., Roth, M. S., Madden, T. J., & Hudson, R. (2015). The effects of social media on emotions, brand relationship quality, and word of mouth: An empirical study of music festival attendees. *Tourism Management, 47*, 68–76.

Hwang, J. S., Liu, C., Li, G., & Park, S. T. (2020). A study on problem solution of culture tourism festival through big data analysis. *International Journal of Emerging Multidisciplinary Research, 4*(3), 20–24.

Kim, T. T., Karatepe, O. M., Lee, G., & Demiral, H. (2018). Do gender and prior experience moderate the factors influencing attitude toward using social media for festival attendance? *Sustainability, 10*, 3509.

Kim, M. J., Lee, C. K., & Jung, T. (2020). Exploring consumer behavior in virtual reality tourism using an extended stimulus-organism-response model. *Journal of Travel Research, 59*(1), 69–89.

Koba, Y. (2020). Internet marketing communication in event tourism promotion. In A. Hassan & A. Sharma (Eds.), *The emerald handbook of ICT in tourism and hospitality* (pp. 149–164). Emerald Publishing Limited.

Koukopoulos, Z., & Koukopoulos, D. (2017). Smart dissemination and exploitation mobile services for carnival events. *Procedia Computer Science, 110*, 24–31.

Lafarge, D. (2017). 'The making of the future: Now': Edinburgh art festival. *Visual Culture in Britain, 18*(3), 401–409.

Lagarense, B. E. S., Hidayah, T., & Abdillah, F. (2018, 26–27 October). Digital technology and Pentahelix Role model for Sport tourism event of IVCA 2018 in Bali. In *Proceedings – 2018 international conference on applied science and technology*. Manado: IEEE, pp. 263–270.

Law, R., Leung, R., & Buhalis, D. (2009). Information technology applications in hospitality and tourism: A review of publications from 2005 to 2007. *Journal of Travel and Tourism Marketing, 26*(5–6), 599–623.

Lee, J., & Lee, C. H. (2019). A study on the analysis of news data for the improvement of local flower festival. *Journal of Industrial Convergence, 17*(4), 33–38.

Lee, W., & Paris, C. M. (2013). Knowledge sharing and social technology acceptance model: Promoting local events and festivals through facebook. *Tourism Analysis, 18*(4), 457–469.

Lee, S. "Shawn", Boshnakova, D., & Goldblatt, J. (2017). *The 21st century meeting and event technologies: Powerful tools for better planning, marketing, and evaluation*. Apple Academic Press.

Lee, W., Xiong, L., & Hu, C. (2012). The effect of Facebook users' arousal and valence on intention to go to the festival: Applying an extension of the technology acceptance model. *International Journal of Hospitality Management, 31*(3), 819–827.

Lexhagen, M., Nysveen, H., & Hem, L. E. (2005). Festival coordination: An exploratory study on intention to use mobile devices for coordination of a festival. *Event Management, 9*(3), 133–146.

Li, J., Ye, Q., Deng, X., Liu, Y., & Liu, Y. (2016). Spatial-temporal analysis on spring festival travel rush in China based on multisource big data. *Sustainability, 8*, 8814.

Li, J., Xu, L., Tang, L., Wang, S., & Li, L. (2018). Big data in tourism research: A literature review. *Tourism Management, 68*, 301–323.

Li, X., Su, X., Hu, X., & Yao, L. (2019). App users' emotional reactions and festival satisfaction: The mediating role of situational involvement. *Journal of Travel & Tourism Marketing, 36*(9), 980–997.

Liere-Netheler, K., Packmohr, S., & Vogelsang, K. (2018). Drivers of digital transformation in manufacturing. In *Proceedings of the 51st Hawaii international conference on system sciences*, pp. 3926–3935.

Lindsey-Mullikin, J., & Borin, N. (2017). Why strategy is key for successful social media sales. *Business Horizons, 60*(4), 473–482.

Liu, C. R., Lin, W. R., Wang, Y. C., & Chen, S. P. (2019). Sustainability indicators for festival tourism: A multi-stakeholder perspective. *Journal of Quality Assurance in Hospitality and Tourism, 20*(3), 296–316.

Lohr, S. (2013). *The origins of 'Big Data': an etymological detective story.* New York Times. Retrieved https://bits.blogs.nytimes.com/2013/02/01/the-origins-of-big-data-an-etymological-detective-story/. Accessed: 2 Dec 2021.

Luxford, A., & Dickinson, J. E. (2015). The role of mobile applications in the consumer experience at music festivals. *Event Management, 19*(1), 33–46.

MacKay, K., Barbe, D., Van Winkle, C. M., & Halpenny, E. (2017). Social media activity in a festival context: Temporal and content analysis. *International Journal of Contemporary Hospitality Management, 29*(2), 669–689.

Mantoro, T., Jaafar, A. D., Aris, M. F. M., & Ayu, M. A. (2011, 7–9 April). HajjLocator: A Hajj pilgrimage tracking framework in crowded ubiquitous environment. In *2011 International conference on multimedia computing and systems.* Ouarzazate, pp. 1–6.

Mariani, M. M., & Giorgio, L. (2017). The "Pink Night" festival revisited: Meta-events and the role of destination partnerships in staging event tourism. *Annals of Tourism Research, 62*, 89–109.

Marine-Roig, E., Martin-Fuentes, E., & Daries-Ramon, N. (2017). User-generated social media events in tourism. *Sustainability, 9*(12), 2250.

McCartney, G., & Ip Si Kei, D. (2018). Art festival attendance: Who attends and why with implications on developing future festival attractiveness. *Journal of Convention & Event Tourism, 19*(3), 268–285.

Mistilis, N., & Dwyer, L. (2000). Information technology and service standards in MICE tourism. *Journal of Convention & Exhibition Management, 2*(1), 55–65.

Mitchell, R. O., Rashid, H., Dawood, F., & Alkhalidi, A. (2013, 20–22 January). Hajj Crowd management and navigation system people tracking and location based services via integrated mobile and RFID systems. In *2013 International conference on computer applications technology (ICCAT).* Sousse, pp. 1–7.

Mizzau, L., Montanari, F., & Massi, M. (2019). Extending a Festival's social atmosphere online: The case of Fotografia Europea. In M. Volgger & D. Pfister (Eds.), *Atmospheric turn in culture and tourism: Place, design and process impacts on customer behaviour, marketing and branding* (pp. 277–289). Emerald Publishing Limited.

Mowafi, Y., Zmily, A., Abou-Tair, D. E. D. I., & Abu-Saymeh, D. (2013, 12–14 November). Tracking human mobility at mass gathering events using WISP. In *Second international conference on future generation communication technologies (FGCT 2013).* London, pp. 157–162.

Ngernyuang, K., & Wu, P.-Y. (2020). Using social media as a tool for promoting festival tourism. *International Journal of Computer Science and Information Technology, 12*(3), 17–32.

Norfolk, L., & O'Regan, M. (2020). Biometric technologies at music festivals: An extended technology acceptance model. *Journal of Convention and Event Tourism, 22*(1), 36–60.

Olya, H., Jung, T. H., tom Dieck, M. C., & Ryu, K. (2020). Engaging visitors of science festivals using augmented reality: Asymmetrical modelling. *International Journal of Contemporary Hospitality Management, 32*(2), 769–796.

Opute, A. P., Irene, B. O., & Iwu, C. G. (2020). Tourism service and digital technologies: A value creation perspective. *African Journal of Hospitality, Tourism and Leisure, 9*(2), 1–18.

Osnes, B. (2010). *The shadow puppet theatre of Malaysia: A study of wayang kulit with performance scripts and puppet designs*. McFarland & Co.

Pan, J., & Lai, J. (2019). Spatial pattern of population mobility among cities in China: Case study of the National day plus mid-autumn festival based on Tencent migration data. *Cities, 94*(967), 55–69.

Quinn, B. (2006). Problematising "Festival tourism": Arts festivals and sustainable development in Ireland. *Journal of Sustainable Tourism, 14*(3), 288–306.

Robertson, M., Yeoman, I., Smith, K. A., & Mcmahon-Beattie, U. (2015). Technology, society, and visioning the future of music festivals. *Event Management, 19*(4), 567–587.

Roemer, M. K. (2007). Ritual participation and social support in a major Japanese festival. *Journal for the Scientific Study of Religion, 46*(2), 185–200.

Roemer, M. K. (2010). Shinto festival involvement and sense of self in contemporary Japan. *Japan Forum, 22*(3–4), 491–512.

Saleh, F., & Ryan, C. (1993). Jazz and knitwear: Factors that attract tourists to festivals. *Tourism Management, 14*(4), 289–297.

Saragih, H. S. (2021). Mining and exploring electronic word-of-mouth from twitter: Case of the Java Jazz festival. *Journal of Hospitality and Tourism Technology, 12*(2), 341–354.

Sterritt, R. (2002, 5–6 December). Towards autonomic computing: Effective event management. In *27th annual NASA Goddard/IEEE Software Engineering Workshop*. Greenbelt, MD, pp. 40–47.

Süli, D., & Martyin-Csamangó, Z. (2020). The impact of social media in travel decision-making process among the Y and Z generations of music festivals in Serbia and Hungary. *Turizam, 24*(2), 79–90.

Talantis, S., Shin, Y. H., & Severt, K. (2020). Conference mobile application: Participant acceptance and the correlation with overall event satisfaction utilizing the technology acceptance model (TAM). *Journal of Convention and Event Tourism, 21*(2), 100–122.

Thomson, A., Proud, I., Goldston, A. L., & Dodds-Gorman, R. (2021). Virtual reality for better event planning and management. In K. Birdir, S. Birdir, A. Dalgic, & D. Toksoz (Eds.), *Impact of ICTs on event management and marketing* (pp. 177–198). IGI Global.

Tofigh, F., Amiri, M., Shariati, N., Lipman, J., & Abolhasan, M. (2020). Crowd estimation using electromagnetic wave power-level measurements: A proof of concept. *IEEE Transactions on Vehicular Technology, 69*(1), 784–792.

Tondo, J. S. F. (2010). Popular religiosity and the transnational journey: Inscribing Filipino identity in the Santo Niño Fiesta in New Zealand. *The Asia Pacific Journal of Anthropology, 11*(3–4), 219–244.

Unurlu, C. (2021). The integration of social media into event tourism. In K. Birdir, S. Birdir, A. Dalgic, & D. Toksoz (Eds.), *Impact of ICTs on event management and marketing* (pp. 69–85). IGI Global.

Uysal, M., & Gitelson, R. (1994). Assessment of economic impacts: Festivals and special events. *Journal of Festival Management and Event Tourism, 2*(1), 3–10.

Van Winkle, C., Mackay, K. J., & Halpenny, E. A. (2017). *Factors affecting mobile device use at festival attractions*. Retrieved from: https://scholarworks.umass.edu/cgi/viewcontent.cgi?article=2107&context=ttra. Accessed 30 July 2021.

Van Winkle, C. M., Bueddefeld, J. N. H., Halpenny, E. A., & MacKay, K. J. (2019). The unified theory of acceptance and use of technology 2: Understanding mobile device use at festivals. *Leisure Studies, 38*(5), 634–650.

Waida, M. (n.d.). Tech Thursday: Everything you need to know about augmented reality event planning. *Social Tables*. Retrieved from: https://www.socialtables.com/blog/event-technology/augmented-reality/. Accessed 30 July 2021.

Wolhuter, S. (2021). How virtual reality is transforming customer experience. *We Are Brain*. Retrieved from: https://www.wearebrain.com/blog/innovation-and-transformation-strategy/virtual-reality-and-customer-experience/. Accessed 30 July 2021.

Wreford, O., Williams, N. L., & Ferdinand, N. (2019). Together alone: An exploration of the virtual event experience. *Event Management, 23*(4–5), 721–732.

Yamin, M. (2019). Managing crowds with technology: Cases of Hajj and Kumbh Mela. *International Journal of Information Technology (Singapore), 11*(2), 229–237.

Yamin, M., Mohammadian, M., Huang, X., & Sharma, D. (2008, 10–12 December). RFID technology and crowded event management. In *2008 International conference on computational intelligence for modelling control & automation*. Vienna, pp. 1293–1297.

Ziyadin, S., Koryagina, E., Grigoryan, T., Tovma, N., & Ismail, G. Z. (2019). Specificity of using information technologies in the digital transformation of event tourism. *International Journal of Civil Engineering and Technology, 10*(1), 998–1010.

Dr Sana Maidullah received a PhD degree in Humanities and Social Science in 2019 from the Indian Institute of Technology, Jodhpur, India. She subsequently joined Lovely Professional University, Punjab, India, where she worked as an Assistant Professor. At present, she is an Assistant Professor in Tourism and Hospitality Management, Indian Institute of Management Sirmaur (IIM Sirmaur), District Sirmaur, Himachal Pradesh, India. Dr Maidullah's principal research interests are in tourism, decision making, consumer behavior, and motivation. She is also a multi-disciplinary design practitioner with experience in tourism products, communication, and environmental design.

Mohammad Rokibul Hossain is currently working as an Assistant Professor of Marketing, Faculty of Business Studies at Premier University, Bangladesh. His area of concentration in Marketing. Hossain's fundamental research interests lie in Tourism Marketing, Customer Engagement, Consumer Behavior, Sustainable Marketing, and Neuromarketing. He published articles on Tourism Marketing, Customer Relationship Management and Agricultural Marketing in different national and international peer-reviewed journals.

Fahmida Akhter is an Assistant Professor of School of Business Administration, East Delta University, Bangladesh. Her area of concentration is Accounting & Information Systems. She started her career as an academic in 2015. Her fundamental research interest lies in the area of Corporate Governance, Corporate Social Responsibility Practices of Firms, Sustainable Environmental Practices of Corporations, Human Resource Disclosure Practices, and other voluntary disclosure practices of Companies and Sustainable Tourism. She's published articles in various national and international peer-review academic journals.

Dr Azizul Hassan is a member of the Tourism Consultants Network of the UK Tourism Society. Dr Hassan has been working for the tourism industry as a consultant, academic, and researcher for over 20 years. His research interest areas are technology-supported marketing for tourism and hospitality, immersive technology applications in the tourism and hospitality industry, and technology-influenced marketing suggestions for sustainable tourism and hospitality industry in developing countries. Dr Hassan has authored over 150 articles and book chapters in leading tourism outlets. He is also part of the editorial team of 25 book projects from Routledge, Springer, CAB International, and Emerald Group Publishing Limited. He is a regular reviewer of a number of international journals.

Chapter 2
Technology Application in Tourism Festivals in Asia: Theoretical Discussions

Emmanuel Sebata and Md. Ruhul Amin Mollah

Introduction

The rapid technological advancements in tourism are turning it into one of the most dynamic industries lately. This development characterized with increased usage of technologies such as internet and various mobile communication devices has led to the emergence of the innovative eTourism (Buhalis & Licata, 2002). This confirms the assertion by Hjalager (2009) that technology is moving into the tourism festival area. Getz (2005) adds that modern festivals have matured with time, lots of original arguments have surfaced and, progressively, they happen to be even more aligned to tourism. Tourism festivals are gradually more staged by several cities globally. Within the past decade, contemporary light events have come into sight not only in Oceania, South America but also in Asia, yet a number of them have already gained international relevance (Giordano & Ong, 2017). Several Southeast Asia economies favor tourism areas and major urban locations via the software business and countrywide tourism, and the tourism sector can be employed as a mechanism to revitalize the local economies through appealing to local and international travelers. Nevertheless, preserving and growing tourism in the long-term requires extended backing for example in terms of physical infrastructure development. By and large, advanced economies revitalize their local markets and protect the identification of their local inhabitants through tourism festival and related events (Kim & Kim, 2013).

E. Sebata
The Sport Business School, Beijing Sport University, Beijing, China

Kyambogo University, Kampala, Uganda

M. R. A. Mollah (✉)
Griffith University, Southport, Australia

Jagannath University, Dhaka, Bangladesh
e-mail: mdruhulamin.mollah@griffithuni.edu.au

A. Hassan (ed.), *Technology Application in Tourism Fairs, Festivals and Events in Asia*, https://doi.org/10.1007/978-981-16-8070-0_2

Since festivals in Asian economies are important and deemed as a mechanism to means to hasten the development of rural areas since they boost local and regional socio-cultural and the financial sustainability in poor localities, yet they also aid in the renaissance of the local economies (Yun & Park, 2014), the authors tourism studies focused on adoption of modern technology in Asia is must be prioritized. This concurs with Luxford and Dickinson (2015) who noted that the advancement of mobile applications has been evident at various tourism festivities and impacting on the leisure experiences. Additionally, in her research, Sigala argues that several research has been centered on the use and impact of technologies in sectors such as food, accommodation, transports, Destination Management Organization (DMOs), tourism attractions and but few research has been done on festivals and related events (Sigala, 2017). Based on this, the need for more studies that can offer a theoretical overview on application and adoption of technology in tourism festivals in Asia is paramount. Moreover, as the COVID-19 pandemic persists and creates a paradigm shift in tourism, researchers in tourism and related academic domains should embrace technology in their studies. For instance, the conditions of COVID-19 (such as physical distancing, lockdowns) have increased the demand for technology usage by both travelers and businesses such as festivals and events (Sigala, 2020). In the same vein, the application of more refined internet technology allows the capture, analysis and utilization of personal data of tourism festival customers; this also enhances online marketing of the festival in general (Pan & Li, 2011), and the combination of innovative technologies and culture commonly encountered in festivals harmonizes the quality of tourism festivals and also produces a well-built appeal (Yang et al., 2011a). There is also a notion that the travel industry is a strong force in furthering the tourism festival development and diversification. The ever-rising demand in the tourism sector has brought about escalated income inflows for several tourism festivals and is accountable for their continued existence as financial feasible projects (Quinn, 2006). Hence, federations in the event and festival industry must aid organizers of tourism festivals and related events to construct "smarter" tourism festivals and market high-tech related services (Cimbaljević et al., 2021).

In brief, many technologies and ICT tools are reshaping the global tourism industry (such as artificial intelligence, smart devices, drones, big data, virtual reality, augmented reality and many other state-of-the-art technological innovations (Sigala, 2017). This applies also to most of the emerging Asian economies such as China, South Korea, Thailand, India, Japan, Bangladesh, Hong Kong, Malaysia and others. Even though many studies have been carried out on tourism festivals in all those countries, to date, there has been no significant and substantial investigations that can offer a theoretical discussion of technological application to tourism festivals in Asia. This study therefore sheds light on the existing theoretical discussions but particularly focusing on the technology application in tourism festivals in the Asian context, an area that has predominantly been understudied.

Tourism Festivals and Trends in Asia Today

Notwithstanding the latest innovations, several tourism festivals are being organized in various communities across the globe, Asia inclusive. Even if the radically changing circumstances comprise of fresh markets, fresh localities, new capitals, new districts and new information and communication systems, numerous individuals still opt to attend big festivals (Fournier, 2019). While leisure pursuits stimulate modern festivals, among the recent developments is improving the festival attraction through inclusion of additional extra events and activities. In a bid to attract bigger audiences, bring in more revenues and achieve globalization, Asian economies have embraced such events commonly known as 'hybrid festivals'. This dynamic process is termed as 'hybridization' (Fournier, 2019, p. 19). Accordingly, Xie and Xu (2004), several festivals such as religious festivals, folk festivals and art festivals among others are hosted in China and other neighboring Asian economies, for example, the international Woodcarving Art Festival in Taiwan (Hsiao & Chuang, 2015).

The Beer Festival at Qingdao is one of the most famous tourism festivals in China. This festival was shaped by a popular Chinese beer brand known as Qingdao beer. This festival offers opportunities for tourism through leisure and entertainment, amusement and related gaming activities or games (Ma & Lew, 2012). In China, festive sport events have become increasingly popular in large cities and they provide a new setting through which participants take part in various sporting activities within a tourism festival environment. In similar vein, in their research done in Korea through employing quantitative model to analyze the effectiveness of festivals, Jung and Kim (2012) note that a country hosting large events such as the Olympic Games, FIFA World Cup and Expos bring on fresh investments through advancing its latest technologies and future development capacity while at the same time growing its exportation potential. The self-determination theory (SDT) is used to understand participants' involvement in these events. Based on this theory, the authors assert that fulfillment of three fundamental psychological needs namely the autonomy need, the relatedness need and the competence need influences the individual motivation and happiness from events and festivals. These events feature recreation, mass participation and celebration experiences in one festival. Sports oriented festivals offer a chance for attendees to take part in relaxation and enjoyment activities in an entertaining ambiance, they as well provide an opportunity to build city identity, and most notable is the 100-Kilometer Hiking Festival organized in Shenzhen (Chen et al., 2021). In Korea, medicine tourism festivals are lately being organized, although large-scale health tourism festivals are under researched (Song et al., 2014). The Loy Krathong Festival is amongst the best water festivities in Thailand regularly staged according to lunar calendar. Water festivals reflect the Thai way of living with their value being perceived in a family, communal, wider societal setting at large and hence they have become a real achievement of festival tourism (Thadaniti, 2014). This explains why Yang et al. (2011a, b) highlights cultural festivals as events with potential to assist festival organizers augment participant profiles, boost attraction levels, and elevate the prospective market in the global tourism market. In Japan, food festivals essentially contribute to local tourism and neighborhood development and also possess huge socio-economic and culture development potential (Kim, 2015). In different parts of India, several queer film festivals are now very fundamental tourism and culture exhibition platforms since 2000s (Mokkil, 2018), while in China, as the drive for creation of new tourism events continues to felt, film, fashion photography, food and tea festivals have turned out to become a common phenomenon in Beijing, Shanghai and throughout different parts of China. The Dragon Boat Festivals originally organized in China has been spread to Korea, Japan and Southeast Asia (Ma & Lew, 2012).

However, the audience for these festivals is not only going to be affected by the ever-increasing number of races, but will be affected by the global COVID-19 pandemic. Thus, Calvo-Soraluze and Blanco (2014) assert that the application of appropriate technologies and modern social media platforms should be a strategy employed to connect with the audience as this also offers a tremendous opportunity that adds to the general visitor experience. In line with this, theories such as the actor-network theory (ANT) can help to appreciate the contribution various human and non-human organs in supporting tourism performances through technology (Tribe & Mkono, 2017). This theory provides room for understanding the function technology plays in enhancing tourism experiences (Jóhannesson, 2005). Unsurprisingly, Calvo-Soraluze and Blanco (2014) assert that technologies and social networking apps have turned into a strong and powerful apparatus in the networking community. A festival tourism conceptual framework is therefore proposed, it outlines four different categories of tourism festivals and these include; global contemporary festivals, national heritage festivals, local contemporary festivals and local heritage festivals putting into consideration the historical and geographical context of their emergence and organization of each of them (Ma & Lew, 2012). Such festivals are deemed as vital in both the tourism sector and in society at large. Such festivals attract tourists to an area; this not only increases revenue but also promotes local industry development. Similarly, knowledge of the cultural beliefs and customs is stirred and amongst locals, unity and identity are promoted (Ngernyuang & Wu, 2020). In this regard, in Vietnam, the social exchange theory and stakeholder theory were employed to examine the perception of local residents about the impacts of tourism and their role in developing tourism (Sinh et al., 2016).

Further, since tourism festivals have previously been analyzed in some literature about tourism and leisure and reported to have potential to provide residents and participants with exceptional and multidimensional leisure time experiences (Van Winkle et al., 2019). More other theoretical perspectives such as the exchange theory can be employed to explain the resident's reactions to such events (Getz, 2008). Nevertheless, several questions must be answered, conventional theories such as theory of social exchange, cognitive constraint and social comparison designed by understanding ordinary individuals ought not to be plugged and played so as to realize the manners of individuals affected by technology (Sigala, 2017). More so, several theories such as game theory can be employed as an instrument to establish various traits that give tourism festivities a key position in the global market.

Correspondingly, the game theory can be used to tourism festivals in several situations to ensure that high quality is attained (Zyl, 2012). In the highly web-driven market environment, technology use in tourism remains a key aspect (Law et al., 2009). According to UNESCO, the justification for application of technology to tourism festivals has been highlighted as follows:

> The overwhelming preference and focus on economic impacts in tourism has precluded exploration important factors such as festivals and technology (UIS-UNESCO, 2015, p. 31).

Areas of Technology Application in Tourism Festivals in Asia

Intriguingly, the travel industry is profoundly abetted by technology advancements and by theoretical innovations that influence the actions of customers and the competition in the business environments. Basing on this school of thought, the development in the socio-economic value of tourism is also partially driven by other key players such as technology not only just the tourism key players (Hjalager, 2015). In relation to this, Lohmann and Netto (2017) explain that the general systems theory allows each of these systems (technology and tourism) either to be analyzed as a whole (the united system), or to be divided into parts to facilitate their understanding and study. From the theoretical perspective, tourism festivals are intangible services offered to tourist putting into consideration the preference of the tourist in terms of both the festival facilities and event day activities. In modern day Korea, a variety of festival related information is channeled to festival goers at a very unprecedented speed with help of information technologies. Take for instance the World Club Dome Korea festival at which the social networking services (SNS) are used to spread information about the event (Kim et al., 2018).

Lately, the diffusion of technology in the tourism business has brought about substantial developments. Thus, unique innovative technologies such as augmented reality (AR) have been employed as an instrument to enhance internet marketing in tourism (Shabani et al., 2019). Again, the use of Artificial Intelligence (AI) and other digital innovations such as machine learning have added more value to the customer virtual and non-virtual experience (Sigala, 2017). E-marketing has also turned into a significant approach in festivals. Therefore, marketing and promotion of tourism festivities through online platforms inspires people to interchange diverse and useful information via SNS and other digital platforms (Kim et al., 2018). Thus, more marketing activities and branding in tourism must be areas of attention in Asian economies particularly Bangladesh (Hassan & Burns, 2014). In order to improve marketing of tourism festivals especially online, festival organizers must forge ideal partnerships with Destination Marketing Organizations (DMO), this can be cost effective since a hot link from the DMO's home page can be used and this directly facilitates overseas marketing of a tourism festival (McKercher et al., 2006). In fact, internet marketing plays a key role in driving the aspirations of tourists at tourism festivals, also, it provides an opportunity for the tourists to make well-versed choices since they have opportunity to explore the various online platforms before making final choices (Pan & Li, 2011; Kim et al., 2018).

In Taiwan, currently tourist festivals are marketed by private tourism investors though they work hand in hand with particular government departments. A number of tourism businesses involved in organizing festivals are by now using information technologies and several media channels as a market tool that helps them to keep in contact with their clients. Note however should be taken that in order to ensure sustainability, these marketing approaches need to be kept up-to-date (Hsiao & Chuang, 2015). The previous two decades have seen technology the evolving technology moving the globe in a more digital phase. Many individuals possess moveable technological gadgets for instance tablets and smartphones and they ensure that these gadgets are well connected to the internet at all times. This demeanor influences tourism but more particularly tourism festivals and other related events (Ngernyuang & Wu, 2020). Technologies like Bluetooth and Wi-Fi data are used to monitor visitor movement at events or festivals (Kalvet et al., 2020). Some of the apps used in these portable devices use Global positioning system (GPS) are part of the recent breed of localization technology that is easily employed to track travelers. The use of such applications can offer substantial and correct information about the travelers such as movement designs. Such information can easily be interconnected with technologies such as geographical information system (GIS), Quantum GIS, Google Earth and many others (Yun & Park, 2014). Consequently, Kim and Kim (2017) identified the task technology fit (TTF) theory as one of the theories that provided a very important theoretical framework for linking mobile technologies to tourism industry. They note that it can be imagined that the connection between mobile technologies and task of festival goers in tourism are absolutely connected to the intent to utilize services such as digital or mobile tourism. Research however shows that tourism festival organizers must make sure that technological applications do not impede with the leisure experience of the customers. More still, festival goers while at their homes can be supported by technologies such as broad-bandwidth and digital subscriber line (DSL) technology such as Asymmetric digital subscriber line (ADSL) as form of data communications technology enables faster digital data transmission at good bandwidths using telephones (Buhalis & Licata, 2002). Such innovative digital technologies are believed to have potential of improving the virtual customer experience which ultimately boosts the overall customer satisfaction. In line with this, Chang and Hsieh (2017) in their study attempted to bridge the gap between the festival organizer and the festival guest basing on the theory of reasoned action. In practice, they explain that this theory helps to meet the satisfaction levels of customers. They add that the identity theory can also be used as a basis for understanding how the festival audience identifies or interacts with the tourism festival or festival organizer.

Further, the stakeholder theory proposed by Reid also offers a key role to both theory and managing of tourism events such as festivals. He reveals that this theoretical approach if used to plan and organize events not only helps to identify and categorize they key stakeholders (among which are the technology service providers), but also helps to assess the needs and wants of the respective stakeholders. This assists festival organizers in monitoring satisfaction levels and relationships. This theory also monitors event stakeholders to gauge festival sustainability (Reid, 2011). Another big trend and influence of technology on tourism festivals is the use of social media. Drawing from the works of Ngernyuang and Wu (2020), it is crystal clear that mobile technology particularly social media platforms provide an opportunity to audiences to interact with the festival service providers; through this the customers have a chance to freely convey their judgments and remarks concerning the events. Their research also identifies social media as one of the potential advertising methods for modern tourism festivals. Pawłowska et al. (2015) outlined some of these social media platforms and these included but not limited to Facebook, YouTube, Instagram, Google+ and Twitter. They also noted digital platforms such as Trip Advisor help to offer an opportunity for exchanging objective follow-up remarks and comments about the tourist destination and related events. Ballantyne et al. (2014) add that the use of social media can help tourism festival organizers to engage attendee through crowd sourcing. Through this, festival attendees can be involved in the process of selection of music and social activities hence facilitating personal engagement and customer satisfaction.

Before 2050, attendees of tourism festivals will be able to use contact lenses to visualize and boost their personal experiences while attending these events (Robertson et al., 2015). In collaboration with security departments such as police, organizers of tourism festivals have explored the application of face recognition technologies as platforms for controlling access, security and safety at during tourism festivals (Crampton et al., 2019). Body scanning technology was also introduced and this involves scanning the whole-body as one of the key safety control measures as opposed to the tradition metal detecting devices. (Hjalager, 2015). Since Asian countries are the most densely populated countries such as Hong Kong, Macau, Singapore, India, China and Indonesia, the adoption and application of innovative technologies for proper security and access control cannot only stimulate improved festival management, but also stimulate the general attendee's leisure experience in the long run. Nevertheless, much as tourism has embraced technological innovations faster, the degree of adoption differs from one event organizer to another. Moving forward, in a bid to make festivals even more "smarter", many economies including those in Asia must devise necessary means of minimizing the limitations surrounding the application of technology (Cimbaljević et al., 2021). This affirms that although technology is assumed to be the cause of progress and improvements in the staging of tourism festivals, the assertion, however that technological application does not negatively impact tourism festivals is quite misleading.

Of late, COVID-19 has created room for embracing the role of technology in the resurgence and rethinking of the tourism-related events, as a result, a paradigm shift is being experience in the tourism is primarily characterized by e-tourism growth. In a bid to counter the COVID-19 pandemic, new technology such as artificial intelligence, mobile technologies, robotics, virtual reality and augmented reality has been adopted by tourism event organizers. Researchers on COVID-19 and tourism should therefore rethink and refocus on the benefits, utilization of such digital technology platforms (Sigala, 2020). In summary, it is clear that tourism festivals in Asia just like it is elsewhere can be influenced by the diffusion of hi-tech innovations or modern technologies and these include but not limited to mobile applications, self-service kiosks or terminals, social media, smart cards, mobile payment, festival-activity forecast, QR codes, online booking, robots, drones, Free Wi-Fi and augmented reality. However, caution should be taken that the speed and magnitude of application of these technologies varies from country to country due to various factors at play.

The Concept of Smart Tourism Festivals

Technology advances are driving every tourism-related supply into a new form known as smart tourism. The degree of smartness nonetheless is influenced by the factors such as the connectivity and components of data which ideally should be "smart" (Sigala, 2017). This innovation has been felt even in tourism festivals in Asia. With technology evolving at a very fast speed, the lifestyles of many individuals in various parts of the globe such as Asia continue to be positively influenced. Society must therefore continue to welcome different hi-tech innovations. Smart tourism is among the thrilling technologies that will shape the tourism sector in the coming decades. The past decade has been characterized by an incredible evolution of information and communication technology and this has drawn a lot of attention from several researchers, technology service providers and many other stakeholders (Ye et al., 2020).

Smart tourism festivals have tremendous value in as far as increasing the quality of leisure experience of festival participants is concerned, they encourage healthy and competitive tourism festival business (Cimbaljević et al., 2021). In Asia and many other parts of the globe, even though the growth of smart tourism events is evidently underway, several key developments that portray well processed, systemized and established smart tourism festivals (such as value creation) are still in infant stages (Gretzel et al., 2015). Festivals with potential to fully adopt modern technology that can enhance the tourist experience of the festival goers are thus referred by this study as smart tourism festivals.

The Opportunity

Tourism festivals have the potential to economically boost regional economies especially during times of monetary devaluation. They provide a very exceptional opportunity for local business ventures to become part and partial of festivals but also to maximize their revenues through tapping from the tourism festival spinoffs (Niekerk & Coetzee, 2011). Following the development of tourism events such as film festivals throughout the past two decades, many economies are bound to continue experiencing a paradigm shift (characterized by increased strategic collaborations with technology service providers, institutions and other key players) in their tourism sectors due to technological advancements (Dawson & Loist, 2018). This explains why the application of technology even among Asian top economies such as China, Japan and South Korea has become part of their tourism development agendas. Accordingly, festivalization has emerged into one of the leading global tourism growth phenomena. Since festivals have the potential to amplify the attendee's good sentiments and sensations, technology adoption by a particular festival planner and or organizer further boosts the tourism experiences and hence the possibility of revisit a similar festival (Kendall et al., 2020).

The application of technology is therefore more likely to facilitate this process through allowing for faster and wider circulation of festival related information and also enhance augmenting marketing and promotional activities. This capitalizes on economic opportunities of tourism festivals. O'Sullivan and Jackson (2002) maintain that tourism festivals create a strong opportunity that can contribute to sustainable economic development since such opportunities have not fully been exploited. This explains why Quinn (2010) argues that whereas festivals such as urban art festivals may be proliferating, one should not necessarily conclude that they are flourishing. This study thus explains that this could be because such tourism festivals have not yet substantially embraced modern technologies.

Giordano and Ong (2017) highlighted that the growth and development of tourism festivals has impact on urban tourism growth, this has facilitated the emergence of tourism cities in different countries across the globe. Modern technology plays a very key role in offering new platforms for individuals in such communities to network and keep in constant online contact with each other; through this they can exchange vital information about their experiences during festivals (Laing & Mair, 2015). On the contrary, this new technology is increasingly disrupting tourism distribution whose basic elements include the tourists, the suppliers and the intermediaries. Basing on the activity theory, even if all these parties are perceived as individual elements in their respective systems, there should be interaction between the three systems (Jorgensen, 2017). This concurs with Sigala (2017) who concluded that the advancements of technologies have interrupted many key players in tourism especially at the stage when new businesses lower the prices as compared to the already existing key players.

As internet continues to grow as the widest and fastest technological revolution in history, tourism festival organizers have the opportunity to promote and advertise their products as a result of the increased accessibility of several mobile technologies that are simple and easy to use within today's real techno-social environment (Calvo-Soraluze & Blanco, 2014). Therefore, this study asserts that there is need to increase awareness of business communities especially among the emerging Asian nations about the opportunities associated with the application modern technologies in meeting the ever changing and increasing diverse needs of tourists.

Limitations

Even though technology embodies a great chance for various categories of tourism and ICT business organizations and key players to collaborate, the collaborative environment for most of their business operations is still limited by various complexities (Buhalis & Deimezi, 2004). Also, however much there is increased emphasis on technology usage and adoption, many tourism service providers are still sluggish in as far as integrating modern technologies in the day-to-day operations of their businesses is concerned. The association between tourism service providers and technology experts is still insufficient; this leaves most of the tourism service providers and managers with inadequate skills and support on how best to handle the application process of technology in their business operations (Law et al., 2009). In the near future, all these stakeholders particularly in Asia need to learn new ways on how to create, maintain and improve their collaboration efforts in the areas of adoption of modern tourism technology. Also, Cimbaljević et al. (2021) noted that due to several internal barriers such as limited finances, staffing challenges, operation related challenges and external obstacles especially market forces, unfavorable laws and policies among others need to be more comprehended by most of the stakeholders in the business of tourism festivals and events. Their study further notes that technology application requires huge infrastructure, management and implementation costs and thus making it difficult to place priority in ICT implementation. Therefore, the assertion by (Hassan et al., 2020) that government needs to identify more with the private investors so as to foster the growth and development of their tourism industry in Asian economies such as Bangladesh is indispensable. The authors of this paper therefore argue that unless both the public and private sectors capitalize on the technological advancements in tourism specifically in festivals, the Asian online tourism and travel industry is likely to continue facing even more stiffer competition especially from its western counterparts such Europe.

The adoption and application of technologies in tourism is being limited by insufficient financial resources, poor telecomputing and geographical obstacles such as forests (Pawłowska et al., 2015). For example, several mobile technology devices such as smartphones and GPS devices are sometimes not accurate due to signal obstruction and disturbance by various natural and artificial obstacles such as skyscrapers (Yun & Park, 2014). In the same line, some biometric technologies used by festival organizers are sometimes not reliable, accurate, and secure, hence they at times compromise with the privacy of festival attendees (Norfolk & O'Regan, 2020). Meanwhile, however much organizations involved in tourism business have realized the need to apply mobile technology in their business operations, there is still a big test of developing of relevant digital marketing content that is suitable for the various information distribution channels (Buhalis & Licata, 2002). This also applies to even to the top Asia economies that are world leaders in technological development such as China, South Korea and Japan.

Conclusion and Future Research

Technology has potential of reshaping tourism festivals in Asia and beyond. In this study, it can be ascertained that advances in technology in tourism are gaining prominence both in theory, policy and practice. Technology in a tourism festival plays an important role in sustainable economic development. It is evident that even though the powerful harmony between technologies and festivals is existent, their application especially in the developing Asian economies has been understudied. If effectively applied, technology in tourism is bound to bring about a new and unique breed of festival known as "smart tourism festivals". Smart tourism aims at development of information and communication infrastructure and capabilities with potential of improving management of tourism festivals and events, enhancing service innovation and boosting leisure experience, and ultimately, improving tourist satisfaction. Due to revitalization, commodification, heritagization and competitiveness of their tourism markets, a drastic transformation from traditional tourism festivals to hybridized festivals characterized by modern hi-tech innovations ranging from augmented reality, mobile applications, GPS to other digital technology is inevitable in Asia. More still, since most studies focus less on tourism festivals, a gap exists between research findings and theory building as well, therefore in future; more effort should be hinged on theory and model building particularly in tourism festival related studies. Clear understanding and adoption of theories such as a stakeholder theory by tourism festival organizers provides a framework to not only tap the efforts of all stakeholders but also to assess their satisfaction among primary stakeholder groups, particularly those in investment, planning, construction, and usage of ICT infrastructure in promotion and development of tourism products and services.

From a theoretical point of view, this paper is a good starting point for understanding some of the specific innovative technologies that can create value addition in tourism festivals. It is thus a first activity in longer project of understanding and analyzing of the application and influence of technology in the digital festivals. The underlying theoretical discussions draws closer the attention of key tourism players in Asian states, NGOs, private investors and multinational funding agencies towards innovation and technological advancement in Asian and global tourism festival markets.

Future theoretical lenses should illuminate through the methodological frameworks of digital marketing technologies and tourism festival planning especially the pre, during and post COVID-19 pandemic period. More studies should also not only extend their scope and explore the application of technology to specific tourism festival domains such as consumption, service provision, but also inquire into cost-benefit analysis of application of technology to tourism festivals particularly in the most underdeveloped Asian economies. More research on adoption, staging and management of smart tourism festivals in Asia is also still required. With sustainability emerging into a major concern in contemporary tourism, more research should also attempt to dissect technology application in tourism festivals towards the realization of the Sustainable Development Goals (SDGs) both in the respective

Asian economies. More broadly speaking, as Asia strives towards globalization, fresh techniques and hypotheses are necessary to examine the connection between technology and tourism festivals in each independent Asian state.

References

Ballantyne, J., Ballantyne, R., & Packer, J. (2014). Designing and managing music festival experiences to enhance attendees' psychological and social benefits. *Musicae Scientiae, 18*(1), 65–83.

Buhalis, D., & Deimezi, O. (2004). E-tourism developments in Greece: Information communication technologies adoption for the strategic management of the Greek tourism industry. *Tourism and Hospitality Research, 5*(2), 103–130.

Buhalis, D., & Licata, M. C. (2002). The future eTourism intermediaries. *Tourism Management, 23*(3), 207–220.

Calvo-Soraluze, J., & Blanco, A. V. (2014). Stimulating attendees' leisure experience at music festivals: Innovative strategies and managerial processes. *Global Journal of Management and Business Research, 14*(2), 2–17.

Chang, K., & Hsieh, T. (2017). From having fun to applause: The study of relationships among festival benefits, festival identity and festival support by viewpoints of the hosts and guests. *Sustainability, 9*(12), 2240.

Chen, F., Naylor, M., Li, Y., Dai, S., & Ju, P. (2021). Festival or sport? Chinese motivations to a modern Urban Hiking event. *SAGE Open, 11*(1), 1–13.

Cimbaljević, M., Stankov, U., Demirović, D., & Pavluković, V. (2021). Nice and smart: Creating a smarter festival – the study of EXIT (Novi Sad, Serbia). *Asia Pacific Journal of Tourism Research, 26*(4), 415–427.

Crampton, J. W., Hoover, K. C., Smith, H., Graham, S., & Berbesque, J. C. (2019). Smart festivals? Security and freedom for well-being in urban smart spaces. *Annals of the American Association of Geographers, 110*(2), 3–9.

Dawson, L., & Loist, S. (2018). Queer/ing film festivals: History, theory, impact. *Studies in European Cinema, 15*(1), 1–24.

Fournier, L. S. (2019). Traditional festivals: From European ethnology to festive studies. *Journal of Festive Studies, 1*(1), 11–26.

Getz, D. (2005). *Event management and event tourism*. Cognizant Communication Corporation.

Getz, D. (2008). Event tourism: Definition, evolution, and research. *Tourism Management, 29*, 403–428.

Giordano, E., & Ong, C. (2017). Light festivals, policy mobilities and urban tourism. *Tourism Geographies, 19*(9), 1–18.

Gretzel, U., Sigala, M., Xiang, Z., & Koo, C. (2015). Smart tourism: Foundations and developments. *Electronics Markets, 25*(3), 179–188.

Hassan, A., & Burns, P. (2014). Tourism policies of Bangladesh – A contextual analysis. *Tourism Planning and Development, 11*(4), 463–466.

Hassan, A., Kennell, J., & Chaperon, S. (2020). Rhetoric and reality in Bangladesh: Elite stakeholder perceptions of the implementation of tourism policy. *Tourism Recreation Research, 45*(2), 1–16.

Hjalager, A. (2009). Cultural tourism innovation systems – The Roskilde festival. *Scandinavian Journal of Hospitality and Tourism, 9*(2-3), 266–287.

Hjalager, A. (2015). 100 innovations that transformed tourism. *Journal of Travel Research, 54*(1), 3–21.

Hsiao, T., & Chuang, C. (2015). The cooperation model between tourism development and traditional culture: New perspectives on regional context. *Journal of Tourism and Cultural Change, 14*(2), 91–106.

Jóhannesson, G. T. (2005). Tourism translations: Actor–network theory and tourism research. *Tourist Studies, 5*(2), 133–150.
Jorgensen, M. T. (2017). Reframing tourism distribution – Activity theory and actor-3 network theory. *Tourism Management, 62*, 312–321.
Jung, B., & Kim, S. (2012). An analysis of festival satisfaction evaluation using multiple regression model EXPO 2012 YEOSU KOREA. *Communications in Computer and Information Science, 350*, 155–156.
Kalvet, T., Olesk, M., Tiits, M., & Raun, J. (2020). Innovative tools for tourism and cultural tourism impact assessment. *Sustainability, 12*(18), 7470.
Kendall, G., Chan, J. H. T., Yeung, M. C. H., & Law, K. K. (2020). Do film festivals attract tourists? *Current Issues in Tourism, 24*(11), 1482–1486.
Kim, S. (2015). Understanding the historical and geographical contexts of food festival tourism development: The case of the Tatebayashi Noodle Grand Prix in Japan. *Tourism Planning and Development, 12*(4), 433–446.
Kim, W., & Kim, H. (2013). Regional development strategy for increasing cultural tourism business in South Korea. *Asia Pacific Journal of Tourism Research, 18*(6), 534–548.
Kim, D., & Kim, S. (2017). The role of mobile technology in tourism: Patents, articles, news, and mobile tour app reviews. *Sustainability, 9*(11), 2082.
Kim, J., Lee, K., Seok, B., Choi, H., & Jung, S. (2018). Competitive factors of electronic dance music festivals with Social Networking Service (SNS) citizenship behaviour of international tourists. *Journal of Psychology in Africa, 28*(2), 116–121.
Laing, J., & Mair, J. (2015). Music festivals and social inclusion – The festival organizers' perspective. *Leisure Sciences, 37*(3), 252–268.
Law, R., Leung, R., & Buhalis, D. (2009). Information technology applications in hospitality and tourism: A review of publications from 2005 to 2007. *Journal of Travel and Tourism Marketing, 26*(5), 599–623.
Lohmann, G., & Netto, A. P. (2017). *Tourism theory: Concepts, models and systems*. CABI.
Luxford, A., & Dickinson, J. E. (2015). The role of mobile applications in the consumer experience at music festivals. *Event Management, 19*(1), 33–46.
Ma, L., & Lew, A. A. (2012). Historical and geographical context in festival tourism development. *Journal of Heritage Tourism, 7*(1), 13–31.
McKercher, B., Mei, W. S., & Tse, S. M. (2006). Are short duration cultural festivals tourist attractions? *Journal of Sustainable Tourism, 14*(1), 55–66.
Mokkil, N. (2018). Queer encounters: Film festivals and the sensual circuits of European cinema in India. *Studies in European Cinema, 15*(1), 85–100.
Ngernyuang, K., & Wu, P. (2020). Using social media as a tool for promoting festival tourism. *International Journal of Computer Science and Information Technology, 12*(3), 17–30.
Niekerk, M. V., & Coetzee, W. J. L. (2011). Utilizing the VICE model for the sustainable development of the Innibosarts festival. *Journal of Hospitality Marketing and Management, 20*(3), 347–365.
Norfolk, L., & O'Regan, M. (2020). Biometric technologies at music festivals: An extended technology acceptance model. *Journal of Convention and Event Tourism, 22*(1), 36–60.
O'Sullivan, D., & Jackson, M. J. (2002). Festival tourism: A contributor to sustainable local economic development? *Journal of Sustainable Tourism, 10*(4), 325–342.
Pan, B., & Li, X. R. (2011). The long tail of destination image and online marketing. *Annals of Tourism Research, 38*(1), 132–152.
Pawłowska, A., Łukasz, M., & Elżbieta, S. (2015). The use of information and communication technologies (ICTs) in tourist information and promotion of the Polish carpathians in foreign tourism markets. *Journal of Travel and Tourism Marketing, 32*(5), 554–577.
Quinn, B. (2006). Problematising 'festival tourism': Arts festivals and sustainable development in Ireland. *Journal of Sustainable Tourism, 14*(3), 288–306.
Quinn, B. (2010). Arts festivals, urban tourism and cultural policy. *Journal of Policy Research in Tourism, Leisure and Events, 2*(3), 264–279.

Reid, S. (2011). Event stakeholder management: Developing sustainable rural event practices. *International Journal of Event and Festival Management, 2*(1), 20–36.
Robertson, M., Yeoman, I., Smith, K., & Mcmahon-Beattie, U. (2015). Technology, society and visioning the future of music festivals. *Event Management, 19*, 567–587.
Shabani, N., Munir, A., & Hassan, A. (2019). E-Marketing via augmented reality: A case study in the tourism and hospitality industry. *IEEE Potentials, 38*(1), 43–47.
Sigala, M. (2020). Tourism and COVID-19: Impacts and implications for advancing and resetting industry and research. *Journal of Business Research, 117*, 312–321.
Sigala, M. (2017). New technologies in tourism: From multi-disciplinary to anti-disciplinary advances and trajectories. *Tourism Management Perspectives, 25*, 151–155.
Sinh, B. D., Nga, V. T., Linh, V. T. H., & Tuan, N. H. (2016). Stakeholder model application in tourism development in Cat Tien, Lam Dong. *Journal of Advanced Research in Social Sciences and Humanities, 1*(1), 73–95.
Song, H., Lee, C., Kim, M., Bendle, L. J., & Shin, C. (2014). Investigating relationships among festival quality, satisfaction, trust, and support: The case of an oriental medicine festival. *Journal of Travel and Tourism Marketing, 31*(2), 211–228.
Thadaniti, S. (2014). Learning the greatness of mother nature festival tourism: A case of Loy Krathong water festival in Thailand. *SHS Web of Conferences, 12*, 1–5.
Tribe, J., & Mkono, M. (2017). Not such smart tourism? The concept of e-lienation. *Annals of Tourism Research, 66*, 105–115.
UIS-UNESCO. (2015). *Festival statistics: Key concepts and current practices, 2009 UNESCO FCS handbook no. 3*. UNESCO Institute for Statistics.
Van Winkle, C. M., Bueddefeld, J. N. H., Halpenny, E. A., & MacKay, K. J. (2019). The unified theory of acceptance and use of technology 2: Understanding mobile device use at festivals. *Leisure Studies, 38*(5), 634–650.
Xie, Y., & Xu, J. (2004). Cultural tourism vs tourist culture: Case of domestic tourism in modern Beijing. *Tourism Recreation Research, 29*(2), 81–88.
Yang, J., Gu, Y., & Cen, J. (2011a). Festival tourists' emotion, perceived value, and behavioral intentions: A test of the moderating effect of festivalscape. *Journal of Convention and Event Tourism, 12*(1), 25–44.
Yang, X., Reeh, T., & Kreisel, W. (2011b). Cross-cultural perspectives on promoting festival tourism – An examination of motives and perceptions of Chinese visitors attending the Oktoberfest in Munich (Germany). *Journal of China Tourism Research, 7*(4), 377–395.
Ye, B. H., Ye, H., & Law, R. (2020). Systematic review of smart tourism research. *Sustainability, 12*, 3401.
Yun, H. J., & Park, M. H. (2014). Time-space movement of festival visitors in rural areas using a smartphone application. *Asia Pacific Journal of Tourism Research, 20*(11), 1246–1265.
Zyl, C. V. (2012). Tourism marketing: A game theory tool for application in arts festivals. *Tourism Economics, 18*(1), 43–57.

Emmanuel Sebata is currently a PhD Candidate at the Sport Business School, Beijing Sport University Beijing, China. He is a Teaching and Research Graduate Fellow in the Department of Sport science, Faculty of Science, Kyambogo University, Kampala, Uganda. He was also selected as an IOC Young Leader 2021-2024. His areas of research interest are: sport tourism, sport economy, Olympics, leisure, event heritage, festivals, and sport and tourism development in developing economies. He completed Master's Degree in Tourism Management biased in sport tourism at Hebei University, China and a Bachelor of Science in Sport and Leisure Management at Kyambogo University, Uganda.

Md. Ruhul Amin Mollah is an Assistant Professor at Jagannath University, Bangladesh. He obtained Master's degree in Tourism Management from Hebei University at China and MBA & BBA from the University of Chittagong at Bangladesh. He is currently a PhD Candidate in the

Department of Tourism, Sport and Hotel Management at Griffith University, Australia. He is also a Teaching and Research Assistant at Griffith University. His research interest includes tourism technology, sport tourism, religious tourism, contemporary management issues and sustainable tourism. He has presented papers in peer-reviewed conferences such as The Council for Australasian Tourism and Hospitality Education (CAUTHE) and The Greater Western Chapter of the Travel & Tourism Research Association (GWTTRA) and also published 10 peer-reviewed articles in reputed journals. His publications appeared in the Journal of Sport & Tourism, Man in India, ABAC journal and several others.

Chapter 3
Tourism Philanthropy: The Role of Tourism Fair, Festival and Event Stakeholders

Bapon Chandra Kuri, Sadia Afrin Ananya, Synthia Islam, and Azizul Hassan

Introduction

Tourism fairs are the only marketing communication instrument that allows you to give important information to prospective customers face to face, as well as present cities and regions' offerings. During trade exhibitions, exhibitors can appeal to all visitors' senses and provide entertainment in a festive atmosphere (Gębarowski & Wiażewicz, 2014). Tourism philanthropy is fundamentally about good global citizenship enjoying, learning, and positively contributing to worthwhile community and conservation projects in tourism destinations (Honey, 2011). The authors argue philanthropy programs need to be initiated in the global society. Tourists and other stakeholders can contribute to residents' well-being through their contribution, consciousness, and commitment. Some argue that tourism suppliers keep a keen eye to create a pipeline between tourists and their business thus helps the business stakeholders to generate more revenue and neglecting social and environmental issues (Frumkin, 2006). As a result, economic greed leads to create unstable socio-cultural and environmental circumstances. In the immediate Coronavirus (COVID-19) situation, there seems to be growing consciousness on rebuilding the moral aspect among citizens to promote both human and planetary health. Moral and environmental education may contribute to bridging the gaps among all pillars of sustainability (i.e., environmental, social, economic, political, and technological). Tourism philanthropy can serve as a positive pathway to promote community peace through tourism. Indeed, in 2008, the survey by MSNBC and Conde Nast Traveler, found the majority of the respondents (1600 respondents) think that they are very or

B. C. Kuri (✉) · S. A. Ananya · S. Islam
Department of Tourism and Hospitality Management, Bangabandhu Sheikh Mujibur Rahman Science and Technology University (BSMRSTU), Gopalganj, Bangladesh

A. Hassan
Tourism Consultant Network, The Tourism Society, London, UK

A. Hassan (ed.), *Technology Application in Tourism Fairs, Festivals and Events in Asia*, https://doi.org/10.1007/978-981-16-8070-0_3

somewhat interested in giving back to local communities when traveling and 62% expressed an interest in volunteer vacations (Lovitt, 2008). The 2009 World Travel Market study identified almost $ 248 million in cash donations raised by 39 companies and organizations over a wide range of periods (Goodwin et al., 2009). The 2008 World Travel Market report tabulated total donations by 36 travel companies of over $6.7 million in 2007, with an almost even 50/50 split donations by the company and their customers, and only 4% contributed by staff. The CMI Green 2009 survey of over 1000 travelers found 83.7% gave between $11 and $500 while traveling (Roth, 2009). For sustainable development, the idea of tourism philanthropy needs to be clear to all by highlighting the benefits and sustainability of travel philanthropy (Novelli et al., 2015). The main objective of this study is to promote global prosperity through adopted actions taken by tourism fairs, festivals, and events stakeholders. Another objective is to elucidate the motivational factors that encourage stakeholders' participation and discuss the strategies to implement a successful tourism philanthropy program.

Background of Tourism Philanthropy in the Perspective of Tourism Fair, Festival and Events

Local festivals are increasingly being used to promote tourism and benefit the economy of their respective regions. This is typically reflected in the amount of public aid they receive. In recent years, the utilization of local festivals as a tool for tourism promotion has acquired widespread traction. While certain long-standing local cultural or religious events have been revived or repackaged as tourism events, new festivals have been created and promoted solely for the goal of attracting new visitors to a city or region. Social dialogues and collective agreements highlight the community's betterment through improving the sustainable standard of living at the destination (Diekmann & Bauthier, 2011). Philanthropy is an act that ensures pure human goodness by donating money, charitable gifts, volunteering (Novelli et al., 2015). Philanthropy and travel philanthropy have been analyzed from namely the traditional, modern, post-modern philanthropy approaches. The practice of the post-modern philanthropy approach includes charity, venture capital, social change, pro-active creative innovation, individual donations, corporate fundraising, volunteering (Novelli et al., 2015). Traditional philanthropy creates an influential relationship between donors and recipients that ultimately helps the local communities and contributes to the broader society (Korf, 2007). Modern philanthropy is the combination of strategic-corporate, social entrepreneurship, and social justice philanthropy. This type of philanthropy helps to accomplish the social purpose by sharing an amount from the profit of the company, encouraging private sectors to invest, forming networks, developing technical skills, the establishment of corporate values (Sulek, 2009) among others. Travel philanthropy gets wider importance in the post-modern philanthropy approach. It indicates the benefit of community and environment directly from the tourism sector (Maathai, 2011; Novelli et al., 2015). Tourism businesses, tourists, and other stakeholders contribute willingly to support the wellness of that tourist zone, communities to bring a sustainable social change. While community support has a direct impact on tangible benefits, it indirectly affects the sustainability of the destination (Than et al., 2020).

From the thoughts of Shepherd (2017), travel philanthropy is where travel is meant to give something back that is useful. She shares the idea of "Philanthropy Plus" where visitors can visit their chosen place and make a direct investment. They can see precisely where their cash is going and the deliverables. Responsible tourism can be promoted due to the practice of travelers' philanthropy which impacts the good travel experience, community development, and travel business (Honey, 2011). The philanthropic values will require fair distribution of benefits and revenue for tourism development (Smith & Duffy, 2003). Both the hosts and guests need to co-create ethical, economic, and aesthetic values with tourism. Climate change, solid waste, and over-tourism have threatened the travel destination, so sustainable tourism with proper measures can eliminate the crisis by building up a mutual bonding with the community and nature (Shasha et al., 2020). Functions related to social development, supporting cultural institutions, and building new infrastructure are among the most essential. A particularly essential role of festival tourism appears to be the ability to spend one's leisure time engagingly, develop one's hobbies, and form new relationships (Dani, 2020).

Philanthropic visitor program at tourist venues ensures the practice of charity, promote responsible behaviors as well as contribute to socio-economic benefits (Lacey et al., 2015). It can attract donations from tourism operators, companies, and individuals (Travel Philanthropy, 2021) promoting a better quality of life for residents and tourists. Strategic philanthropy can be the tool for running tourism businesses more sustainably since its ideology supports synergistic use of organizational resources in a way where the process will simultaneously address stakeholder interests with social and company goals (McAlister & Ferrell, 2002) and contribute to corporate and social responsibility. Long-term partnerships with local communities promote community empowerment and improve environmental wellbeing through the conservation and preservation of natural resources (Alisa & Ridho, 2020).

Methods

The methodological part of this research has approached the Focus Group research method that brings together a small number of individuals for an in-depth discussion regarding the research topic. It has supported the ways of exposing the tourism philanthropy program to ensure community wellbeing through the involvement of stakeholder's financial contribution, a consciousness of sharing talent, and commitment to giving time. All these outputs have been analyzed by the Post Coding Content Analysis method after using 12 respondent's answering sessions. In this method, global prosperity through adopting natural and social measures is explored by using the tourism philanthropy program. This depth information of the study has been provided by the interviews of the experts in the tourism field and their discussion. It has studied that - what are the motivational factors of tourism philanthropy and how the stakeholders can be involved in this program. So the question "How" and qualitative "What" is indicating it as Exploratory research. The focus group research selects those participants who are the stakeholders of the specific area. Respondents of focus groups have true understanding, discuss the facts and play a major role in the relatable area. So the exploratory research with a qualitative approach based on the interviews of the 12 respondents has done the content analysis which is well enough to do a focus group research. Their wider and deeper knowledge has come through the discussion with them. Interview questions are semi-structured and a mixture of open-ended and close-ended questions to get detailed explanations. After the interview session, this research has been gone through the data transcribe part, data coding part, and data analysis part.

The study is exploratory research and depends on the qualitative approach and follows non-probability sampling. The group of samples has been identified by judgment. The survey has recorded interviews with 12 respondents. It has found out the right persons who have matched with the topic and objectives and their deeper knowledge then can depend on the Judgmental Sampling. The judgment of the experts may be a biased one. But the biasedness is ethical and positive to choose the right group and right person as the right sample because of their fruitful contribution in researches and extraordinary involvement in the tourism field. The primary sources are mainly the deeper knowledge and experiences of 12 experts and secondary sources are the documentations, news, and articles of newspapers, journals, and previous researches to have a wider knowledge. The data collection process has been continued through the interview sessions at different specific times of the respondents as per their appointment. They have answered the open-ended and semi-structured 15 questions. Recorded opinions and explanations of questions have been used in analyses. Data has been transcribed and the Post-coding approach has been used according to the objectives. This study has been dependent on open-ended and semi-structured questions. So it was difficult to set any hypotheses. It has not been analyzed by any statistical tools because it is both exploratory and qualitative research. It has used non-statistical tools to analyze the data. Qualitative content analysis has been applied for analyzing data collected from respondents. After coding and analyzing this research has identified the motivational factors and strategy to involve the stakeholders. The study has represented a model of tourism philanthropy program by the flow of input factors, processing factors, and output factors. It has proposed a model where the government, businesses, affluent people, and tourists will support the community and tourist zone through the tourism philanthropy program.

Discussion of Findings

The Basic Theme of the Philanthropy Program

Travelers' philanthropy is tourism businesses and travelers making concrete contributions of time, talent, and treasure to local projects beyond what is generated through the normal tourism business (Honey, 2011). In this research, the researchers use 3C (consciousness, commitment, and contribution) in the replacement of 3 T (Time, Talent, and Treasurer).

Consciousness (Talent)

Every tourist is the peace messenger for any destination. Their knowledge, education, skill, diplomacy, training capability, norms, values, ethics, professionalism, art, tact, and leadership can play a vital role to distribute their unique competencies to community people (Yiannakis & Gibson, 1992). Tourism fairs, festivals, and events stakeholders can contribute the above-mentioned skill to the community people. For example- a doctor and a teacher visit a particular destination to gather knowledge and experiences or relaxation. In the moment of exploring a destination, he may find local host community people and decided to provide free treatment and educational programs to the host people. The doctor provides free treatment to the community people and the school teacher provides teaching to the community people to educate them without any charge. Actually what they did? They visited a destination and at the same time, they contribute their talent to the people through their educational and medical programs.

> We can use our general knowledge and aid initially the community in their emergency moment like natural catastrophe by foods, medical assistance, water, milk, tent, planting trees, disinfect the destination with chemicals, preparing the shelter (Respondent 3).

> …it would be great if we get the chance to help people by using our skills. It is not necessary to do help only in their crisis time. This program will benefit disadvantaged tourist areas throughout the year (Respondent 5).

> Well, if you notice, you can see that indirectly all tourists are contributing to the betterment of the local community through economic growth, social understanding. But when they are doing it deliberately, philanthropy starts. I have seen so many tourists who are environmentally and socially responsible. They are motivating and educating other people to behave in that manner. The ethics that they pursue are incredible which boosts the spirit of tourist behavior (Respondent 9).

Commitment (Time)

The valuable moment the tourist spent with community people, share knowledge and experiences about their lifestyle, culture, heritage, and tradition the way they inherited from generation to generation (Beeton, 2006). The community also illustrates their cultural and traditional value with the tourist in tourism fairs, festivals, and events. Both tourist and the community invest their valuable time to share their feelings and emotions that they are bearing in their heart for a long time and find someone like them to share. This valuable time helps to build up emotional attachment between the tourist and host. This process will heal the community problem and tourists will get motivated to visit again to meet with people at the destination (Piartrini, 2018). This will accelerate the economic multiplier for the community's wellbeing.

> If you consider the situation of ours, the people of Bangladesh are very friendly and hospitable. Knowing about them is like knowing about their culture, tradition, region. An attachment can be ensured easily and they may share their feelings with you (Respondent 2).

> Tourism can construct a sustaining impact in the place by swaying the lives of the local people although I assume there is a chance of having some external effects but undoubtedly it can be altered by adopting an appropriate tourist development strategy (Respondent 8).

> …bonding between tourists and community people enable receiving of tourists in the destination which may lead to a positive tourist experience resulting from the community-tourist interaction (Respondent 12).

Contribution (Treasurer)

This involves financial or physical contributions from tourists to the community people (Butcher & Smith, 2015). They will contribute money to the community people according to their financial stability. They also raise or collect funds from various sources to develop the living patterns of the community and preserve the natural resources at the destination. Physical contribution involves working with community people in the field, firming, carrying goods from home to market, helping them in fishing, collecting honey, voluntarily working in the shop, collecting funds in favor of them, collecting waste, etc. Material contribution involves proving necessary goods (food, water, medical supplies, etc.) to the community people in an emergency.

> …tourists can spontaneously respond to donate money or provide their skills in the destination. Like you can see in this crisis moment of COVID-19, few people are ready physically and mentally to protect the community by providing musk, medical assistance, providing foods, alarm the community as instructed by WHO (Respondent 4).

> A traveler can contribute in his way. If he promotes volunteer tourism around the country and involves other voluntourists by Facebook, advertising, documentary, communication with electric media, posting pictures on social media, tourism philanthropy is eventually created (Respondent 5).

The Motivational Factors of Tourism Philanthropy in Tourism Fairs, Festivals and Events

There is a big question what is the biggest contribution as a tourism stakeholder in society? Tourists are the transferable wealth from one region to another region and are influenced by so many factors. People always try to gain social prestige through their charitable activities.

The Responsibility of a Citizen

A citizen is not just the inhabitant of a country, he or she is the citizen of the world. As a citizen of the world, each person will contribute to social wellbeing by donating financial funds to the community people or giving psychological or physical support in a distressing moment of the country or society.

> …when a tourist is aware of the wider world, they will perform an active role for the betterment of the whole community of the world (Respondent 1).

> The tag line of your philanthropic activities should provide a social message like 'I am a citizen of the world'. The message could even inspire others (Respondent 4).

Push and Pull Factors

Push factor identifies the aspiration, reason for travel and pull factors attract tourists to pick tourism products (Baniya & Paudel, 2016). In the case of tourism philanthropy, the push factors of motivation can emphasize on the tourist that they are bound or compelled to provide supports to the community by giving products or commodities (rice, pulse, blanket, water, medicine), service (voluntarily work with the community, removing waste from the site) or financial funds. Pull factor means donating or contributing funds spontaneously. In case of pull factors, tourists or government provides every sort of facilities in the destination from their core of the heart as part of their social responsibility.

> When we are talking about tourism philanthropy, we have to understand that push and pull factors are working in a different way over here. Sometimes DMO suggests to the travelers that if anyone plants 10 trees at a time, will have 50% ownership of his plantation – this is the example of push factor. Someone is donating spontaneously and swelling the power of tourism for doing good is an example of a pull factor (Respondent 7).

> The determination of push and pull factors is decisive as it assists the stakeholders to recognize the behavior pattern of the travelers. Generally, travelers engage in tourism for relaxation, knowledge gathering, enjoying attractions, and other purposes. Then for the sake of the destination, they assist willingly or involve in different support programs offered by the authority (Respondent 11).

Intrinsic and extrinsic motivation: Intrinsic motivation comes through moral, ethical, and social value to give both psychological and financial support to the community. Intrinsic motivation is nurtured from the early stage of childhood by practicing ethical and environmental education from the parents, teachers, and elders (Locke & Schattke, 2019). For example, doctors visit the destination to provide free treatment, teachers from the various school college provide free teaching, archeologist visits onsite destination and show the community how to preserve and conserve heritage and culture. On the other hand, extrinsic motivation is somewhat rewarded achieving (Locke & Schattke, 2019). Such as fame, grades, and praise that can be achieved through monetary contribution or voluntary work.

> …you cannot force one to engage himself in philanthropic activities. He can be involved willingly or sometimes in gaining attention. In this crisis moment of COVID-19, each country can come close together by supplying medical emergencies and try to make their citizen understand precaution measures to combat COVID-19. The government offers incentive packages, charity programs to give breathing space for the industry to maintain economic stability. A group of students is directly working in the paddy field with farmers, preparing hand sanitizer, collecting funds from various sources, and providing free foods to the poor people. All of these are the outcomes of both motivations (Respondent 5).

The Tradition of Giving

The country that practices offering gifts to others at the crisis moment over the decades as a part of their traditional role. When an individual is simply helped through charity by a commendable person, without any expectation, at the proper time and in the proper place, philanthropy occurs (Mukundananda, n.d.). There is a solid conviction that a gift that is given without any expectation of appreciation or reward is valuable to both the giver and the recipient (Adam, 2004). It is true for the tourism and hospitality industry too. Patrons frequently donate and support tourism without any purpose of having back of returns.

> Yes, one can do humanitarian and charitable jobs for running the practice of tradition though I think it comes more from people's mental satisfaction (Respondent 2).

> The tendency of helping others is the identification of humanity as people are born to serve the planet. Often, the tradition of giving starts from home from where kids can learn about philanthropic activities. Community people, interested bodies often are involved in charitable programs for the development of tourism in Bangladesh (Respondent 10).

Self-Morality and Social Responsibility

People by born learn some social rules and regulations, moral, ethical, social norms and values that can purify the heart of the citizen to build up social responsibility. From here, philanthropic activities may generate among travelers. Moreover, the public expects from a tourism business that it will be involved in community activities and fulfill society's expectations and it must have ensued from their social responsibility (Grigore, 2010).

> Every country has some constitutional rules that suggest how the state will behave with the community and nature positively. A social agreement between the tourism business and society is needed now to create benefits for the global people (Respondent 3).

> Few tourism operators are involved in humanitarian sections by contributing their resources, facilities, donations which is the result of their responsibility to the society where they belong (Respondent 6).

Application of Environmental and Ethical Education

When tourists are conscious and trained from environmental and ethical education, they will apply their ethical morality to society (Tribe, 2002; Grigore, 2010). For example, presenting trees in the wedding ceremony or birthday party three, planting trees at the tourist spot. Grigore (2010) also found a little difference between the two terms including discretionary and ethical responsibilities. Discretionary activities are wanted whereas ethical responsibilities are expected.

> Tourism desires a foundation of ethics and environment that will guide the actions of the tourist and tourism business (Respondent 6).

> An environmentally and ethically sound traveler supports the destination conservation and preservation program. They are very supportive of having a fresh nature and environment-friendly tourist destination as ultimately it is actually for their direct benefit (Respondent 11).

Religious motivation: Every religious theory suggests people doing something good for the surrounding poor forlorn people. These theories influence travelers and tourism operators to engage in charity.

> As all religions encourage charity, so many travelers are inspired by it. Even in few cases, they influence others to donate to make the almighty happy (Respondent 5).

> A major amount of donations comes from inhabitants who are enthused by religions. In the tourism industry, travelers generally want to donate and help the community people even if these people belong to other religions. It ensures peace in the society and represents a better community highlights the image of the tourist place (Respondent 9).

The Ethical and Environment Educational Program

The allure of a fair or festival cannot be replaced by the allure of a location. Festival attractiveness stresses intangible characteristics such as festival mood, interpersonal connection, and so on, whereas destination attractiveness emphasizes concrete elements such as the natural environment, cultural scenic locations, infrastructure, and so on. Intangible factors such as festival mood and interpersonal contact attract tourists, however, these intangible qualities must be brought out through physical features such as cultural emblems and festival events (Li et al., 2020). Students can be the best travelers so that they can teach their families about tourism philanthropy. This program will be conducted all over the country with the inspiration of a philanthropy program. A design of ethical (society) and environmental (nature) educational programs is drawn here, that can be initiated from the ministry of education and other tourism stakeholders.

A probable design of an ethical and environmental education program is given in Table 3.1.

Table 3.1 The design of an ethical and environmental education program

Level	Theoretical	Practical
Early-stage of the childhood	Parents act as a teacher teach primary education and religious study	Storytelling about honesty, morality, and ethical value. Participating in tourism fairs, festivals, and events
Primary level	The theme of environment, The theme of norms, values, and moral behavior. Understanding tourism fairs, festivals, and events, responsibility of a citizen	Discipline, tree plantation, cleaning the classroom, respecting the elders, honesty shop, avoiding copy, generating a good result, visiting a friend who is suffering from illness
Secondary level	The theme of tourism philanthropy (time, talent, treasure), The theme of the ecosystem, Travel and tourism, Tourism resources, Domestic tourism, The good governance Responsibility towards: Society, economic and environment	Arranging tree fair, book fair, field visit to watch the ecosystem, skill development (talent), forming a fund to help the poor and classmates, sharing the foods, books, and pens with classmates, distributing blankets for the poor people in winter, donating money onsite destination, work with the people, educate the host people, respect their culture, follow the dress code, protect and preserve the natural and manmade resources
Higher secondary level	Global tourism, Technology in tourism The constitution of the state	Forming an online donation system to support poor countries
Hon's	Sustainable tourism, eco-tourism, pro-poor tourism, CBT, voluntarism, natural and manmade tourism The psychology of behavior(morality, values, governance)	Arranging seminars in school, college, and varsity to initiate education program in destination, train host community on how to earn money through tourism, teach them how to use technology to attract more tourists, and promote philanthropy program
Masters	Tourism entrepreneurship	Building entrepreneurs from the host community to continue the tourism philanthropy program
PhD	Research-based theory development (Tourism Philanthropy)	Initiating new mode of philanthropy tourism: a sustainable tool of ensuring peace through tourism in the universe

Source: Authors' compilation, 2021

INPUT (Tourism Fair, festivals and Events)	PROCESSING	OUTPUT
Environmental and Ethical Education	Voluntourism	World peace through Tourism
Educate the tourism Stakeholders	Sustainable and ecotourism	The 17 goals of SDG
Educate the Society	Commitment, consciousness, and contribution	Win-win situation, Preservation of nature and culture, Improve the standard of living
Green politics	Community-Based Tourism	Community development

Fig. 3.1 The input, processing, and output of tourism philanthropy. (Source: Authors' compilation, 2021)

Input, Processing, and Output of Tourism Philanthropy on the Occasion of Tourism Fair Festival and Events

Tourism Philanthropy is a driver that will educate the society, tourists, media, citizens, govt. and business about the environmental and ethical issues which will inspire the travelers to visit destinations in a responsible way to contribute their time, talent, and treasure to the community (Honey, 2011). Finally, tourism fairs, festivals, and events facilitate the community to develop their standard of living and the world will gain peace through tourism philanthropy programs. As this program can be continued all over the world, soon it will find the hope of achieving the ultimate 17 goals of SDG. A sum-up of the overall idea of tourism philanthropy has been drawn in Fig. 3.1.

The Role of Tourism Fairs, Festivals, and Events Stakeholders for Global Prosperity

As every destination needs to initiate a tourism philanthropy program, it requires creating a demographical mapping about people like the standard of living, income, education, job, foods and nutrition, health, and children. The tourism fairs, festivals, and events stakeholders can play an imperative role over here.

> Philanthropy in the tourism field needs the participation of people from every sphere who are concerned. It can be travelers, educational institutions, government, tourism operators, etc. (Respondent 1).

> Tourism philanthropy programs may be initiated in that particular tourist destination with the permission of the concerned authority or government. The people who are living below the poverty line needs special attention in the program from the stakeholders like tourism business, tourist, travelers, professionals, host community (Respondent 7).

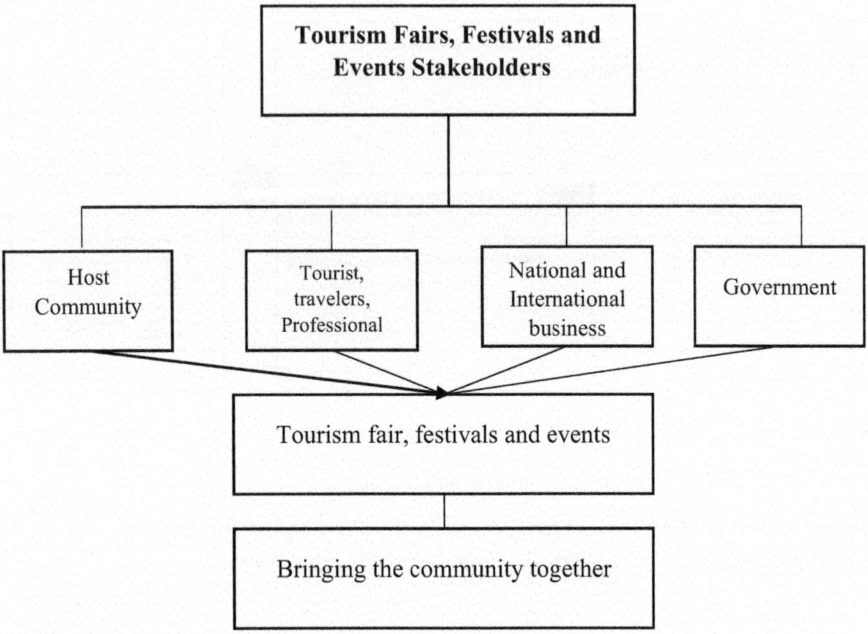

Fig. 3.2 Tourism fairs, festivals, and event stakeholders. (Source: Authors' compilation, 2021)

> …it is the understanding of travelers and tourism operators that instigate themselves to upkeep more about the community than just for earning profits. All operators and companies of tourism can send a share of their profit to assistance programs of the community which may sustenance the environment, public by generating a long-term impact for the betterment of the local civilization (Respondent 12).

The rise in media interest in the area before, during, and after the festival is the most major tourism impact of the fair and festival. The festival's media coverage results in favorable marketing of the area at the regional, national, and even global levels. Tourism marketing authorities in numerous tourist destinations across the world have recognized this potential (Fig. 3.2).

Host Community Philanthropy Towards the Tourist, Nature, and Themselves

Accordingly, residents should be aware of the potential benefits of Heritage festivals, and partners and the government should raise awareness of the importance of cultural and heritage tourism through various cultural events and educational programs. One, conserving and preserving the natural resources and heritage by practicing their inherited traditional activities like knitting muslin saree (a traditional dress of Asian continent), collecting date juice, preparing sweets, nurturing cultural activities, catching fish, honey hunting, running horse or cow cart, arranging boat running competition, etc. Two, Responsibility to Protect (RtoP) means each community will support another community in their distressing moment. Three, sending their children to school to learn education. Five, maximizing tourist satisfaction with memorable experiences by showing art and culture. Six, increasing a fund that is generated from the tourist expenditure to support their community in the crisis moment at the destination. Seven, frequently providing information to the Destination Management Organization (DMO) about their current status of the destination. Eight, promoting tourism resources and community-based tourism to the tourist.

Professional and Non-professional Persons

In this case, all the professional persons will be divided into several groups and then each group will visit every district or province of the tourist zone to provide free service to the community. A doctor may give free medical treatment and a teacher may provide free teaching. An archeologist will demonstrate to the community people how to preserve and conserve the heritage site and the sociologist will teach them how to practice social norms and behavior. On the other hand, a geologist will guide them, how to conserve and preserve natural resources. The business person will tell them how to sell the local products (not originated from animal or coral) to the tourist in a pleasant manner and the technician will illustrate to them how to use technology to promote CBT in the arena. Other celebrities and notable persons will visit the destination to share their feeling and contribute to the community people. A media person will frequently cover all philanthropy-based activities and request the affluent country to donate to the community. Where the layer will tell both tourists and communities about the constitutional or global (UNWTO) laws, rules and regulations to ensure environmental sustainability. Students can be a crucial part of the tourism philanthropy program. They will work with community people by helping them in gardening, tree planting, destination cleaning, teaching to the children, freeing the birds in the destination. This group of students can raise funds from the college or universities. They will share the post about their philanthropy-based activities on social media (FB, Twitter) which will motivate other institutions and society to raise funds and contribute to the community people at the destination. The citizen will take care of the community by taking online classes, medical suggestions, and agro-based farming.

Public-Private Partnership

The success of a trade show is determined by the close collaboration of organizers, exhibitors, and potential attendees. Trade fairs are seen as a valuable tool for promoting economic development, demonstrating innovations, and fostering business relationships and opportunities, and their importance as a source of revenue, employment, and wealth among governments, public administrations, and business sectors cannot be overstated (Guerrero et al., 2020). The government will not solely be able to promote the philanthropy program to the destination. Here, national tourism organizations and non-governmental organizations (PATA, UNWTO, and JICA) must come to support the community people in the crisis moment (natural and human-made disasters).

The Activities of the Government

Festivals can benefit the local community in two important ways when it comes to tourism. First and foremost, people from the local culture feel delighted by worldwide interest in their culture. This boosts local pride and encourages the preservation and promotion of the community's culture. Second, tourists and visitors spend money in the town, which helps to boost the local economy and support restaurants, hotels, and other tourism-related companies (Sahoo & Mukunda, 2020). One, the national and private TV channels will telecast environmental and ethical-based programs in their episodes to promote tourism philanthropy programs all over the world to arise moral values among the citizen. Two, the government will take action to eradicate drug business, frequent accidents, rumors, terrorist attacks, economic recession, food, and medicine adulteration. Three, they must provide immediate medical assistance in the distressed area. Four, the government collects foreign donations from their allied countries to support the philanthropy program in the destination. Five, all the state person, diplomatic, celebrity, notable person, ambassador, and civil society to create public opinion in the international platform to promote tourism philanthropy program. Six, building the bottom line empowerment that means empowering the community, by the way, can able to build up the national image by presenting their culture and heritage. Seven, identifying the donor's country to raise collection and create the government fund for the poor distressed people. Eight, the government must provide the basic needs in the required destination and ensuring safety and security for the community people and the tourist. Nine, ensuring the right justice and fairness for the people who depend on the tourism industry. Ten, providing one-stop service for all tourism stakeholders. Eleven, publishing journals or books on values, ethics, and morality. Twelve, demotivating over tourism as it will destroy the natural carrying capacity of the destination. Thirteen, the human settlement should be in a vertical progression that can accommodate more humans in a single complex thus will help to protect our agro-based land. Fourteen, a family with less than 2 children will be rewarded. Fifteen, initiating plastic point means if a traveler collects plastic from the environment then he will get a bonus point to achieve a reward from the government. Sixteen, school, college, medical center, electricity, road, gas facility, library, games room, playing ground, meeting room, hospitality training institution should be ensured in the destination. Seventeen, taking actions to free the animal from the case or zoo. Eighteen, ensuring a corruption-free society and protect the endangered species of the tourist zone. Nineteen, sending assistance in the emergency. Twenty, raising climate and destination's community wellbeing fund to meet up the 17 goals of SDG.

The Responsibilities of Travelers

On any tourism event, tourists will participate in the program to understand the host community's style and find out their lacking and problems. Then, tourist will turn into a contributor to the local community people. One, tourists will be integrated and be with the community (time) by participating in a local activity (farming, fishing) and sharing their feelings. Two, expert and professional tourists may show the community how to conserve and preserve natural and heritage sites for the future generation. Three, be an eco-tourist and raise charitable funds for the community people. Four, creating public opinion to protect the community people who are deprived of their basic needs through social media and word of mouth. Five, drifters and explores may gain foods from the community people by assisting them in exploring a destination that creates a social bonding between tourists and communities. Six, responsible tourists must give the guideline to the other travelers to free the wild animal at the destination. Seven, promoting not to use or sell animals or sea-made products like coral or snail products. Eight, providing a free education system to the children during their stay. Nine, not spreading unhealthy products like chocolate, chips, candy, ice cream, or any other junk food to the community people. Ten, visiting destination in the emergency moment for example; distribute blanket in the winter season or rebuild the houses of the community people after the calamity. Eleven, not to provide food to wild animals that may cause serious health issues for the animal. Twelve, respecting the culture and maintain the dress code. Thirteen, good-willed travelers can work in an area where the government cannot reach their philanthropy program. Fourteen, removing the ego while working with community people. Fifteen, exchanging knowledge. For instance; the community people may not know how to use internet technology. So the tourist can teach them and on the other hand, the host people may teach the foreign tourist how to cook food in the bamboo a traditional cooking method of ethnic community people. Sixteen, contributing time, talent, and treasure through free working, contributing money, providing gifts and necessary goods, free volunteer service, free teaching, free treatment, training and giving environmental education, etc. Seventeen, staying with a community that will create a mutual bonding between tourists and hosts. Eighteen, enjoying the activities with community people and gather knowledge about nature and society. Nineteen, refraining from carrying and doing illegal activities as well as tangible and intangible littering while traveling a destination. Twenty, being a responsible tourist and never irritating the wildlife on the spot. Twenty-one, planting trees by the name tourist and he will come back 10 years late to see the trees he planted. Twenty-two, supporting a family every month as a family member. Twenty-three, giving them a photo in the departure moment: lifetime memory Twenty-four, developing a website for community people to raise funds. Twenty-five, tourists must not visit a destination with HIV or other contagious diseases (coronavirus). Twenty-six, they must travel the tourist spot without the intention of playing casino, having sex, and consuming drugs. Twenty-seven, stop visiting such a natural area where people should not travel. For example: climbing Mount Everest that brings a self-prestige into society but harmful for the surrounding nature. Twenty-eight, update and alarm the community about the existing and upcoming crisis. Twenty-nine, provide the guidelines and suggestions about health safety measures. Thirty, supplying books and materials on how to protect the community in the crisis moment.

Business Stakeholders

The foundation for the formation of an event organizer is the development of sustainable societal attractions. Service providers and others engaged in the organization of an event through perceived values and positive experiences of event participants promote their service (Mihajlović & Vidak, 2017). The business stakeholders must recruit and contribute local people as a part of Corporate Social Responsibility (CSR). They must arrange regular training secession and demonstrate how to deal with tourists. They also teach the community people about the environmental education program to ensure how to conserve and preserve natural and manmade resources. As a part of the CSR program, every month they can provide $10,000 for a poor family to initiate a small business rather than giving $1000 to 10 families. That means every month each poor family can start a new business to build up their future. After that, the rest of the poor community people will enjoy this same opportunity from the business stakeholders. On the other hand, the hospitality industry must redesign its rules and regulation to distribute its leftover hygienic foods to the community.

The Process of Bringing the Community Together Through Tourism Fairs, Festivals, and Events

Tourism fairs and small events have a big impact on bringing the community together that allows tourism fairs, festivals, and event stakeholders to identify the basic problem of the host community people based on their product selling's, offerings, and overall surroundings.

Develop a Demographic and Geographic Profile

Map the whole country to identify people based on their demographic characteristics and geographic location. Economic breakdown, income, and standard of living are the major factors to be considered to initiate philanthropy by arranging tourism fairs, festivals, and event programs in that area.

Promote Latent Tourism Resources in the Area

Identify the latent tourism resources where community people do not know the importance of these resources. Arranging a tourism fair at that destination can be an influential tool of promotion. Media coverage will play a crucial role to inject tourist flow and generation of income for the local people.

Identify the Poor Local People

One of the major factors to identify the person to whom charitable funds can be distributed based on their product selling at fairs, festivals, and events.

Sources of the Charitable Fund and Mode of Payment

The sources of the fund can be collected from affluent, government, international organizations, business persons, tourists, and others. The method of distribution can be done both directly and indirectly (Bank account, Mobile banking, etc.). Even, the social media platform is another major source (FB, Twitter, documentary) to promote and collect charitable funds.

Identification of Skilled Person

Human Resource Management will take the responsibility to identify the skilled persons who are interested to distribute their skills to the community people.

Voluntary Team

This team will be formed from the several departments of the tourism fair and event organizers. The professionals and nonprofessional persons from the various departments and universities also be part of the voluntary team.

Host Community Team

To empower the community is the ultimate motto of tourism fairs, festivals, and events through philanthropic work. A chain of command will be built in the community inside for ensuring the proper utilization of tourism resources.

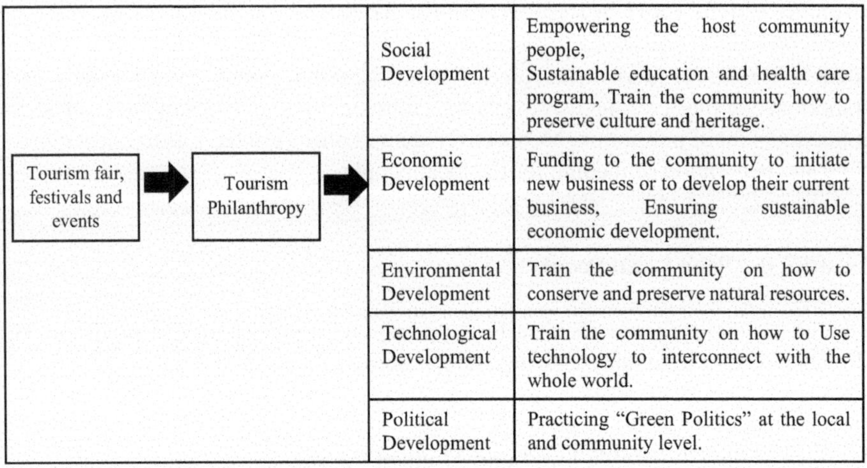

Fig. 3.3 The sustainable development of the community through tourism fairs, festivals, and events. (Source: Authors' compilation, 2021)

Maintain Consistency and Commitment

This continuous tourism fair and event will be held for any specific duration of the year that will help to bring the community together.

Monitoring

The management department of tourism fairs, festivals, and events will monitor the whole process to evaluate the effectiveness and efficiency of philanthropic work (Fig. 3.3).

Feedback

Bringing people from different communities together participation and engagement are essential for any event's success, and a well-organized event, whether high-profile or low-key, can undoubtedly bring the local community together, if only by providing a space and a focal point for activities, but generally much more.

Proposed Model of Tourism Philanthropy Program

There is a need for a tourism philanthropy program where the government, businesses, affluent people, and tourists will support the community and tourist zone. It will work as a roadmap for practitioners, stakeholders, and policymakers of tourism fairs, festivals, and events.

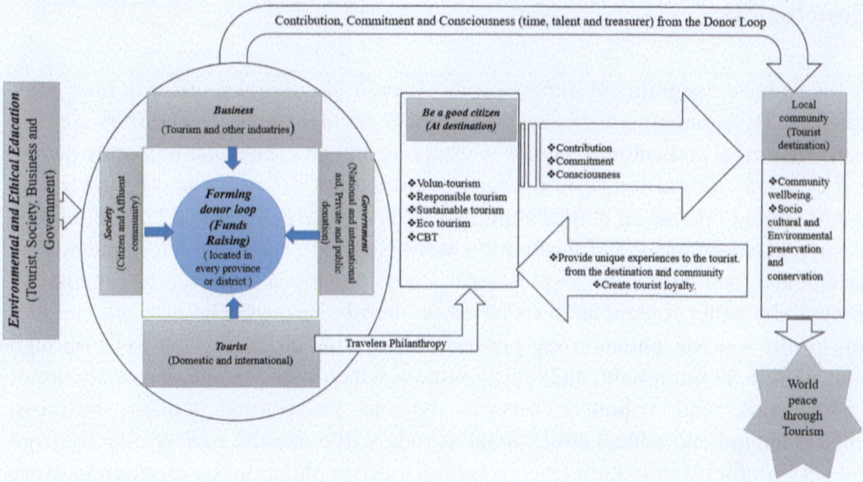

Fig. 3.4 A model of a tourism philanthropy program. (Source: Authors' compilation, 2021)

A proposed model of tourism philanthropy program has been drawn in Fig. 3.4. Forming the donor loop is the ultimate objective of this model to contribute (treasure), consciousness (talent), and commitment (time) to the community. Business stakeholders, government, tourists, and the society (affluent community) will form this donor loop for the welfare of the community. This donor loop will donate or collect donations from various sources (self-donation, internal, external, domestic, national, and international sources). In this case, the government must play a crucial role to gather money from international donations like IDA, JICA, ADB, UNWTO, WTTC, etc. This donor loop will be founded in every district or province of a country and they will provide a donation in a specific period of the month or year. The community will spend this money on their community wellbeing by conserving and preserving their natural and cultural heritage sites. On the other hand, each citizen in the world is a prospective tourist for any particular destination. So, they will responsibly visit the destination to assist and support the community by contributing time, talent, and treasure. Finally, the world will see the hope of ensuring peace through the tourism philanthropy program. This donor loop will keep updating and alarming the community about the upcoming and existing crisis prevailing in the whole world by providing necessary information and advice. They will arrange a demonstration session on how to keep the community protected from any crisis (disease, natural disaster, national conflict, and war, etc.). This donor loop will maintain frequent communication with all the stakeholders to make the philanthropy program successful.

Conclusion

Festivals have a significant impact on the growth of cultural tourism in host communities. By organizing festivals in community settings, festival organizers are now using historical and cultural themes to establish annual events that will draw people and create a cultural image in the host communities. Tourism fairs, festivals, and events should operate an ethical action with good activities and good consequences to support community and destination wellbeing. After getting 7 respondents by judgmental sampling this exploratory study has been represented qualitative research by using content analysis based on their interviews. The motivational factors of the tourism philanthropy program inspire the citizen to 'do well' through contribution, commitment, and consciousness which will generate a new stream of cash, goods, and volunteer services beyond the normal tourism business. Environmental and ethical educational agenda will nurse the society directly from the very root level to mature level to spread tourism philanthropy program in every corner of the world which will facilitate each country to be a welfare state through fulfilling community needs, wants, and demands.

References

Adam, T. (2004). Introduction. In T. Adam (Ed.), *Philanthropy, patronage, and civil society: Experiences from Germany, Great Britain and North America*. Indiana University Press.

Alisa, F., & Ridho, Z. (2020). Sustainable cultural tourism development: A strategic for revenue generation in local communities. *Journal of Economic and Tropical Life Science (JETLS), 4*(2), 47–56.

Baniya, R., & Paudel, K. (2016). An analysis of push and pull travel motivations of domestic tourists in Nepal. *Journal of Management and Development Studies, 27*, 16–30.

Beeton, S. (2006). *Community development through tourism*. Landlinks.

Butcher, J., & Smith, P. (2015). *Volunteer tourism: The lifestyle politics of international development*. Routledge.

Dani, R. (2020). Impact of rural festivals on tourist satisfaction with special reference to Kumaon. *International Journal of All Research Education and Scientific Methods (IJARESM), 8*(7), 370–375.

Diekmann, A., & Bauthier, I. (2011). Tourism 2020–towards the development of a sustainable tourism industry. *Turar Turizm ve Araştırma Dergisi, 1*(1), 29–52.

Frumkin, P. (2006). *Strategic giving: The art and science of philanthropy*. The University of Chicago Press.

Gębarowski, M., & Wiażewicz, J. (2014). Contemporary trade shows as a place of knowledge sharing about tourism products. In M. Gębarowski & J. Wiażewicz (Eds.), *Human capital without Borders: Knowledge and learning for quality of life, proceedings of the management, knowledge and learning international conference* (pp. 25–27).

Goodwin, H., McCombes, L., & Eckardt, C. (2009). *Advances in travel philanthropy: Raising money through the travel and tourism industry for charitable purposes*. WTM Report.

Grigore, G. (2010). Ethical and philathropic responsibilities in practice. *Annals of the University of Petroşani: Economics, 10*(3), 167–174.

Guerrero, J. F., Burgos-Jiménez, J. D., & Tarifa-Fernández, J. (2020). Measurement of service quality in trade fair organization. *Sustainability, 12*(22), 9567.

Honey, M. (2011). *Travelers' philanthropy handbook*. Center for Responsible Travel.
Korf, B. (2007). Antinomies of giving: Moral geographies and post-tsunami aid in Southeast Asia. *Geoforum, 38*(4), 366–378.
Lacey, G., Weiler, B., & Peel, V. (2015). Philanthropic tourism and ethics in charitable organizations: A case study in Central Kenya. *Tourism Recreation Research, 41*(1), 16–25.
Li, J., Dai, G., Tang, J., & Chen, Y. (2020). Conceptualizing festival attractiveness and its impact on festival hosting destination loyalty: A mixed method approach. *Sustainability, 12*(8), 3082.
Locke, E., & Schattke, K. (2019). Intrinsic and extrinsic motivation: Time for expansion and clarification. *Motivation Science, 5*(4), 277–290.
Lovitt, R. (2008). The value of voluntourism. *NBC* NEWS.com. Retrieved from: http://www.nbcnews.com/id/23262573/ns/travel-rob_lovitt_columns/t/value-voluntourism/#.XrqsSmgzY2w. Accessed 11 May 2021.
Maathai, W. (2011). Foreword. In M. Honey (Ed.), *Travelers' philanthropy handbook* (pp. 1–2). Center for Responsible Travel.
McAlister, D. T., & Ferrell, L. (2002). The role of strategic philanthropy in marketing strategy. *European Journal of Marketing, 36*(5/6), 689–705.
Mihajlović, I., & Vidak, M. (2017). The importance of local events for positioning of tourist destination. *European Journal of Social Science Education and Research, 4*(4), 228–239.
Mukundananda, S. (n.d.). Chapter 17: Verse 22 – Bhagavad Gita, The Song of God – Swami Mukundananda. *Holy-bhagavad-gita.org*. Retrieved from: https://www.holy-bhagavad-gita.org/chapter/17/verse/22. Accessed 14 May 2021.
Novelli, M., Morgan, N., Mitchell, G., & Ivanov, K. (2015). Travel philanthropy and sustainable development: The case of the Plymouth–Banjul challenge. *Journal of Sustainable Tourism, 24*(6), 824–845.
Piartrini, P. (2018). The relationship among community based tourism application, community attitude, community empowerment and community life satisfaction. *E-Journal Of Tourism*. https://doi.org/10.24922/eot.v5i2.44048
Roth, T. (2009). The CMIGreen Traveler Study 2009. *CMIGreen Community Marketing, Inc.* Retrieved from: https://communitymarketinginc.com/cmigreen/docs/cmigreentraveler2010v1.pdf. Accessed 14 May 2021.
Sahoo, S. S., & Mukunda, B. G. (2020). The potential of festivals and their contribution to cultural tourism development–a study on Dhanu Jatra in Western Odisha. *Journal of Critical Reviews, 7*(15), 2792–2799.
Shasha, Z. T., Geng, Y., Sun, H. P., Musakwa, W., & Sun, L. (2020). Past, current, and future perspectives on eco-tourism: A bibliometric review between 2001 and 2018. *Environmental Science and Pollution Research, 27*, 23514–23528.
Shepherd, N. (2017). *Travel philanthropy and sustainable tourism – The explorations comp. Explorationscompany.com*. Retrieved from: https://www.explorationscompany.com/about/media-centre/blog/africa/why-does-travel-philanthropy-play-such-an-important-role-within-the-explorations-company. Accessed 14 May 2021.
Smith, M., & Duffy, R. (2003). *The ethics of tourism development*. Routledge.
Sulek, M. (2009). On the modern meaning of philanthropy. *Nonprofit and Voluntary Sector Quarterly, 39*(2), 193–212.
Than, T. T., Kieu, T. P. H., Pham, T. A. D., Van Hoang, T. C., Tran, T. H., Nguyen, H. D., & Dao, T. K. (2020). Impact of community attachment and resident's support on destination sustainability: Evidence from spiritual and community destination in Vietnam. *The Journal of Asian Finance, Economics and Business (JAFEB), 7*(8), 361–369.
Travel Philanthropy. (2021). *Home*. Travelphilanthropy.org.uk. Retrieved from: http://www.travelphilanthropy.org.uk/. Accessed 14 May 2021.
Tribe, J. (2002). Education for ethical tourism action. *Journal of Sustainable Tourism, 10*(4), 309–324.
Yiannakis, A., & Gibson, H. (1992). Roles tourists play. *Annals of Tourism Research, 19*(2), 287–303.

Bapon Chandra Kuri is an Assistant Professor at the Department of Tourism and Hospitality Management, Bangabandhu Sheikh Mujibur Rahman Science and Technology University (BSMRSTU), Gopalganj, Bangladesh. He received BBA and MBA degrees from the Department of Tourism and Hospitality Management, University of Dhaka. He has a good number of articles published in both national and international journals. His areas of research interest are CBT, Rebranding, and identifying employee satisfaction levels. Before joining the university, he was working in the Service Department of Radisson Blue Water Garden Hotel and in the Housekeeping Department of Hotel the Cox Today, Cox's Bazar, Bangladesh.

Sadia Afrin Ananya is an Assistant Professor at the Department of Tourism and Hospitality Management, Bangabandhu Sheikh Mujibur Rahman Science and Technology University (BSMRSTU), Gopalganj, Bangladesh. Ananya earned her Bachelor and Master degrees from the University of Dhaka. As the recognition of her academic feat, she earned Achieved Dean's Honor and Dean's Merit Award 2017 from FBS, DU, and Scholarship from the University of Dhaka. She has research interests in tourism, travel, hospitality, business, marketing, and statistical analysis. She has contributed to a good number of publications that are on the way to be published.

Synthia Islam is now working as an Adjunct Lecturer at the Department of Tourism and Hospitality Management, Bangabandhu Sheikh Mujibur Rahman Science and Technology University, Gopalganj, Bangladesh. She has received her BBA and MBA degrees from Business Administration Discipline of Khulna University with major in Marketing. She has achieved Merit Order Scholarships from Business Administration Discipline of Khulna University. Her main areas of research interest are Sustainable Tourism Marketing, Service Marketing, Entrepreneurship Development, Relationship Marketing, Strategic Marketing, Promotion and Positioning Strategies and Human Resource Management Policies. She is involved in several qualitative and quantitative research projects and has articles in both national and international journals.

Dr Azizul Hassan is a member of the Tourism Consultants Network of the UK Tourism Society. Dr Hassan has been working for the tourism industry as a consultant, academic, and researcher for over 20 years. His research interest areas are technology-supported marketing for tourism and hospitality, immersive technology applications in the tourism and hospitality industry, and technology-influenced marketing suggestions for sustainable tourism and hospitality industry in developing countries. Dr Hassan has authored over 150 articles and book chapters in leading tourism outlets. He is also part of the editorial team of 25 book projects from Routledge, Springer, CAB International, and Emerald Group Publishing Limited. He is a regular reviewer of a number of international journals.

Chapter 4
Pilgrimage and Halal Tourism Event: Application of Technology

Muhammad Khalilur Rahman, Shaharin Akter, Mohammad Mainul Hossain, and Azizul Hassan

Introduction

In just a few decades, rapid technological advances throughout the Muslim world have spilt over into one of the world's most sacred rituals. The Hajj, or Muslim pilgrimage to Mecca, has long become one of the most important events in every Muslim's life (Almuhrzi & Alsawafi, 2017). Many emotions run through this event held, including solidarity among Muslims, repentance for wrongdoing, and insights into Islam's roots. The Hajj is, at its essence, a reinforcement of a spiritual connection between a believer and God (Collins-Kreiner, 2010; Caidi, 2019). Having been stripped of all worldly possessions, staying in tents on Mina's fields, and having to wear only two pieces of cloth. The conclusion of the Hajj in the Valley of Arafat would be to suggest Judgment Day when the encounter with the creator would make them forget all human issues.

Everything that was previously a once-in-a-lifetime trip can now be booked and purchased online with the touch of a keyboard. Hotels and supermarkets intended to compete with tourist sites have replaced dirt-floor accommodations and hazardous street cuisine (Amanullah, 2009). Muslims, who originate from the countries with the highest mobile phone penetration rates in the world, cannot withstand the

M. K. Rahman (✉)
Faculty of Entrepreneurship and Business, Universiti Malaysia Kelantan,
Pengkalan Chepa, Malaysia
e-mail: khalilur.r@umk.edu.my

S. Akter
Faculty of Business Administration, University of Chittagong, Chattogram, Bangladesh

M. M. Hossain
Faculty of Arts and Social Sciences, University of Malaya, Kuala Lumpur, Malaysia

A. Hassan
Tourism Consultants Network, The Tourism Society, London, UK

A. Hassan (ed.), *Technology Application in Tourism Fairs, Festivals and Events in Asia*, https://doi.org/10.1007/978-981-16-8070-0_4

excitement to share the thrill of adventure with near and dear ones by voicemail message and text messages or through SMS. Although globalization has transformed the image of the Hajj, this formerly inaccessible journey was also made accessible to a large number of people who might not have had the chance otherwise. Amanullah (2009) stated that the Hajj authorities are compelled to explore new technical solutions to transportation and structural challenges with ever-increasing annual participation.

Technology is an important aspect of the travel and tourism sector (Pencarelli, 2020) since it aids businesses in their daily operations while also boosting client satisfaction. As a result, hotels, airline companies, cafes, and other businesses must stay updated with the recent technological advancements in the tourism sector. This is incredibly significant in the COVID era when customers' expectations are changing. The current technological developments, and digital technology trends that act as a response to the COVID virus's impact on consumer behavior, which can play a major influence in the travel and tourism industries. At hostels, airlines, and transport networks, technology is employed to give less personalized service while also generating greater income (Ziyadin et al., 2019). Despite its misuse, tourism and travel are heavily reliant on technology, and its application has transformed life much easier (Sigala, 2018). The cell phone has undeniably become an integral component of our daily life. The technological etiquette may hold back technology equipment, the inconvenience of anyone talking too loudly on a mobile phone is well overshadowed by the safety, security, and convenience (Mason, 2006).

The advancement of technology has enabled terrorist activity (Keene, 2011) to target the tourism sector (Rana et al., 2020). Mobile phones have the ability to save lives or destroy bombs; air conditioners help to make living comfortable in hot weather, but they also harm the environment and can carry diseases. The emergence of the internet era helps us to know global weather forecasts, helps international visitors to stay connected with their workplaces, and eliminates the challenge of negotiating time zones, but it may also be used to disrupt airline travel (Ziyadin et al., 2019). For the tourism business (Rahman et al., 2021b), technology has been a mixed blessing since then. It provides numerous benefits while also serving as a means of increasing income and decreasing customer service (Neiterman & Zaza, 2019). The technological world has made air transportation safe and plentiful (Pritchett, 2005), but also made it necessary to undergo long safety lines and daily troubles. This study has attempted to explain the various factors and effects of Coronavirus on the Halal tourism and Halal hospitality markets to improve current understanding in this research field. Halal tourism is booming at a rapid pace. Due to the increasing advantages of Halal products and services given in Halal tourism, the prospects in this sector are seen as very much positive. The main goal of this study is to explore the importance of digital technologies that can improve pilgrimage and tourism events. Technology can be utilized to enhance the Hajj journey more accessible to Muslims around the world while maintaining its spiritual significance.

Technology's Impact on Pilgrimage/Hajj

The face of the Hajj has changed and will continue to change as a result of modernization in technologies, and not even the most conservative segments of the Muslim community appear to be able to stop it (Amanullah, 2009). Muslim scholars must make this transformation into a framework and entrust it with a purpose that is important to a Muslim environment that is more intellectual, knowledgeable and interconnects the Muslim community. The application of technology can make the experience revolutionize (Gretzel, 2011; Othman et al., 2019). It is a duty to the pilgrim community, as well as authorities of Hajj, to supervise it and to be used for maximum spiritual benefit. The Hajj is supposed to guide the believers for Judgment Day, while all of humanity would be brought before Almighty, and pilgrims should embrace the experience as if it were actual. With today's Hajj being completed in comfort and safety, it is easy to be lulled into a state of the holiday instead of worship. Amanullah (2009) stated that the impact of technologies on Hajj as a resilient force hitting an inflexible object, the best thing is perhaps to see if the new Hajj may serve as a spiritual touchstone in a unique and interesting approach. Sending text messages while circumambulating the Kaaba, may appear banal, but it does have a strong influence on the recipients that will last until he or she is unable to personally experience the event.

Impact of Coronavirus on Pilgrimage

Until further notice, Saudi Arabia has restricted all access to Islamic religious sites and locations, including Makkah and Madinah. Umrah has also been suspended until further notice, starting in March 2020. Furthermore, even though the 2020 Islamic Pilgrimage (also known as Hajj) was scheduled to take place in July and August 2020, because of the COVID-19 pandemic (Alam & Uddin, 2020). Hajj was put on hold and was probably going to be cancelled for the year 2020 (based on publicly accessible statistics and information till mid-July 2020). It was not solely a Saudi Arabian decision; other nationalities have also decided that they will not perform Hajj this year (Aljazeera, 2021). In this context, Singapore stated that due to COVID-19, its residents would not attend Hajj this year, as well as those who registered for 2020 Hajj, will be moved forward to 2021 Hajj (Alam & Uddin, 2020).

Rashad and Yaakoubi (2021) indicated that in mid-July 2020, the Saudi Arab authorities declared that a very restricted number of local pilgrims would be allowed to proceed Hajj (around 10,000 pilgrims). Although the Hajj Committee ultimately changed its decision and permitted a small number of pilgrims to perform Hajj in 2020, the numbers were significantly lower than predicted at the beginning of the year. The Edge Markets (2021) reported that taking into consideration the number of pilgrims who take part in Hajj and Umrah every year, cancelling these two most important Islamic rituals will have huge adverse economic and financial consequences for both the Saudi economy and Halal tourism and hospitality companies such as hotels, resorts, market stalls, cafes, jewelry stores, and transportation services.

Many Muslims believe that lock-down due to the COVID-19 is a punishment from Almighty and as a manifestation of God's anger as a result of their and other humans' misdeeds on the planet (El-Gohary, 2020). The holy city of Makkah and Madinah, Mosque, churches and other places of religious worship throughout the world are closed because of the pandemic (Wu & Mursid, 2019). The COVID-19 epidemic is one of the major causes of this. Such closings of two holy cities, mosques and other places of worship have a major negative mental and physiological effect on Muslim people because they never experienced incidents like that before and almost all of them never heard of any similar occurrences leading to such a holiest and religious closure (El-Gohary, 2020). However, this is misleading, as for various reasons Hajj and Umrah have previously been suspended. King Abdulaziz Foundation for Research and Archives (2021) and some other sources stated that Hajj and Umrah were discontinued approximately 40 times during Islam's early days to 2020.

Halal Tourism and Gender Segregation

For every Muslims, travelling (for religious or non-religious purposes) is inextricably related to Islamic Shariah. One of the major causes for this is that every Muslim is required to do Hajj, which must be accomplished in Saudi Arabia's holy towns of Makkah and Madinah (El-Gohary, 2020). As a result, every Muslim (male or female) must go to Saudi Arabia to fulfil one of their religion's pillars. Nonetheless, Muslims are only compelled to participate in Hajj when they are financially able to cover the entire fee and are physically capable of doing so. The Halal Travel idea focuses not just on Muslims' travel activities, but also a wider range of activities (Rahman et al., 2017a). Halal Travel is the purest type of travel activities carried out in full conformity with Islamic Shariah laws (Rahman et al., 2018a).

The author claims that not every Muslim tourist share the same understanding of Halal travel concepts and essentials. The authors imply Muslims are going to disagree in their perception of certain concepts or basic demands as a result of differences in their cultural values, beliefs, dedication level to Shariah Principles (Buisson, 2013; Rahman et al., 2018b), general and spiritual learning level, socioeconomic status, earnings, and other Halal related factors. The Halal travel sector is being driven by Halal tourism trends (Rahman et al., 2019; Rahman et al., 2021a). The growth of solo Muslim women travelers, the increased importance of Halal digital travel and the increasing value of prices include the fact that Muslim travelers predominantly have Halal travel destinations (Rahman & Zailani, 2016; Rahman et al., 2020).

Gender segregation in an academic institution (Hansot & Tyack, 1988; Battour et al., 2019), social functions (Noon & Ayalon, 2018), workplaces, social institutions, and clinics are referred to as physical distinction and prescriptive restriction based on males and females. Because of religious and cultural relationships, gender segregation has been commonly practiced. Islamic segregation laws are thought to preserve modesty in both men and women as a safeguard against temptation, wickedness, and wrongdoing (Rahman et al., 2020). The separation between men and women, for example, while swimming, is one of the key principles of Halal tourism (Blackburn et al., 2002; Rahman et al., 2017b).

Religious Value

Consumers' purchasing behavior is influenced by a variety of socioeconomic, personal, cultural, religious, and political factors. Moreover, based on their religious convictions, individuals from various religious backgrounds may purchase products and services in different ways. A person's attitude is influenced by their religion. Religion is regarded as one of the most powerful social institutions, with a powerful influence on individual and social values, beliefs, and behaviors (Duman, 2012). As a result, religion deploys an important role in the lives of many religious groups, particularly in terms of consumption, purchasing, and travel decisions, which are largely influenced by their religious belief and customs. Individual's tastes differ greatly when it comes to religious law compliance and adherence (Bonne & Verbeke, 2008). However, religiosity is seen as a determining element for purchasing behavior which is defined as the vehemence of a person's attachment with the faith. The impact of religion on food intake is rooted in religion, its application, and the passion with which individuals pursue their religious wisdom (Lada et al., 2009). Consumers' food choices are frequently influenced by religious beliefs as well as social conventions and traditions. Furthermore, it has been discovered that food culture, lifestyle, and religion are all interrelated. In this context, Muslims chose Halal items and services for their everyday needs and activities.

Coronavirus and Its Impact on Halal Tourism and Hospitality

While discussing Halal tourism and hospitality, it is important to note that there are two primary types of Halal tourism and hospitality: religious hospitality (Hajj and Umrah) and other Halal hospitality activities. The demand for religious tourism and hospitality activities (Hajj and Umrah) is inflexible due to the characteristics of these activities and the fact that they are a part of the Muslim religion (El-Gohary, 2020). Furthermore, when performing the Hajj and Umrah, Muslim pilgrims expect to receive more satisfaction than religious fulfilment. In this context, researchers discovered that anticipated organizational benefits and pleasure in assisting others motivate Umrah pilgrims to participate (El-Gohary, 2020).

Some practitioners have begun raising some concerns about the Halal tourism system (Battour et al., 2019) as the starting point for changing its process because of the appalling loss of Halal tourism and hospitality companies and Halal tourism destination due to the effect of the COVID-19 epidemic. Many pessimists predict that it will take a very long time for Halal hospitality to recover from the coronavirus issue. These fears are prompted by the pandemic and the enormous impact on tourism, peoples' movement, and the global tourism sector (El-Gohary, 2020). After the pandemic, the majority of hotels and hospitality companies would struggle to attract Halal travelers, and the expense of doing so would be huge. Some academicians have already begun to claim that it is critical to investigate how the tourism sector will recover after the effect of coronavirus.

Furthermore, travelers' attitudes toward safe transportation nowadays is significantly different from their attitudes before the pandemic time. Most passengers are now more worried about their own and family's safety and security (Duman, 2012). As a result, consumers' security requirements of hotels, tour operators, and hospitality providers will shift dramatically in favor of a much greater level of security. Meeting such security standards will be exceedingly expensive for tourism and hospitality providers, and it will have a significant impact on their competitive advantage and marketing strategy (Battour et al., 2019). The prominent Halal travel trends include the rising importance of pricing as one of the important elements influencing Muslim passengers' decision making and the fact that Halal travel demand is mostly dominated by Muslims (Rahman et al., 2017a). As a result, the inability of hotels and hospitality operators to offer better plans and rates would have a significant detrimental impact on these businesses.

Since Muslim travelers account for the majority of Halal travel demand, any reduction in Muslim pilgrims' earnings would have an impact on Halal travel, tourism, and hospitality (El-Gohary, 2020). Because a significant proportion of Muslim tourists originate from the Middle East and Gulf countries, the present decline in oil and gas prices worldwide would result in a reduction in Muslim travelers' earnings. The worse fact is that Muslim tourists from other regions of the world earn less compared with Gulf Countries. This scenario will deteriorate as a result of the post-COVID-19 economic downturn, leaving Muslim tourists with even less spare money to spend on Halal travel and hospitality. All of these issues could result in a drop in Muslim tourists' purchase intention for Halal travel, tourism, and hospitality.

Technology's Impact on Tourism Events

As far as technological advancements in the tourism sector are concerned, cellphones, smartwatches, and artificial intelligence assistant have all helped to boost the effectiveness of voice search (Nilsson, 2009; Pencarelli, 2020). A rising number of tourism customers, in particular, are utilizing voice search to identify and buy airline tickets, hotels, and travel opportunities. As a result, to cope up with the trend and make a profit from it, it is critical to design your website with voice search in mind. Furthermore, voice control is becoming more common in practical travel experiences. Voice-controlled equipment can be utilized in hotel rooms for controlling lighting and heating or for gathering visitor information without having to speak to an employee. This is likely to spread to other aspects of the tourism business in the future.

One of the most fascinating forms of travel technology is robotics (Launius & McCurdy, 2007), which is continually being advanced. For example, in hotels, robots have been employed to welcome as well as provide information to the visitors after arriving. Some hotels have gone even further with their use, including housekeeping and cargo handling. Robots could eventually play a part in food preparation and serving in restaurants (Lee et al., 2018). They can be used to identify concealed weapons in airports, and some producers are using robots in developing luggage cases that intelligently follow you. Furthermore, travel agencies are employing robotics for screening, trying to make client waiting for times more productive (Launius & McCurdy, 2007). Robotics' use has increased in response to the COVID-19 situation with many other technological innovations in the tourism industry, particularly because they can eliminate physical interaction.

Accepting cashless transactions is another significant aspect of travel technology (Wulandari, 2017). Travel companies will be able to accept transactions considerably more efficiently as a result of this, even if clients do not have access to cash or their credit or debit card. Since it saves time, it can also improve the client experience. Contactless payments began as a convenience, but in the aftermath of the coronavirus pandemic, it becomes a critical component of tourism marketing. Many travelers are in hesitations to handle cash because they are concerned about the transmission of the virus. Wulandari (2017) believed that contactless payments should be used as much as possible to protect the company's employees.

Role of Virtual Reality in Tourism

Virtual reality (VR) is an emerging technology in a variety of fields, but it plays a particularly important role in the tourism business (Adachi et al., 2020). Besides, it enables passengers to explore faraway places from home comfort and can make a difference in whether they finally book or not. VR tours enable clients to experience all from virtual tourist hotels and restaurants to monuments, historic sites and even particular activities (Lo & Cheng, 2020). Interconnectivity and absorption can also assist to get a competitive advantage over its competitors who aren't yet making use of these kinds of technological advances. Most VR tours are already accessible with most popular web browsers to experience the benefits (Williams & Hobson, 1995), making it easier for those who are in hesitations to travel because of the COVID-19 pandemic. In a tourism event, augmented reality can be quite useful. It is comparable to virtual reality in the sense that it enhances rather than replaces a person's actual environment. The fact that it is cheaper than VR and consumers need simply a tablet or smartphone device with Internet connectivity is one of the big positive factors of this particular technological innovation. Those in the tourism sector may significantly improve the client experience by using graphics overlays to provide useful information or simply pure entertainment. Apps, for example, can be used to augment images using filters and effects (Lo & Cheng, 2020). Details about popular destinations can be presented as a consumer points their cellphone at them, offering information at the time when it is most needed.

Cyber Security and Travel Management

As organizations in the industry are constantly in a vulnerable situation to cyber-attacks and other types of information thefts, cybersecurity is a major priority area for people in travel management professions (Chen & Fiscus, 2018). Since they engage a large number of people and also have access to a large amount of client data, travel businesses are a prime target for the aggressor (Paraskevas, 2020). Hacking and malicious software cyberattacks are two of the most common risks in this sector, but because of the modern reliance on data, firms are also vulnerable to human mistake created by their staff. To keep the organization safe, there is a need to invest in cybersecurity training as well as a variety of hardware and software components. It's also essential to stay up to date with legislation and adhere to consumer protection laws.

Internet of Things and Tourism Industry

The Internet of Things (IoT), which incorporates internet-based interconnectivity between everyday devices, permitting them to communicate with other devices, is one of the most interesting emerging travel technology developments (Guo et al., 2014). We are already observing evidence of its involvement in the travel and tourism business, and hopefully, this will go far in the way of development. For example, in hotel rooms, IoT technology may be utilized to provide clients with such a device that links with everything from lighting to heating or air conditioners, enabling them to operate everything from one spot. Meanwhile, in airport terminals, baggage cases with sensors that inform people when they pass by can be introduced (Revfine, 2020). The IoT also offers significant and real-time information about their current condition and working order to immediately benefit IoT devices. This might be important for many of the tourism industries who can replace or repair important gadgets before they stop operating. For instance, if a radiator or a light bulb starts deteriorating, hotel workers can be notified immediately.

The Internet of Things can also be used away from hotels to assist airlines to use a fuel-efficient aircraft or replace components at the proper timing so that both maximum value and security may be achieved. Ivasciuc (2020) indicated that the

IoT eventually includes assigning Internet connectivity to daily gadgets and devices so that they can communicate with each other, and this brings significant advantages for travel operators, including the capability to deliver superior customer experience and optimize internal procedures. Companies in the travel industry may utilize the IoT to send customers location-specific data as well as collect useful data (Car et al., 2019). Messages can be sent to tourists at the most relevant time, based on where they are, by integrating smartphone capabilities with beacon technology or other sensors. For example, this could involve sending notifications about nearby destinations and the best times to visit them, or pointing out nearby public transportation options. The IoT also allows the collection of precise data at different times on the number of persons utilizing certain hotel facilities to optimize staffing levels.

Artificial Intelligence and Tourism Event

In analogy to human intelligence, most definitions tend to explain artificial intelligence. Artificial intelligence is a study of developing smart machinery (Gretzel, 2011; Nilsson, 2009) that involves the term intelligence, which is considered to be able to do complicated tasks, learn from action to particular goals and operate in the environment with forecasting (Gretzel, 2011). Artificial intelligence is being applied in a variety of ways other than in robotics (Nilsson, 2009). The most obvious application in the travel and tourist business is for customer support, with chatbots capable of providing quick responses to problems or inquiries (Hosseini, 2020; Manthiou et al., 2020). It can also learn from consumer interactions continuously. Furthermore, hotels and other tourism companies can employ artificial intelligence to process data accurately and permanently. It can make conclusions about company performance or consumer satisfaction patterns and can even manage inventories smartly.

Conclusion

In comparison to other religions, Islam promotes and inspires Muslims to travel to all corners of the world for a variety of reasons. This is supported in many places by the Holy Quran. Traveling for religious or non-religious purposes is highly connected to the Islamic Shariah for every Muslim. Muslim people are obligated by their religion to travel in order to obtain an education, make a better life, execute religious rituals such as Hajj or Umrah, seeking a cure for illness, enjoy tourism, and so on. This is documented in numerous verses in the Holy Quran (Muslims' religious book) that promote and encourage the Muslim community to travel. Moreover, Islamic Shariah provides Muslim travelers with some privileges, depending on the time and distance they have traveled. The outbreak of coronavirus (COVID-19) has given a lot of global attention and will continue to generate much worldwide attention. The virus has become a pandemic over a relatively short period, leading to huge negative consequences for the health and well-being of the individuals as well as for the world economy, including the tourism and hospitality sectors and other sectors. Coronavirus caused significant changes in customer behavior in a variety of businesses around the world. Such changes will have a huge impact on customer behavior, as well as advertising campaigns and promotion expenditures, and even some big effects on the various marketing mix components and travel industry.

References

Adachi, R., Cramer, E. M., & Song, H. (2020). Using virtual reality for tourism marketing: A mediating role of self-presence. *The Social Science Journal, 58*(1), 1–14.

Alam, S., & Uddin, I. (2020). *Singapore's Muslims to skip Hajj pilgrimage this year*. Retrieved from: https://www.aa.com.tr/en/asia-pacific/singapores-muslims-to-skip-hajj-pilgrimage-this-year/1841944. Accessed 5 June 2021.

Aljazeera. (2021). *Indonesia cancels Hajj again amid concerns over COVID*. Retrieved from: https://www.aljazeera.com/news/2021/6/3/indonesia-cancels-hajj-again-amid-concerns-over-covid. Accessed 5 June 2021.

Almuhrzi, H. M., & Alsawafi, A. M. (2017). Muslim perspectives on spiritual and religious travel beyond Hajj: Toward understanding motivations for Umrah travel in Oman. *Tourism Management Perspectives, 24*, 235–242.

Amanullah, S. (2009). Hajj 2.0-Technology's impact on the Muslim Pilgrimage. *Georgetown Journal of International Affairs, 10*(2), 75–82.

Battour, M., Rahman, M. K., & Rana, M. S. (2019). The impact of PHTPS on trip quality, trip value, satisfaction and word of mouth: Non-Muslim tourists' perspective. *Journal of Islamic Marketing, 11*(6), 1517–1538.

Blackburn, R. M., Browne, J., Brooks, B., & Jarman, J. (2002). Explaining gender segregation. *The British Journal of Sociology, 53*(4), 513–536.

Bonne, K., & Verbeke, W. (2008). Religious values informing halal meat production and the control and delivery of halal credence quality. *Agriculture and Human Values, 25*(1), 35–47.

Buisson, J. (2013). Gender segregation in Islam: Protection or destruction. *Kufa Review, 2*(1), 99–122.

Caidi, N. (2019). Pilgrimage to Hajj: An information journey. *The International Journal of Information, Diversity, & Inclusion, 3*(1), 44–76.

Car, T., Stifanich, L. P., & Šimunić, M. (2019). Internet of things (iot) in tourism and hospitality: Opportunities and challenges. *The 5th International Scientific Conference ToSEE – Tourism in Southern and Eastern Europe 2019 "Creating Innovative Tourism Experiences: The Way to Extend the Tourist Season"*. Opatija: The 16–18th May 2019.

Chen, H. S., & Fiscus, J. (2018). The inhospitable vulnerability: A need for cybersecurity risk assessment in the hospitality industry. *Journal of Hospitality and Tourism Technology, 9*(2), 223–234.

Collins-Kreiner, N. (2010). Researching pilgrimage: Continuity and transformations. *Annals of Tourism Research, 37*(2), 440–456.

Duman, T. (2012). The value of Islamic tourism: Perspectives from the Turkish experience. *ICR Journal, 3*(4), 718–739.

El-Gohary, H. (2020). Coronavirus and Halal tourism and hospitality industry: Is it a journey to the unknown? *Sustainability, 12*(21), 9260.

Gretzel, U. (2011). Intelligent systems in tourism: A social science perspective. *Annals of Tourism Research, 38*(3), 757–779.

Guo, Y., Liu, H., & Chai, Y. (2014). The embedding convergence of smart cities and tourism internet of things in China: An advance perspective. *Advances in Hospitality and Tourism Research, 2*(1), 54–69.

Hansot, E., & Tyack, D. (1988). Gender in American public schools: Thinking institutionally. *Signs: Journal of Women in Culture and Society, 13*(4), 741–760.

Hosseini, S. (2020). *Using a Chatbot to increase tourists' engagement*. Retrieved from: www.theseus.fi/handle/10024/340420. Accessed 12 June 2021.

Ivasciuc, I. S. (2020). Augmented reality and facial recognition technologies. Building bridges between the hospitality industry and tourists during pandemic. *Bulletin of the Transilvania University of Brasov. Economic Sciences. Series, 13*(2), 75–92.

Keene, S. D. (2011). Terrorism and the internet: A double-edged sword. *Journal of Money Laundering Control, 14*(4), 359–370.

King Abdulaziz Foundation for Research and Archives. (2021). *Darah. After an Outbreak of the "New Corona" and the Search for a Cure, "Darah" Tells Events that Stopped the Pilgrimage 40 Times*. Retrieved from: https://www.darah.org.sa/index.php/st-and-rep/darah-events/257-40. Accessed 12 June 2021.

Lada, S., Tanakinjal, G. H., & Amin, H. (2009). Predicting intention to choose halal products using theory of reasoned action. *International Journal of Islamic and Middle Eastern Finance and Management, 2*(1), 66–76.

Launius, R. D., & McCurdy, H. E. (2007). Robots and humans in space flight: Technology, evolution, and interplanetary travel. *Technology in Society, 29*(3), 271–282.

Lee, W. H., Lin, C. W., & Shih, K. H. (2018). A technology acceptance model for the perception of restaurant service robots for trust, interactivity, and output quality. *International Journal of Mobile Communications, 16*(4), 361–376.

Lo, W. H., & Cheng, K. L. B. (2020). Does virtual reality attract visitors? The mediating effect of presence on consumer response in virtual reality tourism advertising. *Information Technology & Tourism, 22*(4), 537–562.

Manthiou, A., Klaus, P., Kuppelwieser, V. G., & Reeves, W. (2020). Man vs machine: Examining the three themes of service robotics in tourism and hospitality. *Electronic Markets, 24*(3), 1–17.

Mason, R. (2006). Learning technologies for adult continuing education. *Studies in Continuing Education, 28*(2), 121–133.

Neiterman, E., & Zaza, C. (2019). A mixed blessing? Students' and Instructors' perspectives about off-task technology use in the academic classroom. *Canadian Journal for the Scholarship of Teaching and Learning, 10*(1), 1–16.

Nilsson, N. J. (2009). *The quest for artificial intelligence*. Cambridge University Press.

Noon, R. B., & Ayalon, L. (2018). Older adults in public open spaces: Age and gender segregation. *The Gerontologist, 58*(1), 149–158.

Othman, B., Harun, A., Rashid, W., & Ali, R. (2019). The impact of Umrah service quality on customer satisfaction towards Umrah travel agents in Malaysia. *Management Science Letters, 9*(11), 1763–1772.

Paraskevas, A. (2020). Cybersecurity in travel and tourism: A risk-based approach. In Z. Xiang, M. Fuchs, U. Gretzel, & W. Höpken (Eds.), *Handbook of e-Tourism* (pp. 1–24). Springer Nature.

Pencarelli, T. (2020). The digital revolution in the travel and tourism industry. *Information Technology & Tourism, 22*(3), 455–476.

Pritchett, A. R. (2005). The air transportation system. *Journal of Aerospace Computing, Information, and Communication, 2*(3), 170–173.

Rahman, M. K., & Zailani, S. (2016). Understanding Muslim medical tourists' perception towards Islamic friendly hospital. *Journal of Investment and Management, 5*(6), 206–213.

Rahman, M. K., Zailani, S., & Musa, G. (2017a). What travel motivational factors influence Muslim tourists towards MMITD? *Journal of Islamic Marketing, 89*(1), 48–73.

Rahman, M. K., Zailani, S., & Musa, G. (2017b). Tapping into the emerging Muslim-friendly medical tourism market: Evidence from Malaysia. *Journal of Islamic Marketing, 8*(4), 514–532.

Rahman, M. K., Zailani, S., & Musa, G. (2018a). Tourists' satisfaction and loyalty intention at Shariah compliant private hospitals in Malaysia. *International Journal of Tourism Sciences, 18*(4), 295–311.

Rahman, M. K., Zailani, S., & Musa, G. (2018b). The perceived role of Islamic medical care practice in hospital: The medical doctor's perspective. *Journal of Islamic Marketing, 9*(1), 2–18.

Rahman, M., Rana, M. S., Hoque, M. N., & Rahman, M. K. (2019). Brand perception of halal tourism services and satisfaction: The mediating role of tourists' attitudes. *International Journal of Tourism Sciences, 19*(1), 18–37.

Rahman, M., Moghavvemi, S., Thirumoorthi, T., & Rahman, M. K. (2020). The impact of tourists' perceptions on halal tourism destination: A structural model analysis. *Tourism Review, 75*(3), 575–594.

Rahman, M. K., Sarker, M., & Hassan, A. (2021a). Medical tourism: The Islamic perspective. In A. Hassan (Ed.), *Tourism products and Services in Bangladesh: Concept analysis and development suggestions* (pp. 87–99). Springer.

Rahman, M. K., Masud, M. M., Akhtar, R., & Hossain, M. M. (2021b). Impact of community participation on sustainable development of marine protected areas: Assessment of ecotourism development. *International Journal of Tourism Research*. https://doi.org/10.1002/jtr.2480

Rana, M. S., Rahman, M. K., Islam, M. F., & Hassan, A. (2020). Globalization effects on tourism marketing in Bangladesh. In A. Hassan (Ed.), *Tourism Marketing in Bangladesh* (pp. 157–171). Routledge.

Rashad, M., & Yaakoubi, A. (2021). *Saudi Arabia considers barring overseas haj pilgrims for second year, sources say*. Retrieved from: https://www.reuters.com/world/india/saudi-arabia-considers-barring-overseas-haj-pilgrims-second-year-sources-say-2021-05-05/. Accessed 2 June 2021.

Revfine. (2020). *4 Ways facial recognition can be used in the hospitality industry*. Retrieved from: https://www.revfine.com/facial-recognition-travel-industry/. Accessed 5 June 2021.

Sigala, M. (2018). New technologies in tourism: From multi-disciplinary to anti-disciplinary advances and trajectories. *Tourism Management Perspectives, 25*, 151–155.

The Edge Markets. (2021). *Prospective pilgrims advised to get ready should haj pilgrimage be allowed this year*. Retrieved from: https://www.theedgemarkets.com/article/prospective-pilgrims-advised-get-ready-should-haj-pilgrimage-be-allowed-year. Accessed 15 May 2021.

Williams, P., & Hobson, J. P. (1995). Virtual reality and tourism: Fact or fantasy? *Tourism Management, 16*(6), 423–427.

Wu, C. H., & Mursid, A. (2019). Loyalty motivations for religious tourism: Indonesian Muslim travelers Umrah participating in Umrah pilgrimage to Mecca, Saudi Arabia. *Tourism Review, 75*(3), 466–478.

Wulandari, N. (2017). Cashless payment in tourism. An application of technology acceptance model. *Journal of Environmental Management & Tourism, 8*(24), 1550–1553.

Ziyadin, S., Koryagina, E., Grigoryan, T., Tovma, N., & Ismail, G. Z. (2019). Specificity of using information technologies in the digital transformation of event tourism. *International Journal of Civil Engineering and Technology, 10*(1), 998–1010.

Muhammad Khalilur Rahman currently works in the Faculty of Entrepreneurship and Business at Universiti Malaysia Kelantan (UMK), Malaysia. He has been actively involved in research activities and published over 55 articles and book chapters in leading service management outlets. He has a wide interest in tourism and service management research which includes medical tourism, eco-tourism, halal tourism, service quality, brand equity, supply chain management, operation management, green and sustainable development. Rahman is a reviewer of Kybernetes, Journal of Hospitality and Tourism Insights, International Journal of Contemporary Hospitality Management, Journal of Islamic Marketing, and BMC Public Health.

Shaharin Akter is currently pursuing her MBA at the Department of Banking and Insurance under Faculty of Business Administration in University of Chittagong, Bangladesh. She is highly passionate about research with a strong desire to learn more. Her current research interests include Determinants of Profitability, Banking and Insurance Sector, Risk and Efficiency Measurement, Sustainability, and Blue Economy.

Mohammad Mainul Hossain is currently pursuing his Master of Philosophy (MPhil) at the Department of Media and Communications Studies under the Faculty of Arts and Social Sciences in University of Malaya, Malaysia. He is a dynamic and highly motivated student who has a strong passion, especially in the research area. His area of research interest is social networking applications, journalism, digital communications, information systems and information technology.

Azizul Hassan is a member of the Tourism Consultants Network of the UK Tourism Society. Dr Hassan has been working for the tourism industry as a consultant, academic, and researcher for over 20 years. His research interest areas are technology-supported marketing for tourism and hospitality, immersive technology applications in the tourism and hospitality industry, and technology-influenced marketing suggestions for sustainable tourism and hospitality industry in developing countries. Dr Hassan has authored over 150 articles and book chapters in leading tourism outlets. He is also part of the editorial team of 25 book projects from Routledge, Springer, CAB International, and Emerald Group Publishing Limited. He is a regular reviewer of a number of international journals.

Part II
Influential Factors

Chapter 5
How Do Social Media Literacy, Psychological Capital and Work Engagement Influence the Employee Morale of the Hospitality and Tourism Industry?

Fatema Johara, Munshi Muhammad Abdul Kader Jilani, and Md. Aftab Uddin

Introduction

The rapid expansion of the novel coronavirus (COVID-19) has led to a significant decline in all socioeconomic activities worldwide (Khan et al., 2021). This epidemic has resulted in substantial shutdowns in all areas of economic operations and caused disruptions in supply chain, resulting in a worldwide wave that impacts all economic sectors which were utterly unpredictable (Assaad & El-adaway, 2021; Smart et al., 2021). Most other activities have also been halted because of strict social distancing concerns and quarantines (Das et al., 2020). On March 8, 2020, the first COVID-19 case was confirmed in Bangladesh. Bangladesh recorded 1,117,310 confirmed cases and 18,125 deaths out of a total population of 164.64 million people (Worldometers, 2021). Delayed in containing the transmission of COVID-19, there are several uncertainties, and many obstacles have been faced by the hospitality and tourism industries. The government's imposement of nationwide lockdown forced the closure of hotels, motels, restaurants, and halt in transportation businesses along with the cancellation of all flights to and from Bangladesh, resulted in

---

F. Johara
Department of Business Administration, Bangladesh Army International University of Science & Technology, Cumilla Cantonment, Bangladesh

M. M. Abdul Kader Jilani
Department of Human Resource Management, Bangladesh Institute of Governance and Management (BIGM), Affiliated to the University of Dhaka, Dhaka, Bangladesh

M. A. Uddin (✉)
Department of Human Resource Management, University of Chittagong, Chattogram, Bangladesh
e-mail: mdaftabuddin@cu.ac.bd

A. Hassan (ed.), *Technology Application in Tourism Fairs, Festivals and Events in Asia*, https://doi.org/10.1007/978-981-16-8070-0_5

significant losses in this sector. It worsened when the Bangladesh government urged the tourism and hospitality industry' to shut down.

While the lockdown has been lifted since June 2020, this industry has become vulnerable and is facing the long-term aftermath of the pandemic. Because of the outbreak of COVID-19, hospitality and tourism have suffered negative consequences such as extended closures and a drop in income, employee turnover, and so on. Consequently, restaurants, like other businesses, have begun to deliver services online. In this regard, social media platforms have become the simplest way to reach customers.

Moreover, organizations related tourism and hospitality need to adopt new enterprise mechanisms to increase value in order to continue their growth, where social media is a digital platform to enhance customer potential and to support tourism products and services (Lin et al., 2020). According to research findings of Chen and Cheng (2019), the frequent spread of fake media news about products and businesses on social media in the modern world is alarming, whereas interactions and easy transmission of information on social media have provided brand leaders with powerful tools to engage more closely with consumers. The hospitality and tourism companies are also not out of this practice. Besides, regardless of the level of technological development in the workplace, hospitality and tourism businesses rely primarily on labor-intensive activities (Adeyinka-Ojo, 2018). Furthermore, social media literacy refers to the technical and cognitive skills that users require to use social media effectively and efficiently for online social interactions and communication (Daneels & Vanwynsberghe, 2017). In addition, age also has a significant impact on internet users' perceptions and behavior (Ye et al., 2019).

Tourism, being a people, and service-oriented industry, is dependent on engaged staff (Tsaur et al., 2019; Zhong et al., 2021). These individuals are critical to the business as they devote a great deal of energy, attention, and intensity to their self-fulfilling work for the pursuit of long-term excellence (Rabiul & Yean, 2021). Furthermore, Schaufeli et al. (2006) pointed out that work engagement (WE) is a term which refers to an employee's positive kind of attitude towards work, and it is a psychological state of mind marked by commitment, vitality, and absorption. Moreover, in a highly competitive world, where adverse changes have accelerated, companies increasingly rely on their employees' strengths and talents (Bakker, 2017). Thus, inspiring employees to build a high level of work engagement is a crucial aspect shaping the firm's employees' morale (EM).

Meanwhile, Kappagoda et al. (2014) also confirmed that psychological capital (PsC) is associated with all aspects of job performance and attitude. Employees with more PsC, for example, will have higher job satisfaction, work contentment, citizenship behavior, and job engagement (Karatepe & Avci, 2017). PsC is frequently cited as an antecedent variable in current tourism research. Paek et al. (2015) argued that employees' job performance, WE, EM, and quality of work-life could all benefit from PsC. Bowles and Cooper (2009) supported the idea that EM is associated with positive well-being based on confidence, usefulness, and purpose. It is defined as an individual's ability to maintain faith and its aims in their business. EM is tied with psychological well-being, WE, and organizational dedication, all of which contribute to increasing the firm's commitment (Pattnaik & Jena, 2020). In line with the previous arguments, we recommend the following research questions for future researchers:

RQ 1. Are social media literacy (SML), PsC, WE, and EM associated in the study contexts?
RQ 2. Are there any indirect effects, such as moderating and mediating, on EM?

In general, the study will make the following contributions in chapters. First, it establishes a novel concrete framework for connecting SML to WE, PsC, and morale. This chapter contributes to the body of knowledge by elucidating the predictors (SML and PsC) and outcomes (EM) of WE. Second, we noticed that past researches have concentrated on SML and employee engagement in the industrial and healthcare sectors (Johara et al., 2021; Tsaur et al., 2019), with little attention paid to EM. Notably, none of the studies established a relationship among PsC, WE and EM. As a result, the current research on the determinants described previously and, their possible direct and indirect effects will contribute to the advancement of preceding concepts of WE and EM.

Third, this study enhances the growing body of research on PsC in hospitality and tourism industry by examining the impacts of work engagement and EM. The current chapter is one of the very few empirical studies to explore the link between PsC and WE, with a specific emphasis on the tourism industry. Finally, we observed a dearth of studies on EM in developing countries (Brunetto et al., 2010; Pattnaik & Jena, 2020). Still, we found no analysis encapsulating social media literacy, PsC, WE, and EM from the hospitality and tourism industry's viewpoint, which would bridge that gap of examining a specific tourism entity.

Literature Review

Social Media Literacy

Social media can be restricted to digital media depending on three specific features: social communication, the recipient, and networking contacts. Social communication is decentralized as users choose their communication channels; the recipient will also be a sender that will increase user content; and social networking contact is collaborative and network-based (Daneels & Vanwynsberghe, 2017). Similarly, literacy in social media means that text, images, devices, and so on can be decoded, evaluated, and transmitted to interact in social media by keeping things private and confidential (Johara et al., 2021). Moreover, Schreurs and Vandenbosch (2020) asserted SML as the technological and cognitive competencies required for using social media to transmit meaningful and efficient messages in social and online communication. Festl (2020) has provided a broad and widespread comprehension of the context of social networking usage that excludes the restriction of using specific platforms or channels describing social interactions among groups in e-world.

Psychological Capital

The concept of PsC is seen as one of the modern management concepts, contributing to the highest performance levels (Durrah et al., 2016). However, efforts are being made to broaden the scope of PsC to include the outside workplace world. Several researchers have developed the PsC construct as the growth of positive organizational behavior. They have demonstrated that the psychological abilities of human resources can be measured, developed, and efficiently managed (Paek et al., 2015). Consequently, the term PsC, which has recently been demonstrated in tourism research, refers to an individual's positively valued situation and can be expressed in four ways: hope, optimism, resilience, and self-efficacy. Firstly, hope implies a well-defined future and goal. Second, optimism links with a favorable future on life events. Third, resilience is defined as the capability to sustain superior performance while recovering from setbacks. Finally, self-efficacy is ultimately a belief in the capability to solve problems (Wirawan et al., 2020).

Work Engagement

WE reveals the state of being involved in an individual's job. WE is a psychological condition in which the person doing the work is directly engaged and enthusiastic about it (Bakker, 2017). Van Wingerden and Poell (2019) identified that engaged employees display their higher energy and self-efficacious belief which reflect their internal locus of control over the events affecting their lives. WE is also essentially a motivational concept in which personal resources are actively allocated for workplace roles (Christian et al., 2011). According to Schaufeli et al. (2002), a positive mood marked by vigor, dedication, and absorption was defined as WE, where vigor features such as high energy levels, mental springiness, and alacrity to invest resources are demanded in a job. Secondly, dedication entails active involvement in work, which entails being enthusiastic, inspired, or proud of one's work while also being challenged by it. Finally, absorption is characterized by concentration, rapid passage of time, and difficulty getting away from work (Bakker et al., 2008; Van Wingerden & Poell, 2019).

Employee Morale

Morale affects an individual's mind and behavior, and its essence has been challenging to define. EM reflects the overall health of the organization. Morale is a relationship between employees and their coworkers and their organization (Arunchand & Ramanathan, 2013). EM represents emotion, satisfaction, and overall workplace attitude (Sania et al., 2015). It denotes excitement, enthusiasm, joy, and a positive attitude toward work, all of which drive productivity, as happy workers are more likely to succeed (Pattnaik & Jena, 2020). EM refers to employees' positive and supportive feelings toward their organization. EM is high when there is work satisfaction, increased labor force, creativity, initiative, a sense of pride in one's organization, and a desire to push group (common) goals ahead of personal goals. It is low when there is a high level of absenteeism, labor turnover, unresolved complaints, and strikes (Linz et al., 2006). Giese and Ruter (1949) discovered that EM is explained by tardiness, absenteeism, and production efficiency. EM is critical because failure to recognize and address the aforementioned factors can result in long-term, complicated, and costly defeatism at work.

Theoretical Framework

Social Media Literacy and Work Engagement

Social media platforms are primary means of raising funds, mobilizing people, exerting pressure on organizations and public institutions, and sharing and debating viewpoints (Kahne & Bowyer, 2019). Moreover, for a long time, media literacy theorists and supporters have claimed that it promotes civic engagement by increasing intellectual learners and promoting self-efficacy (Martens & Hobbs, 2015). Furthermore, practitioners and academics also frequently believe that SML can inspire more significant engagement. It can boost greater engagement and be a necessary tool for virtual continuance. Various authors have discussed the relationship of SML with civic and political engagement (Ashley et al., 2017; Kahne & Bowyer, 2019; Martens & Hobbs, 2015). Therefore, the present chapter assumes that SML influences the WE of hospitality and tourism employees.

H^1. *SML has a positive effect on WE.*

Psychological Capital and Work Engagement

Several researchers, especially Luthans and Youssef (2007), developed the PsC arrangement as an output for positive organizational behavior, and showed that human resource afflictions could be measured, developed, and managed effectively. PsC has demonstrated a significant positive relationship to results or reduced adverse effects (Paek et al., 2015). Previous research has found that employees' PsC is critical to their job success (Tsaur et al., 2019). Different companies need to recruit highly talented workers and inform staff to make their skills more practical to complete work more effectively (Bakker et al., 2011). Because of the growing popularity of evidence-based management, Sweetman and Luthans (2010) argued that understanding the process of employee participation is critical for modern organizations presenting a theoretical design that links PsC to WE by the use of positive emotions. Xanthopoulou et al. (2009) observed optimistic self-efficient employees attempting to create a creative working environment to engage in their activities. A wide range of studies found that PsC directly compressed WE (Paek et al., 2015; Tsaur et al., 2019; Wirawan et al., 2020). Therefore, the study proposed the following hypothesis:

H[2]. *PsC has a positive effect on WE.*

Work Engagement and Employee Morale

EM is a critical aspect of organizational performance (Pattnaik & Jena, 2020). Since the importance of ethics in modern business environments has grown, researchers have looked into employee ethics and morale as outcome variables (Paek et al., 2015). However, morale is an old and well-known construct riddled with incompatible applications and preliminary research (Ivey et al., 2015). Potoski and Callery (2018) stressed that increased employee engagement could boost EM and productivity while decreasing turnover. Moreover, WE is a more powerful precedent because employed workers are able and willing to make significant investments (Ryu & Shim, 2020), where WE only refers to the employee-employer relationship (Lee et al., 2016). However, very few researchers have found the association between WE and EM (Ivey et al., 2015; Paek et al., 2015; Pattnaik & Jena, 2020). Because of that, given this body of evidence and building on the previous discussion, we propose that WE enhances hospitality and tourism EM, leading to the underlined hypothesis:

H[3]. *WE has a positive effect on EM.*

Mediating Effect of Work Engagement

Since its inception, several studies have shown that WE mediates the relationship between organizational factors and results. Karatepe and Olugbade (2016) identified WE as a mediator for at least three reasons. The most important is that it appears to be the most proximate motivating variable to performance outcomes. SML can inspire greater engagement and optimistic self-efficient employees attempting to create a creative working environment to engage in their activities. Furthermore, increased employee engagement can boost EM and productivity while decreasing turnover. Hence, we expected that WE could mediate between SML and EM and, PsC and EM.

H[4]. *WE mediates the influence of SML on EM.*
H[5]. *WE mediates the influence of PsC on EM.*

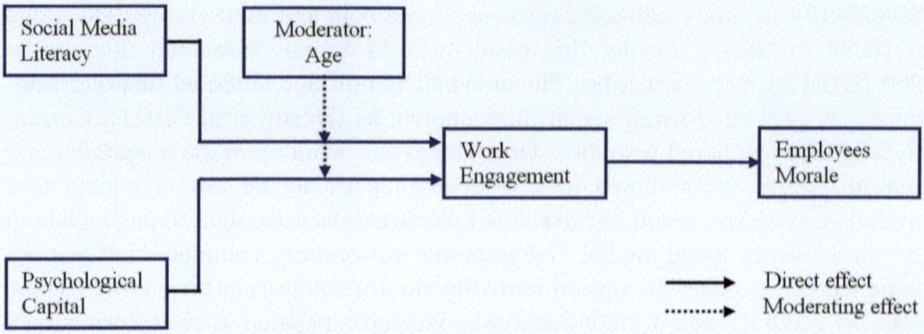

Fig. 5.1 Research model. (Source: Authors' own contributions)

Moderating Effect of Age

Age has been perceived as one of the essential characteristics and indicators in developing an online service (Biswas et al., 2020). Most previous studies looked at the age of various sociodemographic variables. Moreover, in various hospitality and tourism studies, age has been identified as a moderating factor (Biswas et al., 2020; Kang et al., 2018; Kim et al., 2019). The multi-group analysis revealed significant age differences, correlating with previous findings of different age groups and generational groups (Kim et al., 2019). The majority of age-based surveys have concentrated on classification, motivation, limitations, and behavior (Biswas et al., 2020). Furthermore, age is negatively related to internet users' self-efficacy, perceived online community quality, perceived value, and overall behavioral intent (Ye et al., 2019). This study hypothesized that age difference intervenes the influence of SML and PsC on WE. Finally, we developed the following hypotheses:

H^6. *Age moderates the influence of SML on WE because the SML of young adults has a high impact on WE than old adults and vice versa.*

H^7. *Age moderates the influence of PsC on WE because the SML of young adults has a strong effect on WE than old adults and vice versa.*

Figure 5.1 demonstrated that SML and PsC influence WE, and WE impacts EM. Moreover, the moderating effects of age on the influence of SML and PsC on WE are also displayed.

Research Methods

Research Setting

This study employs a deductive reasoning approach to conduct the investigation, and the research design is based on a quantitative research technique. Multi-item scales and self-administered surveys were used. Due to the emergence of

COVID-19, the study collected cross-sectional data and used Google Docs due to social distancing during this pandemic. As social distancing due to the COVID-19 strongly prescribes the snowball sampling technique, this questionnaire was circulated using social media networks (Johara et al., 2021). Current investigations gathered responses from employees working in the hospitality and tourism sector. We gathered data from employees during this changing neo-normal period. As a result, we asked people with internet access to participate in the survey using social media. The response was entirely voluntary, and respondents were also given the option to decline their responses at their will. To provide an accurate and timely response, we have ensured their information's privacy, anonymity, and confidentiality.

Participants' Information

The present study distributed 321 survey questionnaires among the target respondents and, finally, received 158 questionnaires. We used 151 valid questionnaires and dropped seven questionnaires due to incomplete or unmatched data. Of the 151 respondents, the majority of respondents were old adults (90 respondents, 59.6%), male (102 respondents, 32.5%), graduates (56 respondents, 37.1%), and private service holders (72 respondents, 47.7%). Most of the respondents served 5 to 15 years (93 respondents, 61.5%), held middle-level positions (81 respondents, 53.6%) in organizations, and earned family income ranging from BDT. 17,000 to 45,000 (82 respondents, 54.4%). On average, each respondent spent 2.96 h/day on social media.

Measurement Tools

The present study used scales from prior studies. SML, PsC, WE, and EM were measured using the scales of Ak (2020), Tsaur et al. (2019), Saks (2006), and Brunetto et al. (2010), respectively. The PsC was measured with four dimensions (self-efficacy, optimism, hope, and resilience). Sample item includes, 'I can distinguish whether information on the web is trustworthy (SML),' 'I feel confident contributing to discussions about my company's strategy (self-efficacy),' 'I always look on the bright side of things regarding my job (optimism),' 'I can think of many ways to reach my current goals (hope),' 'I usually take stressful things at work in my stride (resilience),' 'I am highly engaged in this job (WE)', and 'There is an atmosphere of trust in this organization (EM)'.

Analytical Technique

The study used structural equation modeling (SEM), SmartPLS version-3, to examine the hypothesize associations. The supremacy of using the PLS-SEM lies in robustness to generate the estimates by integrating both models, i.e., measurement model and structural model (Hair Jr. et al., 2017). Studies showed that this technique is appropriate when the observed model entails moderation and mediation effects (MacKinnon et al., 2002). Moreover, the PLS-SEM is flexible than other SEM and gives accurate results even if the sample size includes 100 or fewer responses (Azim et al., 2019; Fan et al., 2019; Hair Jr. et al., 2017; Johara et al., 2021). The present study also used other statistical tools, such as SPSS 23 and Microsoft Excel 2016, to generate other relevant estimates.

Results

Method Bias

The present study used various measures to prevent method bias. Previous studies reported that the method bias might be a crucial concern when responses for different scales were collected from the same sources (Johara et al., 2021; Podsakoff et al., 2012). Firstly, the respondents were assured confidentiality and anonymity for their responses provided to this study (Kock et al., 2021; Kock, 2015; Min et al., 2016). Secondly, we ran Harman's one-factor test, and the estimates showed that a single factor does not explain more than 50% of total variance (Harman, 1967). Thirdly, following the essence of Bagozzi et al. (1991), we examined the correlation matrix between latent variables, and the results showed that none of their associations exceeds 0.60 (<0.90). Finally, we also discussed the multi-collinearity issue, and estimates showed that none of the variance inflation factors exceeds the minimum range of 5–10, which is mentioned in (Field, 2018, p. 534). Thus, there is no issue with method bias.

Measurement Issues

Despite the study used prior studies' measurement tools, we examined the psychometric properties of the adapted scales for ensuring the validities. For convergent validities, we estimated composite reliability (CR) and the average variance extracted (AVE). Table 5.1 highlighted that the lowest CR and AVE are 0.866 and 0.623, respectively, above the threshold limits (Hair et al., 2020). For discriminant

Table 5.1 Estimates for measurement issues

Variables	1	2	3	4	5	6	7
1. EM	0.803						
2. PsC: Hope	0.422	0.786					
3. PsC: Optimism	0.423	0.510	0.876				
4. PsC: Resilience	0.363	0.502	0.351	0.850			
5. PsC: Self-efficacy	0.273	0.507	0.377	0.471	0.871		
6. SML	0.392	0.214	0.277	0.192	0.204	0.789	
7. WE	0.588	0.367	0.376	0.396	0.303	0.388	0.790
AVE	0.645	0.618	0.767	0.722	0.759	0.623	0.624
CR	0.901	0.866	0.868	0.886	0.904	0.952	0.869
R^2	0.420	–	–	–	–	–	0.280

Source: Authors' own compilation
EM Employee morale, *PsC* Psychological capital, *SML* Social media literacy, *WE* Work engagement, *AVE* Average variance extracted, *CR* Composite reliability

validities, we applied the method of Fornell and Larcker (1981). We found that the square root of a latent variable's AVE is larger than its correlation with other latent variables. Henceforth, the study has no measurement issues with convergent and discriminant validities (Hair et al., 2020).

Hypotheses Testing

Figure 5.2 demonstrated estimates in a structural model that showed the strength of the relationships via path estimates (β) and coefficient of determination (R^2) describes the overall predictability of the model. Scores on both β and R^2 in Fig. 5.2 are satisfactory, as explained in Cohen (1977) and Hair Jr. et al. (2017). We made three sections for testing hypotheses. Section 1 examines direct effects (H_1, H_2, and H_3), section 2 tests mediating effects (H_4 and H_5), and section 3 examines moderating effects (H_6 and H_7).

Assessment of Direct Effects

Table 5.2 showed the estimates of direct effects. Hypothesis 1 predicts that there is a significant influence of SML on WE. Estimates showed that the impact is significant ($p < 0.05$). Thus, hypothesis 1 is supported. In hypothesis 2, the study proposes that PsC affects WE, and results revealed that the effect is also significant ($p < 0.01$). Thereby, hypothesis 2 is supported. Finally, hypothesis 3 postulated that WE significantly influences EM and estimates revealed that the influence is also significant ($p < 0.01$). Thus, hypothesis 3 is also supported.

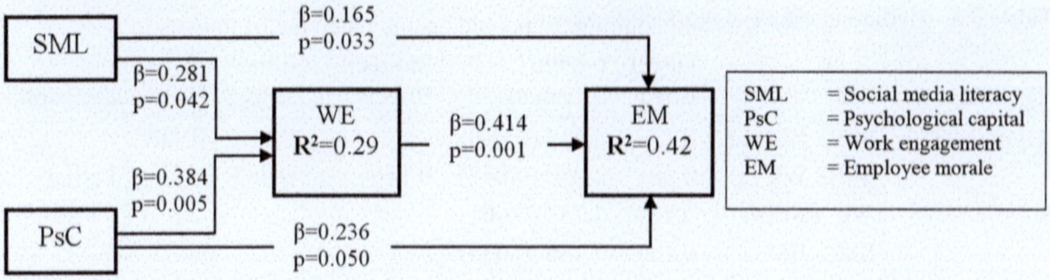

Fig. 5.2 The structural model, along with estimates. (Source: Authors' own contributions)

Table 5.2 Direct effects

Hypothesis	Path relations	β	Standard error	t-statistics	p-Value	Decision
Hypothesis 1	SML → WE	0.281	0.137	2.050	0.042	Supported
Hypothesis 2	PsC → WE	0.384	0.133	2.877	0.005	Supported
Hypothesis 3	WE → EM	0.414	0.128	3.238	0.001	Supported

Source: Authors' own compilation
SML Social media literacy, *EM* Employee morale, *WE* Work engagement, *PsC* Psychological capital

Assessment of Mediating Effects

To examine the mediation effect, we checked the direct effect (c) before and after using mediating variable. The path *c* before using the mediating variable, indirect effect (a) from independent variable to mediating variable, and indirect effect (b) from mediating variable to the dependent variable are also needed to be significant for confirming mediating effect (Azim et al., 2019; Fan et al., 2019). Following the tenet of mediation, if the direct effect (c') after using mediating variable turns insignificant and indirect effects remain significant, the full mediation exists (Uddin et al., 2019). Conversely, suppose both direct and indirect effects are significant after using the mediating variable. In that case, partial mediation might exist if the calculated variance accounted for (VAF) ranges from 0.20 to 0.80 (Hair Jr. et al., 2017). If the VAF score falls behind 0.20, no mediation exists (Hair Jr. et al., 2017).

In hypotheses 4 and 5, we predicted that WE mediates the influence of SML and PsC on EM. In Table 5.3, We observed that the previously significant *c* remains significant for both hypotheses. Thus, full mediation does not exist. The variance accounted for (VAF) was calculated, and Table 5.3 showed that VAF scores for hypothesis 4 is 0.413 and for hypothesis 5 is 0.402, meaning that partial mediation exists. Therefore, both hypothesis 4 and hypothesis 5 are partially supported.

Assessment of Moderating Effects

This chapter conducted a multi-group analysis of the moderating effects of age. Following the method suggested by Lowry and Gaskin (2014), we made two mutually exclusive groups-young adults (61, 40.4%) with an age limit from 18 to 30 years

Table 5.3 Mediating effect of work engagement

Hypothesis	Path	Direct Effect	t-value (p-value)	Indirect Effects (d)	Total Effect	VAF	Decision
Hypothesis 4	SML EM (c)	0.279	2.617 (0.010)			0.413	
	SML WE (a)	0.281	2.050 (0.042)	0.116	0.282		Partial mediation
	WE EM (b)	0.414	3.238 (0.001)				
	SML EM (c′)	0.165	2.146 (0.033)				
Hypothesis 5	PsC EM (c)	0.398	3.254 (0.001)	0.159	0.395	0.402	Partial mediation
	PsC WE (a)	0.384	2.877 (0.005)				
	WE EM (b)	0.414	3.238 (0.001)				
	PsC EM (c′)	0.236	1.975 (0.050)				

Source: Authors' own compilation
SML Social media literacy, *EM* Employee morale, *WE* Work engagement, *PsC* Psychological capital, *d a* time *b*; Total effect = Direct effect+Indirect effect, Variance accounted for (VAF) = Indirect effect/total effect

Table 5.4 Moderating effects of age

Hypothesis	Path Relations	Younger Adults β	SE	P-value	Older Adults β	SE	P-value	Difference in β	P-value	Decision
Hypothesis 6	SML→WE	0.635	0.145	0.000	0.342	0.111	0.002	0.293	0.045	Supported
Hypothesis 7	PC → WE	0.583	0.239	0.015	0.451	0.089	0.000	0.132	0.261	Not supported

Source: Authors' own compilation
[a]*SML* Social media literacy, *PC* Psychological capital, *WE* Work engagement, *SE* Standard error

old, and old adults (90, 59.6%) who are older than 30 years. The results on moderating influences are demonstrated in Table 5.3 and Table 5.4. In hypothesis 6, we hypothesized that SML of young adults has a strong impact on WE than old adults, and results reported in Table 5.4 showed that young adults (β = 0.635, p = 0.000) demonstrated high effects than old adults (β = 0.342, p = 0.002). Additionally, the difference in the effect size between these two groups is significant (β = 0.293, p = 0.045). Thus, hypothesis 6 is supported.

In hypothesis 7, the study also proposed that the influence of PsC on WE varies between young adults and old adults. Table 5.4 also showed that the path coefficients are different between young adults (β = 0.583, p = 0.015) and old adults (β = 0.451, p = 0.000), but the difference in the effect size is not statistically significant (β = 0.132, p = 0.261). Therefore, hypothesis 7 is not supported.

Discussion

Hospitality and Tourism industry employees can deliver advantages by establishing and maintaining quality host-guest relations and developing value in return. Highly motivated and dedicated employees are required to sustain the success of service companies and endeavors. Furthermore, due to mobility restrictions, some businesses have moved their operations online to continue operations. Hotels and restaurants, like these businesses, have begun to deliver food online. In this regard, social media platforms such as Facebook, Twitter, and others are the simplest way to reach customers. Furthermore, age also has a significant impact on the perceptions and behavior of employees. Previous research has shown that older people are less interested in trying new things than younger people.

Also, the organizations' sustainability and growth have been harmed by the abrupt closure due to COVID-19. EM, on the other hand, is suffering as a result. Even though EM is one of the most critical aspects of business operations, the researchers discovered that SML, PsC, and WE are essential predictors of increased EM after conducting a thorough literature review. As a result, this chapter aims to investigate and explore significant relationships between SML and WE, PsC, WE, and WE and EM. Researchers discovered a powerful positive influence among the relationships. This chapter also sought to highlight the role of WE in mediating SML and EM. The results indicated that this relationship was partially mediated.

Furthermore, the researchers attempted to identify the mediating effect of WE on PsC and EM. The researchers discovered partial mediation, just like the previous mediating effect. Furthermore, the results sought to determine the moderating effect of age, assuming that age moderates the influence of SML and PsC on WE. Because young adults' SML has a more significant impact on WE than old adults' and vice versa. Results discovered a considerable development. Furthermore, age moderates the influence of PsC on WE since young adults' SML has a more significant effect on WE than old adults' and vice versa. This assumption's findings were not statistically significant. As a result, this chapter anticipates that employees, particularly older employees, will improve their SML to increase their engagement. As a result, their morale will also rise to cope with this new normal life.

Contributions

This work contributes to the event and tourist management literature in several ways. Logically, the chapter's significance can be divided into two categories: EM and employee engagement in the hospitality and tourism sector. EM has emerged as a critical concern in the event, tourist, and, most importantly, academic domains. This study adds to the body of knowledge about individual morale by proving how employee WE can predict personal enthusiasm. Whenever EM is likely decent, an industry may operate at its total capacity in revenue and profitability. Furthermore, the chapter contributed to the literature on work engagement by establishing a relationship between SML and PsC. Second, our chapter demonstrates how workers' engagement influences EM while explaining that the effects of SML and PsC differ from past research findings. The results mean that managers from hospitality and tourism must refocus their attention away from the direct impact of SML and PsC on EM and toward the mediated effect of WE. In essence, the moderation of age and mediation of WE can better explain EM than directly employing SML and PsC. Finally, our research contributes to a broader knowledge of SML, PsC, and perspectives on employee work engagement as a predictor of EM and, responds to a request for additional research into employees' moral dilemmas. The pandemic's negative experience allowed all stakeholders in hospitality and tourism to cooperate and develop relationships more positively. The current results support this situation as well, demonstrating that SML and technological applications will increase employee efficiency, providing earnings and competitive advantage in the future.

Conclusion

This chapter's findings give empirical support for the pivotal role of SML and PsC. It directly affects WE and EM and has a moderating effect of age on WE. Our findings indicate that, despite the pervasive moderator of age, employees' moral outcomes are more likely to improve when they have higher positive psychological value and job engagement levels. Specific techniques aimed at increasing positive PsC should be devised and made available to employees with varying degrees of inner psychological well-being. Additionally, future research should explore the consequences of tourism-related interventions.

Limitations and Directions for Future Research

Although the present chapter notes numerous contributions and managerial implications, it inherits few constraints that prevent the findings' causality and generalizability. Firstly, the study used cross-sectional data that limits the findings' generalizability. Thus, future researcher researchers are expected to use longitudinal data or mixed-method research to ensure the triangulation of the data. Secondly, the chapter's sample is one of the most significant limitations that also prevents the broader acceptability of the results. Although the present chapter produced results in light of the principles of SEM, future researchers might think of collecting more respondents. Finally, the current research is executed individually, but individual behavior is rarely explained alone by individual-level variables. Thus, a multi-level model enclosing variables from the individual, group, and organizational levels is suggested for future search.

References

Adeyinka-Ojo, S. (2018). A strategic framework for analysing employability skills deficits in rural hospitality and tourism destinations. *Tourism Management Perspectives, 27*(March), 47–54.

Ak, N. Y. (2020). Development of internet literacy self-efficacy scale for pre-service teachers. *Turkish Journal of Education, 9*(2), 179–204.

Arunchand, C. H., & Ramanathan, H. N. (2013). Organizational culture and employee morale: A public sector enterprise experience. *Journal of Strategic Human Resource Management, 2*(1), 1–8.

Ashley, S., Maksl, A., & Craft, S. (2017). News media literacy and political engagement: What' s the connection? *Journal of Media Literacy Education, 9*(1), 79–98.

Assaad, R., & El-adaway, I. H. (2021). Guidelines for responding to COVID-19 pandemic: Best practices, impacts, and future research directions. *Journal of Management in Engineering, 37*(3), 06021001.

Azim, M. T., Fan, L., Uddin, M. A., Jilani, M. M. A. K., & Begum, S. (2019). Linking transformational leadership with employees' engagement in the creative process. *Management Research Review, 42*(7), 837–858.

Bagozzi, R. P., Yi, Y., & Phillips, L. W. (1991). Assessing construct validity in organizational research. *Administrative Science Quarterly, 36*(3), 421–458.

Bakker, A. B. (2017). Strategic and proactive approaches to work engagement. *Organizational Dynamics, 46*(2), 67–75.

Bakker, A. B., Schaufeli, W. B., Leiter, M. P., & Taris, T. W. (2008). Work engagement: An emerging concept in occupational health psychology. *Work and Stress, 22*(3), 187–200.

Bakker, A., Albrecht, S. L., & Leiter, M. P. (2011). Work engagement: Further reflections on the state of play. *European Journal of Work and Organizational Psychology, 20*(1), 74–88.

Biswas, C., Omar, H., & Rashid-Radha, J. Z. R. R. (2020). The impact of tourist attractions and accessibility on tourists' satisfaction: The moderating role of tourists' age. *Geojournal of Tourism and Geosites, 32*(4), 1202–1208.

Bowles, D., & Cooper, C. (2009). *Employee morale: Driving performance in challenging times*. Palgrave Macmillan.

Brunetto, Y., Farr-Wharton, R., & Shacklock, K. (2010). The impact of supervisor–subordinate relationships on morale: Implications for public and private sector nurses' commitment. *Human Resource Management Journal, 20*(2), 206–225.

Chen, Z. F., & Cheng, Y. (2019). Consumer response to fake news about brands on social media: The effects of self-efficacy, media trust, and persuasion knowledge on brand trust. *Journal of Product and Brand Management, 29*(2), 188–198.

Christian, S. M., Garza, S. A., & Slaughter, E. J. (2011). Work engagement: A quantitative review and test of its relations with task and contextual performance. *Personnel Psychology, 64*(1), 89–136.

Cohen, J. (1977). *Statistical power analysis for the behavioral sciences*. Academic.

Daneels, R., & Vanwynsberghe, H. (2017). Mediating social media use: Connecting parents' mediation strategies and social media literacy. *Cyberpsychology, 11*(3). https://doi.org/10.5817/CP2017-3-5

Das, A. K., Abdul Kader Jilani, M. M., Uddin, M. S., Uddin, M. A., & Ghosh, A. K. (2020). Fighting ahead: Adoption of social distancing in COVID-19 outbreak through the lens of theory of planned behavior. *Journal of Human Behavior in the Social Environment, 31*(1–4), 373–393.

Durrah, O., Ahmad, M., Al-Tobasi, A., & A'Aqoulah, A. (2016). The impact of the psychological capital on job performance: A case study on faculty members at Philadelphia university. *International Review of Management and Marketing, 6*(2), 183–191.

Fan, L., Mahmood, M., & Uddin, M. A. (2019). Supportive Chinese supervisor, innovative international students: A social exchange theory perspective. *Asia Pacific Education Review, 20*(1), 101–115.

Festl, R. (2020). Social media literacy & adolescent social online behavior in Germany. *Journal of Children and Media*, 1–23. https://doi.org/10.1080/17482798.2020.1770110

Field, A. (2018). *Discovering statistics using IBM SPSS statistics* (5th ed.). Sage Publications.

Fornell, C., & Larcker, D. F. (1981). Evaluating structural equation models with unobservable variables and measurement error. *Journal of Marketing Research, 18*(1), 39–50.

Giese, W. J., & Ruter, H. W. (1949). An objective analysis of morale. *Journal of Applied Psychology, 33*(5), 421–427.

Hair, J. F., Jr., Hult, G. T. M., Ringle, C. M., & Sarstedt, M. (2017). *A primer on partial least squares structural equation modeling (PLS-SEM)*. Sage Publication.

Hair, J. F., Howard, M. C., & Nitzl, C. (2020). Assessing measurement model quality in PLS-SEM using confirmatory composite analysis. *Journal of Business Research, 109*, 101–110.

Harman, D. (1967). A single factor test of common method variance. *Journal of Psychology, 35*(1967), 359–378.

Ivey, G. W., Blanc, J.-R. S., & Mantler, J. (2015). An assessment of the overlap between morale and work engagement in a nonoperational military sample. *Journal of Occupational Health Psychology, 20*(3), 338–347.

Johara, F., Taher, M. A., & Uddin, M. A. (2021). Social media literacy and COVID-19 awareness for faculty resilience in a moderated mechanism: An empirical study. In B. M. Waseem & A. Emilia (Eds.), *Impact of Infodemic on organizational performance* (pp. 250–269). IGI Global.

Kahne, J., & Bowyer, B. (2019). Can media literacy education increase digital engagement in politics? *Learning, Media and Technology, 44*(2), 211–224. https://doi.org/10.1080/17439884.2019.16011

Kang, J. H., Jang, J. C., & Jeong, C. (2018). Understanding museum visitor satisfaction and revisit intentions through mobile guide system: Moderating role of age in museum mobile guide adoption. *Asia Pacific Journal of Tourism Research, 23*(2), 95–108. https://doi.org/10.1080/10941665.2017.1410190

Kappagoda, U. W. M. R. S., Othman, H. Z. F., & De Alwis, G. (2014). Psychological capital and job performance: The mediating role of work attitudes. *Journal of Human Resource and Sustainability Studies, 2*, 102–116.

Karatepe, O. M., & Avci, T. (2017). The effects of psychological capital and work engagement on nurses' lateness attitude and turnover intentions. *Journal of Management Development, 36*(8), 1029–1039.

Karatepe, O. M., & Olugbade, O. A. (2016). The mediating role of work engagement in the relationship between high-performance work practices and job outcomes of employees in Nigeria. *International Journal of Contemporary Hospitality Management, 28*(10), 2350–2371. https://doi.org/10.1108/IJCHM-03-2015-0145

Khan, A. G., Kamruzzaman, M., Rahman, M. N., Mahmood, M., & Uddin, M. A. (2021). Quality of life in the COVID-19 outbreak: Influence of psychological distress, government strategies, social distancing, and emotional recovery. *Heliyon, 7*(3), e06407.

Kim, W. H., Cho, J. L., & Kim, K. S. (2019). The relationships of wine promotion, customer satisfaction, and behavioral intention: The moderating roles of customers' gender and age. *Journal of Hospitality and Tourism Management, 39*(January), 212–218.

Kock, N. (2015). Common method bias in PLS-SEM: A full collinearity assessment approach. *International Journal of e-Collaboration, 11*(4), 1–10.

Kock, F., Berbekova, A., & Assaf, A. G. (2021). Understanding and managing the threat of common method bias: Detection, prevention and control. *Tourism Management, 86*, 104330.

Lee, Y., Kwon, K., Kim, W., & Cho, D. (2016). Work engagement and career. *Human Resource Development Review, 15*(1), 29–54.

Lin, H. C., Han, X., Lyu, T., Ho, W. H., Xu, Y., Hsieh, T. C., Zhu, L., & Zhang, L. (2020). Task-technology fit analysis of social media use for marketing in the tourism and hospitality industry: A systematic literature review. *International Journal of Contemporary Hospitality Management, 32*(8), 2677–2715.

Linz, S. J., Good, L. K., & Huddleston, P. (2006). Worker morale in Russia: An exploratory study. *Journal of Managerial Psychology, 21*(5), 415–437.

Lowry, P. B., & Gaskin, J. (2014). Partial least squares (PLS) structural equation modeling (SEM) for building and testing behavioral causal theory: When to choose it and how to use it. *IEEE Transactions on Professional Communication, 57*(2), 123–146.

Luthans, F., & Youssef, C. M. (2007). Emerging positive organizational behavior. *Journal of Management, 33*(3), 321–349.

MacKinnon, D. P., Lockwood, C. M., Hoffman, J. M., West, S. G., & Sheets, V. (2002). A comparison of methods to test mediation and other intervening variable effects. *Psychological Methods, 7*(1), 83–104.

Martens, H., & Hobbs, R. (2015). How media literacy supports civic engagement in a digital age. *Atlantic Journal of Communication, 23*(2), 120–137.

Min, H., Park, J., & Kim, H. J. (2016). Common method bias in hospitality research: A critical review of literature and an empirical study. *International Journal of Hospitality Management, 56*, 126–135.

Paek, S., Schuckert, M., Kim, T. T., & Lee, G. (2015). Why is hospitality employees' psychological capital important? The effects of psychological capital on work engagement and employee morale. *International Journal of Hospitality Management, 50*, 9–26.

Pattnaik, L., & Jena, L. K. (2020). Mindfulness, remote engagement and employee morale: Conceptual analysis to address the "new normal". *International Journal of Organizational Analysis*. https://doi.org/10.1108/IJOA-06-2020-2267

Podsakoff, P. M., MacKenzie, S. B., & Podsakoff, N. P. (2012). Sources of method bias in social science research and recommendations on how to control it. *Annual Review of Psychology, 63*, 539–569.

Potoski, M., & Callery, P. J. (2018). Peer communication improves environmental employee engagement programs: Evidence from a quasi-experimental field study. *Journal of Cleaner Production, 172*, 1486–1500.

Rabiul, M. K., & Yean, T. F. (2021). Leadership styles, motivating language, and work engagement: An empirical investigation of the hotel industry. *International Journal of Hospitality Management, 92*, 102712.

Ryu, G., & Shim, D. C. (2020). Developmental leadership, skill development, and work engagement. *International Review of Public Administration, 25*(1), 64–79.

Saks, A. M. (2006). Antecedents and consequences of employee engagement. *Journal of Managerial Psychology, 21*(7), 600–619.

Sania, U., Kalpina, K., & Javed, H. (2015). Diversity, employee morale and customer satisfaction: The three musketeers. *Journal of Economics, Business and Management, 3*(1), 11–18.

Schaufeli, W. B., Salanova, M., Gon Alez-ro, V. A., & Bakker, A. B. (2002). The measurement of engagement and burnout: A two sample confirmatory factor analytic approach. *Journal of Happiness Studies, 3*, 71–92.

Schaufeli, W. B., Bakker, A. B., & Salanova, M. (2006). The measurement of work engagement with a short questionnaire: A cross-national study. *Educational and Psychological Measurement, 66*(4), 701–716.

Schreurs, L., & Vandenbosch, L. (2020). Introducing the social media literacy (SMILE) model with the case of the positivity bias on social media. *Journal of Children and Media, 15*(3), 320–337.

Smart, K., Ma, E., Qu, H., & Ding, L. (2021). COVID-19 impacts, coping strategies, and management reflection: A lodging industry case. *International Journal of Hospitality Management, 94*, 102859.

Sweetman, D., & Luthans, F. (2010). The power of positive psychology: Psychological capital and work engagement. In A. B. Bakker & M. P. Leiter (Eds.), *Work engagement: A handbook of essential theory and research* (pp. 54–68). Psychology Press.

Tsaur, S.-H., Hsu, F.-S., & Lin, H. (2019). Workplace fun and work engagement in tourism and hospitality: The role of psychological capital. *International Journal of Hospitality Management, 81*, 131–140.

Uddin, M. A., Mahmood, M., & Fan, L. (2019). Why individual employee engagement matters for team performance? Mediating effects of employee commitment and organizational citizenship behaviour. *Team Performance Management: An International Journal, 25*(1/2), 47–68.

Van Wingerden, J., & Poell, R. F. (2019). Meaningful work and resilience among teachers: The mediating role of work engagement and job crafting. *PLoS One, 14*(9), 1–13.

Wirawan, H., Jufri, M., & Saman, A. (2020). The effect of authentic leadership and psychological capital on work engagement: The mediating role of job satisfaction. *Leadership and Organization Development Journal, 41*(8), 1139–1154.

Worldometers. (2021). *COVID-19 coronavirus pandemic*. Retrieved from: https://www.worldometers.info/coronavirus/?fbclid=IwAR3ohsw8HMRsdj5gMCmNLDh-j28nkV_0474oJGnrPKS9ac34OU2Ni4K5LQA. Accessed 21 July 2021.

Xanthopoulou, D., Bakker, A. B., Demerouti, E., & Schaufeli, W. B. (2009). Reciprocal relationships between job resources, personal resources, and work engagement. *Journal of Vocational Behavior, 74*(3), 235–244.

Ye, B. H., Barreda, A. A., Okumus, F., & Nusair, K. (2019). Website interactivity and brand development of online travel agencies in China: The moderating role of age. *Journal of Business Research, 99*(September), 382–389.

Zhong, Y., Busser, J., Shapoval, V., & Murphy, K. (2021). Hospitality and tourism student engagement and hope during the COVID-19 pandemic. *Journal of Hospitality & Tourism Education, 33*, 1–13.

Dr Fatema Johara has been working as an Associate Professor at Bangladesh Army International University of Science & Technology. She did her PhD from Universiti Sains Malaysia, Malaysia. Dr Johara also did her Master of Philosophy in Management and MBA in Accounting from the University of Chittagong, Bangladesh. Her research interests lie in talent management, employee engagement, management accounting, strategic management, organizational behavior, and internal auditing. She published a good number of publications in peer reviewed journals and conference proceedings, including in the International Journal of Sustainable Strategic Management, Journal of Sustainability Science & Management, Global Business and Management Research, etc. She has been acting as a regular reviewer and editor for many international conferences and Journals.

Dr Munshi Muhammad Abdul Kader Jilani has been serving as an Assistant Professor at Bangladesh Institute of Governance and Management (BIGM). He completed his Doctor of Philosophy (PhD) in Human Resource Management from the Wuhan University of Technology, China. He also completed an MBA from IISWBM, the University of Calcutta on Human Resource Management. Dr Jilani also completed his BBA and another MBA in International Management from the University of Chittagong, Bangladesh. His areas of research interest are psychological capital, knowledge management, employee resilience, and organizational behavior. He published a good number of articles in peer reviewed journals and conference proceedings, such as the Sustainability Journal, Management Research Review Journal, Journal of Innovation and Sustainability, among others.

Dr Md. Aftab Uddin has been teaching as an Associate Professor in the Department of Human Resource Management, University of Chittagong, Bangladesh 4331. He completed his PhD and MBA majoring in Human Resource Management from the Wuhan University of Technology in China. Previously, Dr Uddin also completed his first MBA (International Management) and BBA (Management) from the University of Chittagong, Bangladesh. Dr Uddin jointly published a book entitled "HRM Cases". He also conducts research on industrial psychology, cognitive psychology, environmental psychology, AI adoption in workplace, which were published in peer-reviewed journals and conference proceedings at home and abroad with different indexations (SSCI/SCIE/CPCI/ESCI/Scopus/ABDC/Scimago/ABI/ABS, etc.).

Part III
Perceptions and Effects

Chapter 6
Protection of Sacred Tourism Sites, Festivals and Events in Bangladesh over Augmented Reality: An Experience of Cultural Heritage Based Tourism

A. H. M. Salimullah and Rakiba Nabi

Introduction

Cultural heritage (CH) depends on the cultural belongings and intangible attributes possess by an ethnic group or society that inherited from overlapping generations (Logan, 2007). Yet the historic cultural resources such as monuments, landscapes, archaeological sites are at risk in many places globally, including Bangladesh. Society does the selection and preservation of the heritage spots. Ancient buildings or monuments with architectural diversity, landscapes, old books, or remarkable arts and crafts are the examples of tangible cultural heritage. Due to the issues of acquisition, ownership of heritage sites and the preservation of tangible cultures related with area or locality often becomes a challenge for the government of a country. Nonetheless, cultural heritage also faces categorical challenges such as climate change, lack of human resources, weak legal and policy framework, modernization, and unplanned urbanization due to the lack of proper awareness and poorest planning. Historically, Bangladesh is disposed with cultural heritage sites. Promoting tourism nationally and internationally will consequently uphold the country's dignity and the country will derive frequent economic benefits from the number of renowned sites and places she inherited.

Several historical tourist places in Bangladesh namely, Sompur Mahavihara in Paharpur, Mahasthangarh in Bogra, the Sundarbans in Khulna (the world's largest mangrove forest), Lalbagh Fort of Dhaka, Shatgombuj Mosque of Bagherhat, Kantazir Temple of Dinajpur, historic city of Sonargaon near Dhaka, Patuakhali, and Sacred places in Sylhet come across many challenges such as infrastructure

A. H. M. Salimullah (✉)
Department of Economics, University of Chittagong, Chattogram, Bangladesh
e-mail: salimullah.economics@cu.ac.bd

R. Nabi
Department of Law, University of Chittagong, Chattogram, Bangladesh

A. Hassan (ed.), *Technology Application in Tourism Fairs, Festivals and Events in Asia*, https://doi.org/10.1007/978-981-16-8070-0_6

degradation, weak planning and management, insecurity, lack of sufficient historic protection laws, and lack of communication as well as transportation, etc. The Sundarbans Reserve Forests (SRF) in Bangladesh comprised an area of 1,40,000 hectares approximately which was inscribed as a UNESCO world heritage site in 1997 has fallen victim to climate change and hence requires resiliency and sustainability measures. The other areas of tourist place are not related to climate action plans, climate mitigation plans, and resiliency plans. Yet the Bangladesh government, including local and district administrations, has adopted several contemporary strategies such as the built realm, energy usage, water usage, diversion and treatment, transportation efficiency, communications, emergency services, natural environment, comprehensive planning, public relations, etc. safeguard historic properties.

Augmented reality (AR) can be a solution to the problems we face in preserving and protecting CH. Superimposing digital content in the real-world environment through AR can be a solution to preserve the authenticity of the prime heritage sites and maintain them to remain as a heritage attraction. Using AR will make the promotion of sustainable tourism and ecotourism easier in the upcoming days.

A Review of the Archaeological Tourism Sites in Bangladesh

Among various unexplored tourism areas, CH tourism is one of the most underutilized tourism in Bangladesh. The lack of care to conserve these sites and the indifference towards promoting such sites deprive the country of reaping the benefits of archaeological tourism. For better conservation and proper utilization of the historical tourist sites in the country, heritage tourism can be divided into three categories of tourism: archaeological, historical, and religious tourism (Hasan & Jobaid, 2014).

There are various micro and macro-level heritage spots in Bangladesh that are yet to be fully explored. Throughout the history, from 200 years of British regime to the Indian subcontinent period and the post 1947 period, Bangladesh witnessed many regimes with different cultural orders. Nevertheless, their cultural practices remained as cultural and archaeological heritage, especially inherited from the ancient medieval period of different Muslim and Non-Muslim rulers and different Buddhist dynasties like the Pala, Dev, Sen and Chandra rulers.

As per the Department of Archaeology, there are approximately 2500 archaeological sites in Bangladesh. Currently, the department maintains 509 heritages. UNESCO has enlisted three heritages as in the world heritage destinations list. Historic Mosque City of Bagerhat and Ruins of the Buddhist Vihara at Paharpur Vihara do have cultural heritage and the Sundarbans as a natural heritage. In addition, there are five sites that are listed in the tentative list of UNESCO for nomination as a world heritage site, Mahasthangarh, and its environs, The Lalmai – Mainamati group of monuments, Lalbagh fort, Halud Bihar, Jaggadala Bihar (UNESCO, n.d.-a). Unfortunately, the list was updated by Bangladesh government in 1999, almost

22 years ago, proving the lack of attention in the conservation of the heritage sites of the country. Some of the important heritage sites in Bangladesh are listed below:

Paharpur Bihar

Between the years 1807 to 1812, Buchanon Hamilton was first noticed this site during a survey in eastern India. India declared this site protected by the Archaeological Survey of India under the Ancient Monuments preservation act, 1904. Department of Archaeology in Bangladesh further excavated this site after the Independence in 1971. It contains 177 monastic cells, and at the center, there is a high pyramidal temple. A clay seal discovered at the monastery ground reveals that Dharmapala Vikramshala, the king of Varendri Magadha, was the first builder of the monastery. In a museum at that site, the excavated discoveries are preserved (Bangladesh Parjatan Corporation, 2014a). Upon the request of the government of Bangladesh in 1973 to preserve this site, UNSECO prepared a master plan in 1983, and this site was therefore, included in the World Cultural Heritage List in 1985. The master plan initiated in 1987 and the whole completion process was conducted in three phases that ended in 2002 (UNESCO, n.d.-b).

Mahasthangarh

Predicted from limestone where the words inscribed is found in 1931, the date when this village was found dates back to the third century. People used it as a city until the eighteenth century. Total amount of land areas is almost 185 hectares. In the center of the city, the citadel is a fort. Excavation on this site discovered coins, ceramics, sculptures and many other artefacts. It is situated at the bank of Karatoya, near the Bogra Rangpur highway and 11 km away from Bogra City. There are several isolated mounds. Local people named them as Khodai Pathar Mound, Mankalir Kunda, Govinda vita temple, Parasumarer Bedi, Jiyat Kunda (Bangladesh Parjatan Corporation, 2014b).

Mainamati

Named after the wife of Manikchandra, a king of the Chandra Dynasty, Mainamati is comprised of 50 ancient Buddhist temple and many Buddhist sites such as Shalban Vihara Kutila Mura, Charpatra Mura, Ananda Bihar, Mainamati Ranir Bedi (The tomb of Mainamati) (Bangladesh Parjatan Corporation, 2014c).

Shalban Vihara

In Cumilla District, a part of the remnants of Mainamati, Shalban Vihara is a vital center reminiscing Buddhism of the seventh–twelfth century. It was the capital of the Dev dynasty. The fourth ruler of the Deva dynasty, Bhava Deva, discovered this site in the middle of the Lalmai Hills Bridge in 168 square meters. The place used to be a lodging place for 115 Monks. It was known as Shalbon Rajar Bari. Later, it was named the Shalbon Bihar or an educative place with residential facilities after discovering the copper artworks. This place represents the Buddhist trinity by the three stupas found side by side (Barua, 2018).

Jaggadala Mahavihara

Probably Rampala, the last emperor of the Pala dynasty, established this site. Many viharas were established during the regime of the Pala dynasty. Among the five great Mahaviharas stood out, one of which was Jaggadala Vihara. UNESCO supported and verified this archaeological site to be a monastery. UNESCO also revealed it as an extensive mound, with 105 m long and 85m' width that represents the archaeological remains of a Buddhist monastery. That hermitage has included terracotta plaques, ornamental bricks, nails, a gold ingot and three stone images of deities (United Nations Educational, Scientific and Cultural Organization, 1999). This site is in the tentative list of UNESCO for the nomination of the world heritage site.

Shait Gombuj Masjid

With 81 domes and 60 stone pillars, the mosque is situated in the Bagerhat district in the Khulna division. The saint-general Khan Jahan Ali built it during the reign of Sultan Nasiruddin Mahmud Shah. It is one of the many mosques built by him in the mosque city of Bagerhat. This site is enlisted in the UNESCO World Heritage list.

Sonargaon

The earliest known capital of Bengal, Sonargaon, was under the Deva dynasty until the thirteenth century. Afterwards, it was controlled by the Delhi sultanate, Bengal Sultanate, the Twelve Bhuiyans and finally the Mughals till the eighteenth century. Afterwards, the British took over, and it was used for trades along with the neighboring Panam Nagar. It was the center of production of the famous Muslin fabric and the cotton-based cloth Khasa. A large part of the population was involved in weaving. In 1975, the veteran Bangladeshi painter Zainul Abedin established a

Bangladesh folk arts and crafts foundation. The world monuments fund placed Sonargaon in the watch list of the 100 most endangered sites in the year 2008 (Bangladesh Parjatan Corporation, 2014d).

Wari Bateshwar

This was a fort city of ancient time. It dates back to 450 BC, and it reminds us of early civilization in Bangladesh. A local school teacher in the Belabo Upazila in Narshingdi District, Hanif Pathan, first discovered the site. Greek geographer, astronomer, mathematician, Ptolemy in his book Geographia mentioned this place as Sounagora or an emporium. He mentioned some other emporia. They are-Arikamedu in India, Mantai of Sri Lanka and Kion Thom of Thailand. Wari Bateshwar is similar to the other place mentioned in his book (Bangladesh Parjatan Corporation, 2014e).

Lalbagh Fort

Mughal emperor Azam Shah started the work of this fort in1678 A.D. and continued for 15 months. Afterwards, Shaista Khan completed this fort during the period of his ruling. In 1684, the daughter of Shaista khan died in the fort, and Shaista Khan saw the fort as a misfortune. For a long time, the place was unoccupied. Until excavation, the fort was seen to have three structures- a place of worship, the tomb of Pari Bibi and the residential quarter known as Dewan-i-Am and the Hammam, but later other features of the structure were found (Bangladesh Parjatan Corporation, 2014f).

Curzon Hall

Lord Curzon established this site on 14 February 1904. It was the location of witnesses of various significant and historical movements such as the Bengali language movement of 1952. It was previously being used as Dhaka College in 1911 and later took up its position as the science division of Dhaka University in 1921 (Banglapedia, 2014).

Suhrawardy Uddyan

From the period of struggle for liberation to the country's Independence, this site speaks for various historical events of the country. In 1969, after got release from the imprisonment the father of the nation Sheikh Mujibur Rahman was accorded the title Bangabandhu by a civic public reception on that ground. On March 7th, in a massive public gathering, Bangabandhu declared the Independence of the country. On December 16, 1971 the nation achieved victory from the then West Pakistan after they surrendered to the allied forces of the country on this ground. This site is currently maintained as a historic park with an eternal fire set up in 1996 which is symbolizing of freedom (Bangladesh Parjatan Corporation, 2014g).

Ahsan Manjil

Popularly known as the pink palace for its color, Ahsan Manjil was a residential Pala (a shovel-shaped or scoop-shaped structure) and headquarter for the Nawabs of Dhaka. In the Mughal era, a landlord of the Jamalpur Pargana, Sheikh Enayetullah, used to reside in a garden house in this place. Later, Sheikh Moti Ullah, son of Enayetullah, sold this palace to the French traders. French merchants carried out their business without paying any taxes that was decreed by the then Emperor Aurangzeb. After several transfers, it was purchased by an established landlord of Dhaka, Khwaja Alimullah. Later his son, Khwaja Abdul Ghani developed the property and titled it Ahsan Manzil, named after his own son, Khwaja Ahsanullah (Bangladesh Parjatan Corporation, 2014h).

The Necessity of Protection of Sacred Tourism Spots

Every nation comes with its unique cultural, historical and religious heritage. Each region has its unique history and culture that tells stories of centuries after centuries of civilizations and societies that had played a crucial role in forming the present identity of that region. Sites that contain the vestiges of our religion, culture and history are not only sacred monuments but also sites for the tourists who are looking for the experience of the culture and history of different nations. Therefore, those heritage sites must be preserved and protected not only for the contemporary but also for the future generations. The memories of millennia worth of cultures, religions, and civilizations can be transmitted properly. The preservation of the sites is also important for generating an economic return and supporting the tourism industry (UKEssays, 2018).

Sacred heritage sites represent the history and culture of a nation. They are the transmitters of age-old practices and values that linked to the identity of a region as well as reflecting the relationship between religious and monumental values of a place (Aulet & Vidal, 2018). They also constitute the architectural values of the area (UKEssays, 2018). The archaeological findings of those sites reflect historical values that relate visible architecture to the important events in history, which are religious, societal or political (Aulet & Vidal, 2018). For example, the ancient history Bengal has seen the rise and fall of different empires such as the Pals, the Sens, the

Guptas, the Sultans, and the Mughals, etc. (Haider, 2015). Places like Paharpur Bihara, Mainamati, Mahasthangarh, Shompur Bihara, and Sonargaon, etc. are therefore, the reflections of vestiges of the glorious past and diverse cultural identity of Bangladesh. Through looking at the archaeological evidence of those sites, we can identify the how rich, and diversified legacy of Bangladesh has in the past (Haider, 2015).

Furthermore, archaeological sites with cultural, historical and religious importance can significantly impact a nation's economy by generating more revenue and other economic benefits with the development of the tourism sector. Tourism industry already accounts for a noteworthy part of the world's economy in boosting foreign exchange, higher Gross Domestic Product, more employment generation, generating governmental revenue, increasing value as well as volume of trade and commercial goods, more exports and imports to create surplus in the current account (Rahman, 2012). Cultural tourism–where a tourist's essential motivation is to learn, experience and consume the cultural attractions in a tourism destination (World Tourism Organization, 2021) forms a significant part of the tourism industry. Tourists who are craving for experiencing nations' culture and history are likely to spend one-third more on average than other tourists (Richards, 2007). In 2007, cultural tourists accounted for 40% of global tourism (Organisation for Economic Co-operation and Development, 2009).

As discussed above, Bangladesh is considered historically as an abundant heritage transmitter of age-old cultural norms and sacred traditions and this economic unit can be considered as a significant revenue and economic benefit earners for the country. Furthermore, many of these places–especially sacred places where people visit to worship and on religious journeys–must also be conserved and protected properly in order to be used by the local populace while being simultaneously explore by the respectful visitors and tourists. There is a desperate need for the protection of the country's sacred tourist spots and heritage sites and measures to balance conservation and use so that these places can be used for the benefit of the present generation and also preserving them for the future generations as well.

The Problems of Maintenance of Cultural Heritage as the Sacred Tourism Spots

Climate Change

Climate change now-a-days, has formed a place for itself on the top of the list of world's problems. There is hardly any aspect of life today that is not affected by climate change. Moreover, our tangible cultural and sacred heritage sites are not exempt from this. Many ancient buildings were built during the periods when maintained specific local climates in mind (United Nations Educational, Scientific and Cultural Organization, 2007). With the climate changing drastically, these buildings and sites are bound to be affected adversely since historic buildings have been more connected to the ground and their surrounding environment. Climate change easily damages archaeological evidence and organic building materials like timber and wood (United Nations Educational, Scientific and Cultural Organization, 2007). Archaeological sites are particularly vulnerable to climate change. In Bangladesh, where there are 19 coastal districts (Ahmad, 2019), the rise of salinity in water is corroding the coastal archaeological sites (Siddique, 2018). Aside from direct physical impact, it also has indirect impacts like fragmentation of population (Dastgerdi et al., 2019), loss of intangible heritage (Henderson & Seekamp, 2018), decline of visitors and disruption of socio-economic activities (United Nations Educational, Scientific and Cultural Organization & United Nations Environment Programme, 2016).

Lack of Government Initiatives

No appropriate measure has yet been taken by the government as required for the preservation of heritage sites. There is a lack of appropriate and adequate heritage infrastructure in place to preserve and protect them. Bangladesh has 2500 archaeological sites, out of which 519 sites are supervised by the authority (Rabbi, 2019). However, most of these are not well represented or well preserved. Many ancient sites have been ruined and demolished, while others are in the same danger due to a lack of conservation measures (Kamol, 2020). During the last decade, around 600 ancient sites and structures were demolished, and due to this the government's apathy has been criticized by the archaeologists and noted as one of the prime reasons for this damage (Ahamad, 2018). Furthermore, the budget allocation is overwhelmingly disproportionate against heritage sites. In 2020, only 0.1% of the national budget was allocated for all cultural activities, including heritage sites (Hossain, 2020).

Lack of Resources and Manpower

One of the main problems of maintaining heritage sites is to finding the right materials for repairing and matching with the original builds (Daniels, 2019). As mentioned above, there are more than 2000 heritage sites in Bangladesh, and most of the time insufficient funds are allocated for their repairment. It is practically impossible to find proper resources to repair and rebuild most of the heritage buildings and sites. Furthermore, there is not enough manpower to maintain these sites either. Each heritage site needs a proportionate amount of workers and caretakers to maintain it. An extreme example is in Khulna, where a number of ninety one (91) sites have fallen into shabbiness due to lack of repairment on time and in some extent totally damaged due to the shortage of manpower when needed to maintain them (Dhaka Tribune, 2015).

Urbanization

Urbanization entails increasing population in and expansion of the cities (Shin, 2015) and shifting the economy towards secondary and tertiary industries instead of primary ones (Chaolin, 2020). It is a multifaceted process that involves migration from rural to urban areas, rural and urban land conversions, reconfiguring settlement spaces, and altering governance and management. (Chaolin, 2020). Urbanization in and of itself is not bad. However, rapid and unplanned urbanization involving haphazard growth, poor communication, planning and insufficient awareness about heritage values can produces several threats against heritage conservation (Kiruthiga & Thirumaran, 2019). This problem is more or less a damn fact in the developing countries, where urbanization occurs at an unprecedented rate (Nandi & Gamkhar, 2013).

Tourism

While tourism is a great tool for cultural exchange and economic benefit, it can also have harmful impacts on host communities and their respective cultural heritage. The most prominent negative impacts include degradation of sites and exploitation of cultural resources (Shahateet, 2017). Tourism experiences are often altered to conform to the tourists' desires, which results in loss of authenticity of local culture and heritage (Shahateet, 2017). When done excessively and with poor planning, tourism development and activities can have an adverse effect on a place by imposing unacceptable levels of changes that threaten its physical characteristics, integrity, and ecological environment of heritage sites (Brooks, 2001). Disrespectful and irresponsible visitors can only exacerbate such damage. Furthermore, there is a possibility that tourism activities may inadvertently result in illicit trading and exploitation of cultural goods (Brooks, 2001). There is a need to find a proper balance that would mitigate these negative impacts while also allowing the tourism industry to flourish properly.

The Legal Barriers for the Protection of Archaeological Sites

Various national legislation and international conventions are in place for the protection of cultural heritages worldwide. However, rapid urban modernization and heritage preservation often conflict with each other, which is the primary barrier to archaeological site preservation. Furthermore, city dwellers often consider modern development over cultural preservation due to the lack of knowledge of the importance of such preservation. As a result, they hinder the government from taking away their interest in the property, which bears a cultural significance. Thus, the clash between collective public benefit and individual interest lies in the heart of barriers to preservation.

Bangladesh is a state party to the Convention concerning the Protection of the World Cultural and Natural Heritage adopted in Paris. It binds each state party to this convention that the duty of ensuring the identification, protection, conservation, presentation and transmission to future generations of the cultural and natural heritage on its territory belongs primarily to that state (United Nations Educational, Scientific and Cultural Organization, 1972). Bangladesh government has been taking many initiatives for the preservation of archaeological heritage. Department of Archaeology, with its divisional offices, enlists different sites to be protected as antiquity. On the other hand, through archaeology, the government, by the official notification, declare any establishment as antiquity by an official gadget. To list a site as protected antiquity, first, the regional office conducts exploration to know the location of the archaeological site, or sometimes the news media or the local people send their team to the site and prepare the documentation of the site. Then, if the site is eligible for protection by the government, it is enlisted (Debnath, 2018).

Urban Study Group, a group of volunteers who works to protect the old heritage building at old Dhaka, claims that there are approximately 2000 eligible buildings to be enlisted as a cultural heritage, but the government has hardly shown any effort to enlist them; till now, there are only 451 protected sites under the archaeology department (Haider, 2019). There have been several demolitions of buildings with architectural heritage values in the capital, Dhaka. For example, the Jahaj Bari, a place that the media claimed to be a city's heritage, was demolished in 2019, and the government did not enlist this site (Morshed, 2019). However, under section 16(1) of the act, the government is liable to protect any immovable antiquity if it apprehends danger to any immovable antiquity. Law and policies have a lot to contribute towards the better preservation of cultural heritage sites. Countries like England, Singapore, the United States bear numerous example where the heritage sites are being preserved, continuing the general use of the properties. In this way, the benefit that used to be derived from the property and the owner does not lose the benefit, and the cultural value of the property also remains intact.

The Antiquities Act renders all the preservation responsibilities to the Department of Archaeology. The department has the authority to declare any site as protected antiquity under section 10 of the Antiquities Act, 1968. Under section 12 of the act, the owner of immovable antiquity can make a written agreement to declare a government officer to be the custodian of antiquity. The owner's rights to the property remain unaltered. However, most of the time, the owners of antiquity may not be interested in such an agreement as they are unaware of the value of preserving heritage. The government have to come up with a better offer to incentivize the owners to make these agreements. The government can include tax rebatement, provide low-interest loans, and moving on complete tax exemptions for heritage organizations and investors, offer grants for heritage protection projects and free consultation for developing business models for historic properties (Morshed, 2019). Overall, the law and policy have to be such that the owners and the people, in general, become engaged in preserving heritage properties.

The Perception of Augmented Reality as a Solution

One of the possible – or plausible – solutions for this problem of preserving the originality and authenticity of heritage sites is Augmented Reality (AR). To preface the potential of this technology in this aspect, first, we need to understand what AR is all about. AR is a kind of technology that allows users to superimpose digital content over a real-world environment (Houston, 2020). It is an experience that enhances the users' experience of the physical world with computer-generated input ranging from audio to video to graphics to GPS overlays and more (Interaction Design Foundation, 2021). This technology became extremely popular in 2016 with Pokémon Go, which allowed users to interact with fantasy creatures named Pokémon through their smartphone screens (Houston, 2020). Simply put, the augmented reality is reality enhanced contrive with interactive digital elements.

Since it became popular, the use of AR in various fields has been increasing day by day. It is used in games like Pokémon Go, training and education, especially medical training, television broadcasts, showing products for selling, etc. (McFadden, 2020). Apart from these, enhanced navigation systems use superimposed image over the actual road to navigate military fighters and troops. Pilots use AR to see height, speed, altitude and a variety of other information, neurosurgeons sometimes use AR to create an image of the brain to help them in surgery, and furniture giants like IKEA uses AR to show how a piece of their furniture would fit in the user's space (The Franklin Institute, 2019).

One of the best potential uses of AR would be to use it to preserve and reconstruct the images and vestiges of heritage sites in their full ancient glory. This technology is already in place in several prominent sites. For example, in Pompeii, Italy, AR technology is used to show the image of ancient civilizations over the ruins, bringing the past to life (The Franklin Institute, 2019). AR can make it possible to reconstruct historically and visualize buildings and monuments that have been or are at risk of being damaged. It can interact with these monuments easier without the risk of damage and can provide virtual tourism opportunities (Noh et al., 2009). In short, AR is an amazing way to bring our heritage to life while improving tourist experiences, bringing e-revisits and earning e-revenues (Flo-Culture, 2020).

The Role of Augmented Reality to Promote Ecotourism and Sustainable Tourism

Tourism has developed with time, often matching the development of technology step by step. Technology has always played a crucial role in developing the tourism industry (Ozkul & Kumlu, 2019). As such, augmented reality can be a really powerful tool to develop and promote sustainable tourism and ecotourism to properly preserve tourism destinations while also maximizing tourist satisfaction and sustaining local and natural authenticity. However, first, we need to understand the concepts of ecotourism and sustainable tourism.

Ecotourism

Ecotourism is defined as a responsible travelling to natural environments that preserve, conserve and observe the wildlife and environment, sustaining the welfare of the local communities, involving the education of both locals and guests (The International Ecotourism Society, 2015). In simpler words, ecotourism is visiting towards mysterious, often threatened location or environments while leaving a minimal impact on the nature. Conservation of the local area's resources, land, forest and preservation of wildlife are important aspects of ecotourism (Pengalaman, 2016). Apart from that, minimizing social and psychological impacts, building cultural and environmental awareness, providing visitors and hosts with positive experiences, providing help for conservation, constructing low impact facilities, etc., are also essential ecotourism principles (The International Ecotourism Society, 2015). While natural ecosystems are usually focused on in discussions of ecotourism, the concept extends to the entirety of an area's environment (Pengalaman, 2016), including cities, attractions and heritage sites. When a tourist leaves, the resources and the local community and environment should be the same as when they arrived.

Sustainable Tourism

Sustainable tourism is "tourism that takes full account of its current and future economic, social and environmental impacts, addressing the needs of visitors, the industry, the environment and host communities" (World Tourism Organization, n.d.). This kind of tourism includes the need-based demand of various types of tourists, increase the benefits of the host communities and reduce the fallacies caused by tourism industry of the destination country without disturbing the ability of those of the future generations while sprouting and dealing onto their tourism activities. (Tourism Notes, 2018). There are three different and equally important aspects of sustainable tourism these are natural, socio-cultural and economic. To make optimal use of environmental resources and to maintain and conserve natural heritage and biodiversity under sustainable tourism, three important principles needs to follow that are-respect and conserve the socio-cultural heritage, admire the traditional values of host communities, and ensure long-term economic benefits to all stakeholders and participants of the industry (World Tourism Organization, n.d.).

Using Augmented Reality to Develop and Promote Ecotourism and Sustainable Tourism

As mentioned above, AR technology can be an amazingly powerful tools to promote ecotourism and sustainable tourism, which could satisfy tourists' needs, increase benefits of the tourism industry, conserve authenticity, and raise benefits of the host communities. Some of the ways which would allow AR to promote ecotourism and sustainable tourism are as follows:

Natural ecosystems like forests and coral reefs are extremely fragile. Large visitor numbers can negatively impact these fragile natural environments and the locality and the local people (Hassan & Ramkissoon, 2021). It is necessary to promote environmental awareness among tourists. AR can be an alternative and innovative way to manage and reduce the flow of excessive tourists while preserving and promoting natural attractions. Many natural attractions are in remote places with poor transportation and accommodation facilities. The existing facilities in these places can sometimes be unconducive to both hosts and visitors. AR can be used in different kinds of natural and forest environments to provide authentic tourist experiences. Examples of using AR to enhance nature tourism are Bushey Norwood, Madagascar, Bath Abbey, Bristol Zoo Gardens, etc. (Forest of Imagination, 2017).

AR can help maximize the tourist experience and entertainment and help provide useful and important information of tourist destinations and respective localities and host communities (Kounavis et al., 2012). This will help raise cultural and environmental awareness and respect among tourists and help them reduce the impact on their destination society-both of which are important elements of ecotourism. Destinations should be tailor-made according to tourists' needs and desires to remain competitive and attractive (Han et al., 2013). The technology of AR can be used for this purpose. AR can enhance tourist attractions or destinations without any kind of change or negative impact on the original form of these destinations.

The aforementioned point is even more applicable in heritage sites and historic places, where the possibilities of damage to these ancient sites are high. Through AR technology, it is possible to document historic objects and buildings for virtual reconstruction, create educational resources to teach culture and history, reconstruct monuments damaged or demolished for visitors, and interact with our heritage without the possibility of damage (Noh et al., 2009). AR can be used as a tourism guide to provide information about the area surrounding their destinations, including restaurants, shopping centers, tour guides, events, excursions, leisure activities, etc. (Ozkul & Kumlu, 2019). This can maximize tourist satisfaction due to easy navigation and enjoyment of their destinations. Examples include Tuscany AR for the region of Tosda and the project Layar for the city of Basel (Ozkul & Kumlu, 2019).

AR can improve the tourist experience at heritage sites by providing immediate access to historical knowledge and images (Molz, 2012). Furthermore, AR provides overlaid information that can help avoid overcrowding physical spaces (tom Dieck & HyungsooJung, 2017), reducing any negative impact such overcrowding may have on fragile heritage sites. AR can be a powerful marketing tool. Successful tourism companies depend considerably on customer feedback (Yin et al., 2021). The opinions and reviews of tourists affect their activities and affair based on the informed choice they make based on their preferences (Yin et al., 2021). Companies advertise these activities to the customers. AR can offer and enjoy these activities, contributing significantly to the development and economic return for the tourism industry.

These are some of the many potential ways to use AR technology to promote and develop ecotourism and sustainable tourism by enhancing tourist experiences while also conserving the socio-cultural authenticity of host communities and the ecology, structure and environment of the tourist destination.

An Assessment of the Impact of the Augmented Reality on Sacred Tourism Places

AR technology is a game-changer (Fingent, 2019) for any industry, including tourism. This concept is even more pertinent in the case of heritage and religious tourism sites, where AR can bring the past to life and show the visitors authentic visual images of these monuments to maximize their experience and satisfaction. This section discusses the impact AR has on these sites and the people that visit these sites. AR is considered a booming industry that plays a significant role in redefining traditional tourism (Ramos et al., 2018). It can increase tourist experience and maximize entertainment and satisfaction by providing instant, relevant information about the destination, making the tourist experience enhanced and high-level. This kind of technology that makes relevant information and technologies to explore tourist spots in no time can significantly impact heritage and sacred sites, which-along with fragile structure and environment-also have a significant amount of cultural and historical value.

There is no doubt that the historical sites have strong architectural, historical and cultural significance; however, their incomplete or fragile physical structure and the absence of indigenous people means that a person needs a considerable amount of knowledge to appreciate this heritage fully (Kerstetter et al., 2001). Moreover, in religious sites and relics, rules and regulations often restrict the use of signs and boards, which can negatively impact these sites (Jung & Han, 2014). Furthermore, tourism has many other negative effects on sacred tourism sites.

AR can be an ideal solution for sustaining tourism at these sites while also reducing the harmful impacts. AR puts multidimensional environment overlaid on the real environment (Tussyadiah et al., 2018), allowing tourists to enjoy the heritage and sacred sites properly while preserving their original structure (Han et al., 2021). Furthermore, the use of AR in these sites can help raise intellectual awareness of architecture, culture and historical events among visitors (Han et al., 2021). Currently, AR is considered an excellent solution to overcome the environmental barriers of visiting heritage sites. It has been particularly praised for its ability to facilitate educational purposes by providing first-hand interaction (Horn, 2006, cited in Jung & Han, 2014) and information depth about heritage sites using virtual space without affecting the physical environment (Jung & Han, 2014).

Mobile AR systems are good tools for navigation and obtaining relevant information about places. This is also applicable to sacred and cultural heritage sites. Navigation was one of the first application systems of AR and remained widely used in commercial tourism (Emmanouilidis, et al., 2013). Using AR to navigate significantly reduces the mental pressure required for self-navigation on tourists and provides opportunities to explore new and unknown surroundings by providing 3D location-based information. Besides, advanced AR systems can help tourists' relive historical places and events by reviving ancient and historic buildings as 3D images placed over the actual structure of the monument (Kourouthanassis et al., 2015), thus increasing visitor experience and entertainment.

As discussed above, AR devices and systems can enhance and maximize tourist satisfaction through various applications. That, by itself, can have long-term effects on heritage sites and sacred tourist spots. Research has shown that this satisfaction is related to destination attitude and intention (Chung et al., 2017). Visitor's satisfaction with AR applications affects the behavioral intention of tourists towards heritage sites significantly (Chung et al., 2017). Tourist satisfaction with AR can bring about a positive attitude among visitors (Buonincontri et al., 2017), resulting in responses such as willingness to pay (Nian et al., 2019) and contribute to the conservation of these sites (Nasser, 2003).

However, despite all the advantages, the use of AR is not without its challenges. There is a risk that the AR- in the context of heritage sites and monuments – may become a "diminished reality" (Barrado-Timón & Hidalgo-Giralt, 2019). People might want to stop interacting with the physical structure altogether in favor of the virtual one created by AR. Furthermore, though the information and application can be tailored according to individual preferences, the invasion of users' privacy and easy access to personal data will challenge the implementation of this technology (Hyman, 2013). Another issue against implementation will be virtual spam and unauthorized advertisements that are allowed by such AR systems and applications from open source networks (Zacharias, 2010, cited in Jung & Han, 2014), which brings with it the risk of digital bugs and viruses as well as theft of personal information and data.

However, despite the risks, the long term benefits from using AR outweigh the drawbacks, and prolonged use and experience are considered to be the solution for overcoming these challenges (Jung & Han, 2014). AR is a perfect and affordable solution for many challenges faced by the cultural and religious tourism industry, especially in developing countries. For Bangladesh, a developing country where heritage sites are abundant, yet most of them are abandoned, the use of AR technology will usher a new era and countless opportunities for sustainable tourism at cultural heritage sites and sacred tourist spots, benefitting the visitors, host communities, the tourism industry as well as the national economy.

Concluding Remarks

The preservation and conservation of the cultural heritage of a country is necessary to safeguard its legacy. Cultural Heritage tourism upholds the cultural diversity of the country and makes people aware of their roots. However, rapid modernization and urbanization, climate change, weak policies put the preservation of cultural heritage at stake. We must ensure a more robust and efficient policy framework to balance modernization and heritage preservation. Laws and policies should also maintain a balance between public benefit and individual interests. Augmented Reality can be one of the most – if not the best – ideal solution for the preservation of our heritage sites. On top of being cost and time effective, vivid presentation of historical incidents or lost cultural heritage through AR may attract more tourist than ever. Not only will this contribute significantly to our economy, but it will also play huge role in preserving our ancient heritage and legacy through various means. Therefore, Augmented Reality is probably one of the most viable solutions to preserve the huge number of heritage sites in Bangladesh for sustainable tourism and eco-tourism.

References

Ahamad, R. (2018). Capital's heritage sites in danger. *New Age Bangladesh*. Retrieved from: http://www.newagebd.net/article/32898/capitals-heritage-sites-in-danger. Accessed 30 May 2021.

Ahmad, H. (2019). Bangladesh coastal zone management status and future trends. *Journal of Coastal Zone Management, 22*(1), 1–7.

Aulet, S., & Vidal, D. (2018). Tourism and religion: Sacred spaces as transmitters of heritage values. *Church, Communication and Culture, 3*(3), 237–259.

Bangladesh Parjatan Corporation. (BPC). (2014a). *Paharpur Buddha Vihara*. Retrieved from: http://www.parjatan.gov.bd/site/page/1c30b619-87ee-40ca-b5bf-db02648l0daa/. Accessed 25 May 2021.

Bangladesh Parjatan Corporation. (BPC). (2014b). *Mahasthangarh*. Retrieved from: http://www.parjatan.gov.bd/site/page/330a32bd-2a86-43f2-b70b-a82cf6fe606e/. Accessed 26 May 2021.

Bangladesh Parjatan Corporation. (BPC). (2014c). *Mainamati*. Retrieved from: http://www.parjatan.gov.bd/site/page/11841889-52f8-4eab-a001-8c837f0545d6/. Accessed 27 May 2021.

Bangladesh Parjatan Corporation. (BPC). (2014d). *Sonargaon – The Old Capital of Bengal*. Retrieved from: http://www.parjatan.gov.bd/site/page/c5614643-0d84-4d48-8b25-bc77dbb3fbf8/. Accessed 27 May 2021.

Bangladesh Parjatan Corporation. (BPC). (2014e). *Wari Bateshwar.* Retrieved from: http://www.parjatan.gov.bd/site/page/4bd003b6-9923-485b-9e91-69aec62ab2a1/. Accessed 27 May 2021.

Bangladesh Parjatan Corporation. (BPC). (2014f). *Lalbagh Fort*. Retrieved from: http://www.parjatan.gov.bd/site/page/ef72190b-25a4-4594-8beb-02ac816c2721/. Accessed 24 May 2021.

Bangladesh Parjatan Corporation. (BPC). (2014g). *Sohrawardy Park*. Retrieved from: http://www.parjatan.gov.bd/site/nolink/. Accessed 27 May 2021.

Bangladesh Parjatan Corporation. (BPC). (2014h). *Ahsan Manjil (The Pink Palace)*. Retrieved from: http://www.parjatan.gov.bd/site/page/0e57d43e-aa04-4bf6-89e7-2fc9a7b7db26/nolink/. Accessed 23 May 2021.

Banglapedia. (2014). *Curzon Hall*. Retrieved from: https://en.banglapedia.org/index.php/Curzon_Hall. Accessed 26 May 2021.

Barrado-Timón, D. A., & Hidalgo-Giralt, C. (2019). The Historic City, its transmission and perception via augmented reality and virtual reality and the use of the past as a resource for the present: A new era for urban cultural heritage and tourism? *Sustainability, 11*(10), 2835.

Barua, K. B. (2018). Shalban Vihara, Comilla, Maynamati, Bangladesh. *Researchgate*. Retrieved from: https://www.researchgate.net/publication/327797189_Shalban_ViharaComillaMaynamatiBanglades. Accessed 27 May 2021.

Brooks, G. (2001). *Heritage at risk from tourism*. Retrieved from: https://www.icomos.org/risk/2001/tourism.htm. Accessed 27 May 2021.

Buonincontri, P., Marasco, A., & Ramkissoon, H. (2017). Visitors' experience, place attachment and sustainable behaviour at cultural heritage sites: A conceptual framework. *Sustainability, 9*(7), 1112.

Chaolin, G. (2020). Urbanization. In A. Kobayashi (Ed.), *International encyclopedia of human geography* (pp. 141–153). Elsevier.

Chung, N., Lee, H., Kim, J. Y., & Koo, C. (2017). The role of augmented reality for experience-influenced environments: The case of cultural heritage tourism in Korea. *Journal of Travel Research, 57*(5), 627–643.

Daniels, S. (2019). Top challenges for preservation of cultural heritage buildings. *Dig This Design*. Retrieved from: https://digthisdesign.net/design-architecture/top-challenges-for-preservation-of-cultural-heritage-buildings/. Accessed 31 May 2021.

Dastgerdi, A. S., Sargolini, M., & Pierantoni, I. (2019). Climate change challenges to existing cultural heritage policy. *Sustainability, 11*(19), 5227.

Debnath, B. K. (2018). Saving our heritage sites. *The Independent*. Retrieved from: https://m.theindependentbd.com/post/146447. Accessed 27 May 2021.

Dhaka Tribune. (2015). *Archaeological sites in Khulna lie unrestored for fund crisis, lack of manpower.* Retrieved from: https://www.dhakatribune.com/uncategorized/2015/05/15/archaeological-sites-in-khulna-lie-unrestored-for-fund-crisis-lack-of-manpower. Accessed 31 May 2021.

Emmanouilidis, C., Koutsiamanis, R. A., & Tasidou, A. (2013). Mobile guides: Taxonomy of architectures, context awareness, technologies and applications. *Journal of Network and Computer Applications, 36*(1), 103–125.

Fingent. (2019). *How is augmented reality reshaping travel and tourism?* Retrieved from: https://www.fingent.com/blog/how-is-augmented-reality-reshaping-travel-and-tourism/. Accessed 4 June 2021.

Flo-Culture. (2020). *How to use augmented reality to enhance onsite experiences*. Retrieved from: https://flo-culture.com/digital/augmented-reality-heritage-experiences/. Accessed 1 June 2021.

Forest of Imagination. (2017). *Augmented reality.* Retrieved from: https://www.forestofimagination.org.uk/tag/augmented-reality/. Accessed 2 June 2021.

Haider, M. H. (2015). Heritage tourism: Opportunities untapped. *The Daily Star.* Retrieved from: https://www.thedailystar.net/lifestyle/special-feature/heritage-tourism-76991. Accessed 21 May 2021.

Haider, S. (2019). *Home*. Retrieved from: http://samaihaider.com/2019/01/23/saving-dhakas-heritage-one-building-at-a-time/. Accessed 28 May 2021.

Han, D. I., Jung, T., & Gibson, A. (2013). Dublin AR: Implementing augmented reality in tourism. In Z. Xiang & I. Tussyadiah (Eds.), *Information and communication Technologies in Tourism 2014* (pp. 511–523). Springer.

Han, S., Yoon, J. H., & Kwon, J. (2021). Impact of experiential value of augmented reality: The context of heritage tourism. *Sustainability, 13*(8), 4147.

Hasan, M., & Jobaid, M. I. (2014). Heritage tourism marketing: Status, prospects and barriers. *IOSR Journal of Business and Management (IOSR-JBM), 16*(5), 40–48.

Hassan, A., & Ramkissoon, H. (2021). Augmented reality application for visitor experiences in nature based tourism. In J. N. Albrecht (Ed.), *Visitor experience management in nature-based tourism* (pp. 118–130). CABI.

Henderson, M., & Seekamp, E. (2018). Battling the tides of climate change: The power of intangible cultural resource values to bind place meanings in vulnerable historic districts. *Heritage, 1*(2), 220–238.

Hossain, M. I. (2020). Budget 2020–2021: Artists, content makers for proper utilization of minimal allocation. *UNB News*. Retrieved from: https://unb.com.bd/category/Special/budget-2020-2021-artists-content-makers-for-proper-utilization-of-minimal-allocation/53560. Accessed 30 May 2021.

Houston, B. (2020). What is augmented reality (AR)? A practical overview. *Threekit*. Retrieved from: https://www.threekit.com/blog/what-is-augmented-reality. Accessed 1 June 2021.

Hyman, P. (2013). Augmented-reality glasses bring cloud security into sharp focus. *Communications of the ACM, 56*(6), 18–20.

Interaction Design Foundation. (2021). *Augmented reality*. Retrieved from: https://www.interaction-design.org/literature/topics/augmented-reality. Accessed 1 June 2021.

Jung, T. H., & Han, D. I. (2014). Augmented reality (AR) in urban heritage tourism. *e-Review of Tourism Research, 5*, 1–6.

Kamol, E. (2020). Bangladesh heritage sites disappearing in absence of list. *New Age Bangladesh*. Retrieved from: https://www.newagebd.net/article/119172/bangladesh-heritage-sites-disappearing-in-absence-of-list-conservation. Accessed 30 May 2021.

Kerstetter, D. L., Confer, J. J., & Graefe, A. R. (2001). An exploration of the specialization concept within the context of heritage tourism. *Journal of Travel Research, 39*(3), 267–274.

Kiruthiga, K., & Thirumaran, K. (2019). Effects of urbanization on historical heritage buildings in Kumbakonam, Tamil Nadu, India. *Frontiers of Architectural Research, 8*(1), 94–105.

Kounavis, C. D., Kasimati, A. E., & Zamani, E. D. (2012). Enhancing the tourism experience through mobile augmented reality: Challenges and prospects. *International Journal of Engineering Business Management, 4*, 1–6.

Kourouthanassis, P., Boletsis, C., Bardaki, C., & Chasanidou, D. (2015). Tourists responses to Mobile augmented reality travel guides: The role of emotions on adoption behavior. *Pervasive and Mobile Computing, 18*, 71–87.

Logan, W. S. (2007). Closing Pandora's box: Human rights conundrums in cultural heritage protection. In H. Silverman & D. F. Ruggles (Eds.), *Cultural heritage and human rights* (pp. 33–52). Springer.

McFadden, C. (2020). Nine powerful real world applications of augmented reality. *Interesting Engineering*. Retrieved from: https://interestingengineering.com/9-powerful-real-world-applications-of-augmented-reality#:~:text=Augmented%20reality%20is%20now%20used,headsets%20or%20augmented%20reality%20glasses. Accessed 1 June 2021.

Molz, J. G. (2012). *Travel connections: Tourism, technology and togetherness in a Mobile world*. Routledge.

Morshed, A. Z. (2019). The calculus of heritage preservation. *The Daily Star*. Retrieved from: https://www.thedailystar.net/opinion/the-grudging-urbanist/news/the-calculus-heritage-preservation-1761766. Accessed 1 June 2021.

Nandi, S., & Gamkhar, S. (2013). Urban challenges in India: A review of recent policy measures. *Habitat International, 39*, 55–61.

Nasser, N. (2003). Planning for urban heritage places: Reconciling conservation, tourism, and sustainable development. *Journal of Planning Literature, 17*, 467–479.

Nian, S., Zhang, H., Mao, L., Zhao, W., Zhang, H., Lu, Y., & Xu, Y. (2019). How outstanding universal value, service quality and place attachment influences tourist intention towards world heritage conservation: A case study of mount Sanqingshan National Park. *China. Sustainability, 11*(12), 3321.

Noh, Z., Sunar, M. S., & Pan, Z. (2009). A review on augmented reality for virtual heritage system. *Lecture Notes in Computer Science, 5670*, 50–61.

Organisation for Economic Co-operation and Development. (OECD). (2009). *Impact of culture on tourism*. OECD.

Ozkul, E., & Kumlu, S. T. (2019). Augmented reality applications in tourism. *International Journal of Contemporary Tourism Research, 2*, 107–122.

Pengalaman, C. (2016). The usage of augmented reality technology in Southeast Asia as interactive tourism information media to achieve sustainable tourism development. *Techraveller*. Retrieved from: https://techraveller.com/the-usage-of-augmented-reality-technology-in-southeast-asia-as-interactive-tourism-information-media-to-achieve-sustainable-tourism-development/. Accessed 2 June 2021.

Rabbi, A. H. (2019). Poorly preserved heritage sites of Bangladesh. *The Business Standard*. Retrieved from: https://www.tbsnews.net/panorama/poorly-preserved-heritage-sites-bangladesh. Accessed 30 May 2021.

Rahman, M. W. (2012). Cultural tourism and Bangladesh: An overview. *Bangladesh Research Publications Journal, 7*(1), 6–15.

Ramos, R. F., Trilles, O. S., Torres-Sospedra, J., & Perales, F. J. (2018). New trends in using augmented reality apps for Smart City contexts. *International Journal of Geo-Information, 7*(12), 478–501.

Richards, G. (2007). *Cultural tourism: Global and local perspectives*. The Haworth Press.

Shahateet, M. (2017). Finding balance: Cultural preservation and tourism. *Chenomics*. Retrieved from: https://chemonics.com/blog/finding-balance-cultural-preservation-tourism/. Accessed 31 May 2021.

Shin, H. B. (2015). Urbanization in China. In J. D. Wright (Ed.), *International encyclopedia of the social & behavioral sciences* (pp. 973–979). Elsevier Ltd.

Siddique, A. (2018). Climate change putting Bangladesh's coastal archaeological sites at risk. *Bengaldelta.net*. Retrieved from: https://bengaldelta.net/climate-change-putting-coastal-archaeological-sites-at-risk/#:~:text=Climate%20change%20putting%20Bangladesh's%20coastal%20archaeological%20sites%20at%20risk,-Rising%20water%20salinity&text=Additionally%2C%20more%20frequent%20natur. Accessed 30 May 2021.

The Franklin Institute. (2019). *What is augmented reality?* Retrieved from: https://www.fi.edu/what-is-augmented-reality. Accessed 1 June 2021.

The International Ecotourism Society. (TIES). (2015). *What is ecotourism?*. Retrieved from: https://ecotourism.org/what-is-ecotourism/. Accessed 2 June 2021.

tom Dieck, M. C., & Jung, T. H. (2017). Value of augmented reality at cultural heritage sites: A stakeholder approach. *Journal of Destination Marketing & Management, 6*(2), 110–117.

Tourism Notes. (2018). *Sustainable tourism*. Retrieved from: https://tourismnotes.com/sustainable-tourism/. Accessed 2 June 2021.

Tussyadiah, I. P., Jung, T. H., & tom Dieck, M. C. (2018). Embodiment of wearable augmented reality Technology in Tourism Experiences. *Journal of Travel Research, 57*(5), 597–611.

UKEssays. (2018). *The importance of heritage conservation*. Retrieved from: https://www.ukessays.com/essays/environmental-sciences/heritage-conservation-is-important-for-identifying-environmental-sciences-essay.php?vref=1. Accessed 21 May 2021.

United Nations Educational, Scientific and Cultural Organization. (UNESCO). (1972). *Convention concerning the protection of the world cultural and natural heritage*. Retrieved from: https://whc.unesco.org/en/conventiontext/. Accessed 20 May 2021.

United Nations Educational, Scientific and Cultural Organization. (UNESCO). (1999). *Jaggadala Vihara*. Retrieved from: http://whc.unesco.org/en/tentativelists/1212/. Accessed 10 May 2021.

United Nations Educational, Scientific and Cultural Organization. (UNESCO). (2007). *Climate change and world heritage*. UNESCO World Heritage Centre.

United Nations Educational, Scientific and Cultural Organization. (UNESCO). (n.d.-a). *Properties inscribed on the World Heritage List (3)*. Retrieved from: http://whc.unesco.org/en/statesparties/BD/. Accessed 20 May 2021.

United Nations Educational, Scientific and Cultural Organization. (UNESCO). (n.d.-b). *Ruins of the Buddhist Vihara at Paharpur*. Retrieved from: http://whc.unesco.org/en/list/322. Accessed 27 May 2021.

United Nations Educational, Scientific and Cultural Organization. (UNESCO) & United Nations Environment Programme (UNEP). (2016). *World heritage and tourism in a changing climate*. UNESCO and UNEP.

World Tourism Organization. (UNWTO). (2021). *Tourism and culture*. Retrieved from: https://www.unwto.org/tourism-and-culture. Accessed 21 May 2021.

World Tourism Organization. (UNWTO). (n.d.) *Sustainable development*. Retrieved from: https://www.unwto.org/sustainable-development. Accessed 2 June 2021.

Yin, C. Z. Y., Jung, T., & Lee, M. Y. (2021). Mobile augmented reality heritage applications: Meeting the needs of heritage tourists. *Sustainability, 13*(5), 2523.

Dr A. H. M. Salimullah is Professor in the Department of Economics at the University of Chittagong, Chattogram, Bangladesh. Dr Salimullah graduated with a BSS (Honors) and MSS degree in Economics from the University of Chittagong in 2003, did a second MA degree in Financial Economics from the Middle Tennessee State University (MTSU) in 2014, and completed his PhD in Econometrics and Quantitative Economics from the same university in 2018. His present areas of interest are Labor Economics, Dynamic Macroeconomics, and Economics of Tourism. He published 10 scholarly articles in national and international journals. Besides, he presented several papers at different national and international conferences. The notable conferences that he attended and presented his research were the Academy of Economics and Finance (AEF) Conference in 2018 at Houston, Texas, and Dr Gregory P. Domin Graduate Conference in 2017 at Columbus State University, Columbus, Georgia. He worked as a reviewer for a journal named *Scientia et Humanitas*, sponsored by the University Honors College at MTSU. He is a proud winner of the Deans' Distinguished Essay Award for his scholarly publication in the Honors College journal of MTSU. Besides, he won important prizes during scholars' week university-wide poster presentation at MTSU. He is chosen as a proud member of the MTSU Chapter of Beta Gamma Sigma due to his scholarly achievement.

Dr Rakiba Nabi is Professor in the Department of Law at the University of Chittagong, Chattogram, Bangladesh. Dr Nabi graduated with an LLB (Honors), and LLM from the University of Chittagong in 2004, did a second LL.M in Human Rights from the University of Hong Kong in 2012-13, and completed her SJD in Indigenous Law from the University of Kansas, the United States in 2020. Her present areas of interest are Human Rights, Refugee Law, Indigenous Law and Policy, Law of International Organization, Tourism and Sustainable Development. She published 10 scholarly articles in national and international journals. She also works in several projects funded by Research Cell, CU, and Law Ministry, Bangladesh.

Part IV
Social Media

Chapter 7
Technology Innovation and Social Media as a Global Platform for Tourism Events

Muhammad Khalilur Rahman, Mohammad Mainul Hossain, Shaharin Akter, and Azizul Hassan

Introduction

Technology is an important part of the life of the consumer, digital health connects with the physical, psychological, spiritual, socioeconomic, professional, technological, economic and environmental components of the well-being of the general public (Stankov & Gretzel, 2021). The predominance of information and communication technology (ICT) in tourism is unsurprising considering that the tourism sector has always been eager to embrace new technologies to improve its operations (Xiang, 2018), often at the expense of the human side of the visitor experience (Stankov & Gretzel, 2020). The expansion in technology in all phases of daily life and the development of a smart growth model have made technology dependency crucial for many tourism providers to give exciting co-creative experiences and excellent service delivery and thus their requirement for sustainable competitiveness (Gretzel et al., 2015). It moreover includes initiatives to incorporate technology into tourism services to promote mindful experiences (Stankov et al., 2020), and to guide to the destination (Kang & Gretzel, 2012). Tourists are using their smartphones to explore and record their special and personal experiences in recent years (Picheta, 2021). Smart tourism development ideas are particularly significant which

M. K. Rahman (✉)
Faculty of Entrepreneurship and Business, Universiti Malaysia Kelantan, Pengkalan Chepa, Malaysia
e-mail: khalilur.r@umk.edu.my

M. M. Hossain
Faculty of Arts and Social Sciences, University of Malaya, Kuala Lumpur, Malaysia

S. Akter
Faculty of Business Administration, University of Chittagong, Chattogram, Bangladesh

A. Hassan
Tourism Consultants Network, The Tourism Society, London, UK

A. Hassan (ed.), *Technology Application in Tourism Fairs, Festivals and Events in Asia*, https://doi.org/10.1007/978-981-16-8070-0_7

aim to address the environmental and social constraints of excessive technological use in smart destinations. These destinations are known as "wise" destinations (Coca-Stefaniak, 2020) because they prioritize sustainability and well-being and use technological solutions to accomplish them.

Nonetheless, the effect of the COVID-19 pandemic on the tourism industry will be diverse in time and space. The ongoing uncertainty of the COVID-19 epidemic has a tremendous and worldwide impact on the economy. It is assumed that the international tourist arrivals will fall by 78%, resulting in a loss of US$ 1.2 trillion in tourism export revenues and direct tourism job cuts of about 120 million (UNWTO, 2020). Adapting to digital platforms may be a solution for some industries while still fighting for survival (Mehrolia et al., 2021). Due to travel limitations and societal isolation, some industries have seen significant failures, making it extremely difficult to survive the pandemic. The United Nations World Tourism Organization (UNWTO) (2020) reported that around 22% decrease in tourist numbers in the first quarter of 2020 compared to the same quarter of 2019, and there exists a threat of 60–80% drop during 2020 (in comparison with 2019), which indicate the potential havoc of COVID-19 pandemic for the global tourism industry. Tourism is one of the industries with the highest labour intensity. So, this downturn in the industry might jeopardize millions of jobs, threatening to reverse progress achieved toward the 2030 agenda for sustainable development (UNWTO, 2020). Technology, on the other hand, is a major influence in allowing the tourism business to be more flexible (Hall et al., 2020). People received a lot of help from technology experts during the COVID-19 pandemic. There are now signs of human replaced by robots, smartphones and tablets apps tracking people's contacts and Big Data analytics anticipating COVID-19 spreading among the people.

On the other hand, social media platforms and Web 2.0 which promotes consumer-generated contents, have created new and inventive product development prospects, allowing tourism businesses to expand their geographic network. Social media has heavily influenced all elements of tourism, resulting in significant shifts in the way tourists plan, consume, evaluate, and promote their experiences. In the tourism industry, the focus is on tourist data management and marketing strategy, and the ability to properly connect users via social networking sites allows service providers to create new business opportunities (Pantano et al., 2017). Social media and new technologies have influenced how people see tourism products and services. Images uploaded online, data collection and recorded has aided researchers in developing and defining new ways of interpretation and analysis of a location to better understand tourist dynamics. Sharing data has enabled the creation of a link between the real and the virtual worlds, as well as the ability to learn about users' interests and behaviour (Bertacchini et al., 2018). Tourists' interests and favoured activities are reflected in images shared on the web. Tourists share their information on latitude and longitude coordinates using the technology (Sun et al., 2015).

Flickr, Facebook, Twitter, wikis, blogs and other technological platform services are among the top social networking websites. Tourism operators may utilize social media to attract and build interest among prospective consumers by using it as a means of engaging with them (Roque & Raposo, 2016). The value of social media as a promotional tool has been well demonstrated by several studies (Chen et al., 2017; Amaro et al., 2016). According to the literature, effective social media tourist marketing includes expressing both the destination's benefits and its attractiveness (Buhalis, 2000). Companies have been found to employ a variety of communication channels in the context of social media as a worldwide platform, including traditional advertising media (Jun & Oh, 2015) and social networking sites such as Facebook, Twitter, and YouTube (Lee et al., 2014). This study has aimed to explore technology innovation and social media as a global platform for tourism events.

Technology Innovation in the Tourism Industry

Technology plays a significant role in increasing flexibility in the tourism industry (Hall et al., 2020). The proper use of ICT is essential because the data is the backbone of the tourist sector. The travel sector is quickly adopting the entire information technology system. At the retail site, intangible tourist services cannot be physically examined before purchase. They are typically bought before they would be used and away from the place of consumption. Intangible tourism services rely on representations provided by the travel & tourism industries to attract customers (Shehzad et al., 2019). The key to meeting tourist satisfaction is accurate information about users' needs. As a result, digital platforms serve as the data backbone which works as an influential factor for the travel business. Supply and demand growth make ICT application an indispensable collaborator and so plays an important role in the advertising, promoting, delivery and management of tourism (Adeola & Evans, 2020).

Artificial intelligence, robotics, and automated technologies can help businesses to save money, boost solvency, and increase flexibility. Robotics would assist people in maintaining social distance (Assaf & Scuderi, 2020; Thomas & Chopra, 2020) because technology allows individuals to interact without having to meet in person. As a result, technology enables handling pandemic related issues including identifying travelers, detecting COVID-19 scenarios, tracing connections, and assuring online education for students (Hall et al., 2020). People are willing to modify their opinions using the technology. People nowadays have begun to overlook privacy concerns to achieve a greater technological advantage (Stankov et al., 2020). Gretzel et al. (2020) have formulated the "six transformative e-tourism research pillars" for proactively employing IT resources for both short- and long-term objectives to bring about changes in e-tourism.

Technology Combats on COVID-19 in the Tourism Industry

To combat the impact of the COVID-19 pandemic, technology would be one of the innovative and sustainable solutions in the tourist industry and reopening tourism and economic activities (i.e., mobility tracking applications, robotic artificial intelligence (AI) services, digital health passport and identity control, social isolation and overcrowding control technologies, quick and real-time decision making, and the sterilization of publicly available services, body temperature measurement). While technology is regarded as a solution for our COVID-19 driven needs to standardize monitoring, protect health and wellbeing, and collect and analyze personal information for quick decision-making. Although COVID-19 pandemic research in the field of tourism will not be able to block technological advancements, it should examine and reset their objectives, structures and functionalities, interpretations, and applied ethics. Technologies are made up of distinct functionalities, the growth and manifestation of which are controlled by the institutional logics that govern how technologies are created, developed, and employed (Zuboff, 2015).

The effect of COVID-19 pandemic research in the field of tourism could examine and enhance our knowledge and technical capacities to gather, analyze, and then utilize the information for accurate understanding, forecasting, managing, and influencing people's behaviour. For example, tourists and employees' behavior are a method of generating money and controlling the market (Zuboff, 2015). Technologies are very much a facilitator, a driver for innovation and creativity, and an instrument for building tourism resilience in times of crisis (Hall et al., 2020). The COVID-19 epidemic has emphasized the importance of technology in the rehabilitation and continuation of tourism while reinforcing existing e-tourism paradigms. To address the effect of the COVID-19 pandemic in tourism consequences, competitive dynamics and acceptance of smart marketing of tourism services, artificial intelligence, robots, and other digital breakthroughs are now expedited. The effect of COVID-19 on the tourism industry should rethink and restructure the goals, uses, and techniques of technological advancements that have a substantial impact on how our communities and economy are transforming, and how tourism is practised, controlled, and developed with the effect of COVID-19 pandemic.

Social Media and Its Role in Tourism

Although social networking is highly significant in all parts of life and society, as well as in all sectors, it is especially important in tourism. Xiang and Gretzel (2010) stated that travelers are more likely to find sources of information on social media. Travelers are seeking information online through the search engine on social media platforms. During the trip, tourists are commonly used to generate information and interpret them and enjoy their desired locations (Fotis et al., 2011; Rahman et al., 2017a), as social media is widely accessed via smartphones by the users. Social networks sites can also be used for recreational purposes. Tussyadiah and Fesenmaier (2009) believed that social media travel content encourages fantasizing, stimulates dreams, evokes nostalgia, and transports users to exotic locations. Social media provides almost trial-oriented adventures, while simultaneously serving the desires for immediate escape online. The consumption of vacation content shared by others could, however, potentially contribute to what is called "Facebook envy" (Taylor, 2017).

Furthermore, although many social media users upload vacation related information to assist someone and share their opinions with friends and family, many often say that they enjoy the process (Gretzel & Yoo, 2014). Kim and Fesenmaier (2017) revealed that such an interaction entertained the past vacation experience and also led to more positive assessments and performed a psychologically important role. Di Pietro et al. (2018) highlighted the function of social media use as an entertainment way during leisure times for traveling. Using social media platform, people can change their flights for travel destinations. Social media marketing can promote digital tour packages (Mkwizu, 2020). Digital marketing is divided into seven categories such as social media advertising, instant messaging, content marketing, search engine optimization, web platforms, e-mail marketing, and pay-per-click (Yasmin et al., 2015). Key criteria that influence successful digital advertising include providing customers with confidence that they can use social media sites as recommended by tourism associations (Bala & Verma, 2018). Furthermore, several investigations discovered that the structure of tourism marketing and advertising via social media was time-consuming for most companies considering the psychological benefits of the internet such as social and economic reasons (Canovi & Pucciarelli, 2019).

Social Media and Tourism Research

Social media plays a significant role in many aspects of *tourism*, especially in information search and decision-making behavior, information search for travel and *tourism destinations,* tourism promotion, decision-making behavior, and in focusing on best practices for interacting with consumers. Leveraging off social media to market tourism products has demonstrated to be an amazing methodology. Social media encourages people to regular communication and frequently share knowledge and GPS location to user-generated material. Social media networking is related to the large scale data paradigm. Social media data is a treasure for research because of its inherent increased visibility, availability, and durability. Web data extraction techniques allow the collection of vast amounts of data from social networking sites. The database fits the social media platform's provider, it is widely regarded as publicly available and accessible for reasonable use in educational research including travel and tourism studies. Data extraction places a high demand on the social media platform's server, which is normally prohibited and technically forbidden (e.g., by narrowing the timeframe and the number of servers accessed).

Data analysis in social media research necessitates a distinct mind-set (Sloan & Quan-Haase, 2017) as data is abundant, but it is also extremely complex. For instance, Twitter data reflects people's perception given the many distinct user types (from customers to influencers, corporations and automated bots), the specific qualities of the content creator, and the different motivations that motivate people to write. According to Xiang et al. (2017), the reliability of data represents be an important topic that needs additional consideration.

Social Media Activism and Tourism

Pre, during, and after travel, social media serves a variety of purposes. It is used by travelers to find, arrange, communicate, and discuss their experiences and memories, and on the other hand, businesses use it to attract, identify, and anticipate their customers (Leung et al., 2013). The use of social media in social activism was highlighted by Mkono and Tribe (2017). Gretzel (2017) explained that the tourism environment exposes itself to social media involvement because it provides higher levels of involvement and high-risk scenarios when the tourists are separated from their current location. Gretzel (2017) further pointed to social media activity driven by other participants in the tourist business and cited incidents of workers' reporting of suspicious activities of tourism corporations as well as objections against overtourism by the local people. When targeting specific establishments, the influence of social media activity on tourism is likely to be greater than when attempting to enact change in society. The responsibility of hotel managers to preserve their brand image online was addressed by Gössling and Hall (2019), although it was tied to the issue of oligopoly in online bookings rather than as a response to social action.

Challenges and Possibilities of Social Media Technology in Tourism Events

In organizational day-to-day business processes, the potential and challenges that the digital world presents for tourism-related sectors are evident (Leung et al., 2013; Sigala et al., 2012). Tourists have become co-manufacturer, co-contributors, co-promoters, and co-customers of travel experiences when they use social media networking (Sigala et al., 2012). Such advances in Web2.0 technologies had a significant effect on customer psychology. Social media platforms have a key role in several topics relating to tourism, in particular consumer preferences, advertising and information sharing/interacting with consumers of tourism and management of corporate activities and practices (Bilgihan et al., 2016; Law et al., 2014). One of the major issues with e-tourism and digital marketing is the absence of face-to-face interaction. These difficulties have been criticized by previous researchers specifically social media marketing. Typically, many people prefer face-to-face communication, also known as personal engagement, since they assume that they will be able to speak with store workers who sell the products and have the options to examine and see the goods. In terms of sale or purchase, using social media advertising provides the opportunity of accessing an endless worldwide audience. Moreover, opportunity of social media marketing is very resourceful. The online distribution of services and products is undoubtedly cost effective and may achieve its goals at a very low cost (Poon & Jevons, 1997). A further problem for customers is lack of trust. Because consumers require faith to assist them in overcoming consumer issues that are recognized as a major development issue for the growth of social media marketing. Therefore, in tourism and hospitality study, the expectations of social media marketing with respect to trust are becoming increasingly important.

Policy Implication of Social Media Technology in Tourism

Tourism providers need to embrace proper ways and execute initiatives to develop and re-design the activities and processes of businesses (Rana et al., 2020; Rahman et al., 2021a) by taking advantage of opportunities and addressing new digital challenges (Law et al., 2014). Tourism service providers must develop effective plans and employ appropriate management activities (Battour et al., 2019; Rahman et al., 2021b). Two basic actions are needed: (1) understanding social media because it is vital for tourist companies to understand the use of these social networking platforms; and (2) understanding how to efficiently apply electronic strategic management and social media marketing (Benckendorff et al., 2014; Law et al., 2014). In the tourist industry, acceptance and usage of social media in the virtual world, strongly suggests that tourist companies have to take new strategies to their management and marketing operations and execute them. Tourist enterprises need to integrate social media into their management and digital marketing efforts to take full use of social networking sites. Knowledge and practice implications have demonstrated that stakeholders in the tourism sector should consider the digital act by utilizing digital marketing through the development of new capacities in using content and mobile promotion as the modern technology focused on advertising tourism to promote the digital marketing strategy. In a sense of organizing vacations via a search engine, social networking sites have become an important element of the online tourism arena. Using social networking sites to pursue a variety of options may require substantial effort. As a result, tourist industry stakeholders such as merchants and administrations should concentrate their attention on social media to increase their economic sustainability and capabilities. Through social networking sites, the tourism department may optimize all data, covering special occasions, destinations, and travel experiences that are shared by the users by pictures or posts. In addition, the tourism department must guarantee that problematic issues such as inadequate data and slow response are reduced because that will lead to better comments by the consumers on the online review websites (Rahman, 2019; Rahman et al., 2017b, 2020). Furthermore, the tourist ministry may build improved norms and regulations regarding social media for expanding the market. It could lead to the development of services based on the preferences of travelers.

The strategic administration and improvement of modern trade organizations are influenced by information and communication technology (ICT). The use of ICT transforms the key tactical strategies of businesses globally (Kumar & Kumar, 2020). By modifying their productivity, diversification, operating expenditures, and responsiveness, ICT transforms the strategic position of tourism companies and organisations (UNWTO, 2020). Digital technology has sparked a drastic transformation in the travel and tourist industry's distributional and operational activities. According to Briedenhann and Wickens (2004), human resource training and education are critical for tourist service innovation management, and ICT will enable the tourist industry to better understand current progress and fantasize for the years ahead. To serve the target market, increase profitability, improve infrastructure, productivity, and manage sustainable performance in tourism marketing, the tourism sector should recognize and employ modern technology. Tourism destinations will be defined by a combination of better management, strategy, advertising, and smart use of digital technology for the tourism industry's success.

Conclusion

The influence and impact of social media on the travel planning process have been studied in this chapter. In particular, the use, effect, and consequences of customer reviews and assessments in all phases of the customer's travel and experiences were examined through social media, and the impact of social media on tourist behaviour was investigated. The study showed afterwards the role of social media on the tourism customer preferences that affect advertising, selling, service quality, transportation, expertise, rebranding and the tourism supplier reputations, as well as its strategic and operational management and function. The interpretations in this chapter demonstrate the understanding social networking sites is essential not just for understanding contemporary tourism, but also for the tourism industry's success. Social networking sites may have an impact on activities of tourism and stakeholders such as vocational training, field-based instruction, experimentation, real-world experience, and technological innovation-based knowledge. To establish a sustainable tourism sector, information technology-related companies and tourism industry stakeholders such as travel agencies, accommodation providers, device manufacturers, and resorts should collaborate and communicate continuously. The tourism sector can generate critical information, which is a vital strategic and important instrument for the tourism industry. The use of information and communication technology can open up new opportunities for data analysis in the tourism industry. To develop the tourism business and assist people in the upcoming times, a gradual shift from information systems to reflective knowledge systems is required.

References

Adeola, O., & Evans, O. (2020). ICT, infrastructure, and tourism development in Africa. *Tourism Economics, 26*(1), 97–114.

Amaro, S., Duarte, P., & Henriques, C. (2016). Travelers' use of social media: A clustering approach. *Annals of Tourism Research, 59*, 1–15.

Assaf, A., & Scuderi, R. (2020). COVID-19 and the recovery of the tourism industry. *Tourism Economics, 26*(5), 731–733.

Bala, M., & Verma, D. (2018). A critical review of digital marketing. *International Journal of Management, IT & Engineering, 8*(10), 321–339.

Battour, M. M., Rahman, M. K., & Rana, M. S. (2019). The impact of PHTPS on trip quality, trip value, satisfaction and word of mouth: Non-Muslim tourists' perspective. *Journal of Islamic Marketing, 11*(6), 1517–1538.

Benckendorff, P. J., Sheldon, P. J., & Fesenmaier, D. R. (2014). Social media and tourism. In P. J. Benckendorff, P. J. Sheldon, & D. R. Fesenmaier (Eds.), *Tourism information technology* (pp. 120–147). CABI.

Bertacchini, F., Giglio, S., Gabriele, L., Pantano, P., & Bilotta, E. (2018). New technologies for improving tourism students training. In L. G. Chova, A. L. Martínez, & I. C. Torre (Eds.), *EDULEARN18 proceedings of the 10th international conference on education and new learning technologies* (pp. 4155–4162).

Bilgihan, A., Barreda, A., Okumus, F., & Nusair, K. (2016). Consumer perception of knowledge-sharing in travel-related online social networks. *Tourism Management, 52*, 287–296.

Briedenhann, J., & Wickens, E. (2004). Tourism routes as a tool for the economic development of rural areas–Vibrant hope or impossible dream? *Tourism Management, 25*(1), 71–79.

Buhalis, D. (2000). Marketing the competitive destination of the future. *Tourism Management, 21*(1), 97–116.

Canovi, M., & Pucciarelli, F. (2019). Social media marketing in wine tourism: Winery owners' perceptions. *Journal of Travel & Tourism Marketing, 36*(6), 653–664.

Chen, F.-W., Guevara Plaza, A., & Alarcon Urbistondo, P. (2017). Automatically extracting tourism-related opinion from Chinese social media. *Current Issues in Tourism, 20*(10), 1070–1087.

Coca-Stefaniak, J. A. (2020). Beyond smart tourism cities–towards a new generation of "wise" tourism destinations. *Journal of Tourism Futures*. https://doi.org/10.1108/JTF-11-2019-0130

Di Pietro, L., Mugion, R. G., & Renzi, M. F. (2018). Heritage and identity: Technology, values and visitor experiences. *Journal of Heritage Tourism, 13*(2), 97–103.

Fotis, J., Buhalis, D., & Rossides, N. (2011). Social media impact on holiday travel planning: The case of the Russian and the FSU markets. *International Journal of Online Marketing (IJOM), 1*(4), 1–19.

Gössling, S., & Michael Hall, C. (2019). Sharing versus collaborative economy: How to align ICT developments and the SDGs in tourism? *Journal of Sustainable Tourism, 27*(1), 74–96.

Gretzel, U. (2017). Social media activism in tourism. *Journal of Hospitality and Tourism, 15*(2), 1–14.

Gretzel, U., & Yoo, K.-H. (2014). Premises and promises of social media marketing in tourism. In S. McCabe (Ed.), *The Routledge handbook of tourism marketing* (pp. 513–526). Routledge.

Gretzel, U., Sigala, M., Xiang, Z., & Koo, C. (2015). Smart tourism: Foundations and developments. *Electronic Markets, 25*(3), 179–188.

Gretzel, U., Fuchs, M., Baggio, R., Hoepken, W., Law, R., Neidhardt, J., & Xiang, Z. (2020). e-Tourism beyond COVID-19: A call for transformative research. *Information Technology & Tourism, 22*, 187–203.

Hall, C. M., Scott, D., & Gössling, S. (2020). Pandemics, transformations and tourism: Be careful what you wish for. *Tourism Geographies, 22*(3), 577–598.

Jun, J., & Oh, K. M. (2015). Framing risks and benefits of medical tourism: A content analysis of medical tourism coverage in Korean American community newspapers. *Journal of Health Communication, 20*(6), 720–727.

Kang, M., & Gretzel, U. (2012). Effects of podcast tours on tourist experiences in a national park. *Tourism Management, 33*(2), 440–455.

Kim, J., & Fesenmaier, D. R. (2017). Sharing tourism experiences: The posttrip experience. *Journal of Travel Research, 56*(1), 28–40.

Kumar, N., & Kumar, R. R. (2020). Relationship between ICT and international tourism demand: A study of major tourist destinations. *Tourism Economics, 26*(6), 908–925.

Law, R., Buhalis, D., & Cobanoglu, C. (2014). Progress on information and communication technologies in hospitality and tourism. *International Journal of Contemporary Hospitality Management, 26*(5), 727–750.

Lee, H., Wright, K. B., O'Connor, M., & Wombacher, K. (2014). Framing medical tourism: An analysis of persuasive appeals, risks and benefits, and new media features of medical tourism broker websites. *Health Communication, 29*(7), 637–645.

Leung, D., Law, R., Van Hoof, H., & Buhalis, D. (2013). Social media in tourism and hospitality: A literature review. *Journal of Travel & Tourism Marketing, 30*(1–2), 3–22.

Mehrolia, S., Alagarsamy, S., & Solaikutty, V. M. (2021). Customers response to online food delivery services during COVID-19 outbreak using binary logistic regression. *International Journal of Consumer Studies, 45*(3), 396–408.

Mkono, M., & Tribe, J. (2017). Beyond reviewing: Uncovering the multiple roles of tourism social media users. *Journal of Travel Research, 56*(3), 287–298.

Mkwizu, K. H. (2020). Digital marketing and tourism: Opportunities for Africa. *International Hospitality Review, 34*(1), 5–12.

Pantano, E., Priporas, C.-V., & Stylos, N. (2017). 'You will like it!' using open data to predict tourists' response to a tourist attraction. *Tourism Management, 60*, 430–438.

Picheta, R. (2021). *New Zealand tells travelers to stop copying other people's travel photos*. Retrieved from: https://edition.cnn.com/travel/article/new-zealand-tourist-photos-campaign-scli-intl/index.html. Accessed 4 June 2021.

Poon, S., & Jevons, C. (1997). Internet-enabled international marketing: A small business network perspective. *Journal of Marketing Management, 13*(1–3), 29–41.

Rahman, M. K. (2019). Medical tourism: Tourists' perceived services and satisfaction lessons from Malaysian hospitals. *Tourism Review, 74*(3), 739–758.

Rahman, M. K., Zailani, S., & Musa, G. (2017a). Tapping into the emerging Muslim-friendly medical tourism market: Evidence from Malaysia. *Journal of Islamic Marketing, 8*(4), 514–532.

Rahman, M. K., Zailani, S., & Musa, G. (2017b). What travel motivational factors influence Muslim tourists towards MMITD? *Journal of Islamic Marketing, 8*(1), 48–73.

Rahman, M., Moghavvemi, S., Thirumoorthi, T., & Rahman, M. K. (2020). The impact of tourists' perceptions on halal tourism destination: A structural model analysis. *Tourism Review, 75*(3), 575–594.

Rahman, M. K., Masud, M. M., Akhtar, R., & Hossain, M. M. (2021a). Impact of community participation on sustainable development of marine protected areas: Assessment of ecotourism development. *International Journal of Tourism Research*. https://doi.org/10.1002/jtr.2480

Rahman, M. K., Sarker, M., & Hassan, A. (2021b). Medical tourism: The Islamic perspective. In A. Hassan (Ed.), *Tourism products and Services in Bangladesh: Concept analysis and development suggestions* (pp. 87–99). Springer.

Rana, M. S., Rahman, M. K., Islam, M. K., & Hassan, A. (2020). Globalization effects on tourism Marketing in Bangladesh. In A. Hassan (Ed.), *Tourism Marketing in Bangladesh: An introduction* (pp. 157–171). Routledge.

Roque, V., & Raposo, R. (2016). Social media as a communication and marketing tool in tourism: An analysis of online activities from international key player DMO. *Anatolia, 27*(1), 58–70.

Shehzad, K., Liu, X., Rauf, A., Arif, M., Mazhar, S., Sohail, N., & Amin, W. (2019). Revolutionising tourism development in China: An effective role of ICT and Western Silk Road project. *Asia Pacific Journal of Tourism Research, 24*(9), 965–977.

Sigala, M., Christou, E., & Gretzel, U. (Eds.). (2012). *Social media in travel, tourism and hospitality: Theory, practice and cases*. Ashgate Publishing.

Sloan, L., & Quan-Haase, A. (Eds.). (2017). *The SAGE handbook of social media research methods*. SAGE Publications Ltd.

Stankov, U., & Gretzel, U. (2020). Tourism 4.0 technologies and tourist experiences: A human-centered design perspective. *Information Technology & Tourism, 22*(3), 477–488.

Stankov, U., & Gretzel, U. (2021). Digital Well-being in the tourism domain: Mapping new roles and responsibilities. *Information Technology & Tourism, 23*(1), 5–17.

Stankov, U., Filimonau, V., & Vujičić, M. D. (2020). A mindful shift: An opportunity for mindfulness-driven tourism in a post-pandemic world. *Tourism Geographies, 22*(3), 703–712.

Sun, Y., Fan, H., Bakillah, M., & Zipf, A. (2015). Road-based travel recommendation using geo-tagged images. *Computers, Environment and Urban Systems, 53*, 110–122.

Taylor, D. G. (2017). *(Don't you) wish you were here? Narcissism, envy, and sharing of travel photos through social media*. Paper presented at the Nineteenth Academy of Marketing Science World Marketing Congress. Paris: The 19–23rd July, 2017.

Thomas, A., & Chopra, M. (2020). On how big data revolutionizes knowledge management. In B. George & J. Paul (Eds.), *Digital transformation in business and society* (pp. 39–60). Palgrave Macmillan.

Tussyadiah, I. P., & Fesenmaier, D. R. (2009). Mediating tourist experiences: Access to places via shared videos. *Annals of Tourism Research, 36*(1), 24–40.

UNWTO. (United Nations World Tourism Organization). (2020). *2020: A year in review*. Retrieved from: https://www.unwto.org/covid-19-and-tourism-2020. Accessed 4 June 2021.

Xiang, Z. (2018). From digitization to the age of acceleration: On information technology and tourism. *Tourism Management Perspectives, 25*, 147–150.

Xiang, Z., & Gretzel, U. (2010). Role of social media in online travel information search. *Tourism Management, 31*(2), 179–188.

Xiang, Z., Du, Q., Ma, Y., & Fan, W. (2017). A comparative analysis of major online review platforms: Implications for social media analytics in hospitality and tourism. *Tourism Management, 58*, 51–65.

Yasmin, A., Tasneem, S., & Fatema, K. (2015). Effectiveness of digital marketing in the challenging age: An empirical study. *International Journal of Management Science and Business Administration, 1*(5), 69–80.

Zuboff, S. (2015). Big other: Surveillance capitalism and the prospects of an information civilization. *Journal of Information Technology, 30*(1), 75–89.

Dr Muhammad Khalilur Rahman currently works in the Faculty of Entrepreneurship and Business at Universiti Malaysia Kelantan (UMK), Malaysia. He has been actively involved in research activities and published over 55 articles and book chapters in leading service management outlets. He has a wide interest in tourism and service management research which includes medical tourism, eco-tourism, halal tourism, service quality, brand equity, supply chain management, operation management, green and sustainable development. Rahman is a reviewer of Kybernetes, Journal of Hospitality and Tourism Insights, International Journal of Contemporary Hospitality Management, Journal of Islamic Marketing, and BMC Public Health.

Mohammad Mainul Hossain is currently pursuing his Master of Philosophy (MPhil) at the Department of Media and Communications Studies under the Faculty of Arts and Social Sciences in University of Malaya, Malaysia. He is a dynamic and highly motivated student who has a strong passion, especially in the research area. His area of research interest is social networking applications, journalism, digital communications, information systems and information technology.

Shaharin Akter is currently pursuing her MBA at the Department of Banking and Insurance under Faculty of Business Administration in University of Chittagong, Bangladesh. She is highly passionate about research with a strong desire to learn more. Her current research interests include Determinants of Profitability, Banking and Insurance Sector, Risk and Efficiency Measurement, Sustainability, and Blue Economy.

Dr Azizul Hassan is a member of the Tourism Consultants Network of the UK Tourism Society. Dr Hassan has been working for the tourism industry as a consultant, academic, and researcher for over 20 years. His research interest areas are technology-supported marketing for tourism and hospitality, immersive technology applications in the tourism and hospitality industry, and technology-influenced marketing suggestions for sustainable tourism and hospitality industry in developing countries. Dr Hassan has authored over 150 articles and book chapters in leading tourism outlets. He is also part of the editorial team of 25 book projects from Routledge, Springer, CAB International, and Emerald Group Publishing Limited. He is a regular reviewer of a number of international journals.

Part V
Destination Branding and National Image

Chapter 8
Technology Innovation as a Destination Branding Tool for Festivals and Events Tourism

Muhammad Khalilur Rahman, Md. Sohel Rana, Shaharin Akter, and Azizul Hassan

Introduction

Tourism destinations attract visitors by offering some unique and alluring spots, locations, cultures, modernizations, history and historical relics, religious and cultural rituals, events like sports, conferences, and trade fairs. Some destinations are very famous globally for tourism activities as they offer these experiences and events for many tourists across the world (Rana et al., 2020; Rahman & Zailani, 2017). People usually get to know the details of these destinations' offerings by the brand value and information integration in the online sources. The concept of destination branding is the result of intense growing competition in the tourism industry. Destination branding is also referred to as destination marketing incorporating a package of tourism offers, experiences and events through an innovative tourist's friendly approach. The development of destination branding is always a complex task. It provides a positive image of the destination in the visitors' minds. Moreover, tourists develop their expectations based on prior experience (Rahman et al., 2021a), word of mouth, media reports, advertising, and other sources (Rahman et al., 2017a, 2018a). Tourists usually can make differences between destinations through the tourism destination brand image.

M. K. Rahman (✉)
Faculty of Entrepreneurship and Business, Universiti Malaysia Kelantan,
Pengkalan Chepa, Malaysia
e-mail: khalilur.r@umk.edu.my

M. S. Rana
Faculty of Business and Accountancy, University of Malaya, Kuala Lumpur, Malaysia

S. Akter
Faculty of Business Administration, University of Chittagong, Chattogram, Bangladesh

A. Hassan
Tourism Consultants Network, The Tourism Society, London, UK

A. Hassan (ed.), *Technology Application in Tourism Fairs, Festivals and Events in Asia*, https://doi.org/10.1007/978-981-16-8070-0_8

The purpose of destination branding is to promote any particular destination among the tourists of many countries (Rahman et al., 2019) by offering a lucrative package of tourism products and services in the most sophisticated manner (Rahman et al., 2018b, 2021b). Consumer products and other forms of services are significantly less multifaceted than destinations. According to positioning theory, a short message focusing on one or a few brand associations is required to reach the minds of busy customers (Pike, 2005). Destination branding is developed in order to highlight the tourism treasures of a country. It is widely linked with the marketing activities of tourism products and services in a broad spectrum. In recent years, tourism destination marketing is now widely regarded as an important part of destination management (Blumberg, 2005; Rahman et al., 2017b, 2020). Some studies regard destination marketing as a type of "market-oriented strategic planning", and hence as a strategic approach tourism destination marketing is related to the development concept rather than just a promotional tool. Moreover, branding of any product and service also requires a development process which is often more than a marketing or a management tool. The development process of any products and services demands innovation and technological integration.

Although much of the focus on innovation is on new products and technology, all innovation is founded on questioning existing assumptions and ways of thinking (Moscardo, 2008). Innovation can take various shapes, but they all have three things in common: creativity, a problem-solving mindset, and a new way of thinking (Moscardo, 2008). These dimensions of innovation often refer to technological integration. According to Hjalager (1997), there remains a propensity to link innovation solely with creativity or the development of new products, which limits our perception of a sector's potential for innovation. Perdomo-Ortiz et al. (2006) have concluded that innovation is best defined as a dynamic capability based on patterns of thought and action that allows a company to change its ways of doing business regularly to improve its effectiveness. All innovation is founded on new ways of thinking, which are frequently linked to creative problem-solving strategies.

Globalization's impact on global tourism has heightened competition among places in search of the best positioning (Rana et al., 2020). In this environment, events have become an important component of destination tourist development, attracting visitors, encouraging investment, improving image, and boosting the local economy. Hosting events has become a key component of cities' and regions' strategies for attracting tourists and investments (both local and foreign), developing the tourism industry, and improving the competitiveness of any particular destination (Getz, 2008; Rahman, 2014). Events are usually one-time celebrations that are planned and staged in a specific location by several public or private organizations and may include a variety of activities. Events are a type of tourism offer in which the content of the events is linked to the specific resources of the location where they are held in order to attract potential visitors, and where a variety of tourism programs relating to natural and other tourist resources and values are organized. Events are unique and appealing happenings that occur infrequently and for a set time, allowing participants to engage in social contact and transact in ways that are not possible in ordinary life (Jago & Shaw, 1998). The unique appeal of event tourism stems from a desire to live a life experience that is unlike any other tourism experience available in other countries.

The examination of various interrelated aspects such as natural and cultural resources, infrastructure, services, institutional image, promotional materials, travel evaluations, and available events, among others, is required to provide a comprehensive analysis of a tourism destination (Tyrrell & Johnston, 2001). A special event either local or international provides the destinations with an opportunity to brand the tourism sector of any country. The attractiveness and pomposity of events will certainly create a brand image of that particular destination among the visitors. However, such attractiveness of the events becomes much entertaining and joyous when it shows its features of uniqueness than other tourism activities. For example, the new year celebration in Dubai and London is much aspiring as these destinations prepare and design the new year event celebration for the visitors in such a way by integrating sophisticated technologies and innovation that ensures satisfaction to the visitors. The integration of technology and innovation makes sure an immaculate attractiveness of the events. This chapter focuses on the importance of technology and innovation integration in destination branding especially when it comes to festivals and events tourism. Further, this chapter discusses the challenges and solutions for using digital technologies in festivals and event tourism.

Technology Innovation and Tourism Destination

Any industry striving to reach its full potential must engage in creative behaviors, and tourism is no different. Tourism might be a powerful driver for fostering and spreading innovation. However, research on tourism innovation is inadequate, scant and scattered (Hjalager, 2005) and there is still a gap explaining the importance of tourism innovation in practice. Despite the development of tourism innovation literature in recent years, the focus on the use of information and communication technology (ICT) as an innovative strategy towards sustainable tourism for destinations is still quite limited. In the advancement of a greener low-carbon economy, technology and innovation play a significant role, and technology and innovation are key factors in ensuring sustainability (Bartlett & Trifilova, 2010). The use of ICT has been recognized as necessary for innovation. ICT, according to Hjalager (2005), is a key accelerator for tourism innovation. Racherla et al. (2008) claimed that the potential of ICT to link innovative tourism planning and development has not been taken on destinations branding. For the tourism sector, ICT could be a productive endeavor.

For the tourism business, technology innovation is an innovative, problem-solving strategy and a unique line of thinking. The analytical typology for sustainable tourism innovation has been proposed by Hjalager (1997) which covers product innovation dimensions, conventional process innovation, information process innovation, management innovation and institutional innovation. Product innovation entails modifying, integrating, or creating a new tourism service or product where the uniqueness appeals to tourists. The development of new types of ICT for visitors would be an innovation in ICT that would make their activities more sustainable during all stages of the trip. Process innovation refers to increasing the degree of performance by redesigning production and delivery processes to save on manufacturing inputs like labor (Hjalager, 1997). This type of innovation occurs when technology is utilized to manage a destination's ecological, socioeconomic, and public resources. The use of ICT for information management is referred to as process innovation for information handling. Through the effective maintenance and coordination of environmental information at the destinations, ICT can play a vital role in sustainable tourism. Management innovation applies to innovative management practices that alter old authority structures, create new roles, collaborate in new ways, and empower employees (Rahman et al., 2021a). As ICT changes the way destinations are now operated, new roles in Destination Management Organizations (DMO) will be generated.

Technological Impacts on the Event Industry

Technologies have been a major contributor to the growth, advancement and viability of numerous events and tourism industries, and have acquired dominance (Koba, 2020). Using technologies in the events sector results in more professional planning, better trustworthy performance, and superior transportation, as well as the advantages of globalization. According to Bowdin et al. (2012), developments in technological communication, information linkage and social media, globalization, and other advancements have influenced event marketing and created a global chance to access new markets. Technology is always changing and improving, and the events business is continuously adapting to these changes. While technology enables the digitalization of different aspects of event planning and management, it also offers the events sector both possibilities and difficulties.

In regular activities, technology has a huge impact on the tourism and events industry. It can be observed throughout the stages, the satisfaction and interaction of the participants, greater awareness and the limiting of anti-social behavior due to security and expanded communication routes (Washington et al., 2012). Technological innovation has improved equipment and machinery, which revolutionized the way events are carried out. Technology has advanced the events and provided a level playing field for events of all sizes and budgets. The event sector has become customer-driven and it is of utmost importance to have access to information from anywhere and at any time. The key trend coming from the literature (Bowdin et al., 2012) is the increased usage of online ticketing for events and tourism locations.

Festivals and Events Tourism Reach Digital Maturity

As we enter a new decade, academic research of festivals and events continues to flourish in the tourism sector. The significance of regional and international festivals and events has been studied all over the world, and while festival failures remain a risk for producers. The existing research has proved the importance of events to persons, organizations, and societies (Ryan et al., 2020). The information and communication technologies (ICTs) are used before, during and after the on-site interaction during festivals and events (Koba, 2020). By incorporating hardware, software, NetWare, and HumanWare into the event, ICTs have transformed communication in all life domains (Poushter, 2016). The use of ICT is frequent both in intrinsic and extrinsic events because technology involves both the backstage and the front (Vinnervik, 2020). However, the effects on the events, participants, organizers and other players because of the continuing development of ICT are rarely known. In event studies, ICT integration into festival management, planning, advertising, logistics, and risk mitigation has earned little attention, and it requires a better knowledge of supply and demand perspectives. Conceptual and theoretical improvement in this well-established but ever-changing field will lay the groundwork for future education, application, and development.

Bossey (2020) has investigated "availability in all sectors", arguing that equality in festival contexts requires more attention. As a result, the study examines the present level of connectivity at UK festivals, as well as technological advancements that could improve components of the festival experience and execution. The researchers gathered a supply-side perspective on accessibility and diversity in festival environments by surveying festival employees. Bossey (2020) explains the social model of disability, which emphasizes the causes of elimination and encourages us to think about how we can improve access. According to the researchers, numerous facilities are integrated into festival settings to improve access, but there are still many obstacles to overcome. Festival audiences, according to Bossey (2020), are open to technology since festivals are typically the site of one-of-a-kind experiences, and this gives a chance for organizers to collaborate with customers to co-create ways to increase access. Although not all approaches can or should be digital, the writers acknowledge that technology should be taken into account. The importance of attending the activities was mentioned, but many persons with disabilities continue to be excluded. The authors propose that they collaborate with their audiences to ensure that live events are accessible and integrated.

At festivals and events, ICT is improving access and inclusivity. Dinis et al. (2020) examined accessibility as well, but with an emphasis on social media, to develop a methodology to enable event organizers to measure the accessibility of their social networking sites for people with hearing and vision difficulties. The researchers convincingly claim that providing access to social media is critical for facilitating various people's involvement in the event. The study examines a variety of social media channels utilized by a large event and emphasizes the importance of ICT in engaging a frequently underserved community. The results reveal that communication advantages can be lost without careful consideration of how ICT is utilized. More research is needed to understand how ICT can help some people participate while alienating others (Eusébio et al., 2020). As we reach the Fourth Industrial Revolution, also known as Industry 4.0, the situation of technology in the events industry is examined. Ryan et al. (2020) explain the importance of digital technology, assess industry experts' opinions and academics on a social networking site to understand the learning outcomes of the events. The researchers can explain the meaning of Event 4.0 and the digital maturity of event tourism. Tourism event planners can strive to improve technological interactions with the consumers and with other organizations.

Effect of Technology on Event Tourism Management

With the advent of digitization, their perceptions and strategies in the area of management have shifted dramatically. In the digital age, adopting ICTs involves transforming procedures in the tourism business into a digital environment (Zhou et al., 2019). With ICTs, companies can develop and strengthen their managerial functions with improved technology. As a result, event management becomes more manageable. In the sphere of event management, ICTs also have an impact on competition strategies and influence how people think about competition. In a competitive market, the efficiency of management tactics is critical to the success and long-term viability of events. The instruments to be utilized for event management are selected based on event dimensions, product offerings and cost parameters. The impact of digitalization has resulted in the digital transformation of the corporate environment, as well as any firm that has adapted to technology. In the digital age, tourism companies can be managed through ICT integration (Lawrence & Tar, 2018). Management and marketing techniques have transformed as a technological orientation in event management. The efficient digital transformation of tourist companies with technological infrastructure allows them to deal with the changing order's possibilities and challenges. Companies should integrate technology's possibilities and challenges to regulate, coordinate, monitor and use processes in terms of human resources more favorably by focusing on marketing demands. Digitalization enables ICTs for achieving and developing the primary advantages of successful events organization, enhancing the quality of care of the event's potential customers, targeting and identifying and planning a competitive strategy (Ryan et al., 2020). The organization can identify the top market trends through competitive strategies to realize and improve the core competencies of the company in terms of performing work and increasing the quality of life of its people. In a competitive climate, the importance of marketing communications is vital to the success and long-term viability of event and tourist management.

Application of ICTs and Well-Being of Tourism

Consumers are increasingly looking for ways to accomplish their well-being through travel experiences. Well-being with the use of sports apps is an essential subject for further exploration in tourism studies, both from a view to good results for users and as a management concern for providers of tourism. Adeola and Evans (2020) provide readers with useful theoretical and practical implications for the tourism sector in exploring psychological and norm-based determinants that affect the consumer's intention in the field of sports applications. As healthy conditions and well-being come hand in hand, the tourism sector must be equipped to provide ways of enhancing access and safe health tourism for customers. In this regard, Rivera et al. (2021) suggest blockchain technology as a consumer-friendly healthcare tourism service. They also investigate the potential of medical tourism (Rahman, 2019; Rahman et al., 2021c) by improving information asymmetry, accessibility and the credibility provided by ICTs in building on the well-known advantages of blockchain technology.

Avoiding rapidly rising environmental toxic substances could produce good results on subjective well-being. Consumers are typically ignorant of common contaminants, such as artificial lights, that they are already exposed to. As a result, Lee et al. (2019) highlight the need for dark skies and proper ICT use during the evening hours to improve mental health and well-being. They recommend using astronomical tourism as a novel strategy for combating light pollution risk while on vacation. In their research, Baldwin et al. (2021) highlighted how a simulated, immersive tourist environment might be used to test tourism's capacity to relieve stress and improve mood. The report emphasizes the need for data triangulation and testing rigor. It also makes a significant contribution to the expanding academic literature on tourism experiments (Viglia & Dolnicar, 2020). This also points to the prospect of using technology to recreate effective multi-sensory experiences in tourism events, which improve their well-being more efficiently than traditional leisure activities, such as watching Television. Three basic guidelines for tourism industry are being developed by Stankov and Gretzel (2021), which include the need to take a look at tourist digital welfare, to establish new regulations within the field of tourism and to build strategies for the development of new products and services.

Challenges the Use of Digital Technologies

Events take place for a multifarious purpose. The nature of one event is separate from the other. Events can be classified according to their nature, philosophy and activities. Events are mainly related to cultural celebration, political and state programs, arts and entertainment, business and trade, education and scientific functions, sports and competitions. Apart from these, events can be private as well. The diversified nature of events often makes it challenging to incorporate the use of digital technologies. Often events that have international exposures are telecasted live globally which requires broadcasting technologies to set up at the event spot. Such technological establishment depends on the availability of broadcasting or technological expertise which involves huge costs. Nevertheless, information and communication technology (ICT) has been gaining utmost priority and deep importance in our daily lives which creates both positive and negative spillover effects (Gretzel & Stankov, 2021). Moreover, as technology is incorporated backstage and front-of-house, both utilitarian and hedonistic uses of ICT are common at events. Little is known about the effects of ICT evolution on events, participants, volunteers, and other stakeholders.

In the present context of tourism diversity, ICTs have become critical for various tourist industry stakeholders to effectively provide and offer tourism experiences (Xiang, 2018; Xiang & Fesenmaier, 2020). However, the extensive use of ICT, on the other hand, can drastically interrupt tourism experiences and put visitors' and others' well-being goals in jeopardy (Stankov & Gretzel, 2021). Festival audiences are generally opened to technology since festivals are typically the site of one-of-a-kind experiences, and this gives a chance for organizers to collaborate with audiences to co-create solutions to improve access. While not all solutions can or should be digital, therefore, it is acknowledged that technology should be taken into account. Digital technologies and the internet are among the ten global issues recognized by the World Economic Forum. The accessibility to sophisticated technology is challenging for many nations. For example, Finland, have made Internet access a right of citizenship, but access does not imply inclusion. On the other hand, access to the internet and modern technologies is scarce in many underdeveloped countries. Moreover, the safety and security of ICT and technological intervention should be considered another global challenge.

To address these difficulties, careful theoretical research and analysis of the merits and drawbacks of using ICT will be required. Changes and advancements in ICTs have given the tourism industry a boost and challenge. Consumers in the tourism industry use the digital age for information marketing, supplying goods and services, managing consumer communication channels, and giving convenience and experience. ICTs are employed in the selection of a tourism destination, as well as in the planning and determination of alternative tourism types. Many advances, referred to as digital innovation and revolution, pose challenges for organizations. Numerous digital platforms, rising data prevalence, and other factors make it tough for firms to sell events in the digital era (Akel, 2021). It has to deal with the difficulties of monitoring and controlling businesses in increasingly complicated and fast-changing marketplaces.

Management and marketing events have countless potentialities along with challenges in the digital world. Several aspects have produced advantages or challenges, from company to client, from marketing leadership and negotiation to service, from conventional to digital (Akel, 2021). More expense and poor efficiency have generated possibilities for efficient marketing in conventional marketing communications. In contrast to traditional communication tactics, event marketing integrates digital marketing techniques for enhancing consumers' participation in marketing and increasing customers' awareness of events (Camilleri, 2018). Increasing the number of respondents in the event would be the same as increasing service awareness.

The business sector's and government's digital transformations have been important tactics for adopting and using technology breakthroughs. Digital tourist companies' technological innovations are essential for value-producing events. In the business sector, the fundamental and developed capacities of disciplined digital followers make a significant difference. The events that lasted in the conventional tourism sector that existed in the olden days have been substituted by the modern tourism business (Akel, 2021). New technology creates new utility, new ways of life, and, consequently, new tourism offerings. The added value and advancements that modern technology brings to the industry should be assessed from both a corporate management and marketing standpoint.

The experience of consumers is a greater sense of possibilities for their journeys to destinations (Vada et al., 2020). At the same time, ICTs have quickly permeated customer life and become essential for the efficient supply and provision of tourism experiences by different participants in the tourism sector (Xiang, 2018; Xiang & Fesenmaier, 2020). The tourism management team and marketer apply a number of technological inventions to enhance customer benefits by making tourism experiences fun, engaging and individualized to benefit from the growth of ICTs (Mizrachi & Gretzel, 2020; Stankov et al., 2019). More and more applications are also being implemented that enable travelers to be more physically, intellectually engaged and comfortable to take advantage of the beneficial effects of holidays and tourism activities (Stankov et al., 2020).

Solutions and Suggestions

Tourism companies should devote resources to engage a large audience and to carry out a successful tourism event. All industry players should use the most effective event management and marketing techniques in accordance with their operational processes and implementation strategies. Managerial and commercial events should be employed for market competitiveness and long-term sustainability. Given the short life of the events, the management and marketing methods affected by technological change and development are essential not to be overlooked. Tourism, which was once considered a labor-intensive industry, has evolved into an information and technology-intensive industry. It is difficult to bring out marketing and operations events apart from technologies in an atmosphere where the focus must be exclusively on client requests and demands, as well as service and experience quality. The regularity and trustworthiness of management and marketing events are some of the elements to examine in all situations. Only regularity and dependability can produce efficiency. Industry 4.0 companies should focus their efforts to sell their events and managing them through ICTs. Entrepreneurs must extend their manufacturing and advertising networks through the online platform to enhance their participation in Industry 4.0 to ICT development. As a result, Industry 4.0 elements like the internet of things (IoT), and Smart Factories require suitable infrastructures for event management and promotional events.

Conclusion

The digital era adoption of ICTs implies the transition of tourism procedures into a digital world. The impact of digitization has resulted not only in businesses that have adapted to technologies but also in the digitalization of the corporate environment. Tourism and marketing businesses link all of their activities with information and communication technologies to achieve a competitive edge and to adjust to digital changes. Digitalization benefits the entire economy by increasing efficiency, employment, and revenue. Numerous initiatives have been undertaken, including advancements in information and communication technologies, the medical field, administration, and communications. ICT has had a huge impact on macroeconomic and social structure in general, rapidly touching practically all sectors. Because of the influence of information and communication technologies on tourism, the modern technological state has made tourism a fairly information-intensive industry.

ICTs are a valuable instrument in the management and utilization of event marketing, and also in the evaluation of events, the development of methodologies, and the systematization of actions. And in a digital era, it delivers ICTs intending to achieve and develop the key advantages of an event that is effective in arranging events, improving the quality of life of the target group, aiming at a competitive advantage, and emphasizing marketing strategies. Event management, event marketing, communications, and economic effects should all be taken into account when evaluating ICTs, and events should be planned accordingly. Furthermore, the management and marketing events in the tourism sector through the use of digitalization, potentialities and difficulties created by this digitalization process are the main aspects of this study. In the digital age, technological breakthroughs and advancements in the management and commercialization of events provide numerous opportunities. Opportunities include the facility of communications, productivity, cost reductions, service quality, accessibility and creativity.

ICTs are becoming much more popular in the modern world in tourism marketing strategies, transaction costs are decreasing, and web transactions are contributing to efficiency. Furthermore, ICTs not only increase efficiency but also minimize error rates and manufacturing costs. Additionally, ICTs eliminate geographical and physical limits, eradicate the perception of distance, and provide options such as border removal. As a result, mass communication has been able to create large events that are aimed at the entire world. ICTs have overcome national borders, allowing for rapid industry management and the ability to conduct e-commerce. For all types of technological procedures to be conducted in the management and marketing of tourism events, infrastructure is necessary. As a result, e-knowledge sufficiency is essential for information and communication technologies and the quality of products and services in tourism event management.

References

Adeola, O., & Evans, O. (2020). ICT, infrastructure, and tourism development in Africa. *Tourism Economics, 26*(1), 97–114.

Akel, G. (2021). Management and marketing events in a digital era: Opportunities and challenges. In K. Birdir, S. Birdir, A. Dalgic, & D. Toksoz (Eds.), *Impact of ICTs on event management and marketing* (pp. 105–122). Hershey, PA.

Baldwin, J., Haven-Tang, C., Gill, S., Morgan, N., & Pritchard, A. (2021). Using the Perceptual Experience Laboratory (PEL) to simulate tourism environments for hedonic wellbeing. *Information Technology & Tourism, 23*, 45–67.

Bartlett, D., & Trifilova, A. (2010). Green technology and eco-innovation: Seven case-studies from a Russian manufacturing context. *Journal of Manufacturing Technology Management, 21*(8), 910–929.

Blumberg, K. (2005). Tourism destination marketing–A tool for destination management? A case study from Nelson/Tasman Region, New Zealand. *Asia Pacific Journal of Tourism Research, 10*(1), 45–57.

Bossey, A. (2020). Accessibility all areas? UK live music industry perceptions of current practice and information and communication technology improvements to accessibility for music festival attendees who are deaf or disabled. *International Journal of Event and Festival Management, 11*(1), 6–25.

Bowdin, G., Allen, J., Harris, R., McDonnell, I., & O'Toole, W. (2012). *Events management* (3rd ed.). Routledge.

Camilleri, M. A. (2018). *Travel marketing, tourism economics and the airline product*. Springer.

Dinis, M. G., Eusébio, C., & Breda, Z. (2020). Assessing social media accessibility: The case of the Rock in Rio Lisboa music festival. *International Journal of Event and Festival Management, 11(1)*, 26–46.

Eusébio, C., Silveiro, A., & Teixeira, L. (2020). Website accessibility of travel agents: An evaluation using web diagnostic tools. *Journal of Accessibility and Design for All (JACCES), 10*(2), 180–208.

Getz, D. (2008). Event tourism: Definition, evolution, and research. *Tourism Management, 29*(3), 403–428.

Gretzel, U., & Stankov, U. (2021). ICTs and well-being: Challenges and opportunities for tourism. *Information Technology & Tourism, 23*, 1–4.

Hjalager, A. M. (1997). Innovation patterns in sustainable tourism: An analytical typology. *Tourism Management, 18*(1), 35–41.

Hjalager, A. M. (2005). Innovation in tourism from a welfare state perspective. *Scandinavian Journal of Hospitality and Tourism, 5*(1), 46–62.

Jago, L. K., & Shaw, R. N. (1998). Special events: A conceptual and definitional framework. *Festival Management and Event Tourism, 5*(1-2), 21–32.

Koba, Y. (2020). Internet marketing communication in event tourism promotion. In A. Hassan & A. Sharma (Eds.), *The Emerald handbook of ICT in tourism and hospitality* (p. 149). Emerald Publishing Limited. https://doi.org/10.1108/978-1-83982-688-720201010

Lawrence, J. E., & Tar, U. A. (2018). Factors that influence teachers' adoption and integration of ICT in teaching/learning process. *Educational Media International, 55*(1), 79–105.

Lee, T. D., Park, H., & Lee, J. (2019). Collaborative accountability for sustainable public health: A Korean perspective on the effective use of ICT-based health risk communication. *Government Information Quarterly, 36*(2), 226–236.

Mizrachi, I., & Gretzel, U. (2020). Collaborating against COVID-19: Bridging travel and travel tech. *Information Technology & Tourism, 22*(4), 489–496.

Moscardo, G. (2008). Sustainable tourism innovation: Challenging basic assumptions. *Tourism and Hospitality Research, 8*(1), 4–13.

Perdomo-Ortiz, J., Gonzalez-Benito, J., & Galende, J. (2006). Total quality management as a forerunner of business innovation capability. *Technovation, 26*(10), 1170–1185.

Pike, S. (2005). Tourism destination branding complexity. *Journal of Product & Brand Management, 14*(4), 258–259.
Poushter, J. (2016). Smartphone ownership and internet usage continues to climb in emerging economies. *Pew Research Center, 22*(1), 1–44.
Racherla, P., Hu, C., & Hyun, M. Y. (2008). Exploring the role of innovative technologies in building a knowledge-based destination. *Current Issues in Tourism, 11*(5), 407–428.
Rahman, M. K. (2014). Motivating factors of Islamic tourist's destination loyalty: An empirical Investigation in Malaysia. *Journal of Tourism and Hospitality Management, 2*(1), 63–77.
Rahman, M. K. (2019). Medical tourism: Tourists' perceived services and satisfaction lessons from Malaysian hospitals. *Tourism Review, 4*(3), 739–758.
Rahman, M. K., & Zailani, S. (2017). The effectiveness and outcomes of the Muslim-friendly medical tourism supply chain. *Journal of Islamic Marketing, 8*(4), 732–752.
Rahman, M. K., Zailani, S., & Musa, G. (2017a). Tapping into the emerging Muslim-friendly medical tourism market: Evidence from Malaysia. *Journal of Islamic Marketing, 8*(4), 514–532.
Rahman, M. K., Zailani, S., & Musa, G. (2017b). What travel motivational factors influence Muslim tourists towards MMITD? *Journal of Islamic Marketing, 8*(1), 48–73.
Rahman, M. K., Zailani, S., & Musa, G. (2018a). The perceived role of Islamic medical care practice in hospital: The medical doctor's perspective. *Journal of Islamic Marketing, 9*(1), 2–18.
Rahman, M. K., Zailani, S., & Musa, G. (2018b). Tourists' satisfaction and loyalty intention at Shariah compliant private hospitals in Malaysia. *International Journal of Tourism Sciences, 18*(4), 295–311.
Rahman, M., Rana, M. S., Hoque, M. N., & Rahman, M. K. (2019). Brand perception of halal tourism services and satisfaction: The mediating role of tourists' attitudes. *International Journal of Tourism Sciences, 19*(1), 18–37.
Rahman, M., Moghavvemi, S., Thirumoorthi, T., & Rahman, M. K. (2020). The impact of tourists' perceptions on halal tourism destination: A structural model analysis. *Tourism Review, 75*(3), 575–594.
Rahman, M. K., Masud, M. M., Akhtar, R., & Hossain, M. M. (2021a). Impact of community participation on sustainable development of marine protected areas: Assessment of ecotourism development. *International Journal of Tourism Research*. https://doi.org/10.1002/jtr.2480
Rahman, M. K., Yaacob, M. R., Khalique, M., & Adeiza, A. (2021b). Analyzing tourists' intention to revisit Islamic tourism destinations. *Middle East Journal of Management, 1*(1), 1–21.
Rahman, M. K., Sarker, M. M., & Hassan, A. (2021c). Medical tourism: The Islamic perspective. In A. Hassan (Ed.), *Tourism products and services in Bangladesh: Investment and development perspective*. Springer.
Rana, M. S., Rahman, M. K., Islam, M. F., & Hassan, A. (2020). Globalization effects on tourism marketing in Bangladesh. In A. Hassan (Ed.), *Tourism marketing in Bangladesh: An introduction* (pp. 157–171). Routledge.
Rivera, R., Amorim, M., & Reis, J. (2021). Technological evolution in grocery retail: A systematic literature review. In *2021 16th Iberian Conference on Information Systems and Technologies (CISTI)* (pp. 1–8). IEEE.
Ryan, W. G., Fenton, A., Ahmed, W., & Scarf, P. (2020). Recognizing events 4.0: The digital maturity of events. *International Journal of Event and Festival Management, 11*(1), 47–68.
Stankov, U., & Gretzel, U. (2021). Digital well-being in the tourism domain: Mapping new roles and responsibilities. *Information Technology & Tourism, 23*(1), 5–17.
Stankov, U., Filimonau, V., & Slivar, I. (2019). Calm ICT design in hotels: A critical review of applications and implications. *International Journal of Hospitality Management, 82*, 298–307.
Stankov, U., Filimonau, V., & Vujičić, M. D. (2020). A mindful shift: An opportunity for mindfulness-driven tourism in a post-pandemic world. *Tourism Geographies, 22*(3), 703–712.
Tyrrell, T. J., & Johnston, R. J. (2001). A framework for assessing direct economic impacts of tourist events: Distinguishing origins, destinations, and causes of expenditures. *Journal of travel Research, 40*(1), 94–100.
Vada, S., Prentice, C., Scott, N., & Hsiao, A. (2020). Positive psychology and tourist well-being: A systematic literature review. *Tourism Management Perspectives, 33*, 100631.

Viglia, G., & Dolnicar, S. (2020). A review of experiments in tourism and hospitality. *Annals of Tourism Research, 80*, 102858.

Vinnervik, P. (2020). Implementing programming in school mathematics and technology: teachers' intrinsic and extrinsic challenges. *International Journal of Technology and Design Education*. https://doi.org/10.1007/s10798-020-09602-0

Washington, M. C., Okoro, E. A., & Thomas, O. (2012). Intercultural communication in global business: An analysis of benefits and challenges. *International Business & Economics Research Journal (IBER), 11*(2), 217–222.

Xiang, Z. (2018). From digitization to the age of acceleration: On information technology and tourism. *Tourism Management Perspective, 25*, 147–150.

Xiang, Z., & Fesenmaier, D. R. (2020). Travel information search. In Z. Xiang, M. Fuchs, U. Gretzel, & W. Höpken (Eds.), *Handbook of e-tourism* (pp. 1–20). Springer.

Zhou, Q., Gao, P., & Chimhowu, A. (2019). ICTs in the transformation of rural enterprises in China: A multi-layer perspective. *Technological Forecasting and Social Change, 145*, 12–23.

Dr Muhammad Khalilur Rahman currently works in the Faculty of Entrepreneurship and Business at Universiti Malaysia Kelantan (UMK), Malaysia. He has been actively involved in research activities and published over 55 articles and book chapters in leading service management outlets. He has a wide interest in tourism and service management research which includes medical tourism, eco-tourism, halal tourism, service quality, brand equity, supply chain management, operation management, green and sustainable development. Rahman is a reviewer of Kybernetes, Journal of Hospitality and Tourism Insights, International Journal of Contemporary Hospitality Management, Journal of Islamic Marketing, and BMC Public Health.

Md. Sohel Rana is a PhD researcher in the Faculty of Business and Accountancy, University of Malaya, Malaysia. He holds a bachelor's degree in business administration and an MBA in Management Studies from the University of Rajshahi, Bangladesh. His area of expertise in research includes Microfinance & Financial Inclusion, Islamic Microfinance, Poverty and Development Economics, Productivity & Efficiency, Tourism, Halal Tourism, Insurance, Environmental Economics, etc. He has already published a number of book chapters, journal articles in the Web of Science and Scopus indexed journals and several editorial opinions in newspaper. He has also reviewed several journal articles. He has the experience of working as a full time Researcher at the University of Malaya, Malaysia. He also holds six years of industrial experience in financial institutions and export-oriented companies in Bangladesh.

Shaharin Akter is currently pursuing her MBA at the Department of Banking and Insurance under Faculty of Business Administration in University of Chittagong, Bangladesh. She is highly passionate about research with a strong desire to learn more. Her current research interests include Determinants of Profitability, Banking and Insurance Sector, Risk and Efficiency Measurement, Sustainability, and Blue Economy.

Dr Azizul Hassan is a member of the Tourism Consultants Network of the UK Tourism Society. Dr Hassan has been working for the tourism industry as a consultant, academic, and researcher for over 20 years. His research interest areas are technology-supported marketing for tourism and hospitality, immersive technology applications in the tourism and hospitality industry, and technology-influenced marketing suggestions for sustainable tourism and hospitality industry in developing countries. Dr Hassan has authored over 150 articles and book chapters in leading tourism outlets. He is also part of the editorial team of 25 book projects from Routledge, Springer, CAB International, and Emerald Group Publishing Limited. He is a regular reviewer of a number of international journals.

Chapter 9
Technology Application in Tourism Event, Education and Training for Making a Nation's Image

Md Rabiul Alam, Md Rashed Khan Milon, Muhammad Khalilur Rahman, and Azizul Hassan

Introduction

None of the phases of human world are left untouched from the blessing of the advancement of information technology. The entire world is being changed with the sharp touch of technology. Like other phases, tourism is not an exception in this regard. Tourism sector is emerging very fast (Rana et al., 2020; Rahman et al., 2021a, b, c) that in many respects portrays the enriched image of a nation with the commotion of information technologies. The Industrial Revolution 4.0 (IR 4.0) has brought a revolutionary change in people's life in terms of adopting technology in all their functions. Hence, the improvement of new technologies has brought a great diversity in the tourism and it is seen in tourism offer as well as tourism demand (Januszewska et al., 2015) also proposed to the hoteliers incorporating IT eagerly in their endeavors to develop the quality of service. If the hoteliers can improve the IT knowledge of their employees, the quality of service can be achieved tremendously. But according to Buick (2003), hospitality usually cannot keep pace up for want of knowledge in ICT.

Mobile as well as web 2.0 applications have been developed with modern technologies and these things are being used for modernizing the market. Buyer sites

M. R. Alam
Faculty of Education, University of Malaya, Kuala Lumpur, Malaysia

M. R. K. Milon
Department of English, Port City International University, Chattogram, Bangladesh

M. K. Rahman (✉)
Faculty of Entrepreneurship and Business, Universiti Malaysia Kelantan, Pengkalan Chepa, Malaysia
e-mail: khalilur.r@umk.edu.my

A. Hassan
Tourism Consultants Network, The Tourism Society, London, UK

A. Hassan (ed.), *Technology Application in Tourism Fairs, Festivals and Events in Asia*, https://doi.org/10.1007/978-981-16-8070-0_9

seem to assuming an important part in every public relation as well as advertising events to spread a lot of purchasers all over the world. The web-based media is playing an important role in pre-trip travel arranging, decision taking, as well as for sharing inferred information and its scope is expanding day by day. The reception of ICT is additionally offering devices to cluster and cooperate for globalization as well as formation of a critical mass (Khatri, 2018). So, the IT sector has a great impact on the tourism industry with the change of the competitive area.

Now-a-days, tourism industry cannot work for a day without the help of ICT. The travel service providers, tourists as well as travel agencies use ICT to minimize the communication among themselves. Therefore, ICT has become an essential part for tourism sector, and anyone can see it in various tourist websites of the service providers as well as hotels and motels. At present, the tourists prefer booking reservation and other activities in online instead of booking face to face or over phone. The travelers prefer visiting the websites of their desired hotels or tourist places (Rahman, 2019; Rahman et al., 2017a), compare the advantages with others of same status, and then confirm their bookings by using their credit card, debit card or e-wallet. All these have fueled the adoption of technology in tourism industry. As a result, the use of e-tourism is being increased.

Taking the utmost necessity of e-tourism into account, the aim of this book chapter is to evaluate the contemporary changes, improvement and acceptance of ICT in tourism sector and how it changes a nation's image overnight. For this, the chapter focuses on the published articles in tourism-related journals which mainly deals with the application of ICT in tourism industry. Based on those, it will shed light on the necessity of ICT and event education for flourishing e-tourism, the drawbacks that the tourism industry faces currently, and the feasible solutions that the policymakers can take into consideration to boost this sector by keeping pace with the competitive world.

The Necessity of ICT in Tourism Industry

The tourism industry is actually an information related industry. The tourists would need subtleties of the region (Rahman et al., 2017b), the urban areas, the places of interest, the climate (Rahman et al., 2021a, b, c) and the vehicle choices etc. prior to arranging and settling the tour. There is a developing desire for personalized visits (Rahman et al., 2020) and a person has the adaptability to thoroughly customize each piece to suite their enjoying and furthermore progressively need data during the excursion (Goecke, 2020; González-Serrano & Talón-Ballestero, 2020). Now-a-days, people of the world do not have any ideal opportunity to go through with their families and companions and so, the travel industry is one such amenity to invest quality energy with family or companions. The greatest worry prior to finishing is that tourists cannot pre-test the visit or request discount if the excursion is not up to their assumptions. As a result, right, reliable, proper, and relevant data is needed for taking a dynamic decision.

A significant cause of using Information and Communication Technology in traveling industry is the result of immaterialness of travel industry items. An individual cannot notice and examine it before going to visit. So, this is reliant upon outlines as well as imageries to help travelers choose (Afaneh, 2021). Furthermore, a spot is steady, the individual requires to visit to encounter the item and they cannot buy it. No similarity is found in two spots, which indicates the uniqueness of it. The information of a spot is unquestionably not quite the same as that of an adjoining place. This prompts variety among the various areas. Finally, individuals will in general cover different areas on a solitary outing and not simply visit a solitary area. This prompts a lot of changes of the areas, the chance of backup ways to go, the request for visit, method of transport, sorts of convenience and food. With such a large amount varieties and potential outcomes, the need of ICT is unquestionably a key necessity (Goecke, 2020). This is simply concerning one partner, for example the traveler. If anyone thinks of the other partners like the destination suppliers, the sellers, the vehicle suppliers, the drivers, the traveler area management, the staff of these places of interest – the circumstance will require the great role of Information and Communication Technology (Afaneh, 2021).

Vendors face a similarly overwhelming assignment of understanding their clients who have a place with various nations, religion, pay group, having shifted interests, expectations and mindfulness. Their decisions and their sentiments contrast so does their correspondence styles. ICT gives a summed-up stage to incorporate these assorted prerequisite arrangements of decisions and alternatives for and of the partners in this competitive travel industry. The importance of the application of ICT in travelling industry can be observed from different angles. ICT is coordinated in the travel industry to enhance the related service activities. The major activities in the travel industry like convenience, transportation, correspondence and so forth are upheld by the ICT for the benefits of host countries as well as the voyagers (Goecke, 2020; Afaneh, 2021). Thus, the basic reasons shrouded with ICT are discussed below:

Necessity of Data

The web-based media, mobile and ICT on the travel industry is broadly utilized for data search, sharing and exchanging. As it is seen from the research, there are a good number of studies that have highlighted the importance of ICT in tourism industry such as tradeshows for data trade (Han & Verma, 2014). Duan et al. (2016) said about the essential data hotspot for buyers as well as business on lodging administration execution; Kang and Namkung (2015) revealed the role of ICT in sharing data. Besides, Benckendorff (2009) showed wikis in information based advanced economy for communitarian learning and Okazaki et al. (2015) expressed the role of ICT for data access in travelling arrangement.

Moreover, a change is noticed in the approach of web 2.0 on a very basic level in the traveler data search as well as destination decision measure (Lo et al., 2011); new technologies offer simple admittance to a lot of the travel industry data. As per

Jung et al. (2015), the expanded accessibility of cell phone and smart phone devices has changed the travel industry through accessing data by the tourists when they visit. According to Murphy et al. (2016), web seeking tool and personal computer are the much time utilized data as well as gadget sources for seeking as well as reservation. Subsequently, an answer for the data need in the travel industry is generally given by ICT. The review tracked down a lot of articles identified with the essential purpose behind information need.

Advancement and Advertising

The travel industry is a cutthroat industry. There is a major contest inside the business, and numerous objections are rivaling every other to draw in the guests (Khatri, 2018). In this situation, advancement, as well as marketing of the travel industry and hospitality items and administrations, is vital. For the destination improvement, the executives are significant marketing elements of the travel industry promotion (Khatri, 2018). Liu and Mattila (2016) examined the best approach to promote innovation-based accommodation administrations. Besides, Kim et al. (2016) showed educational media (literary versus pictorial) has critical impacts with respect to the effect of mental distance on voyagers favored limited time data. Again, Choi et al. (2018) prove that ICT empowers multilateral correspondence designs in terminus presentation. Later on He et al. (2018) show that different enterprises are using Augmented Reality (AR) as an advertising instrument. Information and Communication Technologies' usages are being abused yet as the proper application of ICT has a great impact on people's behavior that is reflected in their decision-making attitude.

Application of ICT in Tourism and People's Behavior

ICT study on the behavior of the people and it has become a cache affair in traveling sector. Various studies are found on the matter like the tourists travel thinking and their adopting of internet (Xiang et al., 2015); diversification of networking shapes in various traveling shareholders in virtual reality (Ying et al., 2016); tourist operators' impact on the customers' pleasure (Caber et al., 2013); traveling advertising learners' beliefs as well as knowledge of speech delivering (Sanders & LeClus, 2009); the factors of voyagers' adopting of traveling data in case of smartphone (No & Kim, 2013).

ICT is utilized in numerous conduct and execution contemplates. For instance, anticipating programs and buyers of lodging sites (Wu et al., 2013); anticipating representatives' selection conduct of new ICT (Cheng & Cho, 2011); creating mental outcomes on a traveler's discernments (Choi et al., 2018). Besides, as per Vu et al. (2020), ICT is used for deciding the traveler exercises and the fleeting inclinations for every action at various destinations with new kind of information called setting referred to web-based media information; catching the developments and voyaging examples of sightseers at a huge scope; analyzing the factors of buying departures from minimal expense transporter sites (Escobar-Rodriguez & Carvajal-Trujillo, 2014); tracking down the online buy goal of country the travel industry administrations (Martin & Herrero, 2012) and examining use conduct of lodging data framework (Kim et al., 2010). Moreover, the investigation of people's conduct with ICT appears as an expanding marvel in the traveling sector. Likewise, voyagers' utilization of the web for trip arranging as well as the way of embracing the web after some time (Xiang et al., 2015); dissecting the systems administration examples of various travel industry partners on the internet.

Multi-facet Benefits and Outcomes of ICT Application in the Tourism Industry

The full-fledged inclusion of ICT in tourism industry brings myriad benefits through which both the service providers and receivers get benefitted that ultimately contributes to the national and sustainable development of a country. The overall and sustainable development of a country helps enhance its image in the global arena (Rahman et al., 2021a, b, c). Berne et al. (2012) showed that the rebuilding interaction of the travel industry area and of the dispersion channels brought about by Information and Communication Technology can change the force locations in the channels. Moreover, Leung and Law (2012) inspected IT usages, the selection of electric information trade in hotel frameworks, as well as the executives help for Information Technology divisions that have huge impact in activity measure. Besides, Leung and Law (2012) considered present day advanced mechanics technology which can increase laborers' psychological limit as well as supplanting direct work. E-airline service provided by SST means Self-Service Technology (Lee, 2013) is likewise influencing the working cycle of this tourism sector.

According to Lee et al. (2013), the traditional relationship among hotels as well as travel planners as the colleagues of business have been altered by the advancement of internet. As per Vogt (2010), Customer Relationship Management (CRM) is generally utilized in the traveling sector industry, with unwaveringness plans keeping clients returning as well as traveling sites yielding an enormous size of e-exchanges. Besides, tourist kit that makes client unwaveringness by using smart innovation to collect precise traveler information (Angeloni, 2016). Information and Communication Technology can help for building up congenial networks which form a physical as well as virtual atmosphere for the conveyance of administrations to create customer requests (Breukel & Go, 2009); smart traveling is viewed as an individual traveling emotionally supportive network inside the setting of data services (Li et al., 2017). Notwithstanding, CRM framework, when tended to by utilizing arising Big Data methods for effective customer profiling, is as yet in its beginning phases (Talón-Ballestero et al., 2018).

Furthermore, a great technological change is noticed in the diversity of services in tourism sector (Araña et al., 2016). E-ability of guest data centers (D'Ambra & Mistilis, 2010; Lyu & Hwang, 2015); technology for water use in destination (Sun & Hsu, 2019); utilization of enlisted explorers' biometric frameworks in air traveling as an ingenuity as well as safety (Morosan, 2011); e-lineation as a particular type of estrangement in Information and Communication Technology empowered the travel industry (Tribe & Mkono, 2017); and geo-based innovation set up encounters (Tussyadiah & Zach, 2012) have brought a revolutionary development in modern tourism industry. Apart from these, convenience of webcast visits as an explanatory media (Kang & Gretzel, 2012); travelers' versatility utilizing Global Positioning System (Afaneh, 2021); aerial computerized symbolism in mix with LiDAR means Light Detection and Ranging information as well as GIS means Geographic Information System planning and investigation (Yang et al., 2012) are the latest received creative process in the travel industry that are being applied for the improvisation of tourism sector.

Many other scholars have studied the application of ICT in tourism from different perspectives. Paris (2012) gives some fascinating bits of knowledge in current hiker culture as well as the proceeding with intermingling of actual traveling with ICT. Divisekera and Nguyen (2018) considered the development and contributions to the travel industry and discovered one significant info is IT. Aldebert et al. (2011) featured the force of advancement action in the travel industry and gave proof about the fast development of the travel industry just as the generous effect of ICT on this advancement. In one investigation, Liu and Nijkamp (2019) presented that although the inbound travel industry might be another and amazing main impetus for local development, while the impact of inbound travel industry on mechanical advancement has all the earmarks of being more fragile than that on friendly advancement.

Besides, administrative and social process stress on the legitimate, the social and ecological piece of the business. ICT work with the consistence to social and administrative process in most of the cases. Fuchs et al. (2010) pointed out that most of the present ICT sway considers dismiss infrastructural, hierarchical and natural factors ordinarily answerable for fruitful e-business reception. As per Chiabai et al. (2011), we cannot ignore the use of e-investment for reasonable social and travel industry board. Wu and Chen (2016) conducted a research in this case and showed that satellite pictures or far-off detecting information empower the exploration of eco-natural alterations. Sharing practices via online media show up as important explanations of friendliness and enthusiastic help (Munar & Jacobsen, 2014).

All these benefits and outcomes brought by the inclusive application of ICT in tourism industry contribute to enhancing a nation's image from various perspectives like in value making as well as competitive merits. The creation of value is related to exercises through which an institution can build the worth of goods or services to buyers. Reasonable merits empower the firm to make prevalent incentive for its clients which of clients receive from its clients. Value is made with bringing down price like Díaz-Chao et al. (2016) broke down fresh co-imaginative wellsprings of firm work usefulness or by embracing separation as Yepes (2014) contemplated the eatery tablet lists with light, shading as well as another realistic highlights that indicated a huge relationship to its apparent impact on the decision of food as well as furthermore for appeal positioning. A fitting dissemination of the ICTs in this area can further develop the social and monetary effects show what these variables mean for the limit of business sectors and objections to meet and associate adequately in a worldwide the travel industry atmosphere (Minghetti & Buhalis, 2009). Cabiddu et al. (2013) recommended that administrators that accomplish unrivaled execution as far as appropriating esteem do as such in view of prevalent vital fit with the targets of the worth co-creation drive, cooperative energy with different individuals from the organization, and IT availability to lead business electronically.

Then in terms of competitive merit, Lin (2016) expressed present ASP means Application Service Provider which is an arising sort of data innovation re-appropriating administration that offer an exhaustive of data innovation services for SMEs means Small and Medium-sized Enterprises to acquire data abilities as well as keep up with upper hand. Essentially, aircraft organizations are increasingly using e–commerce for advertising their products as well as to improve the loyalty of their customers (Kim et al., 2009); the development in computerized channel appropriation has permitted air terminals to acquire upper hand through inventive contributions to travelers (Straker & Wrigley, 2018). Number of guests is another motivation to build the productivity and upper hand of the objective. In this way, Taiwanese National Park Headquarters utilize progressed data advances to draw in explorers, scientists, and different guests (Tsai et al., 2010). The talk in the advanced age is coordinated to upper hand by which the travel industry can maintain and draw in expanding traveler streams.

Challenges Faced by the Tourism Industry

Though Tourism sector is expanding greatly with the touch of globalization, it faces some challenges due to economic crises caused by various uncertainty like the outbreak of COVID-19 pandemic and some influenza viruses. Besides, this sector is hindered due to ignoring the use of ICT in a full swing. Regard to this, the challenges are drawn below.

Firstly, the challenge faced by tourism is DMS means Destination Management System. It is a computer related ability that is a combination of PCs, mobiles, pages, organizing apparatuses, on the web and disconnected booking methodology, customer demands and administrations that are anticipated from the Destination Management Organization (DMO) to make the area positive for the traveler. DMS keeps up with systems administration between the tourists, the neighborhood public and native assets which impact the upper hand of the location (Collins et al., 2003). The framework capacities by promoting and showcasing the areas by providing data, pictures, recordings on exercises and offices the objective gives. So, the government as well as the policy makers should think of the use of DMS to change the image of their country.

Secondly, the tourists hesitate to take decision in traveling. They do not have any idea regarding Decision Support System (DSS). It is a robotized emotionally supportive network intended to help strategy implementers use data, information, data set, innovation, systems administration and information constructions to perceive and resolve issues and take improved choices through communication (Power, 2008; Rahman & Hassan, 2021a). An ordinary illustration of the utilization of it is TIS means Tourism Information Systems. It contains information stockrooms that handle the travel industry area related information and helps destination chiefs in receiving valuable as well as compelling choices. The more mainstream TIS frameworks in activity are Illinois Tourism Network, Canadian Tourism Exchange, Australia, TourMIS, Austria and DECIPHER. The travel industry data framework helps in keeping up with and refreshing the information base of the customers, clients, items and administrations through market overviews, criticisms and exposure/promoting drives in a joint effort with neighborhood governments and nodal the travel industry associations.

Thirdly, the tourist has a little idea regarding GIS means Geographical Information System. It is actually a modernized framework for social occasion, saving, recovering, planning and surveying locational information from the genuine area. The public uses the framework to get the subtleties of a specific area (Duran et al., 2004). Normally the questions are with respect to what should be done, the spots for visiting, the offices which are accessible, the spots for remaining as well as eating, the climate situations and so forth. It is accomplished with the help of Web-based GIS or Information Kiosks situated in well-known tourists' regions. The other client of the GIS is the authoritative or the administration side. They incorporate the public authority or managerial workplaces, nearby or local associations, bunch administrators or individual administrators. They normally ask the framework to consider the client profile, spot of beginning, their financial foundations, their requests and inclinations.

Fourthly, the tourists do not get the idea of the visiting countries or environment. Actually, they are failed to understand Environment Management Information System (EMIS). This framework is computerized which is utilized by partners to comprehend the regular, natural and climate components of a site. EMIS have a huge scope of uses; a portion of the eminent ones are squander global positioning frameworks, discharge observing frameworks, task plan and booking frameworks, synchronizing authorizations and accreditation, guiding expense/advantage investigation and choosing elective materials. Environmental change is perhaps the most troublesome difficulties that humanity needs to look in the upcoming years (Afaneh, 2021). Since the transformation is unalterable, it converted basic to foster sound transformation strategies to deal with future changes in environment framework. Intergovernmental Panel for Climate Change (IPCC) showed zero tolerance to environment pollution and suggested ways to adjust the environmental alteration by legitimate administration of environment related information. Thus, ICT plays a significant part to this motivation. Climate and environment impact the fascination of a place of interest and the exercises that are coordinated there. Most travelers will in general check the climate conditions prior to confirming a visit – the odds of precipitation/snowfall, the greatest hotness, the base temperature and so forth the specialist co-ops utilize the gauges to help the business acquire money related advantages through better arranging, timing, the board and taking care of dangers and dangers related with terrible climate.

Fifthly, tourism sector faces a problem of Location Based Services (LBS). LBS deals with the frameworks that give the utilization the recent area of the traveler given by his cell phone and so give him the pertinent data of the area dependent on his necessities. This area is normally the geographical area. The normal LBS frameworks incorporate vehicle following, traffic as well as stream the board, leaving, calamity the executives and crisis administrations, route, peace and lawfulness organization, site-explicit charging and versatile commerce (Afaneh, 2021). Another benefit of LBS is that it gives constant and precise data. Accompanied by the progression of WWW means World Wide Web, correspondence information as well as programming usages, sightseers utilize virtual guides from Wikimedia, Google Earth and so forth.

Sixthly, tourism sector faces a hindrance of Intelligent Transport Systems (ITS). Its aim is to get, break down, evaluate and coordinate sensor information utilizing network innovation to accomplish traffic productivity. The normal ITS innovations concerning the travel industry are course route plans, mechanized auto situating framework, armada organization plans and vehicle clog control frameworks (Sheldon, 1997). ITS assists the driver with exploring, track down the most streamlined course by keeping away from traffic as well as utilizing a few time, the briefest remoteness for arriving at the destination; by trains, transports as well as planes it tends to utilize for taking care of and advancing armada the executives, mechanized tagging and tickets and momentum traffic data; and for people on foot, it is utilized for co-organizing traffic light, cautioning and mindfulness while going across streets.

Seventhly, most of the host countries do not have the knowledge of Global Positioning System (GPS) that is a worldwide space-based radio route framework (Afaneh, 2021). It makes areas as well as courses for drivers, accordingly, assist in decreasing vehicle mileage as well as fuel utilization by making substitute streamlined courses and relates genuine versus proposed courses.

Eighthly, tourism sector faces a hindrance in Computer Simulation. Simulation addresses the thought – learning by doing practically with the assistance of a PC framework by first structure a model and afterward working the model. The reproduced rendition is proposed to depict how it ought to act in real situation as well as is helpful in checking the working of different elements as well as its interchanges among the diverse partners that might be hard for exact perception. The strategies for displaying- dynamic, discrete or stochastic. PC reproductions end up being a useful administration help that grants customized experiments that permit accepting the working of the business and consequently give an accommodating instrument to the board choice help (Buchta & Dolničar, 2003). In case of travel industry, many reproductions either stresses on (i) macroeconomic highlights, predicting worldwide the travel industry developments based on gathered indicators or (ii) showing a microsystem focused on functional references to work on the assistance or the objective.

Ninthly, most of the developing countries cannot uphold their Virtual Tourism properly to the travelers. It gives a modernized travel feel of the genuine experience of visiting areas using innovation without the difficulty of venturing to every part of the distance or investing energy and cash. However, the interest for the travel industry will consistently remain and will be advanced to more online. With AR (Augmented Reality), travelers require interface with the web to touch the way of life, history, custom, celebrations and so on of a location. AR is compelling for objections which have (i) overcome their conveying limit (ii) are vulnerable to be harmed by the travel industry (iii) in lieu of prohibited exercises such as chasing (iv) to instigate sightseers mindful of the issues of an area prior to arranged visit (v) for unsafe experience games for instance stone climbing, remote ocean plunging, high height skiing, sky jumping and so on. Virtually the travel industry will decrease the contamination levels at these areas and furthermore the contamination caused to arrive at the spot by flight or vehicle. Augmented Reality Web Design is considered as an award winning 360-degree all-encompassing online visits site page. The upsides of 360 virtual visits give a viable method to pitch the office, area, and items straightforwardly to likely clients (Goecke, 2020; Afaneh, 2021).

Tenthly, Community Informatics is the obstacle in tourism sector. The main aim of it is to foster society support in choices that impact the local area. It includes a bunch of philosophy that are fixated on planning and utilizing ICT for improving local area improvement through individual, cultural, instructive, monetary and financial turn of events (Beaton, 2004). There are a few fields where ICT works on the worth of local area life. These are the progression of majority rule impact, development of social as well as manageable economy, strengthening of gatherings like minorities as well as fortifying of the privileges of the person.

Suggested Policies to Improve the Tourism Industry

The Enterprises' sites ought to be redesigned and refreshed with the furthest down the updated technologies to build the business probability. The governments and the endeavors ought to energize and prepare the business as far as using the web-based media. The administrations should see that the high velocity Internet accessibility and moderateness to arrive at every person and costs of the PC frameworks ought to go down further. The mindfulness for the stakeholders towards Google search ought to be given. Endeavors ought to coordinate all divisions through ICT. Fast web offices ought to be given by the public authority and force disappointments ought to be stayed away from. The public authority ought to support the improvement of versatile business (e.g., advancement, installments and so forth,) and ought to be coordinated inside the travel industry. Moreover, the authority should train up the concerned people like drivers, hotel management etc. for the better service of the tourists. If the authority can ensure the mentioned uses of ICT in tourism industry, it will only motivate the foreign tourists but also the local visitors. In the meantime, diversification is seen in tourism. Moreover, the concerned authority should focus on event and tourism education that will attract the young generation to invest their time and energy in promoting this sector. An educated generation will easily come to understand how to adopt technology properly in this sector along with realizing its possible threats that will make them gauge the measures to surmount those threats and challenges (Afaneh, 2021). So, the host countries should not only concentrate on traditional tourism but also follow their competitors. Similarly, changing the image of a nation, every wing of possibility in tourism should be considered. If the concerned authority can ensure the mentioned steps dealing with the application of ICT in tourism industry, a country is thought to experience economic development (Rahman et al., 2021b; Rahman & Hassan, 2021b) that will eventually augment its overall image to the global platform. So, it is high time to take the stated steps into consideration to make modern tourism into a booming industry so that it can augment the image of a nation in the global arena besides contributing to the sustainable development of a country.

Conclusion

Since the advancement of technology dreams the world as a global village with no boundaries, the tourism industry is going to be a booming sector day by day that plays a very vivid role in contributing to the economic development of a country as well as enhancing a nation's image to the global arena by introducing the country's prolonged history, culture, heritage, rituals, religious beliefs and so forth across the globe through travelers. Hence, it is imperative to contemplate on the developmental phase of this sector very minutely. Regard to this, mass technological application in all the aspects of tourism industry has been thought to be a way forward. Therefore, this chapter has illustrated the necessity of technology application in tourism event, education and training, and its benefits and contribution to enhance a nation's image, challenges of technology adaptation in tourism sector along with, most importantly, policy recommendations to surmount those constraints so that this sector might be a vital player to promote wellbeing of its population and its overall image to the global stage. The proper management with the inclusion of advanced technology can make tourism a booming sector that will subsequently help the country achieve the sustainable development goals.

References

Afaneh, S. M. (2021). E-tourism, its prospects "scopes" and impact on the development of tourism industry (the Jordanian experience). *PalArch's Journal of Archaeology of Egypt/Egyptology, 18*(4), 1897–1911.

Aldebert, B., Dang, R. J., & Longhi, C. (2011). Innovation in the tourism industry: The case of Tourism@. *Tourism Management, 32*(5), 1204–1213.

Angeloni, S. (2016). A tourist kit 'made in Italy': An 'intelligent' system for implementing new generation destination cards. *Tourism Management, 52*, 187–209.

Araña, J. E., León, C. J., Carballo, M. M., & Moreno Gil, S. (2016). Designing tourist information offices: The role of the human factor. *Journal of Travel Research, 55*(6), 764–773.

Beaton, B. (2004). The K-Net story: Community ICT development work. *The Journal of Community Informatics, 1*(1), 1–15.

Benckendorff, P. (2009). Evaluating wikis as an assessment tool for developing collaboration and knowledge management skills. *Journal of Hospitality and Tourism Management, 16*(1), 102–112.

Berne, C., Garcia-Gonzalez, M., & Mugica, J. (2012). How ICT shifts the power balance of tourism distribution channels. *Tourism Management, 33*(1), 205–214.

Breukel, A., & Go, F. M. (2009). Knowledge-based network participation in destination and event marketing: A hospitality scenario analysis perspective. *Tourism Management, 30*(2), 184–193.

Buchta, C., & Dolničar, S. (2003). Learning by simulation-computer simulations for strategic marketing decision support in tourism. *International Journal of Tourism Sciences, 3*(1), 65–78.

Buick, I. (2003). Information technology in small Scottish hotels: Is it working? *International Journal of Contemporary Hospitality Management, 15*(4), 243–247.

Caber, M., Albayrak, T., & Loiacono, E. T. (2013). The classification of extranet attributes in terms of their asymmetric influences on overall user satisfaction: An introduction to asymmetric impact-performance analysis. *Journal of Travel Research, 52*(1), 106–116.

Cabiddu, F., Lui, T. W., & Piccoli, G. (2013). Managing value co-creation in the tourism industry. *Annals of Tourism Research, 42*, 86–107.

Cheng, S., & Cho, V. (2011). An integrated model of employees' behavioral intention toward innovative information and communication technologies in travel agencies. *Journal of Hospitality and Tourism Research, 35*(4), 488–510.

Chiabai, A., Paskaleva, K., & Lombardi, P. (2011). e-Participation model for sustainable cultural tourism management: A bottom-up approach. *International Journal of Tourism Research, 15*(1), 35–51.

Choi, Y., Hickerson, B., & Kerstetter, D. (2018). Understanding the sources of online travel information. *Journal of Travel Research, 57*(1), 116–128.

Collins, C., Buhalis, D., & Peters, M. (2003). Enhancing SMTEs' business performance through the Internet and e-learning platforms. *Education and Training, 45*(8/9), 483–494.

D'Ambra, J., & Mistilis, N. (2010). Assessing the E-capability of visitor information centers. *Journal of Travel Research, 49*(2), 206–215.

Díaz-Chao, Á., Miralbell-Izard, O., & Torrent-Sellens, J. (2016). Information and communication technologies, innovation, and firm productivity in small and medium-sized travel agencies: New evidence from Spain. *Journal of Travel Research, 55*(7), 862–873.

Divisekera, S., & Nguyen, V. K. (2018). Determinants of innovation in tourism evidence from Australia. *Tourism Management, 67*, 157–167.

Duan, W., Yu, Y., Cao, Q., & Levy, S. (2016). Exploring the impact of social media on hotel service performance: A sentimental analysis approach. *Cornell Hospitality Quarterly, 57*(3), 282–296.

Duran, E., Seker, D. Z., & Shrestha, M. (2004). Web based information system for tourism resort: A case study for side/manavgat. *Proceedings for XXth International Society for Photogrammetry and Remote Sensing*. Istanbul, Turkey, July, pp. 12–23.

Escobar-Rodriguez, T., & Carvajal-Trujillo, E. (2014). Online purchasing ticket for lowcost carriers: An application of unified theory of acceptance and use of technology (UTAUT) model. *Tourism Management, 43*, 70–88.

Fuchs, M., Höpken, W., Föger, A., & Kunz, M. (2010). E-business readiness, intensity, and impact: An Austrian destination management organization study. *Journal of Travel Research, 49*(2), 165–178.

Goecke, R. (2020). Advanced web technologies and E-tourism web applications. In Z. Xiang, M. Fuchs, U. Gretzel, & W. Höpken (Eds.), *Handbook of e-tourism* (pp. 1–32). Springer.

González-Serrano, L., & Talón-Ballestero, P. (2020). Revenue management and E-tourism: The past, present and future. In Z. Xiang, M. Fuchs, U. Gretzel, & W. Höpken (Eds.), *Handbook of e-tourism* (pp. 1–28). Springer.

Han, H. S., & Verma, R. (2014). Why attend tradeshows? A comparison of exhibitor and attendee's preferences. *Cornell Hospitality Quarterly, 55*(3), 239–251.

He, Z., Wu, L., & Li, X. R. (2018). When art meets tech: The role of augmented reality in enhancing museum experiences and purchase intentions. *Tourism Management, 68*, 127–139.

Januszewska, M., Jaremen, D. E., & Nawrocka, E. (2015). The effects of the use of ICT by tourism enterprises. *Service Management, 2*(16), 65–73.

Jung, T., Chung, N., & Leue, M. C. (2015). The determinants of recommendation to use augmented reality technologies: The case of Korean theme park. *Tourism Management, 49*, 75–86.

Kang, M., & Gretzel, U. (2012). Effects of podcast tour on tourist experience in a national park. *Tourism Management, 33*(2), 440–455.

Kang, J.-W., & Namkung, Y. (2015). Restaurant information sharing on social networking sites: Do network externalities matter? *Journal of Hospitality and Tourism Research, 40*(6), 739–763.

Khatri, I. S. H. W. A. R. (2018). Innovation research in tourism business: A review from two decades of studies. *Journal of Tourism, 19*(1), 1–13.

Kim, H. B., Kim, T. T., & Shin, S. W. (2009). Modeling roles of subjective norms and eTrust in customers' acceptance of airline B2C eCommerce websites. *Tourism Management, 30*(2), 266–277.

Kim, T., Suh, Y. K., Lee, G., & Choi, B. G. (2010). Modelling roles of task-technology fit and self-efficacy in hotel employees' usage behaviours of hotel information systems. *International Journal of Tourism Research, 12*(6), 709–725.

Kim, J., Kim, P. B., Kim, J. E., & Magnini, V. P. (2016). Application of construal-level theory to promotional strategies in the hotel industry. *Journal of Travel Research, 55*(3), 340–352.

Lee, L. Y. (2013). Hospitality industry web- based self-service technology adoption model: A cross- cultural perspective. *Journal of Hospitality and Tourism Research, 40*(2), 162–197.

Lee, H. A., Denizci Guillet, B., & Law, R. (2013). An examination of the relationship between online travel agents and hotels: A case study of Choice Hotels International and Expedia. com. *Cornell Hospitality Quarterly, 54*(1), 95–107.

Leung, R., & Law, R. (2012). Evaluation of hotel information technologies and EDI adoption: The perspective of hotel IT managers in Hong Kong. *Cornell Hospitality Quarterly, 54*(1), 25–37.

Li, Y., Hu, C., Huang, C., & Duan, L. (2017). The concept of smart tourism in the context of tourism information services. *Tourism Management, 58*, 293–300.

Lin, S. (2016). The critical success factors for a travel application service provider evaluation and selection by travel intermediaries. *Tourism Management, 56*, 126–141.

Liu, S. Q., & Mattila, A. S. (2016). Using comparative advertising to promote technology- based hospitality services. *Cornell Hospitality Quarterly, 57*(2), 162–171.

Liu, J., & Nijkamp, P. (2019). Inbound tourism as a driving force for regional innovation: A spatial impact study on China. *Journal of Travel Research, 58*(4), 594–607.

Lo, I. S., McKercher, B., Lo, A., Cheung, C., & Law, R. (2011). Tourism and online photography. *Tourism Management, 32*(4), 725–731.

Lyu, S. O., & Hwang, J. (2015). Are the days of tourist information centers gone? Effect of the ubiquitous information environment. *Tourism Management, 48*, 54–63.

Martin, H. S., & Herrero, A. (2012). Influence of the user's psychological factors on the online purchase intention in rural tourism: Integrating innovativeness to the UTAUT framework. *Tourism Management, 33*(2), 341–350.

Minghetti, V., & Buhalis, D. (2009). Digital divide in tourism. *Journal of Travel Research, 49*(3), 267–281.

Morosan, C. (2011). Voluntary step toward air travel security: An examination of travelers' attitudes and intensions to use biometric systems. *Journal of Travel Research, 51*(4), 436–450.

Munar, A. M., & Jacobsen, J. K. S. (2014). Motivations for sharing tourism experience through social media. *Tourism Management, 43*, 46–54.

Murphy, H. C., Chen, M. M., & Cossutta, M. (2016). An investigation of multiple devices and information sources used in the hotel booking process. *Tourism Management, 52*, 44–51.

No, E., & Kim, J. K. (2013). Determinants of the adoption for travel information on smartphone. *International Journal of Tourism Research, 16*(6), 534–545.

Okazaki, S., Campo, S., Andreu, L., & Romero, J. (2015). A latent class analysis of Spanish travelers' mobile internet usage in travel planning and execution. *Cornell Hospitality Quarterly, 56*(2), 191–201.

Paris, C. M. (2012). Flashpackers: An emerging sub-culture? *Annals of Tourism Research, 39*(2), 1094–1115.

Power, D. J. (2008). Understanding data-driven decision support systems. *Information Systems Management, 25*(2), 149–154.

Rahman, M. K. (2019). Medical tourism: Tourists' perceived services and satisfaction lessons from Malaysian hospitals. *Tourism Review, 74*(3), 739–758.

Rahman, M. K., & Hassan, A. (2021a). Tourists experience and technology application in Bangladesh. In A. Hassan (Ed.), *Technology application in the tourism and hospitality industry of Bangladesh*. Springer.

Rahman, M. K., & Hassan, A. (2021b). Tourism as an element for economic growth in Bangladesh: Investment analysis for product and service development. In A. Hassan (Ed.), *Tourism in Bangladesh: Investment and development perspectives*. Springer.

Rahman, M. K., Zailani, S., & Musa, G. (2017a). What travel motivational factors influence Muslim tourists towards MMITD? *Journal of Islamic Marketing, 8*(1), 48–73.

Rahman, M. K., Zailani, S., & Musa, G. (2017b). Tapping into the emerging Muslim-friendly medical tourism market: Evidence from Malaysia. *Journal of Islamic Marketing, 8*(4), 514–532.

Rahman, M., Moghavvemi, S., Thirumoorthi, T., & Rahman, M. K. (2020). The impact of tourists' perceptions on halal tourism destination: A structural model analysis. *Tourism Review, 75*(3), 575–594.

Rahman, M. K., Masud, M. M., Akhtar, R., & Hossain, M. M. (2021a). Impact of community participation on sustainable development of marine protected areas: Assessment of ecotourism development. *International Journal of Tourism Research*. https://doi.org/10.1002/jtr.2480

Rahman, M. K., Rana, M. S., & Hassan, A. (2021b). Development and investment in core niche tourism products and services in Bangladesh. In A. Hassan (Ed.), *Tourism in Bangladesh: Investment and development perspectives*. Springer.

Rahman, M. K., Sarker, M., & Hassan, A. (2021c). Medical tourism: The Islamic perspective. In A. Hassan (Ed.), *Tourism products and services in Bangladesh: Concept analysis and development suggestions* (pp. 87–99). Springer.

Rana, M. S., Rahman, M. K., Islam, M. K., & Hassan, A. (2020). Globalization effects on tourism marketing in Bangladesh. In A. Hassan (Ed.), *Tourism marketing in Bangladesh: An introduction* (pp. 157–171). Routledge.

Sanders, D., & LeClus, M. (2009). All weird up but not plugged in: An evaluation of tourism and leisure marketing students' expectations and experiences of lecture podcasting. *Journal of Hospitality and Tourism Management, 16*(1), 94–101.

Sheldon, P. J. (1997). *Tourism information technology*. CAB International.

Straker, K., & Wrigley, C. (2018). Engaging passengers across digital channels: An international study of 100 airports. *Journal of Hospitality and Tourism Management, 34*, 82–92.

Sun, Y. Y., & Hsu, C. M. (2019). The decomposition analysis of tourism water footprint in Taiwan: Revealing decision-relevant information. *Journal of Travel Research, 58*(4), 695–708.

Talón-Ballestero, P., González-Serrano, L., Soguero-Ruiz, C., Munoz-Romero, S., & Rojo-Álvarez, J. L. (2018). Using big data from customer relationship management information systems to determine the client profile in the hotel sector. *Tourism Management, 68*, 187–197.

Tribe, J., & Mkono, M. (2017). Not such smart tourism? The concept of e-lienation. *Annals of Tourism Research, 66*, 105–115.

Tsai, W. H., Chou, W. C., & Lai, C. W. (2010). An effective evaluation model and improvement analysis for national park websites: A case study of Taiwan. *Tourism Management, 31*(6), 936–952.

Tussyadiah, I. P., & Zach, F. J. (2012). The role of geo-based technology in place experience. *Annals of Tourism Research, 39*(2), 780–800.

Vogt, C. A. (2010). Customer relationship management in tourism: Management needs and research applications. *Journal of Travel Research, 50*(4), 356–364.
Vu, H. Q., Li, G., & Law, R. (2020). Cross-country analysis of tourist activities based on venue-referenced social media data. *Journal of Travel Research, 59*(1), 90–106.
Wu, S., & Chen, Y. (2016). Examining eco-environmental changes at major recreational sites in Kenting National Park in Taiwan by integrating SPOT satellite image and NDVI. *Tourism Management, 57*, 23–36.
Wu, E. H., Law, R., & Jiang, B. (2013). Predicting browsers and purchasers of hotel websites: A weight-of-evidence grouping approach. *Cornell Hospitality Quarterly, 54*(1), 38–48.
Xiang, Z., Wang, D., O'Leary, J. T., & Fesenmaier, D. R. (2015). Adapting to the internet: Trends in travelers' use of the web for trip planning. *Journal of Travel Research, 54*(4), 511–527.
Yang, B., Madden, M., Kim, J., & Jordan, T. R. (2012). Geospatial analysis of barrier island beach availability to tourists. *Tourism Management, 33*(4), 840–854.
Yepes, M. F. (2014). Mobile tablet menus: Attractiveness and impact of nutrition labelling formats on millennials' food choices. *Cornell Hospitality Quarterly, 56*(1), 58–67.
Ying, T., Norman, W. C., & Zhou, Y. (2016). Online networking in the tourism industry: A webometrics and hyperlink network analysis. *Journal of Travel Research, 55*(1), 16–33.

Md Rabiul Alam is a PhD candidate at the Faculty of Education, University of Malaya, Malaysia. Prior to continue his higher study at the University of Malaya, he had been serving as a faculty in the Department of English, Asian University of Bangladesh for nearly eight years. Besides his study at the University of Malaya, he has been working as a Graduate Research Assistant (GRA) since 2018. His research interest includes TESOL, teacher education, financial education, tourism development, digital financial inclusion and so forth. Alam has published a few of his works in reputed peer-reviewed journals.

Md Rashed Khan Milon has been serving as an Assistant Professor and the Chairman at the Department of English, Port City International University, Bangladesh. He mostly teaches Applied Linguistics courses, English literature and also supervises thesis students both at the undergraduate and graduate level of study. Besides, Milon is a PhD candidate at the Department of English, Jagannath University, Bangladesh. He has published a few of his papers in reputed peer-reviewed journals. Milon's research areas are Teacher Education, Second Language Acquisition, ELT Methodology, Teaching Techniques, Gender issue and Diversity in Education.

Dr Muhammad Khalilur Rahman currently works in the Faculty of Entrepreneurship and Business at Universiti Malaysia Kelantan (UMK), Malaysia. He has been actively involved in research activities and published over 55 articles and book chapters in leading service management outlets. He has a wide interest in tourism and service management research which includes medical tourism, eco-tourism, halal tourism, service quality, brand equity, supply chain management, operation management, green and sustainable development. Rahman is a reviewer of Kybernetes, Journal of Hospitality and Tourism Insights, International Journal of Contemporary Hospitality Management, Journal of Islamic Marketing, and BMC Public Health.

Dr Azizul Hassan is a member of the Tourism Consultants Network of the UK Tourism Society. Dr Hassan has been working for the tourism industry as a consultant, academic, and researcher for over 20 years. His research interest areas are technology-supported marketing for tourism and hospitality, immersive technology applications in the tourism and hospitality industry, and technology-influenced marketing suggestions for sustainable tourism and hospitality industry in developing countries. Dr Hassan has authored over 150 articles and book chapters in leading tourism outlets. He is also part of the editorial team of 25 book projects from Routledge, Springer, CAB International, and Emerald Group Publishing Limited. He is a regular reviewer of a number of international journals.

Part VI
Cases

Chapter 10
Technological Adaptation in Tourism Events, Fairs, and Festivals: Way to a Revolutionary Transformation in Bangladesh

Kaniz Kakon

Introduction

In the reality of the present-day, our everyday lives are inescapably moving faster to the adaptation of technology, in every way, to persist in a speedy-phase dynamic world. One of the positive impact of this is having a stronger infrastructure in the tourism industry of an emerging country like Bangladesh. Bangladesh has amusing cultural uniqueness of over hundreds of years. And the celebration and carrying on this cultural exclusivity emerge in the forms of different fairs, festivals, and events. So in the tourism sector of Bangladesh, different kinds of cultural celebrations have enormously become a special interest of present-day tourists. Fairs, Festivals, and Events in a city are organized in different cities to attract tourists, having a promotion of the city along with the culture in a wider perspective, the purpose of entertainment to local inhabitants (Cudny et al., 2012). These celebrations allow getting people together in one specific place to inspire each other and continue traditions that have been around for many generations. And this increasing attention to these fair, festival and event-based tourism is significantly influenced by the adaptation of technological use in every mode of the tourism sector of Bangladesh. Not long ago, in Bangladesh, technological use was rarely that familiar to the tourists, and the tourism industry was not that much flourished with the technological usage in promoting, hosting, facilitating, and also designing the events. At present, tourism has become an extensive part of life and the use of technology even became an evident and influential part of its operation through e-tourism as well as smart tourism. As a result, the interest of tourists in attending a festival or fair, or any cultural event has reached new heights that have never before been conceived of in Bangladesh. This

K. Kakon (✉)
Department of Philosophy, College of Arts and Sciences (CAAS), IUBAT—International University of Business Agriculture and Technology, Dhaka, Bangladesh
e-mail: kanizkakon@iubat.edu

A. Hassan (ed.), *Technology Application in Tourism Fairs, Festivals and Events in Asia*, https://doi.org/10.1007/978-981-16-8070-0_10

study is devoted to analyzing the radical development of Bangladeshi tourism based on some case studies.

Events, Fairs, and Festivals in Bangladesh

The long history of Bangladeshi culture developed over the centuries and comprises the cultural multiplicity of several social groups of Bangladesh. Fairs and festivals are an indispensable part of Bangladeshi social life along with the cultural tourism in Bangladesh. The various event-based festivals both minor and major are held all through the year and being celebrated regularly to enthuse one another and carry on traditions that have been around for many generations. A conceptual definition of cultural tourism is given by Richards (1996), as persons go to cultural attractions outside of their usual home to obtain fresh knowledge and experiences to meet their cultural requirements.

Fair is as reported in 1911 Encyclopaedia Britannica as: "The most commonly accepted derivation of the word "fair" is from the Latin feria, a name which the church borrowed from Roman custom and applied to her own festivals" (Wikisource, n.d.).

Festivals are celebrated by organizing or centering to any event or fair. Following Getz (2008, 2010), tourism, where the aim is to go to any festivals, is considered as a part of event tourism and events play the role of an important motivator in the development of tourism.

Quinn (2010) narrated that the exceptional culture presented throughout the events are from time to time offered the main attraction for the tourists, and stated that, Arts festivals were of little concern to policymakers until the last few decades, except for those who were explicitly entrusted with culture. Even back then, arts festivals were regarded as a "poor relation" of cultural activities, typically seen as a first step toward the establishment of more sustained, year-round, 'productive' cultural activities in policy terms. This isn't the case anymore. Since the late 1980s, arts festivals have been on the rise.

In the history of Bangladeshi culture, the Bengal Renaissance of the nineteenth century brought radical changes including social, cultural, psychological, and intellectual revolution (Banglapedia, n.d.). The contribution of noted Bangladeshi scholars from different cultural sectors like literature, music, dance, art and craft, drama, film, folklore and folktale, philosophy, and even in religion, have played a significant role in the research of their development and also celebrating them in the mode of different kind of event, fairs, and festivals. And that tradition has added so many new events of celebration even in the modern time. Consequently, a wide variety of festivals are celebrated in Bangladesh all over the year. Among them a small number of prominent and extensively celebrated events, fairs, or festivals of Bangladesh can be generally categorized into the following types (Fig. 10.1).

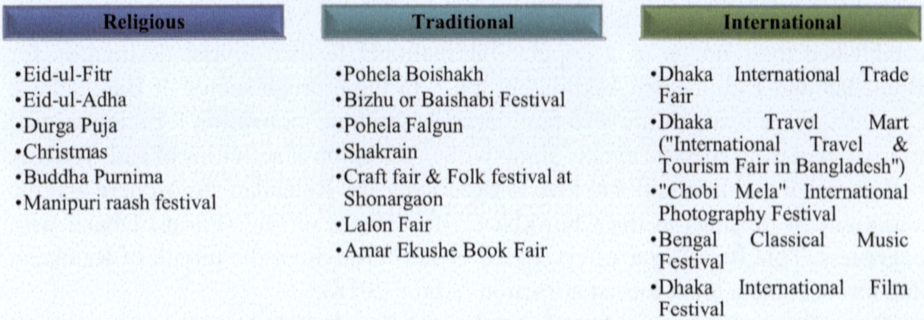

Fig. 10.1 Some of the major fairs, events, and festivals celebrated in Bangladesh. (Source: The author)

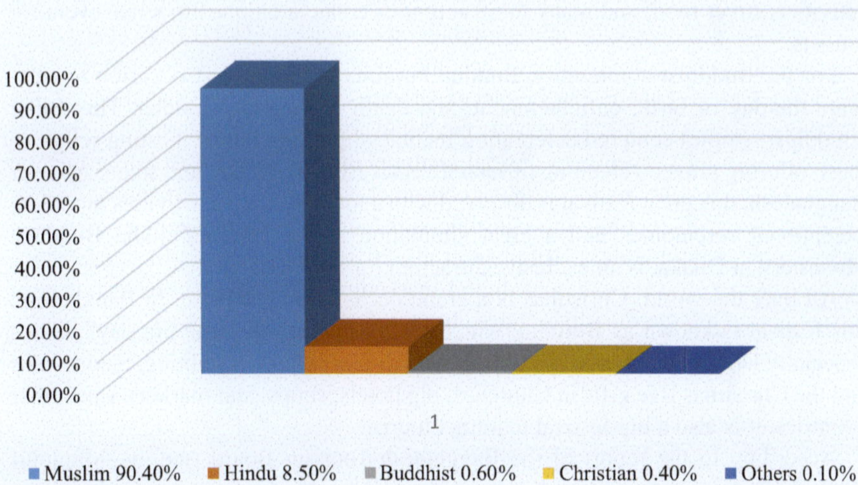

Fig. 10.2 Religion-based population in Bangladesh (2011). (Source: Bangladesh Bureau of Statistics, 2011)

Religious Events and Festivals

In the constitution of Bangladesh, a very significant feature is communal harmony and it has been present in Bangladeshi culture over the years. As per, Bangladesh Government official census result of the 2011 census (Fig. 10.2).

People from all these religious communities feel free to celebrate their religious festivals with social integrity and diligence in Bangladesh.

In the entire Muslim world including the Muslim majority country Bangladesh, two biggest and widely popular festival is Eid-ul-Fitr and Eid-ul-Adha. Eid-ul-Fitr followed by month-long fasting in Ramadan and Eid-ul-Adha by sacrificing domestic animals as Qurbani. Both of the Eid festivals are complemented with unlimited happiness and the festival of great social gathering. Many cities around the Muslim world like Dubai, Kuala Lumpur, Istanbul, Fes, Cairo, Mecca, Manila has already established their image as a popular international festival tourist destination for Ramadan and Eid-ul-Fitr (Askmjareb, n.d.). In the tourism sector of Bangladesh, currently, these festivals are also familiarized as unique motivation for international tourists centering specific events along with other cultural activities of Eid. Tourists both national and international love to experience the Ramadan atmosphere and the variations of cuisines at the Chawkbazar Iftar Bazar of Old (Puran) Dhaka with eagerness. This Iftar Bazar offers a vibe of fair throughout the month of Ramadan and has become a great tourist attraction (Alam, 2018).

Durga Puja is the biggest holy festival of the Bangladeshi Hindu community and being celebrated around 32,000 Mandaps (festival pavilions) all over the country every year. During the five-day Durga Puja festival, the thematic designs and stunning lighting of the Mandaps, demonstration of colors, music, rituals, art, cultural activities, street food, and many more activities create a unique attractive event for tourists.

For the Buddhist community, Buddha Purnima is the biggest occasion to celebrate the day of birth, enlightenment, and death of Gautam Buddha. This Bihar (Buddhist temple) centered celebration includes lightning lamps, hosting religious flags, offering prayers, chanting of sacred verses from Tripitaka, and so on. All over Bangladesh, this great festival is observed with a lot of attractive activities including exceptional ceremonies and a great simulation at the Dharmarajikha Buddhist Monastery in Dhaka, is an exclusive attraction for the tourists.

All over the world, Christmas is a grand occasion to celebrate. In Bangladesh, this festival is known as Borodin, which means in Bengali- "the big day". Cities especially Dhaka becomes very colorful with different kind of lighting, Santa Claus and the Christmas Eve gifts in Churches, big hotels, shops, and markets. Like other countries, it is also a big festival to attract tourists.

According to the report of the Bangladesh Tourism Board (n.d.-b), Manipuri raash festival is a very colorful indigenous religious festival among the Manipuri tribal community in Bangladesh. To rejoice in the divine love of Hindu Goddess Radha and Krishna this festival is celebrated in many places of Bangladesh. Among all Monipuri Royal Palace at Madhabpur and Adampur of Komalganj Upazilla in Moulvibazaar District of Bangladesh organizes it majestically since 1769. This celebration includes tribal dance, drama, music on Radha-Krishna, and a grand fair where a wide variety of products are being brought to sell like, Manipuri handicrafts, household items, toys, foods, and many more. Thousands of exciting visitors irrespective of caste and creed, including a good number of international tourists, assemble in the place to see the fair every year.

Traditional Events and Festivals

The vibrant celebration of the first day of the Bengali Calendar is called the Pohela Baishakh. Beyond the difference of caste, creed, religion it is the biggest cultural festival for all Bangladeshi people for several centuries. It is widely celebrated all over the country, but in Dhaka, it is celebrated grandly. Some of the exciting highlights of this celebration are the grand cultural show of Ramna Park in the very early morning, Mongal Shovajatra (Boishakhi parade), Boishakhi fair, and a variety of sports or events like; puppet show, bullfight, boat race and many more. It is a big attraction for tourists from both home and abroad.

Bizhu or Baisabi is the main social and cultural festival of the Chakma community living in the Chittagong hill tracts in Bangladesh. This festival is also celebrated by the Marma community as Shangrai. This festival is celebrated for three consecutive days from Bengali Calendar months including the last 2 days of Chaitra and the first day of Baisakh. Bizhu celebration centered on agricultural activities on the hills with the first rain to highlight the ethnic unique cultural heritage. This festival includes different cultural programs, traditional tribal dance and music, Jolkeli (customary activities based on water), and so forth. Warmhearted welcoming access for the people from every community to join this colorful celebration opens up the opportunity to share and cherish social bonds together among all participants. Many tourists from home and abroad gather at the festival area every year.

People of Bangladesh, particularly the youth, welcome the arrival of spring – the "Pohela Falgun" in a dazzling style. On this occasion especially the capital city Dhaka put on a festive look with a wide range of events including chorus songs, folk dance, folk music, the dance of indigenous people, recitation of poems, drama, musical concerts, and procession organized by several socio-cultural organizations and educational institutions. As red and yellow are the representative colors of spring, people prefer to wear dresses of these two primary colors in celebrating the day. Thousands of visitors or tourists flocked to celebrate the festival and enjoy various events (The Daily Star, 2011).

On the occasion of Poush Sangkranti (the end of Bengali Calendar month "Poush"), one of the ancient and a remarkable traditional annual Kite festival is celebrated massively in the oldest part of Dhaka every year. The main attraction of this festival is that people fly kites of different varieties, shapes, designs, and colors all day long, especially in the afternoon, from the rooftops around the entire area along with enjoying the delightful tastes of Shiter Pitha (cake in winter) with near and dear ones. A minimum of eight to ten kites being flown from each building, followed by other features like music, cultural program, fire breathing, laser show, fireworks at the evening time make this iconic event a great joyful attraction for tourists both from home and abroad. It is the festival of colors, excitements, smiles, and hospitality of the people of Old Dhaka (Afrin, 2019).

Every year, Bangladesh Folk Art and Craft Foundation arranges a month-long "Craft Fair & Folk Festival" at its premises in Sonargaon, Narayanganj District. In this exceptional fair around 200 used to set up in displaying and selling folk artworks and handicrafts, traditional foods, dresses, ornaments from almost every corner of the country to introduce the visitors to the core culture of the country. In this initiative, the authority also arranges cultural programs including folk and ethnic performances, every evening through this entire festival period providing a complete picture of Bangladeshi culture. Moreover, some other attractions of this festival are rural games, the exhibition on rural life, bioscope, boat driving, puppet show, cockfighting, and Pitha Utsab (traditional cake festival). A huge number of visitors from home and abroad come every day to have an exceptional experience of this delightful festival (Asha, 2017).

The Lalon Academy, in association with the Ministry of Cultural Affairs of Bangladesh, organizes every year a three-day-long memorial festival called the Lalon Mela on the occasion of the death anniversary of Baul Lalon Shah (a Bangladeshi mystic philosopher, songwriter, singer, and social reformer) at his shrine at Chheuria village in Kushtia district. Around four to five lakh people from home and abroad usually gather at this majestic event to have the experience of grand village fair, take part in the discussion of his songs and philosophy, enjoy the passionate performance of Lalon songs that satisfy the souls of the visitors (Bangladesh Tourism Board, n.d.-b).

The biggest annual national book fair of Bangladesh as well as one of the largest book fairs in South Asia, the Amar Ekushey Book Fair is organized by Bangla Academy in February, remembering the sacrifice of the martyrs who died on the 21st February 1952 in a protest shouted for the establishment of Bangla as the state languages. This month-long book fair, which takes place at Bangla Academy courtyard and the adjacent historic Suhrawardy Udyan is a hub of books from different genres and subjects. Around 400 publishing house participate in this fair with their new and old publications displaying in their stalls for sell and the fair always witness a huge crowd all through the month. Moreover, the host Bangla Academy also organizes literary and cultural events on each day of the fair. The uniqueness it holds, it is such an international book fair which held on the background of the tragic history of sacrifice for mother language and literature (Wikipedia, n.d.).

International Events and Festivals

The Dhaka International Trade Fair, the largest trade exhibition and fair in Bangladesh since 1995, is a jointly organized event by the Export Promotion Bureau and the Ministry of Commerce, Bangladesh. This month-long event is held in Dhaka annually to present the exportable quality of Bangladeshi products and also the foreign participants set their stalls to display their state-of-the-art, products, technology to the native producers and importers. Every day thousands of visitors both foreign and local find it very joyful to find their preferred products among hundreds of pavilions. Consequently, this enthusiasm creates a festive environment in the city (The Korea Post, 2021).

The Dhaka Travel Mart is the first and one of the biggest travel fairs in Bangladesh, started its journey in the year 2002 and is being held annually regularly. The Bangladesh Monitor, a travel publication, organizes this excellent event where major travel, tourism, and hospitality industry in Bangladesh and abroad exhibit their products and services along with different kinds of promotional offers. Many countries like India, Nepal, Bhutan, Sri Lanka, Thailand, Singapore, Malaysia, Cambodia, Philippines, Myanmar, Malaysia are involved in the event for increasing the quality of the event. The thousands of visitors find this event special and helpful as they like to know the latest trends in the tourism sector (The Financial Express, 2019).

Chobi Mela, Asia's largest international photography festival is a biennial event organized by Drik Picture Library Ltd. and Pathshala South Asian Media Institute. This exciting festival of world photography, which usually lasts for 2 weeks, hosts several exhibitions, online talks, podcasts, radio projects, and also workshops by photo practitioners from all around the world (Chobi Mela, n.d.).

The World's largest classical music festival in terms of the number of performers on a single stage, audience capacity, and span, is being hosted almost every year by Bengal Foundation in Dhaka with the title "Bengal Classical Music Festival". Since 2012, along with Bangladeshi classical music icons the performances of many influential legendary musicians from the Indian subcontinent and other parts of the world provides an irreplaceable experience of heart-wrenching classical music to thousands of national and international audiences at this five-day long event (Bengal Classical Music Festival, n.d.).

Since 1992 the biggest film festival of Bangladesh and one of the most attended festivals in Dhaka, Dhaka International Film Festival (DIFF) is being organized by Rainbow Film Society. This is one of the most prestigious film events in South Asia is devoted to promoting a healthy cine culture in Bangladesh. By this event, Bangladeshi artists, filmmakers, producers, and directors have an opportunity to interact and create networks with the international film community. Nowadays, it has become a must-see event for tourists who want to discover the film culture of Bangladesh as well as wish to enjoy the globally acclaimed films (Dhaka International Film Festival, n.d.).

From the description above the basic features of some of the notable events, fairs, and festivals of Bangladesh might be felt like that, generally, those are organized primarily focusing the native visitors but gradually they have attracted the foreign tourists to a noticeable extent significant tourist appealing power. But, for their unique nature, all those have both national and international significance from the tourist point of view in their way and the tourist reputation has further enhanced progressively. According to the statistical data provided by the World Bank (n.d.), since 2013, the number of international tourist arrivals in Bangladesh has progressively increased in the following manner (Fig. 10.3).

Impact of Technology in Boosting the Event, Fair, and Festival Tourism in Bangladesh

In terms of product, process, and also information and communication technology, fairs- festivals- events of Bangladesh are changing their activities in more attractive ways day by day. In previous years the arrangements were very basic. But in present time, use of colorful led lights, laser show, music arrangement with better quality sound arrangements, central air conditioning for indoor venue-based events, mobile

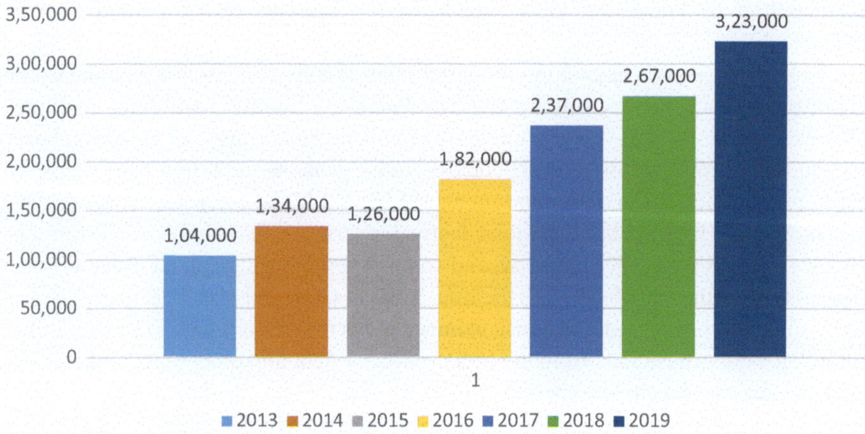

Fig. 10.3 Number of international tourist arrivals in Bangladesh (2013–2019, in millions). (Source: The World Bank, n.d.)

charging facility, digital map of the event places, digital video display, Global Positioning System (GPS) tracker for the vehicle at the event places and much more product technology has been introduced. Process technology has also been implemented in the management of fair, festival, and event spots, marketing, and branding them. And the most revolutionary effect is of Information and communications technology (ICT) in boosting up the popularity of these three kinds of tourism in Bangladesh. Even the introduction of digitalized event management tools has made it easier for event managers to do their work more efficiently. Form 2008 Bangladesh has started a journey for digitalization in every sector by 2021 with the slogan "Digital Bangladesh". With this vision now the ICT facilities are at the doorsteps of the general people of the country. The greatest use of innovative technology and the internet are closely reciprocal to each other (Giannopoulos, 2004). And the use of ICT and smart devices or more specifically, the use of the internet and mobile phones has been increased dramatically in the last decade in Bangladesh. According to the data provided by Bangladesh Telecommunication Regulatory Commission (BTRC) (n.d.-a), the total number of Mobile Phone subscribers has reached 176.41 Million and the total number of Internet Subscribers has reached 120.95 Million at the end of June 2021. So, domestic tourists, as well as international tourists traveling to Bangladesh, are also enjoying all the internet-based facilities in tour planning and the execution of that with more developed infrastructures. But this development has come progressively. According to the data provided by Bangladesh Telecommunication Regulatory Commission (BTRC) (n.d.-b), the total number of Internet Subscribers in Bangladesh since 2013 is as follows (Fig. 10.4).

And as we can see from Fig. 10.3, even the number of international tourist arrivals in Bangladesh has also increased since 2013. On October 12, 2012, the 3G operation in Bangladesh was commercially launched (Rahman, 2013). The outcome of the introduction of the 3G network in the tourism industry is visible from 2013.

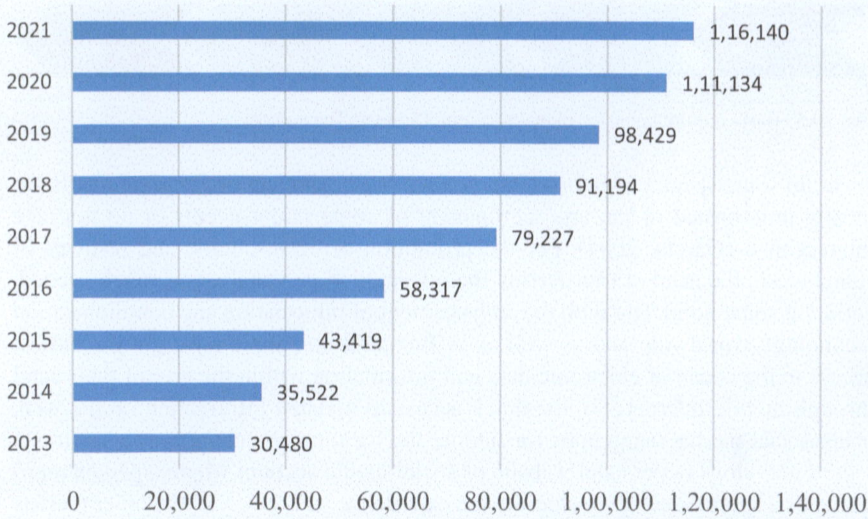

Fig. 10.4 Total number of Internet Subscribers in Bangladesh (2013–2021, in Millions). (Source: Bangladesh Telecommunication Regulatory Commission, n.d.-b)

Then overtime on February 19, 2018, Bangladesh started the 4G network service to ensure faster and improved mobile internet (The Daily Star, 2018).

Tourist satisfaction is influenced by the assurance of few aspects related to the personal quality of life, especially those which support the infrastructure of the destination they are making a plan to go, for example, transport, means of accommodation, catering and cuisine, places of leisure, prices, the conservation of tourist attractions, and especially, the efficient flow of information (Correia et al., 2013). Additionally, health, safety, and security are also very important. And the success of a fair, festival, and event is replicated in the increasing interest of people in participation. This participation is followed by proper planning of a tourist. So, these things are the very basic requirements for planning a tour with enthusiasm as presented in Fig. 10.5.

With the demand of the present time tourism industry, acceptance of technological tools and use of ICT in all these phases were evident in increasing the tourism involvement in Bangladesh. The choice of any planned tourist destination mostly and at the very first stage, depends on the image they form within their mind by following the available information lane, and by this, they can form at least an expectation of what awaits. Usually needed information can be assembled by tourists in streets, squares, and by utilizing mobile phones, laptops, and televisions (Gursoy & McCleary, 2004). But tourists of present days, want access to diversified information in the least time-consuming ways. Due to poor information and communication infrastructure tourists in some cases lose interest in attending some really attractive

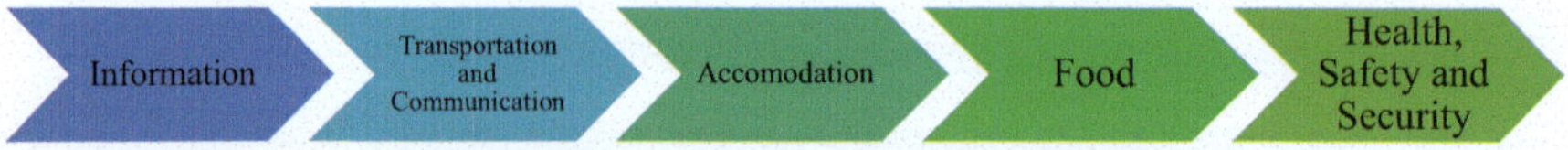

Fig. 10.5 Basic requirements for planning a tour. (Source: The author)

event. In some specific situations, the general public or even mainstream media are mostly uninformed of the arts' component of some major events or do not take interest in it (García, 2004). For the promotion of events, fairs, and festivals of Bangladesh, the need of transferring the information personally was much needed, earlier at some point. But with the advancement of information and communication technology worldwide and as well as in Bangladesh, people have got the easiest access to the ocean of electronic data and information within the grip of their hand through mobile internet. So, the quick access to updated information on the web, using social media, using apps for finding the location, watching numerous travel videos and vlogs (a personal website or social media account where a person regularly posts short videos), considering the experience and review of the other tourists have played a prodigious role in increasing the branding of these fairs, events, and festivals among the local and international tourists.

To any kind of traveler, the key qualities of choosing the mode of transportation are cost, comfort, and traveler flexibility (Hanna & Drea, 1998). By using web 2.0 technology tools and services in the transportation sector of Bangladesh the wide selection of online-based vehicle service industries such as app-based ride-sharing, online travel tickets booking, etc. has introduced a new dimension to the tourist transport and communication facility all over the country in getting into a tourist event. This kind of advancement in the transportation system reduces the amount of effort, time, and cost needed for mobility from one place to another (Yu et al., 2017). Because of the handiness of smartphones and extensive access to the internet connection in them, different app-based services have increased radically in Bangladesh. According to a report published in UNB News (2020), in between 2014 and 2020 a good number of ride-sharing apps and e-ticketing apps has been introduced in Bangladesh for all mode of transportation. So, it has become helpful to identify the nearest vehicles within the shortest probable time.

Accommodation is a ground requirement of all moods of tourism, and consequently, a suitable choice of accommodation is required with the purpose of improved development of the tourism industry. As a result, like other countries of the world, business trends, and ways of the accommodation sector have been developed a lot in recent years in Bangladesh (Travel Daily News, 2014). Even tourists nowadays are much more independent from travel agencies or tour operators to book their accommodation for their trip because of easy online booking facilities than previous years. Tourists coming to visit fairs, festival, and events in Bangladesh, are also nowadays enjoying technology-driven facilities in their hotels and resorts, which includes: technology-oriented check-ins to their accommodations, room service, and other customer-oriented digital interactions, smart room access systems, Wi-Fi infrastructure for seamless internet. Nowadays, consequently, the comfort level of the tourists is increasing in Bangladesh for this kind of service optimization and development. Which is also playing a vital role in the increasing number of tourists.

Food is a basic physical necessity and has always been an important component of tourism for the substantial part it plays in the overall experience of a tourist. In every fair and festival of Bangladesh, stalls and food corners with diversified food can be seen as a major attraction of the tourists. Even at the time of their visit to the event place, finding out the nearest restaurant by using Google map and also see their menu online before heading to that place made the decision making easy of a tourist in the choice making of food around the visited place. Platform offering mobile app-based food order and delivery, with the services like real-time tracking and scheduled delivery, has brought incredible revolution and accessibility in the Bangladeshi food scene. Local delivery services were always there, but the rise of digital technology ensured comfort by routing the orders from the tourists as customers to restaurants. In July of 2013, the country's first online food delivery service, HungryNaki initiated this service in Bangladesh, and later on several other local companies, along with a few global chains, joined the business like Foodpanda, Uber eats, Pathao Food, Shohoz Foods and few more (Muntasir, 2019). The Decision-making process of all these enterprises is very much data-driven and has invested a good portion for APIs (Application Programming Interfaces) to make the customer experience better.

In terms of the safety and security of the tourists in Bangladesh, tourist police always play a major role. But with the advancement of telecommunication technology and the introduction of the hotline or toll-free numbers, tourist service and satisfaction have improved in Bangladesh simply by being available to their single call. Bangladesh government has also introduced a single National emergency service number and which is 999, the number to call in case of any emergency. Even the medical facilities are now just a phone call away. Furthermore, in almost all events, fairs, and festivals- the law enforcement body sets weapon and bomb detectors at the entry point and closed-circuit television cameras at different points, addressable smoke detectors at indoor events to ensure the security of the tourists from any human-created danger. And even mobile internet, GPS, and Google Maps can locate the position of the tourist from getting lost in the crowd. So, now a day's tourists are more relying upon participating in any festival or event in terms of safety.

Challenges and Implications

With the introduction of the speedy digitalization process by the vision- "Digital Bangladesh", Bangladesh is already focusing on the creation of a global market for all the industries with diversified innovation. With this vision, Bangladesh Parjatan Corporation (BPC), has taken many steps on branding the fairs, events, and festivals, increasing the public awareness in collaboration with the a2i (Access to Information), launching YouTube channels, accepting complaints and feedbacks of the tourists through Grievances redress system (GDS) from any part of the country, solving the problems of the tourists with the prompt response with digital initiatives (Howlader, 2018). Despite notable attainments in the recent past, the position of Bangladesh, as a supplier and consumer of ICT services is far behind many countries. So, it is a matter of challenge to maximize the benefit of these initiated services of both government and private initiators in the tourism sector.

Furthermore, changes in technological implementation both internal and external sides of the event-fair –festival industry can destroy the tourism experience. Excessive use of or fascination with technology has the potential to destroy the true essence of enjoying these tourism arrangements with their core value and nature. According to Stipanuk (1993), reliance on technology to preserve the tourist experience opens the door for technology to potentially contribute to the ruin of the experience if the technology fails to protect the experience. Therefore, the use of technology in fairs-festivals-events, more specifically the traditional one should be done sensibly.

Conclusion

Technological transformation in tourism can make the country have easy access to tourists from home and abroad. Fairs, events, and festivals of Bangladesh were always with their charm and significance to attract the fixed kind of tourists from the beginning. But the Internet is an advanced technology of communication that minimizes the time and spatial obstructions and creates seemly access to the community. Tourism companies of Bangladesh are now more concentrated on offering web-based services providing full packages consist of airlines, transportation, accommodations, recreation activities, foods, and other facilities to the tourists through the mobile space. ICT has extended the business scope of travel agencies by maintaining direct communication with partner service providers and sponsors. Even their marketing policy has also shifted greatly on webs and apps. And the positive impact of all this technological adaptation in generating the tourists' interest and participation in events-fairs-festivals of Bangladesh is visible from the points, cases, and information discussed. Still, there are lots of areas left in this sector where more featured and extensive use of all these technological support could be used to enhance this revolutionary effect. However, in developed countries already the adaptation of 4th industrial Revolutionary tools is being introduced to the tourism sector. As a result, events- fairs-festivals of Bangladesh are still staying back in the race of attracting foreign tourists as expected from this revolt. How can this be resolved and is it possible to apply these upgraded tools practically within the prevailing infrastructure of the festival tourism industry in Bangladesh, this can be the point of further analysis. For reaching a comprehensive conclusion the need for quantitative data based on a detailed field study, rather than only case studies, is evident.

References

Afrin, S. (2019). *Tales of Shakrain. Dhaka Tribune*. Retrieved from: https://www.dhakatribune.com/feature/2019/01/12/tales-of-shakrain. Accessed 1 Aug 2021.

Alam, H. (2018). Chawkbazar: The original iftar market in Dhaka. *The Daily Star.* Retrieved from: https://www.thedailystar.net/city/chawkbazar-the-original-iftar-market-dhaka-1578430. Accessed 1 Aug 2021.

Asha, A. (2017). Folk festival, fair draw huge attention. *The Independent.* Retrieved from: https://www.theindependentbd.com/post/78583. Accessed 4 Aug 2021.

Askmjareb. (n.d.). *Popular travel destinations for Ramadan and id al-Fitr*. Retrieved from: https://www.askmjareb.com/en/ramadan/travel-ramadan-eid-al-fitr/. Accessed 1 Aug 2021.

Bangladesh Bureau of Statistics (BBS). (2011). *Population and Housing Census-2011* (p. xiii). Retrieved from: http://203.112.218.65:8008/WebTestApplication/userfiles/Image/National%20Reports/Union%20Statistics.pdf. Accessed 4 Aug 2021.

Bangladesh Telecommunication Regulatory Commission (BTRC). (n.d.-a). *The total number of Mobile Phone Subscribers in Bangladesh.* Retrieved from: http://www.btrc.gov.bd/telco/mobile. Accessed 1 Aug 2021.

Bangladesh Telecommunication Regulatory Commission (BTRC). (n.d.-b). *The total number of Internet Subscribers in Bangladesh.* Retrieved from: http://www.btrc.gov.bd/telco/internet. Accessed 2 Aug 2021.

Bangladesh Tourism Board (BTB). (n.d.-a). *Maanipuri raash festival.* Retrieved from: https://beautifulbangladesh.gov.bd/event/38. Accessed 1 Aug 2021.

Bangladesh Tourism Board (BTB). (n.d.-b). *Lalon Fair*. Retrieved from: https://beautifulbangladesh.gov.bd/event/36. Accessed 2 Aug 2021.

Banglapedia. (n.d.). *Bengal Renaissance*. Retrieved from: https://en.banglapedia.org/index.php/Bengal_Renaissance. Accessed 1 Aug 2021.

Bengal Classical Music Festival. (n.d.). *Research.* Retrieved from: https://bengalclassicalmusicfest.com/research/. Accessed 1 Aug 2021.

Chobi Mela. (n.d.). *Home*. Retrieved from: https://chobimela.org/. Accessed 1 Aug 2021.

Correia, A., Kozak, M., & Ferradeira, F. (2013). From tourist motivations to tourist satisfaction. *International Journal of Culture, Tourism and Hospitality Research, 7*(4), 411–424.

Cudny, W., Korec, P., & Rouba, R. (2012). Resident's perception of festivals – The case study of Lodz. *Sociológia – Slovak The Sociological Review, 44*(6), 704–728.

Dhaka International Film Festival. (n.d.). *Festival history*. Retrieved from: http://www.dhakafilmfestival.org/festival-history/. Accessed 1 Aug 2021.

García, B. (2004). Urban regeneration, arts programming and major events. *International Journal of Cultural Policy, 10*(1), 103–118.

Getz, D. (2008). Event tourism: Definition, evolution, and research. *Tourism Management, 29*, 403–428.

Getz, D. (2010). The nature and scope of festival studies. *International Journal of Event Management Research, 5*, 1–47.

Giannopoulos, G. A. (2004). The application of information and communication technologies in transport. *European Journal of Operational Research, 152*(2), 302–320.

Gursoy, D., & Mccleary, K. (2004). An integrative model of tourists' information search behaviour. *Annals of Tourism Research, 31*(2), 353–373.

Hanna, J. B., & Drea, J. T. (1998). Understanding and predicting passenger rail travel: On empirical study. *Transportation Journal, 38*(1), 38–46.

Howlader, M. Z. H. (2018). Digital technology powers tourism industry. *The Financial Express*. Retrieved from: https://www.thefinancialexpress.com.bd/views/digital-technology-powers-tourism-industry-1537805524. Accessed 1 Aug 2021.

Muntasir, B. (2019). Meteoric rise of online food business. *Dhaka Tribune*. Retrieved from: https://www.dhakatribune.com/business/2019/05/26/meteoric-rise-of-online-food-business. Accessed 1 Aug 2021.

Quinn, B. (2010). Arts festivals, urban tourism and cultural policy. *Journal of Policy Research in Tourism, Leisure & Events, 2*, 264–279.

Rahman, S. (2013). The fuss with 3G. *The Daily Star*. Retrieved from: https://www.thedailystar.net/news/the-fuss-with-3g. Accessed 1 Aug 2021.

Richards, G. (1996). *Cultural Tourism in Europe* (p. 24). CABI.

Stipanuk, D. M. (1993). Tourism and technology: Interactions and implications. *Tourism Management, 14*(4), 267–278.

The Daily Star. (2011). *Pohela Falgun celebrated*. Retrieved from: https://www.thedailystar.net/news-detail-174151. Accessed 1 Aug 2021.

The Daily Star. (2018). *Bangladesh enters 4G era*. Retrieved from: https://www.thedailystar.net/frontpage/bangladesh-enters-4g-internet-service-era-on-february-19-2018-1534357. Accessed 1 Aug 2021.

The Financial Express. (2019). *Dhaka Travel Mart kicks off in city*. Retrieved from: https://thefinancialexpress.com.bd/public/trade/dhaka-travel-mart-kicks-off-in-city-1553166459. Accessed 1 Aug 2021.

The Korea Post. (2021). *Thousands of people visit the Dhaka International Trade Fair every day from all parts of the world*. Retrieved from: https://www.koreapost.com/news/articleView.html?idxno=23128. Accessed 1 Aug 2021.

The World Bank. (n.d.). *International tourism, number of arrivals – Bangladesh*. Retrieved from: https://data.worldbank.org/indicator/ST.INT.ARVL?end=2019&locations=BD&start=1995&view=chart. Accessed 1 Aug 2021.

Travel Daily News. (2014). *Bangladesh: Types of accommodation*. Retrieved from: https://www.traveldailynews.com/post/2.-bangladesh-%3A-types-of-accommodation%2D%2D51074. Accessed 3 Aug 2021.

UNB News. (2020). *Best ride sharing apps in Bangladesh*. Retrieved from: https://unb.com.bd/category/Tech/best-ride-sharing-apps-in-bangladesh/59407. Accessed 2 Aug 2021.

Wikipedia. (n.d.). *Ekushey Book Fair*. Retrieved from: https://en.wikipedia.org/wiki/Ekushey_Book_Fair. Accessed 1 Aug 2021.

Wikisource. (n.d.). *1911 Encyclopaedia Britannica/Fair*. Retrieved from: https://en.wikisource.org/wiki/1911_Encyclop%C3%A6dia_Britannica/Fair. Accessed 1 Aug 2021.

Yu, B., Ma, Y., Xue, M., Tang, B., Wang, B., Yan, J., & Wei, Y. M. (2017). Environmental benefits from ridesharing: A case of Beijing. *Applied Energy, 191*, 141–152.

Kaniz Kakon is an Assistant Professor, Department of Philosophy, College of Arts and Sciences, IUBAT-International University of Business Agriculture and Technology, Dhaka, Bangladesh. After receiving her Bachelor of Arts and Master of Arts degrees in Philosophy from Jahangirnagar University, Savar, Dhaka, she started her career as university faculty. As in her philosophy classes, she gets students from multi-disciplinary departments, her research interests further developed in other studies similarly. She has published research articles and conference proceedings on different issues. She became interested in the zone of Tourism to concentrate on Philosophical Issues in Tourism.

Chapter 11
Promoting Catamaran Tourism in Bangladesh: The Challenges of Blue Economy

Rakiba Nabi and A. H. M. Salimullah

Introduction

As part of marine tourism considered as the fastest-growing industry, the catamaran tourism is indeed an evolving sector that gradually become a prominent industry in the world economy. Due to its rapid expansion, this growing sector has been emerged and contributed significantly in Bangladesh since last decade. The geo-political location of Bangladesh and the presence of the Bay of Bengal on the entirety of the Southern part act as an added benefit to bolster catamaran tourism. This small country has copious leeway to safeguard its huge marine resources and ensure sustainable marine environment.

The integration of catamaran tourism and cruise tourism into our marine tourism industry is nowadays a contemporary approach to step down marine pollution. The country has a huge potential, blessed with gorgeous seascapes and ample marine resources, for catamaran tourism that would boost our tourism industry. But the industry is to be expanded sustainably so that our marine and coastal resources may be utilized and protected at the same time. To this end, measures are needed to be taken in compliance with the concept of Blue Economy. This paper explores the possibilities of introducing catamaran tourism and the ways to develop it in our country in a sustainable manner to expand our tourism industry and contribute to the reinforcement of the national economy.

R. Nabi
Department of Law, University of Chittagong, Chattogram, Bangladesh

A. H. M. Salimullah (✉)
Department of Economics, University of Chittagong, Chattogram, Bangladesh
e-mail: salimullah.economics@cu.ac.bd

A. Hassan (ed.), *Technology Application in Tourism Fairs, Festivals and Events in Asia*, https://doi.org/10.1007/978-981-16-8070-0_11

Marine and Coastal Tourism

Marine and coastal tourism are inextricably connected with each other to create an intricate and growing tourism sector. Marine and coastal tourism are the largest segments of the tourism industry. They also involve the greater portion of economic activities taking place in the sea (Tegar & Gurning, 2018). Marine tourism represents 5% of the world economy and makes up 6–7% of global employment (UNWTO, 1995, cited in Bhuiyan et al., 2020). The UNWTO has related marine tourism to SDG-14, which has pinpointed sustainable use and development of marine resources and ensures economic benefit for host countries through tourism (Bhuiyan et al., 2020).

The marine tourism industry does not only bring out the sea and water-based activities, but also integrates a diverse range of businesses. These pursuits sort from small one-person operations to large multi-national corporations. An even larger and diverse group of businesses are indirectly linked with marine tourism, such as – coastal resorts, equipment suppliers and rentals, island ferry services, artists, souvenirs, and even rubbish collectors (Orams, 2002). Government agencies also play an important role in the monitoring and management of marine tourism. Furthermore, non-profit groups, as well as environmental groups, are also ‘stakeholders’ in this industry (Orams, 2002). In fine, it is a large and diverse industry to protect and promote sustainable marine environment.

While traveling to coastal areas for recreational purpose survived for time immemorial (thousands years), the origin of modern marine tourism can be traced back to sea-based tourism in late eighteenth-century Europe (Walton, 2012; Orams, 2002). Though the undertakings of coastal tourism continued mostly on land, endowed with social gatherings, parks, shopping, theatres, gambling houses, etc., it was not until the development of technology, a development that started in the 1950s, enriched with incredible inventions and facilities such as – scuba diving, surfboards and sailboards, underwater accommodations, passenger-carrying, and recreational vessels, submarines, navigational aids etc. (Orams, 2002) – that marine tourism developed into what it is today.

While marine tourism has existed since ancient times, its development as a commercial industry is quite recent. However, it is now the fastest-growing segment of the tourism industry. There is an emerging trend in Bangladesh as well for its prolific marine resources and its associated riverine environment. Tourism is a rising industry in Bangladesh. It accounted for 4.4% of the total GDP in 2015 and there was 1208 million USD worth of international tourism expenditure (LightCastle Analytics Wing, 2020a). The government allocated 34 billion takas in FY2019–20 in Civil Aviation (Biman) and Tourism Ministry, demonstrating that the government is now giving more importance to the tourism industry (LightCastle Analytics Wing, 2020b). However, despite the government allocating more importance to the tourism industry, it has been unable to utilize very little of its resources and the huge tourism market (LightCastle Analytics Wing, 2020b).

As Bangladesh is situated in a coastal region, marine tourism usually constitutes a huge part of that. Bangladesh has a huge potential for marine tourism which has not been developed to its full extent due to lack of proper initiative. With the right steps, it is possible to develop our marine and coastal areas sustainably with a focus on ecological balance, immaculate marine ecosystem, conservation and proper utilization of natural resources (Bhuiyan et al., 2020). Devising and implementing marine tourism become a significant part of the tourism industry and play an exclusive role in strengthening our economy. One powerful approach to start the process would be to utilize the concept of catamaran tourism and integrate it properly into our marine tourism industry.

The Growing Significance of Launching Catamaran Tourism

The maritime zones of Bangladesh have been determined pursuant to the Territorial Waters and Maritime Zones Act, 1974. Bangladesh also ratified the 1982 Convention on the Law of the Sea (UNCLOS-III) in 2001, under which the Bay of Bengal has been divided into three zones. Bangladesh has jurisdiction of 12 nautical miles (nm) of territorial waters, 200 nm of exclusive economic zones (EEZ), and 350 nm of the continental shelf from Bangladesh baseline (Belal, 2013).

Furthermore, there are currently 405 rivers running through different areas of Bangladesh (Bangladesh Water Development Board, n.d.). A major part of Bangladesh resides in the Ganges-Brahmaputra-Meghna (GBM) delta (Sieghart & Rogers, 2015). The Ganges (called the Padma in Bangladesh), the Brahmaputra (called the Jamuna in Bangladesh), and the Meghna all intersect with each other before flowing into the Bay of Bengal (Nations Encyclopedia, n.d.).

Bangladesh's coastline is fragmented due to rivers and streams of the delta region. In Chattogram, however, the coastline is the uninterrupted shore of Cox's Bazar. A section of Khulna covers the coastline region from the western border to Padma-Meghna bay and it is called the Sundarbans, the largest mangrove forest in the world (Nations Encyclopedia, n.d.).

In accordance with UNCLOS-III, Bangladesh has a total sea area of 2,07,000 square kilometers, nearly 1.4 times the total land area (Belal, 2013). And that is only sea area except other water bodies. Needless to say, Bangladesh has abundant places of attractions to be utilized for the inception of catamaran industry, an evolving concept to protect and promote marine tourism.

Bangladesh has a huge promising future for marine tourism, specifically for catamaran tourism, as it is surrounded by the Bay of Bengal on one side. As a result, she has an abundance of marine and riverine resources to employ and exploit. The country officially entered the cruise tourism industry in 2017, when the ship Silver Discoverer visited Maheshkhali and the Sundarbans twice carrying 162 tourists in total from different nations. The first trip generated 35lakh taka in foreign currency (The Daily Star, 2017). More recently, the same ship came to Mongla Port from Chennai with 61 passengers with a plan to visit various places like Dangmai Dolphin

Sanctuary, Sundarbans Harbaria Forest Office, Harmal canal, Tikona Island, Hironpoint etc. in 2019 (Dhaka Tribune, 2019). Even though this is a step towards flourishing catamaran tourism, Bangladesh is debilitated to utilize its full potential to explore this profitable industry suitably. There have been only a few attempts at wide-scale operations of tourism to pull in foreign tourists and currency. Catamaran tourism offers an opportunity to do that by utilizing our marine and riverine pots and resources.

The Evolving Concept of Catamaran Tourism

A catamaran is a powered water vessel with twin hulls, usually consisting of a superstructure or deck that connects the hulls (Merriam-Webster, n.d.). The word "catamaran" comes from the Tamil word, "kattumaram", which means two logs tied together (Discover Boating, n.d.). This name comes from the fact that its design is based on a raft of two logs connected by planks, a design that had been used in the Indonesian archipelago and Polynesia and Micronesia. The catamaran was first developed in the latter half of the twentieth century for recreation and sports. (Britannica, 1998).

The modern form of the catamaran was first developed and produced in the 1950s (Britannica, 1998). And while catamaran vessels have been around for a while, it is only recently that we have seen an increasing growth of these twin-hulled vessels. Now, there are many types of catamaran vessels all around the world being used for different purposes, such as – cruises, sail yachts, leisure crafts, ferries etc. (Sahoo et al., 2008).

Catamaran tourism is a part of marine tourism that exists in many countries around the world. Nowadays, the catamaran is a preferred mode of vehicle for adventure-loving tourists as it has several advantages such as more deck and space for passengers, stability on the water, high speed and great maneuverability, comfortable space, and facilities for passengers etc. (Discover Boating, n.d.). Catamaran tourism includes a wide range of marine activities such as snorkeling, windsurfing, fishing, observing marine mammals and birds, cruising on ship and ferry industry, scuba diving, sea kayaking, sailing, motor yachting, visiting fishing villages and lighthouses.

Recently, leisure cruising has been expanding rapidly (De Cantis et al., 2016). It is one of the fastest-growing tourism industries. The cruise industry offers different types of cruises to a diverse range of tourists- from boutique ships carrying fewer than 200 passengers to mega-ships carrying between 2400 and 5000 passengers (Foroozanfar, 2014). Even in Mauritius, cruising is being further developed where 50 vessels are expected to arrive in Mauritius by 2021 (Fakun, 2018). Megaships have been coined as floating resorts and are equipped with endless onboard amenities and entertainment attractions, including zip lines, rollercoasters, water parks, and golf simulators (Ketz et al., 2019). The best catamaran vessels can easily make 250-mile voyages, offer great performance, carry an amazing load, and have layout and facilities that can be easily utilized by individuals, making it a favorite of sailors and passengers (Wade, 2021).

The Introduction of Catamaran Tourism in Bangladesh

Marine and water bodies are ample in Bangladesh. It is undoubtedly acknowledged that Bangladesh is blessed with a lot of attractions related to these bodies as well. A wide range of marine attractions ranging from mangrove forests- the Sundarbans to the deep-sea marine attractions- swatch of no ground, the world's longest sea beach is there. Much of the spots are left out requiring proper utilization. These entire marine environments with lots of natural resources justify the foundational basis of setting in motion the catamaran tourism here.

Bangladesh lags behind to inaugurate the profitable trend of Catamaran tourism. While catamaran tourism does exist in Bangladesh, it has been introduced very recently and, thus, did not have the time to grow and become widespread like many other countries. Many spots with the scope of introducing catamaran tourism are remarkable in terms of their growing significance on our economy and marine environment.

Cox's Bazar

Cox's Bazar, an epicenter of marine tourism, is a district in southeastern Bangladesh. It is the longest sea beach (around 120 km or 75 miles) attracts millions of tourists every year around Bangladesh and overseas. This tourist spot is the granary of natural beauty, mingled with wide accessibility, and affordable or luxurious service, facilities (Bhuiyan et al., 2020). In addition to many other tourist destinations like St. Martin's Island and Teknaf sea beach, the main city of Cox's Bazar has an entire, fulfilling tourist destination all by itself. Some of the main fascinations of Cox's Bazar are Cox's Bazar beach, Laboni point, Himchori, Inani Beach, Marine Drive, the Burmese Market etc. Foreign tourists can spend days exploring the entirety of this small but vibrant city.

The marine activities available here include surfing, swimming, water sports, kayaking, boat tours etc. Coastal activities include shopping local handicrafts, delicious seafood, horse riding, beach walking, sunbathing etc. Many resorts, restaurants, hotels, and tour operators provide services in Cox's Bazar, making the place more lucrative than any other tourist spots in Bangladesh (Bhuiyan et al., 2020). In fact, in 2015, Cox's Bazar was visited by 1.5 million tourists (Amin, 2016). The gentle slope of the uninterrupted beach of Cox's Bazar can now host 85,000 daily (The Independent, 2019).

The first-ever cruise ship in Bangladesh, named the "MV Bay 1", was launched in Cox's Bazar- St. Martin route in December 2020 (Hussain, 2020). Its previous name was the "Salvia Maru" and was imported from Japan (Hussain, 2020). It goes directly to St. Martin's Island from Cox's Bazar and Chattogram. The luxury five-star cruise has modern amenities like Swimming Pool, Conference Hall, Presidential Suite, International Restaurant, Twin Bed Cabin, cabin, and general seats numbering around two thousand (Travel Mate, 2021b). This ship has 50 VVIP cabins. The cruise is 118 m in length and 17 m in width. It can travel at a speed of around 20 nm/hour with an average capacity of roughly 6200 tons. A jetty was constructed in Cox's Bazar Marine Drive to ensure the ship's smooth operation (Hussain, 2020).

The "MV Karnaphuli Express" is another cruise ship from the "Karnafuli Ship Builder Ltd.", a Chattogram-based company that was also responsible for MV Bay 1 (Travel Mate, 2021b). The cruise ship starts Nuniachara BIWTA Ghat behind Cox's Bazar airport and goes via the Bankkhali River to the Bay of Bengal before making the journey to St. Martin's Island by sea parallel to Cox's Bazar Marine Drive (MV Karnaphuli Express, n.d.). It is a 4-story ship that has 3 decks and 17 luxury rooms with the capacity to house 510 passengers at a time (Travel Mate, 2021a). These cruise ships endowed with numerous facilities are not properly maintained for strengthening catamaran events.

Cox's Bazar is already endowed with a plethora of facilities that make it a charming tourist destination. However, there is lack of proper overall infrastructure and security as well complementary leverages or activities that are essential to foreign tourists, such as scuba diving, snorkeling, mammal watching, boat rides etc. Introducing these facilities would boost up international catamaran tourism here. Furthermore, creating jetties and docks with proper facilities and service providers for catamaran vessels, so that catamaran vessels can dock and stay overnight, is a must to integrate catamaran tourism. People from all over the country visit Cox's Bazar but there is a lack of cultural exchange and exhibitions that would highlight local culture and heritage, making it an adorable place for those looking to experience the unique cultures of the country. Around the catamaran tours organized, several events such as mammals watching, sightseeing, beach-walking, kayaking could be some of the soothing events for the tourists to make the overall tour enjoyable. Finally, Cox's Bazar can be turned into an appealing spot for catamaran tourism through marketing and promoting properly to the international tourist community like what Bangladesh Tourism Board did for the Cricket World Cup 2011 as a host nation (LightCastle Analytics Wing, 2020b).

Teknaf Sea Beach and St. Martin's Island

Teknaf, an Upazila of Cox's Bazar, is another nerve center for the tourists. It is dubbed after the river Naf, placed on its banks and shares Bangladesh's border with Myanmar. The Teknaf sea beach, while not as prominent as Cox's Bazar beaches or St. Martin's, is visited by a fair amount of tourists. Teknaf is usually treated as a shortstop and the gateway to St. Martin's Island (Wikitravel, n.d.).

St. Martin's Island is the only coral island in Bangladesh, remarkable for its scenic beauty and unique features. Its land area is only 8 square km and as such, it is very easy to navigate (Lonely Planet, 2019). Over the last decade or so, it has become an increasingly acclaimed tourist spot. November to April is the main tourist season because of good weather (Wikitravel, n.d.) and tourists can book a trip from Cox's Bazar or Chattogram. Its core temptations are the existence of its coral reefs, natural and distinctive beauty, the local handicrafts, and the unique cuisine which is almost completely dominated by local seafood.

The present design to introduce catamaran tour in St. Martin is mundane and prosaic. The green line was originally a company providing land transportation services by bus. However, recently it has extended its services over water as well through Green Line Waterways. The "MV Green Line 1" was built in 2013 and launched soon after. It is a passenger ferry that operates on the Teknaf- St. Martin's Island route (Metacentre, 2013). Since then, several similar passenger vessels were ventilated along the same route. But the industry is not flourishing up to the remark.

The existence of catamaran tour demands the development of associated events as well. To fully experience the coral reef, scuba diving and snorkeling is a must. Currently, there is only one diving operation in St. Martin's Island termed as the Oceanic Scuba Diving Service that operates all kinds of diving activities and services in St. Martin's and adjacent areas (bddivers, 2020). Increasing such organizations and operations with increased employees and service providers, providing more cost-effective services, and ensuring international level guidance and instruction from professionals would make scuba diving in St. Martin's Island a pleasant activity, which would help entertaining tourists interested to observe the underwater ecosystem of the only coral island of Bangladesh. St. Martin's Island is a small island that one can walk around in a day. The presence of service providers, facilities and recreational activities not only on the beaches but interspersed throughout the entirety of the island would attract tourists from all over if they are marketed and advertised through the proper and relevant channels. There are trips to Chwera Dwip, which is an unpopulated extension of St. Martin's Island. However, these are usually one-person operations without any proper infrastructure or security measures in place. In addition, one more thing that could be added to St. Martin Island's charm would be to build an aquarium to preserve and exhibit a part of the coral reef for those who want to experience it without going underwater. St. Martin's Island already has the groundwork laid; now it will only need further development to become a desirable destination for catamaran vessels internationally.

The Sundarbans

The Sundarbans, the largest mangrove forest in the world, which is located in the Ganges-Brahmaputra-Meghna (GBM) Delta, specifically the Khulna division is the most mesmeric tourist destination. The area is distinguished for its wide range of fauna like 260 species of birds and endangered species such as the Royal Bengal

Tiger and the Indian Python (United Nations Educational, Scientific and Cultural Organization, 1997). Furthermore, its natural beauty, special marine-based flora and fauna, rivers, creeks, estuaries, mudflats, and tidal flats also make up major parts of its charm (United Nations Educational, Scientific and Cultural Organization, 1997).

This eminent tourist spot offers river cruise or safari to the forest via Khulna. Inside Sundarbans itself, marine and coast-based tourist activities are available. There are eco-parks, wildlife sanctuaries, hiking services, reservation centers, mammal watching etc. that make the Sundarbans a charming tourist spot. There are morning boat tours, watching riverine dolphins that are native to Sundarbans, coastal sightseeing, visiting wildlife sanctuary etc. The Sundarbans is home to the lesser-known Jamtola Beach, situated deep inside the forest beside the Bay of Bengal (Nijhoom Tours, n.d.). Unlike other overcrowded beaches, it is the cleanest and quietest beach in Bangladesh and is a very attractive, albeit less visited than others, a tourist spot (Nijhoom Tours, n.d.). However, despite all of these, inception of catamaran tourism remains relatively ominous due to lack of transport, other facilities such as safety and security, and accessibility (United Nations Educational, Scientific and Cultural Organization, 1997).

However, Viking Yacht is a Bangladesh-based-company that provides riverside tours around Bangladesh by a variety of boats including luxury yachts and cabin cruisers (Viking Yacht, n.d.). Many cruises from Dhaka and Khulna provide river tours of the Sundarbans and an example is the "MV Wave" (GetMyBoat, n.d.). A journey around riverine and marine Bangladesh would show a variety of ships and cruises providing services to tourists.

The Sundarbans is another appropriate choice for catamaran tourism through both river and sea. Small cruises and other vessels already make the trips by the river. There is a need for an overall interconnected and collaborative structure that would allow catamaran tourism into the Sundarbans. Increased construction of docks, providing services and proper security measures, appointment of professional guides and instructors, ensuring proper facilities and accommodation for tourists, among many other things, would boost cruise tours into the Sundarbans. Jamtola Beach in the Sundarbans is the perfect place for catamaran tourism. Building docks for catamaran vessels, ensuring facilities, recreational activities and service providers, ensuring accommodation and infrastructure for safe, overnight stays on board would create a safe and splendid place for catamaran vessels to visit and stay in the Sundarbans.

Potenga and Parki Beaches in Chattogram

Chattogram is a coastal district situated just beside the Bay of Bengal and the largest international seaport in Bangladesh. While not as prominent as Cox's Bazar, Chattogram is home to few beautiful -beaches and water bodies that are mostly visited by locals and natives but which can be turned into a massive tourist attraction for foreign tourists with the inception of catamaran events. The major beaches here are Patenga Beach, Parki Beach and Naval Beach (TripAdvisor, n.d.-a). The lesser-known beaches are Kattoli Beach, Bashbaria, Bashkhali, Guliakhali etc. (Nobi & Majumder, 2019). A systematic study could be made to assess which beaches are more conducive to implement the catamaran tour. The location, natural beauty, marine resources of Patenga beach, Parki and Naval beach validate the initiation of catamaran events there.

Kuakata Beach

Kuakata is a rare scenic spot in Patuakhali district in southernmost Bangladesh, distinct for being home to a panoramic sea beach. It slopes gently into the Bay of Bengal. This virgin sandy beach is lined with coconut trees and home to migratory winter birds, making it a place of serene natural beauty (Bangladesh Parjatan Corporation, n.d.). It is one of the rarest places which offers a full view of sunrise and sunset over the Bay of Bengal. Locally known as "Shagor Konya" (Daughter of the Sea), Kuakata is a holy place of pilgrimage for both Hindu and Buddhist communities and has unique local customs and culture from its Rakhine tribal citizens (Informationbd.org, n.d.). With its rare, natural beauty and unique culture, it is a place that can and does attract tourists with ease. Therefore, there is ample opportunity to inaugurate catamaran tour in and around Kuakata beach.

No such attempt has noticeable till date. Green Line has launched the Green Line Water Bus that provides passenger services through Dhaka-Barisal waterways and has a capacity of carrying 600 passengers (Bangladeshfact.com, 2017). Dhaka River Cruise is similar to the Dhaka Dinner cruise that provides daylong tours from Buriganga to Meghna (Bangladesh Expeditions, n.d.). A mega step could be taken like MV 1 to promote catamaran tourism in Kuakata.

This beautiful place can easily become a beautiful cruise destination if proper infrastructure is put in place. Catamaran vessels can dock here, allowing tourists to enjoy the natural beauty of the place. The best time for a tour to Kuakata would be winter when the weather is mild and migratory winter birds add to the charm of the place. Furthermore, Kuakata has a unique cultural heritage due to the existence of the Rakhine tribe who first discovered it and due to being a holy place of pilgrimage for Hindus and Buddhists. The Buddhist Temple, which has been dated a hundred years of age, is a notable historical site (Bangladesh.com, n.d.-a). To make Kuakata an even more appealing destination for catamaran tours, locals can be encouraged to co-operate with relevant organizations to set up a cultural exchange and exhibitions for visiting foreign tourists.

Coastal Islands

There are 19 coastal districts in Bangladesh, including Chattogram and Cox's Bazar (Ahmad, 2019). These are birthplace to many lesser-known, beautiful islands that have the potential for becoming massive tourist attractions acting as breakthrough for the industry. The islands number around 70 (Ahmad, 2019). Some of the major islands are Bhola Island, Dal Char, and Manpura Island in Bhola district of Barisal division, Dublar Char in Khulna division, Swarna Dwip and Hatiya in Noakhali district, Sandwip and Urirchar in Chattogram district, Kutubdia, Sonadia, Maheshkhali in Cox's Bazar district and Chera Dwip, an uninhabited island which is an extension of St. Martin's Island.

Other Spots

There are many other water bodies worthy of a visit in Bangladesh. Hakaluki Haor, and Hail Haor most beautiful wet marshland situated in Moulovi Bazar of Sylhet, is one of them. Madhobkunda waterfall, the largest waterfall in Bangladesh also situated in Moulovi Bazar, is another tourist spot. Other places include Tanguar Haor of Sunamganj district, Mari River of Jaflong in Sylhet, Sangu River in Bandarban, Madhabpur Lake in Sreemangal, Durga Sagar in Barisal, among many others. These places are expedient to introduce catamaran tourism even in short distances and deserve a careful consideration and scrutiny.

The "Dhaka Dinner Cruise" is a riverine cruise service and a unique tourism service within Bangladesh. It was developed with the help of Bangladesh Parjatan Corporation. It provides ships of several sizes with varying capacities for passengers ranging from 20 to 150. These ships usually travel through Buriganga-Meghna or Muriganga-Sitalakkha route. They offer short dinner cruises and daylong cruises along with special events like annual events, picnics, corporate outings, conferences, social get-together etc. (Dhaka Dinner Cruise, n.d.).

There is much room for the development of catamaran events in Bangladesh. Catamaran tourism and cruising is now a promising and propitious business in the tourism industry that must be capitalized on. To harvest all the benefits of cruising, we must comprehensively expose it to the people. Such an extensive task requires collaborative planning and it brings us the challenges of balancing the economic, social, and cultural impacts relating to the benefit of such tourism. Otherwise, we might miss the huge opportunity to boost the economy by utilizing all the resources available.

Tours of the Bay of Bengal

The Bay of Bengal forms the southern border of Bangladesh and many coasts, beaches and islands, including Cox's Bazar, St. Martin's Island and Kuakata, are nestled against it. It is the largest bay in the world and is counted among the world's 64 largest marine ecosystems (Bangladesh.com, n.d.-b). It is a prime location for long trips of catamaran vessels such as – cruises, yachts, sailboats etc. It is also a good place for marine wildlife watching due to the existence of diverse species such as sea snake *Kerlia jerdonii,* Oliver Ridley sea turtle, the Bryde's whale, different kinds of dolphins etc. (Bangladesh.com, n.d.-b). Furthermore, its large and diverse marine ecosystem also makes it a major location for underwater tourism activities.

As the Bay of Bengal surrounds Bangladesh on the south, encompassing the coastal areas, catamaran or cruise tour of the Bay of Bengal along the border of Bangladesh would facilitate a tour of the entirety of coastal Bangladesh, together with all the unique sceneries, facilities, recreations, and culture that it has to offer. There are many ports in Bangladesh alongside the bay for trade and tourism purposes (Bangladesh.com, n.d.-b). These ports can be enhanced and used for large-scale tours of the southern border of Bangladesh through the Bay of Bengal. To implement such a nationwide operation, collaborative measures between big cruise corporation as well as government and non-government organizations are required to ensure wide-scale communication and cooperation among ports, proper facilities, accommodations and security measures for tourists, cooperation with local and national businesses connected to coastal and marine tourism as well as a proper infrastructure in place for the smooth operation of these tours.

Tours in the Rivers of Bangladesh

Bangladesh is a riverine country, situated in the biggest delta in the world with more than 400 rivers dotting its map. While the tours of the Bay of Bengal would take tourists around the borders of Bangladesh, riverine tourism would allow foreign tourists to explore throughout the country as they travel the rivers. It is one of the best ways to experience the beauty and culture of Bangladesh. Riverine tourism is a huge prospective industry in Bangladesh. One-third of foreign tourists' express interest in touring the rivers of Bangladesh (Ali, 2013). Despite all of this, the riverine tourism industry has declined due to negligence from the government and relevant authorities, decaying infrastructure and navigability, and lack of security and maintenance (Nahar, 2019).

Catamaran tourism would be a great way to revive river tourism by carrying large numbers of tourists on cruises through the rivers of Bangladesh. Most river cruises now run under private operations and are overcrowded and unsafe (Ali, 2013). Furthermore, navigating the rivers is difficult for these private operations and there is lack of proper mapping of river routes (Nahar, 2019). If the relevant authorities take proper measures, catalyzing these private operators while also boosting inland infrastructure, security, facilities, accommodation and recreation, the rivers of Bangladesh would be a major location for cruises and catamaran vessels to tour. Furthermore, maintenance of rivers to reduce pollution as well as monitoring and maintaining docks and jetties and ensuring proper facilities to welcome tourists at riverside ports would boost tourism as well.

Apart from these, many other water bodies have much potential for catamaran tourism and are visited by locals frequently. These are – Karnafuli River, Kaptai Lake, Bhatiyari Lake, Boga Lake, Sangu River etc. (TripAdvisor, n.d.-b). These are all beautiful places that have untapped potential but lack of proper initiatives result in potential wasted for boosting the catamaran tourism industry.

Prospects of Catamaran Tourism in Bangladesh

Catamaran tourism is a very useful way to use the natural marine resources of a country most productively and efficiently. As previously discussed, Bangladesh has a wide range of marine attractions, such as Cox's Bazar, St. Martin's Island, Kuakata, and the Sundarbans. Much of them are not exploited properly, for instance- Nijhum Dwip, Burir Char, Bhola Island, Hatiya, Manpura Island, Maheshkhali Island, Kutubdia, Kattoli, Bashbaria, Sonadia and so on. Such indifference to a large number of tourist locations means missing out a great opportunity of deriving economic benefit for the country. Keeping the following prospects in mind, we must develop catamaran and marine tourism in all these pleasant marine locations to reap all the benefits of existing physical natural resources.

Influencing GDP Growth

Catamaran tourism can make a huge impact on GDP consistent with the development of the contributing factors of GDP such as- infrastructure development, new business development, acceleration of the supply chain of industrial products (Papageorgiou, 2016), enhancing employment opportunities, poverty reduction. Finally, the positive impacts on these factors will show a cumulative effect on the GDP. The host country can also generate tax revenues from the cruise ship industry. In 2017, Tourists from 95 different nations visited Maheshkhali Island in Cox's Bazar and the Sundarbans, the largest mangrove forest in the world, between February 22 and 24 on a ship named Silver Discover. Around BD Tk. 35 lakhs in foreign currency on that tour were earned (The Daily Star, 2017). According to the latest report of the World Travel and Tourism Council, the contribution of tourism to total GDP in the year 2019 was 3% (WTTC, 2020). With the advent of the proper utilization of the underdeveloped marine attractions, this percentage is expected to rise and thereby marking higher GDP in the upcoming years. If the concept of catamaran tourism can be applied in its standard form and integrated into our tourism industry, it will strengthen our GDP growth.

Enhanced Employment Opportunities, Poverty Reduction and a Developed Lifestyle

As the catamaran tourism industry flourishes, the local wholesaler and retailer business will expand gradually. Thus, the employment opportunity upsurges significantly. Sometimes, it gives the local population an alternative job other than exploiting natural resources to earn their livelihood. There are formal jobs in hotels and motels, cafes & restaurants, shopping malls at the tourist site, transport, tourist guide, tourism agency, local handicrafts etc. Besides, a large number of local people find their livelihoods in the informal sector, such as rickshaw driving, hawkers, selling homemade food and non-food items, garbage collecting, unprofessional children singer, camera operator, and so on (Hossain & Wadood, 2020).

Tourism activities have provided employment opportunities for lots of young and unemployed people in Cox's Bazar. More than 10,000 tour guides and more than 200 travel and tourism organizations are working in tourism services in Bangladesh (Airways Office, 2019). There are many jobs, which require minimal skills to join. As per the report of WTTC tourism sponsored 2.9% of the total jobs in 2019 (WTTC, 2020). Proper utilization of all the marine attractions may lead to a rise in this employment rate in the coming days in Sundarbans, Kuakata and other important places of marine attractions.

Furthermore, catamaran tourism itself is a potential source of drastically improved enhancement opportunities, as many entities, industries and service providers, such as – ship crew, port operators and handlers, caterers, excursion operators, local communities and market, cruise industry associations, ship suppliers and storage facilities, etc., have to be involved to ensure the smooth operation of the ships and tours (UNWTO, 2016). Enhanced employment opportunities, infrastructural and other developments will cause a poverty reduction in the locality.

Environmental Prospects

The concept of ecotourism and sustainable tourism must be kept in mind to reap the environmental benefit from tourism. Ecotourism is measured against four standards (Hetzer, 1965):

- Minimum Environmental impact
- Minimum impact on and maximum respect for the host culture
- Maximum economic benefit for the host country
- Maximum recreational satisfaction for the participating tourist

Government must create awareness among the local people to preserve the very reason for which the tourists are here, that is the natural attractions. Also, tourists have to be mindful of safeguarding the environment of the tourist spots. In this regard Orams (2002) argued that the government agencies can ensure responsible tourist behavior by incentivizing eco-friendly behavior and vice-versa. In addition, a portion of the profit of this industry must be used by the government to renovate the environment by reducing the hazards caused by the tourism activity, for instance, restoration of damaged reef fragments, removing rubble, reattaching living corals. In this way, we can avoid over-tourism, ensure sustainable tourism (i.e., eco-tourism). Applying the concept of ecotourism to catamaran and cruise tourism would minimize the negative environmental impacts.

Socio-cultural Benefits

The socio-cultural impacts of tourists can change the character, family value systems, relationships, behavior, safety, moral conduct, and traditional ceremonies of communities (Brida & Zapata, 2010). Tourism helps to exchange culture among nations. Bangladesh is a land of cultural diversity. Besides the cultural practices in the marine tourist destinations, native arts and crafts such as different handicrafts, clay-based pottery, Shital Pati, Jamdani etc. can be promoted to the tourists and thereby preserve the cultural value of the nation. Catamaran tourism is mainly targeted towards foreign tourists, which produces great opportunities for the intermixing of cultures and the displaying of our unique cultural heritage and history. This branding will create a unique identity for our country around the world.

Challenges of Catamaran Tourism in Bangladesh

It is to be expected that, like any other industry, the industry of marine tourism encounters many challenges. If we focus particularly on catamaran or cruise tourism, the challenges increase significantly. This part attempts to outline some of the major challenges this industry faces.

Environmental Impacts

Catamaran vessels, like every other invention, have both negative and positive impacts. When it comes to the environment, the impacts usually lean towards negative. A cruise ship can be seen as floating hotels with a range of facilities similar to land-based resorts and the higher number of passengers results in more waste production and hazardous emission from waste incineration (Johnson, 2002, cited in

Lamers et al., 2015). A cruise with a capacity of around 2800 people is estimated to produce 4.25 million liters of black water (human waste) and greywater (wastewater from the kitchen, bath, laundry, etc.) (Lamers et al., 2015). Certainly, the fuels needed to mobilize cruise ships and other catamaran vessels also result in the emission of a range of water and air pollutants. In coastal regions, these pollutants, such as – nitrate, sulfate and other volatile matters, can cause significant land, air and water pollution, endangering the environment and the health of the locals (Lamers et al., 2015).

In Bangladesh, pollution is already a severe problem that is directly or indirectly responsible for 26.6% of all deaths in Bangladesh (Global Alliance on Health and Pollution, 2018). Therefore, large-scale production of catamaran vessels and the development of the catamaran tourism industry would have to be preceded by initiatives to reduce the negative impacts such a venture would have on the environment. For this, concepts of ecotourism and blue economy for sustainable development should be introduced and implemented.

The Adversity of COVID-19 Pandemic

The COVID-19 pandemic has adversely affected the tourism industry like every other aspect of the world economy. Bangladesh is not an exception. At the beginning of the pandemic, UNWTO had predicted a loss of USD 470 million for the Bangladesh tourism industry in 2020 and the Pacific Asia Travel Association's (PATA) Bangladesh Chapter had predicted employment loss of 0.3 million people in the tourism industry (LightCastle Analytics Wing, 2020a) Similarly, the cruise industry has not been spared from the impacts of Covid-19, as cruise tourism requires many people to be housed nearby, allowing the spread of the virus. In February 2020, Carnival-owned Princess Cruises had to put 3700 passengers in quarantine (Richter, 2020). Similar pandemic-related incidents occurred in the cruise tourism industry, resulting in dozens of deaths and over 1500 COVID-19 cases within June 2020 (Richter, 2020). Considering the health hazards and the severe repercussions of the Pandemic on our national economy, it would be tough to get back to the functional catamaran tourism industry.

Lack of Sufficient Government Initiatives

While the tourism industry has been getting more attention from the government in recent times, it is not nearly enough. World Tourism and Travel Council predicted that investment in tourism would increase to 69.8 billion takas in 2022 (Afroz & Hasanuzzaman, 2018). However, this is a global report in which the specifics of how, when, and where it will be invested are unclear. Moreover, this statistic is now most certainly untrue and overestimated due to the raging pandemic. Besides, all tourism policies are usually targeted towards already existing tourist organizations and facilities, (Afroz & Hasanuzzaman, 2018) which does not leave much room for the development of new ones.

Another obstacle is the visa policy of the government, which puts many restrictions on tourists, making them reluctant to visit. Lifting visa restrictions would result in an increased influx of foreign tourists from Europe, the USA, and other western countries (Alam, 2018). Furthermore, Bangladesh, being a riverine country, has a great potential for riverine cruising and catamaran tourism. One-third of foreign tourists are interested in river tours (Ali, 2013). However, this potential is wasted due to the negligence of proper authorities and ministries and bureaucratic "red tapism" (Nahar, 2019).

Lack of Befitting Infrastructures

This is perhaps the most important issue when it comes to any kind of tourism in our country. The poor infrastructure, lack of facilities, and transport are problems that contribute to the reluctance of international tourists to come to Bangladesh. Tourists cannot access international currency through most ATMs, which is an essential part of foreign tourism (Alam, 2018). Bangladesh ranked 133rd among 140 countries in Tourism Service Infrastructure and is the worst out of all Asian countries (Calderwood & Soshkin, 2019). Furthermore, local infrastructures like airports and service providers can be unsupportive and even deceitful. Bangladesh ranks 111th in the Air Transport Infrastructural Index (Calderwood & Soshkin, 2019). There is also no proper infrastructure in place to ensure the safety and security of the tourists and western countries issue warning against Bangladesh travel for reasons like terrorism, crime and kidnapping etc. (U.S. Department of State, 2021).

Furthermore, despite the lack of proper infrastructure, poor accommodation and facilities, lacking health and hygiene standards as well as problems with ICT readiness, Bangladesh's tourism cost tends to be on the higher end of the spectrum compared to its neighbors with better infrastructure and facilities (Calderwood & Soshkin, 2019). It is no secret that service providers in Bangladesh try to overcharge foreign tourists. This would be even more applicable for catamaran and cruise vessels, which are luxurious and costs a lot to produce and maintain. There is a need for the development of affordable services so that catamaran tourism would appeal to tourists.

Launching Blue Economy with Catamaran Tourism

Blue economy seeks to ensure whether the marine resources are utilized to the full extent and whether they are used keeping in mind the declining ocean health and the overuse of other maritime resources. Blue economy can secure the maximum possible contribution of maritime resources to the growth of an economy, the maximum benefit of the coastal population while ensuring that the policies are compatible to preserve the very nature that is the reason for the derivation of such benefits. A country with an intent to implement the blue economy faces the biggest challenge while ensuring oceanic sustainability. The policies have to confirm both the maximization of benefits and the preservation of the resources.

Blue Economy and Blue Tourism in Bangladesh

The term blue economy was first introduced at the Rio +20 UN Sustainable Development Conference. (United Nations, n.d.) Different organizations defined the blue economy in different ways. According to the Centre for Blue Economy, blue economy includes three related but distinct meanings- the overall contribution of oceans to economies, the need to address the environmental and ecological sustainability of the oceans, and the ocean economy as a growth opportunity for both developed and developing countries (Middlebury Institute of International Studies, n.d.). Tourism is one of the bests among various diverse components of the blue economy that pull in more employment and more GDP growth for a country.

As per the World Bank (2017), an activity is qualified as a component of the blue economy if it provides a social and economic benefit for current and future generations; restore, protects, and maintains biodiversity, productivity, resilience, core functions, and intrinsic value of marine ecosystems. It is based on clean technologies, renewable energy, and circular material flow that will reduce waste and promote recycling of materials.

Blue economy seeks to anticipate the impacts on the coastal ecosystems and marine resources resolve those issues while policymaking. This concept enables the least developed countries to implement sustainable development goals. Conservation and sustainable use of the oceans, seas, and marine resources for sustainable development is what goal no. 14 is about (Upadhyay & Mishra, 2020). It provides a low carbon and resource-efficient path to economic growth for these countries and reduces poverty (World Bank, 2017). The concept of a blue economy is the only way to devise a circular economy out of marine resources.

Principles of Blue Economy and Blue Tourism

The World Wildlife Foundation provided some principles of a sustainable blue economy (World Wide Fund for Nature, 2015). A sustainable blue economy is a maritime-based economy that provides socio-economic benefits for current generations, restores, protects, and maintains the diversity, productivity, resilience, core functions, and intrinsic value of marine ecosystems is based on clean technologies, renewable energy, and circular material flows. A sustainable blue economy is governed by public and private processes that are inclusive, well informed, precautionary and adaptive, accountable and transparent, holistic, cross-sectoral, and long-term innovative and proactive. The public and private sector must set clear, measurable, and internally consistent goals and targets for sustainable blue economy access and communicate their performance on these goals and targets.

There must be a level of economic and legislative playing field that provides the blue economy with adequate incentives and rules. The use of marine space and resources must be planned, managed, and effectively governed, applying inclusive methods and the ecosystem approach. They must recognize that the maritime and land-based economies are interlinked and that many of the threats marine environments may face, originate on land. They must actively comply with activities such as sharing information, knowledge, best practices lessons learned, perspectives, and ideas, to realize a sustainable and prosperous future for all.

Introducing Blue Economy in Bangladesh

To expedite the process of SDG fulfillment by 2030, pragmatic steps have to be taken to ensure the sustainable development of the ocean and derivation of benefit from the ocean. Several incomplete initiatives mark the failure in the implementation of the blue economy in Bangladesh. The first international workshop on the Blue Economy in 2014 and the second one in 2017 was held in Dhaka. Honorable Prime Minister in the workshop emphasized on the workshop that for the economic uplift of the country, poverty alleviation, ensuring food and nutrition security, combating climate change impacts Blue Economy could play an important role. Underlining Blue Economy as a window of opportunity for development, the Prime Minister resolved to turn the Bay of Bengal into an economic development hub. She re-affirmed Bangladesh's commitment to conservation and balanced development of natural resources keeping the integrity of environmental and bio-diversity aspects while pursuing development for the people of the country (Alam, 2019). A brief overview of the organizations related to the blue economy will help us assess the progress of its realization.

Organizational Progress and Present Scenario of Blue Economy, Blue Tourism and Catamaran Tourism in Bangladesh

Several organizations are working in the field of tourism in Bangladesh such as Bangladesh Parjatan Corporation, Blue Economy Cell, and Bangladesh Tourism Board Tour Operators Association etc. Bangladesh Parjatan Corporation's functions are forming human resource for the tourism sector, incentivizing the public and private sector to join in the infrastructure as well as tourism-related projects, stimulating ecotourism and improving the tourism sector such as making new tourist spots, arranging safe hotels, motels, cottages, restaurants, picnic spots, rent a car etc. (Bangladesh Parjatan Corporation, 2019). Since 2016, after 4 years of establishment, the blue economy cell, formed under the Mineral and Energy Resource Division, is hardly contributing to realizing the blue economy concept. Due to shortage of manpower and low budgetary allocation, the Cell comes to a standstill. Its functions are only limited to holding meetings. Tourism was included in the 26 identified sectors of the blue economy in the workshop of 2014 and 2017 but any initiative seems to be far away (Hasan, 2021).

Bangladesh Tourism Board was formed as the national tourism organization with the vision and mission of developing Bangladesh as a tourism spot under the Tourism Board Act of 2010. Till date, there has been a little contribution from this organization other than building up the Beautiful Bangladesh website (Bangladesh Tourism Board, 2018). Tour Operators Association Bangladesh is a private body formed in 1992 to overcome the problems faced by different tour agencies in Bangladesh. If this association is brought under the concept of blue economy, the tourist agencies may become one of the key components to realize blue tourism (Tour Operators Association of Bangladesh, n.d.). The impecunious progress towards the blue economy, blue tourism, and catamaran tourism is conspicuous in Bangladesh. Cruise tourism, though introduced in a small scale in Cox's Bazar and Sundarban, needs to be devised on a larger scale as it becomes one of the most profitable industries around the world as a component of blue tourism and thereby shall contribute towards the blue economy. Without the proper functions of these organizations and implementation of the coastal and marine tourism opportunity will leave a gap in GDP and the overall development of the country.

Challenges to Blue Economy and Sustainable Catamaran Tourism

Bangladesh with omnifarious prospects of the blue economy is yet to reap all the benefits from the components of the blue economy. Several challenges work as main barriers to apply the principles of blue economy. The problems of marine tourism commensurate with that of catamaran tourism and thus, contrary to the sustainable development of blue economy and blue tourism.

Climate Change

Marine and coastal tourism are especially vulnerable to climate change, as it is entirely dependent on the marine environment. The rising sea level due to climate change is one of the biggest long-term threats to the marine tourism industry, which depends on the infrastructure above sea level (Attri, 2018). Moreover, climate change has an adverse effect on the marine ecosystem, particularly critical coastal ecosystems like coral reefs, estuaries, and wetlands (Claussen, 2002). The losses of these ecosystems and damage from flooding would also result in erosion of beaches as well as infrastructure and facilities on land, thus reducing their value as a tourist destination as it directly or indirectly affects coastal and marine activities (Moreno & Amelung, 2009).

Bangladesh is one of the most vulnerable countries to climate change. The impact of climate change on water bodies will be critical in Bangladesh (International Union for Conservation of Nature, n.d.). Bangladesh is a disaster-prone country, with 26% of land subjected to flooding annually and cyclones hitting once every 3 years on average, due to climate change (Hussain & Hoq, 2018). The sea-level rise is another concern. The Bay of Bengal creates storm surges, which can make sea-level rise enough to destroy low-lying coastal areas (Hussain & Hoq, 2018). The Sundarbans can be drowned with a 1-meter rise in sea level (World Bank, 2000) and St. Martin's Island can be destroyed as the coral reef ecosystem is dying out due to climate change (Hussain & Hoq, 2018). If this continues, it will be impossible to preserve the marine environment and ecosystem while properly sustaining tourism. It is quite clear that sustainable tourism and the blue economy are impossible without first tackling the problem of climate change.

Pollution

Pollution of water bodies is another huge problem that stands against the blue economy and sustainable tourism. This is a major problem in Bangladesh where water pollution is massive. Marine and coastal pollution, specifically, involves human-made contaminating products that enter the ocean and the sea. In Bangladesh, marine and coastal pollution mainly occur due to direct discharge from ships and marine activities, indirect discharge from coastal activities, waste oil from industries and ships, discharge from rivers, industrial waste, drainage from ports etc. (Kabir, 2010). With increased industrialization as well as increased marine activities, the pollution keeps proliferating precipitously. It is not an ideal situation to sustain any kind of catamaran tourism.

Furthermore, tourism itself is a major contributor to pollution. Every year, 8 million tons of plastic end up in the oceans (One Planet Network, 2020). Marine and coastal tourism, being the largest segment of the tourism industry, is a major contributor to this (One Planet Network, 2020). Furthermore, cruise and catamaran vessels themselves are also contributing to pollutions with waste discharge, chemicals and fuels, emission of pollutants in marine and coastal environments etc. (Lamers et al., 2015). Untreated effluents of cruise ships are also another contributor (Baldwin, 2018). Even a medium-sized cruise ship is reported to create huge amounts of pollution. And tourists themselves are another contributor to pollution. Some discharge wastes into the water and recreational activities, such as careless boating, diving, snorkeling, etc., have damaged and disturbed marine ecosystems and aquatic animals (Coastal Wiki, n.d.). Therefore, the expansion of sustainable catamaran tourism has to be done to ensure minimal pollution to comply with the policies of the blue economy.

Over-Exploitation of Resources

The protection of marine and coastal resources has to be executed effectively. Otherwise, safeguard turns into exploitation and ultimately becomes destructive to our resources. Pollution is just one example of it. Unmanaged tourism, unnecessary interference with natural marine resources and aquatic flora and fauna, over-fishing, haphazard tourism development, excessive water extraction by coastal resorts and other infrastructure etc. all result in severe damage to and even loss of marine and coastal resources (Hall, 2001).

Even seemingly harmless activities like whale watching can have an impact on the natural environment, habitat, and habits of these animals, which can ultimately lead to reducing their survival rates (British Broadcasting Corporation, 2011). One example in Bangladesh itself is the coral reef of St. Martin's Island. The increasing pollution is naturally eating away at this fragile ecosystem. However, the biggest threat the coral reefs confront is the direct extraction of corals, which declined 24% of the previous coral population on St. Martin's Island (Avro, 2019). It is estimated that 67% of the coral reef has been damaged so far (Avro, 2019) and this damage will continue if the exploitation of this fragile and much-needed ecosystem continues.

It is obvious that to expand and develop tourism we need to make use of our coastal and marine resources. However, trade-off between use and protection should be maintained (Hall, 2001). The very definition of sustainable development is to use without exploiting natural, cultural, and other tourist resources in the present generation so that they can be preserved for the next generation (Angelevska-Najdeska & Rakicevik, 2012). Therefore, we need to find ways to retrieve economic benefits from these resources while protecting and preserving them.

Lack of Initiative and Collaboration

Though the concept of Blue Economy, and sustainable tourism as a part of it, has been introduced in Bangladesh, there has not been much practical progress in this field. There is lack of policy related to the blue economy and lack of focus on marine-related fields at higher policy-making levels (Islam et al., 2018). Bangladesh has no legal instrument that sets the standards for utilization of marine resources in compliance with the principles of the Blue Economy. There is a need for an effective administrative and legislative framework, proper institution and infrastructure, research institutions, and establishment of relevant technology etc. to make substantial progress in the field of Blue Economy (Alam, 2019). However, no appropriate step has yet been taken to this effect.

The Blue Economy Cell (BEC) of Bangladesh was established in 2017 under the Energy and Mineral Resource Division (EMRD) (Hasan, 2021). It was established to ensure and manage collaboration between different ministries to explore and utilize maritime resources for economic benefit (Hasan, 2021). However, apart from some meetings, no real work has been done. Shortage of workforce, funds and focus on this field has made this institution fall into decay and no development towards blue economy has been done in the past 4 years (Hasan, 2021). Finally, dearth of efficient leadership and bureaucracy as well as absence of collaboration between private and public sectors towards investment and development of the blue economy is another major pitfall when it comes to implementing the blue economy and sustainable development (Islam et al., 2018).

Lack of Human Resources

The development of any given economic sector is impossible without human resources. For the proper development of any sector, there is a need for skilled and expert workforce engaging on different relevant fields developing the sector. The deficiency of skilled human resources has been identified as one of the key challenges to the development of the blue economy (Alam, 2019). Since there is lack of focus on sustainably utilizing marine resources, there is shortage of people interested or willing to work in this field. The dearth of knowledge, technology, and expert workforce is one of the major obstacles in maximizing the benefits of the blue economy (Islam et al., 2018). We have shortage of skilled labor that would assess and determine what maritime assets are available for utilization (Islam et al., 2018).

To give one example, the BEC was formed with 25 temporary posts with an additional secretary (Hasan, 2021). However, so far, only 13 of those posts have been filled and no effort has been made to expand the organization to include more skilled and organized labor force (Hasan, 2021). Lacking manpower is the main problem in this institution. Expansion and integration of expert and skilled human resources in this field, who would co-ordinate and work to develop the blue economy in marine economic sectors, including sustainable marine tourism are a must.

Conclusion

The evolving notion of catamaran tourism is a significantly paramount approach to flourish marine and coastal tourism. Given the growing importance on the necessity of introducing catamaran events as part of sustainable marine resources protection, several governmental as well as private organizations come forward to design and devise catamaran tours in the places of attractions such as Cox's Bazar, Sundarban, Kuakata (Barisal) etc.

Considering the pitfalls of marine and catamaran tourism, Bangladesh particularly the Blue Economy Cell, Bangladesh Parjatan Corporation etc. requires involving actively to facilitate and strengthen the functions of catamaran tours along with organizing, embellishing and coordinating different aspects of activities complying with the fundamental principles of blue economy.

As a riverine country with many coastal sites, the approach of catamaran events could be a forceful, vigorous and dynamic mechanism to bolster the sustainable marine as well as catamaran tourism. Now, the evolution of catamaran tours is at its nascent stage. Attracting tourists (national and international) necessitates implementation of basic rules such as good organization skills, good food, and neat and clean environment, stable and strong catamaran boats, and good residential facilities, supportive appliances to entertain tourists both in and outside the boats. Preconditions to successful catamaran tourism should be maintained and implemented effectively. The responsibility of Bangladesh government is to adopt pragmatic approaches and demarcate functions among different departments and implement those to sustain the golden rule of blue economy and blue tourism.

References

Afroz, N. N., & Hasanuzzaman, M. (2018). Problems and prospects of tourism in Bangladesh Bandarban District case. *Global Journal of Management and Business Research, 12*(23).

Ahmad, H. (2019). Bangladesh coastal zone management status and future trends. *Journal of Coastal Zone Management, 22*(1), 1–6.

Airways Office. (2019). Retrieved from: https://airwaysoffice.com/top-10-travel-agency-in-bangladesh/. Accessed 13 Apr 2021.

Alam, J. (2018). Problems and prospects of tourism industry in Bangladesh: A case of Cox's Bazar tourist spots. *International Journal of Science and Business, 2*(4), 568–579.

Alam, M. K. (2019). *Blue economy – Development of sea resources for Bangladesh*. Ministry of Foreign Affairs (MoFA). Retrieved from: https://mofa.gov.bd/site/page/8c5b2a3f-9873-4f27-8761-2737db83c2ec/OCEAN/BLUE-ECONOMY%2D%2DFOR-BANGLADESH?fbclid=IwAR3JpPf5-8IEvnb2h_EXumtW5RfwgFwKH898qyzGJV5aruRu71YJyz6zOAU. Accessed 26 Apr 2021.

Ali, T. (2013). River ways and tourism. *The Daily Star.* Retrieved from: https://www.thedailystar.net/news/river-ways-and-tourism. Accessed 15 Apr 2021.

Amin, M. R. (2016). Domestic tourists' mindset towards responsible tourism management: A case study on Cox's Bazar, Bangladesh. In *Proceedings of the Asia Tourism Forum 2016 – The 12th Biennial Conference of Hospitality and Tourism Industry in Asia*. https://doi.org/10.2991/atf-16.2016.1

Angelevska-Najdeska, K., & Rakicevik, G. (2012). Planning of sustainable tourism development. *Procedia – Social and Behavioral Sciences, 44*, 210–220.

Attri, V. N. (2018). *The role of marine tourism in IORA: The pathways ahead*. Paper presented at the 3rd Tourism Experts Meeting for the Establishment of the IORA Core Group on Tourism. Durban, 6th–8th May, 2018.

Avro, A. N. (2019). Coral reefs of Bangladesh are under threat. *The Independent.* Retrieved from: https://m.theindependentbd.com/printversion/details/207802. Accessed 25 Apr 2021.

Baldwin, G. (2018). How does tourism impact the ocean: Water pollution*? Sea Going Green.* Retrieved from: https://www.seagoinggreen.org/blog/waterpollution/. Accessed 25 Apr 2021.

Bangladesh Expeditions. (n.d.). *Dhaka river cruise.* Retrieved from: http://expeditions-bd.com/tour/dhaka-river-cruise/. Accessed 14 Apr 2021.

Bangladesh Parjatan Corporation (BPC). (2019). *Vision & mission.* Retrieved from: http://www.parjatan.gov.bd/site/page/68c2d54a-112f-4804-8624-9180bf59df98/. Accessed 18 Apr 2021.

Bangladesh Parjatan Corporation (BPC). (n.d.). *Kuakata.* Retrieved from: https://web.archive.org/web/20110719081254/http://www.bangladeshtourism.gov.bd/beach_kuakata.php. Accessed 13 Apr 2021.

Bangladesh Tourism Board (BTB). (2018). *Background of Bangladesh Tourism Board.* Retrieved from: http://www.tourismboard.gov.bd/site/page/039c81b0-e8d4-4627-8f3e-f6c4a711b7e7/. Accessed 20 Apr 2021.

Bangladesh Water Development Board (BWDB). (n.d.). *Rivers information.* Retrieved from: https://www.bwdb.gov.bd/rivers-information. Accessed 12 Apr 2021.

Bangladesh.com. (n.d.-a). *Enjoy the tranquility of Kuakata.* Retrieved from: https://www.bangladesh.com/blog/enjoy-the-tranquility-of-kuakata. Accessed 21 Apr 2021.

Bangladesh.com. (n.d.-b). *Discover the beautiful Bay of Bengal.* Retrieved from: https://www.bangladesh.com/blog/discover-the-beautiful-bay-of-bengal/. Accessed 21 Apr 2021.

Bangladeshfact.com. (2017). *Greenline water bus schedule, ticket price and booking counters.* Retrieved from: https://bangladeshfact.com/greenline-water-bus/. Accessed 14 Apr 2021.

Bddivers. (2020). *Oceanic scuba diving service.* Retrieved from: http://www.bddivers.org/?fbclid=IwAR1LGjhL-Gpv3RhoY84jij7syTlIh0TEAB3fgfHsLOc8V96-6baDHQOj-W4. Accessed 21 Apr 2021.

Belal, A. S. M. (2013). *Maritime boundary of Bangladesh: Is our sea lost?* Bangladesh Institute of Peace and Security Studies (BIPPS).

Bhuiyan, A. H., Darda, A., Habib, W., & Hossain, B. (2020). *Marine tourism for sustainable development in Cox's Bazar, Bangladesh.* Asian Development Bank Institute. Retrieved from: https://www.adb.org/sites/default/files/publication/616761/adbi-wp1151.pdf. Accessed 14 May 2021.

Brida, J. G., & Zapata, S. (2010). Cruise tourism: Economic, socio-cultural and environmental impacts. *International Journal of Leisure and Tourism Marketing, 1*(3), 205–226.

Britannica. (1998). *Catamaran.* Retrieved from: https://www.britannica.com/technology/catamaran. Accessed 14 Apr 2021.

British Broadcasting Corporation (BBC). (2011). *Who, what, why: Is whale watching harmful to whales? BBC.* Retrieved from: https://www.bbc.com/news/magazine-14107381#:~:text=So%20does%20it%20harm%20the,whales%2C%20putting%20everyone%20at%20risk. Accessed 25 Apr 2021.

Calderwood, L. U., & Soshkin, M. (2019). *The Travel & Tourism Competitiveness Report 2019-travel and tourism at a tipping point.* World Economic Forum.

Claussen, E. (2002). *Coastal and marine ecosystems & global climate change: Potential effects on U.S. resources.* Center for Climate and Energy Solutions. Retrieved from: https://www.c2es.org/document/coastal-and-marine-ecosystems-global-climate-change-potential-effects-on-u-s-resources/#:~:text=Climate%20change%20is%20likely%20to,and%20oxygen%20to%20marine%20organisms. Accessed 24 Apr 2021.

Coastal Wiki. (n.d.). *Impact of tourism in coastal areas: Need of sustainable tourism strategy.* Retrieved from: http://www.coastalwiki.org/wiki/Impact_of_tourism_in_coastal_areas:_Need_of_sustainable_tourism_strategy. Accessed 25 Apr 2021.

De Cantis, S., Ferrante, M., Kahani, A., & Shoval, N. (2016). Cruise passengers' behavior at the destination: Investigation using GPS technology. *Tourism Management, 52*, 133–150.

Dhaka Dinner Cruise. (n.d.). *About: Dhaka Dinner Cruise.* Retrieved from: https://dhakadinnercruise.com/about-2/. Accessed 14 Apr 2021.

Dhaka Tribune. (2019). *Bangladesh makes a splash on the international cruise circuit.* Retrieved from: https://www.dhakatribune.com/feature/travel/2019/01/27/bangladesh-makes-a-splash-on-the-international-cruise-circuit. Accessed 19 Apr 2021.

Discover Boating. (n.d.). *Catamaran boats: Types, uses & activities*. Retrieved from: https://www.discoverboating.com/buying/boat/catamarans. Accessed 14 Apr 2021.

Fakun, P. N. (2018). *Cruise travel: The new craze for Mauritians*. Defimedia.info. Retrieved from: https://defimedia.info/cruise-travel-new-craze-mauritians. Accessed 10 Apr 2021.

Foroozanfar, M. H. (2014). The future of cruise ship tourism industry: The challenges of cruising. *Journal of Social Issues & Humanities, 2*(7), 213–224.

GetMyBoat. (n.d.). *155' long full AC luxurious ship MV wave for cruising Sundarban, Bangladesh*. Retrieved from: https://www.getmyboat.com/trips/3a2xq31K/. Accessed 14 Apr 2021.

Global Alliance on Health and Pollution (GAHP). (2018). *Bangladesh & pollution: Health impact, economic impact, injustice, and solutions*. Retrieved from: https://www.pureearth.org/wp-content/uploads/2018/11/Bangladesh-Pollution-Health-and-Economic-Impacts-The-Lancet-Report.pdf. Accessed 10 Apr 2021.

Hall, C. M. (2001). Trends in ocean and coastal tourism: The end of the last frontier? *Ocean and Coastal Management, 44*(9), 601–618.

Hasan, S. (2021). Bangladesh's blue economy cell falters; no progress in 4 yrs. *United News of Bangladesh*. Retrieved from: https://unb.com.bd/category/Bangladesh/bangladeshs-blue-economy-cell-falters-no-progress-in-4-yrs/64715. Accessed 26 Apr 2021.

Hetzer, N. (1965). *Environment, tourism, culture*. UNNS, Reported Ecosphere. Ecosphere, pp. 1–3.

Hossain, B., & Wadood, S. N. (2020). Potential unexplored? Tourism and economic growth of Bangladesh. *Journal of Tourismology, 6*(1), 10–11.

Hussain, A. (2020). First-ever cruise ship launched on Cox's Bazar-St Martin route. *Dhaka Tribune*. Retrieved from: https://www.dhakatribune.com/bangladesh/nation/2020/12/20/first-ever-cruise-ship-to-launch-on-cox-s-bazar-st-martin-route. Accessed 12 Apr 2021.

Hussain, M. G., & Hoq, E. (2018). Impacts of climate change on coastal and marine fisheries resources in Bangladesh. *Researchgate*. Retrieved from: https://www.researchgate.net/publication/329590623_Impacts_of_Climate_Change_on_Coastal_and_Marine_Fisheries_Resources_in_Bangladesh. Accessed 24 Apr 2021.

Informationbd.org. (n.d.). *Kuakata sea beach*. Retrieved from: https://sites.google.com/site/informationbdcom/travel/kuakata-sea-beach. Accessed 13 Apr 2021.

International Union for Conservation of Nature (IUCN). (n.d.). *Climate change and water in Bangladesh: Information brief*. IUCN.

Islam, M. K., Rahaman, M., & Ahmed, Z. (2018). Blue economy of Bangladesh: Opportunities and challenges for sustainable development. *Advances in Social Sciences Research Journal, 5*(8), 68–178.

Kabir, M. H. (2010). Coastal and marine pollution. *The Daily Star*. Retrieved from: https://www.thedailystar.net/news-detail-135561#:~:text=The%20coast%20of%20Bangladesh%20is,coastal%20and%20marine%20environment%20vulnerable.&text=Marine%20pollution%20also%20arises%20from,mineral%20exploration%20and%20production%20activities. Accessed 25 Apr 2021.

Ketz, D., Ketz, R., & Jennings, C. (2019). Managing the impacts of cruise ship tourism. *106 Group*. Retrieved from: https://106group.com/wp-content/uploads/2019/02/Managing-the-Impacts-of-Cruise-Ship-Tourism.pdf. Accessed 11 Apr 2021.

Lamers, M., Eijgelaar, E., & Amelung, B. (2015). The environmental challenges of cruise tourism: Impacts and governance. In C. Hall, S. Gössling, & D. Scott (Eds.), *The Routledge handbook of tourism and sustainability* (pp. 430–439). Routledge.

Light Castle Analytics Wing. (2020a). *COVID-19 & tourism: A tragic end to the thriving industry?* Retrieved from: https://www.lightcastlebd.com/insights/2020/06/covid-19-tourism-a-tragic-end-to-the-thriving-industry#:~:text=The%20travel%20and%20tourism%20industry,among%20the%20South%20Asian%20countries. Accessed 13 Apr 2021.

Light Castle Analytics Wing. (2020b). *Tourism: A possible new driver for the economy of Bangladesh*. Retrieved from: https://databd.co/stories/tourism-a-possible-new-driver-for-the-economy-of-bangladesh-9357. Accessed 14 Apr 2021.

Lonely Planet. (2019). *St. Martin's Island*. Retrieved from: https://www.lonelyplanet.com/bangladesh/chittagong-division/st-martins-island. Accessed 13 Apr 2021.

Merriam-Webster. (n.d.). *Catamaran.* Retrieved from: https://www.merriam-webster.com/dictionary/catamaran. Accessed 14 Apr 2021.

Metacentre. (2013). *Green Line 1.* Retrieved from: http://metacentre.com.bd/greenline1.html. Accessed 14 Apr 2021.

Middlebury Institute of International Studies. (n.d.). *Our history and methodology.* Center for Blue Economy. Retrieved from: https://www.middlebury.edu/institute/academics/centers-initiatives/center-blue-economy/about/history. Accessed 20 Apr 2021.

Moreno, A., & Amelung, B. (2009). Climate change and coastal & marine tourism: Review and analysis. *Journal of Coastal Research, 56*, 1140–1144.

MV Karnaphuli Express. (n.d.). *MV Karnaphuli Express.* Retrieved from: http://www.mvkarnaphuliexpress.com/. Accessed 14 Apr 2021.

Nahar, K. (2019). 'Riverine tourism' potentials wither. *The Financial Express.* Retrieved from: https://thefinancialexpress.com.bd/trade/riverine-tourism-potentials-wither-1548044387. Accessed 15 Apr 2021.

Nations Encyclopedia. (n.d.). *Bangladesh.* Retrieved from: https://www.nationsencyclopedia.com/geography/Afghanistan-to-Comoros/Bangladesh.html. Accessed 12 Apr 2021.

Nijhoom Tours. (n.d.). *Sundarban Mangrove Forest: 8 things to do for the best experience.* Retrieved from: https://nijhoom.com/sundarbans/. Accessed 14 Apr 2021.

Nur Nobi, M., & Majumder, M. (2019). Coastal and marine tourism in the future. *Journal of Ocean and Coastal Economics, 6*(2), 12–31.

One Planet Network. (2020). *Tourism's plastic pollution problem.* Retrieved from: https://www.oneplanetnetwork.org/sustainable-tourism/tourisms-plastic-pollution-problem. Accessed 25 Apr 2021.

Orams, M. (2002). *Marine tourism development impacts and management.* Routledge.

Papageorgiou, M. (2016). Coastal and marine tourism: A challenging factor in marine spatial planning. *Ocean and Coastal Management, 129*, 44–48.

Richter, H. (2020). Cruising post-COVID-19: Lessons and challenges for the cruise ship industry. *Sustainalytics.* Retrieved from: https://www.sustainalytics.com/esg-blog/cruising-post-covid-19-lessons-and-challenges-for-the-cruise-ship-industry/. Accessed 14 Apr 2021.

Sahoo, P. K., Mason, S., & Tuite, A. (2008). Practical evaluation of resistance of high-speed catamaran hull forms – Part II. *Ships and Offshore Structures, 3*(3), 239–245.

Sieghart, L., & Rogers, D. (2015). Bangladesh: The challenges of living in a delta country. *World Bank Blogs.* Retrieved from: https://blogs.worldbank.org/endpovertyinsouthasia/bangladesh-challenges-living-delta-country. Accessed 12 Apr 2021.

Tegar, D., & Gurning, R. O. S. (2018). Development of marine and coastal tourism based on blue economy. *International Journal of Marine Engineering Innovation and Research, 2*(2), 128–132.

The Daily Star. (2017). *Bangladesh enters cruise tourism.* Retrieved from: https://www.thedailystar.net/business/bangladesh-enters-cruise-tourism-1368913. Accessed 19 Apr 2021.

The Independent. (2019). *Tourism sector prospects bright.* Retrieved from: https://theindependentbd.com/post/181751. Accessed 13 Apr 2021.

Tour Operators Association of Bangladesh (TOAB). (n.d.). *About TOAB.* Retrieved from: https://www.toab.org/index.php?action=page&page_id=6. Accessed 19 Apr 2021.

Travel Mate. (2021a). *MV Karnafuly Express: About, ticket booking & updated pricing.* Retrieved from: https://www.travelmate.com.bd/mv-karnafuly-express/. Accessed 14 Apr 2021.

Travel Mate. (2021b). *MV Bay One Cruise: About, ticket booking & updated pricing.* Retrieved from: https://www.travelmate.com.bd/mv-bay-one-cruise/. Accessed 14 Apr 2021.

TripAdvisor. (n.d.-a). *Beaches in Chittagong division.* Retrieved from: https://www.tripadvisor.com/Attractions-g667480-Activities-c61-t52-Chittagong_Division.html. Accessed 13 Apr 2021.

TripAdvisor. (n.d.-b). *Bodies of water in Chittagong division.* Retrieved from: https://www.tripadvisor.com/Attractions-g667480-Activities-c57-t162-Chittagong_Division.html. Accessed 13 Apr 2021.

U.S. Department of State. (2021). *Bangladesh travel advisory.* Retrieved from: https://travel.state.gov/content/travel/en/traveladvisories/traveladvisories/bangladesh-travel-advisory.html. Accessed 15 Apr 2021.

United Nations (UN). (n.d.). *Blue economy concept paper*. Retrieved from: https://sustainabledevelopment.un.org/content/documents/2978BEconcept.pdf. Accessed 23 Apr 2021.

United Nations Educational, Scientific and Cultural Organization (UNESCO). (1997). *The Sundarbans.* Retrieved from: https://whc.unesco.org/en/list/798/. Accessed 13 Apr 2021.

Upadhyay, D. K., & Mishra, M. (2020). Blue economy: Emerging global trends and India's multilateral cooperation. *Maritime Affairs: Journal of the National Maritime Foundation of India, 16*(1), 30–45.

Viking Yacht. (n.d.). *About us – Viking Yacht.* Retrieved from: https://vikingyachtbd.com/about-us/. Accessed 14 Apr 2021.

Wade, D. (2021). *12 Best catamaran sailboats.* Retrieved from: https://www.lifeofsailing.com/post/best-catamaran-sailboats. Accessed 14 Apr 2021.

Walton, J. K. (2012). Tourism: Definition, history, types, importance, & industry. *Britannica.* Retrieved from: https://www.britannica.com/topic/tourism. Accessed 10 Apr 2021.

Wikitravel. (n.d.). *Saint Martins Island.* Retrieved from: https://wikitravel.org/en/Saint_Martins_Island. Accessed 13 Apr 2021.

World Bank (WB). (2000). *Bangladesh – Climate change and sustainable development*. WB.

World Bank (WB). (2017). *The potential of the blue economy: Increasing long-term benefits of the sustainable use of marine resources for small island developing states and coastal least developed countries.* Retrieved from: https://openknowledge.worldbank.org/bitstream/handle/10986/26843/115545.pdf?sequence=1&isAllowed=y. Accessed 21 Apr 2021.

World Tourism Organization (UNWTO). (1995). *Technical manual: Collection of tourism expenditure statistics*. UNWTO.

World Tourism Organization (UNWTO). (2016). *Sustainable cruise tourism development strategies–tackling the challenges in itinerary design in South-East Asia*. UNWTO/APTEC.

World Travel & Tourism Council (WTTC). (2020). *Bangladesh 2019 key data.* Retrieved from: https://www.statista.com/statistics/233223/travel-and-tourism%2D%2Dtotaleconomic-contribution-worldwide./ Accessed 21 Apr 2021.

World Wide Fund for Nature (WWF). (2015). *Principles for a sustainable blue economy.* Retrieved from: https://wwf.panda.org/?247477/Principles-for-a-Sustainable-Blue-Economy. Accessed 22 Apr 2021.

Dr Rakiba Nabi is Professor in the Department of Law at the University of Chittagong, Chattogram, Bangladesh. Dr Nabi graduated with an LLB (Honors), and LLM from the University of Chittagong in 2004, did a second LL.M in Human Rights from the University of Hong Kong in 2012–2013, and completed her SJD in Indigenous Law from the University of Kansas, the United States in 2020. Her present areas of interest are Human Rights, Refugee Law, Indigenous Law and Policy, Law of International Organization, Tourism and Sustainable Development. She published 10 scholarly articles in national and international journals. She also works in several projects funded by Research Cell, CU, and Law Ministry, Bangladesh.

Dr A. H. M. Salimullah is Professor in the Department of Economics at the University of Chittagong, Chattogram, Bangladesh. Dr Salimullah graduated with a BSS (Honors) and MSS degree in Economics from the University of Chittagong in 2003, did a second MA degree in Financial Economics from the Middle Tennessee State University (MTSU) in 2014, and completed his PhD in Econometrics and Quantitative Economics from the same university in 2018. His present areas of interest are Labor Economics, Dynamic Macroeconomics, and Economics of Tourism. He published 10 scholarly articles in national and international journals. Besides, he presented several papers at different national and international conferences. The notable conferences that he attended and presented his research were the Academy of Economics and Finance (AEF)

Conference in 2018 at Houston, Texas, and Dr Gregory P. Domin Graduate Conference in 2017 at Columbus State University, Columbus, Georgia. He worked as a reviewer for a journal named *Scientia et Humanitas*, sponsored by the University Honors College at MTSU. He is a proud winner of the Deans' Distinguished Essay Award for his scholarly publication in the Honors College journal of MTSU. Besides, he won important prizes during scholars' week university-wide poster presentation at MTSU. He is chosen as a proud member of the MTSU Chapter of Beta Gamma Sigma due to his scholarly achievement.

Chapter 12
Information and Communication Technology Application in Tourism Events, Fairs and Festivals in India

Saroop Roy B. R.

Introduction

Information and Communications Technologies (ICT) plays a major role in organizing events, fairs and festivals, which are part and parcel of tourism industry. The developments in ICT has paved way to make changes in the way event organizers interacts with the participants. The introduction of event management tools has simplified things for event managers to make the planning exercise efficient. These technological developments help the planners in saving time, planning the event cost-effectively, facilitates better communication, helps in event ticketing and registration and ensures event security. The event management softwares arrange data for the event planners so that they can immediately start working on the key components of the event, thereby saving time. The event planning tools also help to track how much money is spent on each event by the organizer. This helps in comparing the money spent on previous similar events and to analyze where one can make cuts in future events to make it cost effective. Since everyone part of the event works on the same platform, it is easy for them to communicate and collaborate with one another. Linking with social media platforms makes it easier for participants to share their feedback to event planners. The social media platform also works as a great tool for advertising an event (Bronson, 2018). Introduction of online booking applications has helped in booking tickets for events on various mobile apps and websites. Technological advancements have also helped in easing the registration process for events and even to select seats of their choice. Digital payment platforms have made money transfer easy for booking the events and do the registration process. Security is another aspect that needs to be addressed very carefully. A lot of

S. R. B. R. (✉)
Centre for Responsible Tourism, Kerala Institute of Tourism and Travel Studies (KITTS), Thiruvananthapuram, Kerala, India
e-mail: sarooproy@kittsedu.org

A. Hassan (ed.), *Technology Application in Tourism Fairs, Festivals and Events in Asia*, https://doi.org/10.1007/978-981-16-8070-0_12

personal data is gathered as part of the event and it is important that the collected data does not reach the public domain. Technology plays a key role in protection of this data. In addition to this, people security is also important. Equipments like retina and fingerprint scanners, body scanners, metal detectors are being widely used to ensure safety of the participants of the events (Sharma, 2019).

The current decade has witnessed development of more novel technologies in ICT and the global event industry has in fact capitalized on it to improve their operational efficiencies and to meet the demands of the customers. Many new trends in ICT like AI, Augmented Reality (AR), Virtual Reality (VR), Mobile Event Apps, Projection Mapping, Live Translation, Blockchain Technology, Charging Tables, Wearable Technology, Drones, Live Streaming have been extensively used now a days in event management for marketing purposes, operational efficiency as well as to enhance the level of participant satisfaction. The following section looks in detail on the theoretical aspects of ICT and its application in events, festivals and fairs with a focus on Indian scenario.

Artificial Intelligence

AI is a technique for teaching a computer, a computer-controlled robot, or software to think intelligently in the same way as intelligent humans do (Tutorialspoint, n.d.). AI is widely used in events to deliver highly personalised experience even in conferences and festivals with more than thirty thousand attendees. AI is often used to analyse data from attendees' social media profiles and propose that they attend conference sessions, look at new goods, and so on. Using an AI plug-in, such as Grip, in conjunction with an event app is one way to apply AI to events (Saini, 2019). Telangana, an Indian state, has effectively employed AI to monitor crowds during Asia's largest yearly tribal celebration, the "Medaram Jatara". Authorities were able to use the technology to track changes in crowd density in a given region over time and take corrective steps such as personnel deployment, public announcements, and crowd rerouting to less congested places to avert risks such as stampedes (Telangana Today, 2021).

Chatbots

A Chatbot is artificial intelligence software that can mimic a natural language discussion (or chat) with a user via messaging apps, mobile apps, websites, or the telephone (Expert System, n.d.). In event management, Chatbots help both participants as well as event organisers. Chatbots can be designed to provide venue details, show booking dates, confirm conference registration, allocate seats, sending alerts and reminders as well as respond to queries of participant etc. (Makerobos, n.d.). As the event industry is getting global, Chatbots can be used to make conversations in the participant's preferred language. Thus, a multilingual Chatbots can eliminate the need for representatives from across globe to be available round-the-clock to answer queries (Maruti Techlabs, n.d.. It helps the staff of event companies to work efficiently, promote and sell tickets with discounts on certain dates, collect catering orders without human involvement and also helps the organisers to fine tune future events by getting user feedback (Makerobos, n.d.). Chatbots can also be used to promote event's merchandise by facilitating its purchase by the participants, thus serving the purpose of E – (vent) commerce. In addition, this, Chatbots can create meaningful engagements with participants by conducting polls and surveys during the event. With this, the event organisers could gain insights about popular sessions and hold real-time contests, thereby underpinning two-way dialogue between participants and experts at the event. In the case of events spread across large area, Chatbots help the participants to navigate from one place to another by providing details of the location (Maruti Techlabs, n.d.). Even a few years ago, event bots were becoming popular in India. A virtual assistant was utilised during the 20th E-governance conference in India in 2017 to direct guests based on their profile in terms of who to meet, which booth to visit, and so on (Sheth, 2017).

Facial Recognition

The technique of recognizing or verifying an individual's identification by looking at their face is known as facial recognition. Patterns based on a person's face features are captured, analyzed, and compared using this technique (Thales, n.d.). Facial recognition is being currently used at the registration desk to reduce time in getting data of the attendees and assigning badges to them. It also helps to handle the attendees at numerous check points, thereby ensuring more security for the event. The FR technology also helps in tracking accurate numbers of participants as well as in providing real time information to security personnel during big events/festivals.

NEC India has collaborated closely with the Andhra Pradesh State Government on Facial Recognition Systems (FRS) by integrating with CCTV surveillance cameras installed by the Vijayawada City Police at key locations during the prestigious religious festival of "Krishna Pushkaram", which takes place once every 12 years at the river Krishna and attracts 50 million pilgrims. By monitoring and scanning photos of suspects and criminals, the FRS system may enable real-time criminal identification to aid front-line policing operations (Security Today, 2016).

The Delhi Police Authority deployed approximately 30 face recognition cameras using superior facial recognition technology to enable police identify terrorists and criminal elements during the Republic Day festivities in 2019. The quality of the facial recognition cameras was such that if any visitor had a 70% face match with any accessible image of terrorists or criminals in the database, the system would transmit an alarm to the control centre (Choudhury, 2020).

Facial recognition system was used to provide accurate headcounts, track patterns of movement, and measure time spent between distinct points of pilgrims during the 2020 “Magh Mela”, an annual Hindu pilgrimage and festival that draws millions of people in a single day and assembles at Prayagraj, where the Ganges, Yamuna, and Saraswati rivers converge in the north Indian state of Uttar Pradesh. People frequently carried large bags on their heads as they navigated the mela throng, thus this did not produce the desired outcomes (Jha, 2020b).

Emotional Recognition

Emotion Recognition Technology combines psychology, human expressions, and artificial intelligence to identify and recognize human emotions on a person’s face automatically. In real time, the facial analysis algorithm can detect many emotional states such as happy, sorrow, disgust, surprise, rage, fear, and so on (Sightcorp, n.d.). By using this technology, the business can get a feedback of their products or services in real time. The technology aids event planners in determining how participants feel and interact with seminars, entertainment, and other aspects of the event. By just aiming a camera at attendees, it allows for continuous study of their emotions during the session. Although this is being used by certain advanced countries, it has not find a place in India yet.

Augmented Reality and Virtual Reality

AR aids in the real-time integration of digital data with the user’s environment. It works with the current environment and adds additional information on top of it (WhatIs.com, n.d.-a). VR is a software-created artificial world that is presented to the user in such a way that the user suspends disbelief and accepts it as a genuine environment (WhatIs.com, n.d.-b). There has been an increased use of AR and VR by the event industry in the past few years for enhancing the delegate experiences as well as for marketing and promotional campaigns.

For physical events, AR technology can assist attendees in finding their way to the event location. AR can help participants avoid getting lost in pavilions at conferences and summits by assisting with event venue navigation. Participants may provide stronger presentations backed by statistical data and concept screening with the aid of AR-based event applications. When combined with a little gamification, AR technology can increase participant engagement. With the aid of AR technology, event planners may involve participants in activities such as digital treasure hunts and the setup of competition stations (Thomas, 2020). AR is comparatively easer to implement than VR as a mobile device with AR capabilities can be easily used to leverage the technology. Device-based AR can be a cost-effective option since it only requires a little investment in hardware, as it can be used to interact with AR material using a smartphone or tablet with AR capabilities. This has got a wider reach as most people carry smartphones and delegates only have to open the AR app to use the function. Wearable AR devices like Microsoft Hololens and Google Glass can overlay virtual information onto the real world, thereby providing the most immersive AR experience for the participants. The most advanced form of AR solutions is the Transparent AR-capable screens, which uses Organic Light Emitting Diode (OLED) display for projecting a three-dimensional environment over a 2D screen, thereby providing the participants a 360-degree immersive experience without a smart glass or mobile phone (Joshi, 2020).

VR provides a better experience than AR as participants can experience a 360-degree immersive and engaging experience of a dissimilar place altogether without physically being there. The advantages of VR over AR include better engagement, better graphical environment and minimal distractions. Participants can get a virtual tour of operational components of a production unit or a warehouse using VR devices, get a virtual demo when businessmen showcase their products using VR devices, interact with other participants remotely using VR devices, or virtually attend training sessions using a live account. VR also allows venue management to display the insides of their space to event vendors and planners. In India, Goa-based Avinash Kumar has been organizing the EyeMyth Media Arts Festival since 2016, which focuses on immersive new media ranging from code-based design to art that encompasses VR and AR, in addition to interactive experiences like gaming and performance art (Shankar, 2019).

Mobile Event Apps

A mobile application is software that is meant to operate on a mobile device, such as a tablet computer and smartphone (Techopedia, n.d.). A mobile event app is a smartphone software designed to improve the experience of attendees at a live event. The major features of mobile event app include providing digital event agenda, speaker profiles, sponsor pages, one-to-one messaging, getting participant profiles, social media integrations etc. (Bizaboo, n.d.). These apps not only facilitate meeting with others but helps participants identify other participants whom they want to meet and network with. As discussed in case of Chatbots, the mobile event apps are also able to conduct polls and surveys and can create meaningful engagement of the participants. The most widely used Event Apps in India include Hubb, Aventri, Evenesis, EventsWallet, Meeting box etc. (Sourceforge, n.d.). The Eventival Mobile App has been widely used to enhance the film festival experience in the country like the International Film Festival Kerala (IFFK) by helping the participants to browse the films and screening schedule, create their own screening schedule, get any new information about festival news like itineraries and who is there for the event (Eventival, n.d.).

Projection Mapping

Buildings, stages, runways, and even water may be transformed into dynamic 3D displays using projection mapping, a video projection method that manipulates lighting onto a surface. As a result, these surfaces become a canvas, with images projected onto them, providing a pleasant light and illusion experience (Business News Daily, 2020). Projection mapping is becoming more popular as a medium for video trends at events to give a more engaging experience for guests through visual stimulation (EE News Desk, 2020).

Creative Factory, a Mumbai-based visual content firm, created a stunning projection mapping for the GVK Sangeet Ceremony in Hyderabad in 2017. With close to 25,000 × 1080 pixels used for mapping, this was India's largest projection mapping experience for a wedding. For the guest entrance, 9 distinct themes with themed music and lights were designed that changed every 2 min. They also created a dynamic mapping experience lasting four and a half minutes, as well as an hour of content for artist acts and graphics for the Disc Jockey (DJ) (Jain, 2017). In March 2020, India was set to host emrg2020, the world's first public festival of light and new media arts, co-organized by Artillume, a platform for promoting and fostering experimental new media arts, and Delhi Tourism. The event's objective was to provide artists from all across the country a chance to showcase their ability and originality via more than 40 immersive and interactive light art installations and projection mapping presentations (Art Reveal, 2020). Due to the severity of the COVID epidemic at the time, the program had to be postponed (Jha, 2020a).

Live Translation

The advancement of live translation technology, which allows for the provision of information and services in a variety of languages, aids in the expansion of international events into underserved non-English speaking areas. In a country like India where more than 19,500 dialects or languages are spoken as mother tongues, the development of such technologies is quite significant for millions of people to get exposed with most of the seminars and conferences taking a digital mode with the current global pandemic situation. This technology, which allows rapid translation feeds through mobile devices, earpieces, regular dial-in service, and other means, is being used by an increasing number of event management businesses across the world. The technological problems associated with the usage of Wi-Fi facilities in providing these services will be reduced with the advent of 5G technology. The Indian government is actively developing AI technology that can translate several vernacular Indian languages in real time, allowing communication amongst people who do not speak the same language (The Week, 2021). Microsoft's Translator service now supports 10 Indian languages, including Gujarati, Malayalam, Marathi, Kannada, and Hindi, as well as Urdu, Tamil, Bengali, and Telugu. In India, these technologies will be heavily used in event management in the future years (Faisal, 2020).

Blockchain Technology

Blockchain is a technology that publicly stores all transactions between parties that a particular network produces. The technology got its name Blockchain as each register gets encrypted and grouped into blocks that form a chain. The main features of Blockchain technology is that the data is decentralised, immutable and secured (Cancelas, 2018). At present, the Blockchain technology is being widely used in event management. The technology allows a delegate's ID and transactional information for following them during the event. The system does not need to know who the delegate is at the time of validation using this technique. When a delegate purchases something and pays with their Blockchain ID, the seller is notified that a transaction has occurred and that the correct amount of money has been handed to them. Access control and other information transfers are in the same boat. At the time of transaction, Blockchain eliminates the need for a person's identifying information. In India, three students from Malla Reddy Engineering College for Women in Hyderabad devised a method that uses Blockchain technology to make voting secure, efficient, and easy by removing hurdles such as waiting to cast a vote and standing in lines. The system was able to provide its users with a high level of security by making tampering with votes impossible or non-existent (Yolmo, 2019).

Charging Tables

Charging Tables are provided for attendees of a conference to charge their networking devices especially the smartphones, tablets etc. While attending conferences, delegates take notes, record, take photographs etc., which drain out the batteries of their networking devices. The attendees feel nervous when they see their battery charge running down if they do not have a power bank. Event planners can apply this opportunity for improving the attendee experience by providing charging tables as a value added service and thereby build trust with the attendees. Charging tables offer a concrete area for the attendees to congregate and establish networking with other participants. For the event management companies, incorporating logos into charging table wrap is a great way to expose their brand to attendees (Incharged, 2017). Technology has progressed to the point that wireless battery charging is now possible, allowing any battery-powered device to be charged simply by placing the appliance near a wireless power transmitter or a designated charging station (Renesas, n.d.).

Wearable Technologies

Wearable technology has acquired a lot of traction in event management over the last few years, and it is going to be one of the most prominent event technologies in the next years. Radio Frequency Identification (RFID) is widely used in wearable technology and in event industry, its functions include event ticketing attendee mapping etc. The RFID attendee tracking systems helps for storing useful information for various types of events like conferences, trade shows, competitions, exhibitions, corporate events etc.

One of the most widely used RFID wearable technology is the RFID Badges. The RFID badges contain small chip put on the attendee's wristbands or name tags for giving them a boundless experience. Also, the touch points in the RFID and Near Field Communication (NFC) helps to collect data personalising data of the attendees. The RFID badges also manage the check in for event as well as the sessions and thus provides a balance in the session areas on the number of audience (Inc.Outlook, 2019). In India, there are a number of companies that produce RFID badges for conferences. The Identium Tech solutions provide RFID badges that can automatically track in and out of delegates, the number of delegates present at the venue at a given point of time, trigger alerts to organisers informing them when an important delegate has arrived or left, display personalised welcome messages for each delegate when they enter venue etc. (Indiamart.com, n.d.).

Drones

Unmanned aerial vehicles (UAVs) or unmanned aircraft systems (UASes) may be remotely controlled or fly autonomously using on-board sensors and GPS, as well as software-controlled flight plans in their embedded systems (Earls, 2019). Other than flying, drones may provide event security and surveillance, execute search and rescue operations, analyze crime scenes, and captivate audiences with breath-taking displays, among other things (Inc.Outlook, 2019). Drone videography has become a frequent occurrence in India, at both private and public events. A drone buzzed over the 900-plus crowd during the 2017 Bangalore Comedy Festival, which was hosted at UB City. Before then, the drones covered events including the Echoes of Earth Festival, Bangalore Music Week, Art Bengaluru 2016, and the Embassy Group's Republic Day celebration (Balram, 2017).

The D-4 drone technology was utilized to provide security during the 2021 Republic Day parade on Rajpath. The Defence Research and Development Organisation's (DRDO) D-4 drone technology can quickly identify close-range assaults (Mint, 2021). With the rising popularity of drones, the "Drone Festival of India" was held in New Delhi in January 2020 to debate global drone regulations, technical developments in drones, operations and traffic management in the Drone Ecosystem, and the drone industry's future (Chaudhary, 2019).

Live Streaming

When video is broadcast via the Internet in real time rather than being recorded and saved, it is referred to as live streaming. It distributes a video clip in little chunks via the internet by sending a few seconds of the movie at a time. The benefit of live streaming is that the client devices do not need to download the full video before they can begin playing it (Cloudflare, n.d.). Globally, many sports events, conferences and festivals are reached to a larger audience using live streaming.

Festivals are increasingly being streamed live in India, as event organisers recognise its worth. This may be seen in the One Plus Katy Perry show on YouTube and the Sunburn live on Zee5 aired live in 2019. The rising trend of live streaming live music, sports, and other events is excellent news for fans who have been unable to obtain tickets to their desired events. For example, 18.6 million viewers watched the live stream of the Indian Premier League (IPL) final on the Hotstar website and app in 2019 (The Sherp, 2020). Bharat Bhavan along with Department of Culture, Government of Kerala and other cultural forums has launched a multi-media mega streaming named "Mazhamizhi" for 65 days starting from 28th August 2021 to showcase various art forms of Kerala in order to give local artists a platform to perform as well as to support them financially (Mathrubhumi, 2021).

Conclusion

Event Management Technologies have revolutionised the experience of participants drastically in the last one decade. Most of the technologies developed were for enhancing the participant satisfaction as well as improving the efficiency of project management for physical events. But with global pandemic COVID -19, there has been a drastic shift from organising physical events to online platforms as most of the seminars have turned to webinars. Even big international conferences have become live streaming exercises from different countries. As most people have adopted to such online platforms, which is cost-effective, the chances for going back to offline platforms will take some time taking into consideration of the economic crisis prevailing in countries like India. New technological developments would now be mostly focusing on strengthening such online events.

References

Art Reveal. (2020). *India to witness the first ever public festival of Light & New Media Arts – emrg2020*. Retrieved from: https://www.artrevealmagazine.com/india-to-witness-the-first-ever-public-festival-of-light-new-media-arts-emrg2020/. Accessed 15 Sept 2021.

Balram, S. (2017). Bengaluru's latest obsession: Drones at private and public functions. *The Economic Times*. Retrieved from: https://economictimes.indiatimes.com/magazines/panache/

bengalurus-latest-obsession-drones-at-private-and-public-functions/articleshow/57532706.cms?from=mdr. Accessed 20 July 2021.
Bizaboo. (n.d.). *Mobile event and conference apps: The 2020 guide*. Retrieved from: https://blog.bizzabo.com/event-apps-guide#event-app. Accessed 25 July 2021.
Bronson, R. (2018). The role of technologies in event management. *StartUs Magazine*. Retrieved from: https://magazine.startus.cc/role-technologies-event-management/. Accessed 24 July 2021.
Business News Daily. (2020). *Projection mapping: What it is and why businesses love it*. Retrieved from: https://www.businessnewsdaily.com/10527-projection-mapping-what-is-it.html. Accessed 25 July 2021.
Cancelas, A. (2018). What we can expect from Blockchain in the tourism industry. *We are marketing*. Retrieved from: https://www.wearemarketing.com/blog/what-we-can-expect-from-blockchain-in-the-tourism-industry.html#:~:text=Blockchain%20and%20tourism%20have%20the,information%20to%20the%20different%20firms. Accessed 31 July 2021.
Chaudhary, V. (2019). Drone festival of India 2020 – India's largest UAV conference. *Times Next*. Retrieved from: https://timesnext.com/drone-festival-of-india-2020-indias-largest-uav-conference/. Accessed 25 July 2021.
Choudhury, A. (2020). How facial recognition is becoming the Indian government's new best friend. *Analytics India Mag*. Retrieved from: https://analyticsindiamag.com/how-facial-recognition-is-becoming-the-indian-governments-new-best-friend/. Accessed 3 June 2021.
Cloudflare. (n.d.). *What is live streaming? | How live streaming works*. Retrieved from: https://www.cloudflare.com/en-in/learning/video/what-is-live-streaming/. Accessed 23 June 2021.
Earls, A. R. (2019). Drone (UAV). *Tech Target*. Retrieved from: https://internetofthingsagenda.techtarget.com/definition/drone. Accessed 26 July 2021.
EE News Desk. (2020). *10 Event Tech solutions that have become indispensable for event planners*. Retrieved from: http://everythingexperiential.businessworld.in/article/10-Event-Tech-solutions-that-have-become-indispensable-for-event-planners-/09-01-2020-181760/. Accessed 20 July 2021.
Eventival. (n.d.). *The world's number one film festival software*. Retrieved from: https://www.eventival.com/. Accessed 10 Sept 2021.
Expert System. (n.d.). *Chatbot: What is Chatbot? Why are Chatbots important?* Retrieved from: https://www.expertsystem.com/chatbot/. Accessed 10 July 2021.
Faisal, M. (2020). Microsoft Translator adds 5 more Indian languages: How to use real-time translation. *The Indian Express*. Retrieved from: https://indianexpress.com/article/technology/tech-news-technology/microsoft-translator-supports-5-new-indian-languages-how-to-use-real-time-translation-6365594/. Accessed 24 July 2021.
Inc.Outlook. (2019). *How technology is shaping the event management industry?* Retrieved from: https://www.incoutlook.com/2019/11/20/how-technology-is-shaping-the-event-management-industry/. Accessed 20 June 2021.
Incharged. (2017). *10 Ways charging tables can benefit your event*. Retrieved from: https://incharged.com/10-ways-charging-tables-can-benefit-event/. Accessed 25 July 2021.
Indiamart.com. (n.d.). *Identium tech solutions*. Retrieved from: https://www.indiamart.com/identem-tech-solution/rfid-conference-badges.html. Accessed 22 June 2021.
Jain, S. (2017). Creative factory creates stunning projection mapping experience for GVK Sangeet in Hyderabad. *EventFaqs.com*. Retrieved from: https://www.eventfaqs.com/news/ef-13424/creative-factory-creates-stunning-projection-mapping-experience-for-gvk-sangeet-in-hyderabad. Accessed 2 June 2021.
Jha, L. (2020a). Coronavirus: Media and entertainment events hit across India. *Mint*. Retrieved from: https://www.livemint.com/industry/media/coronavirus-media-and-entertainment-events-hit-across-india-11583408794798.html. Accessed 15 Sept 2021.
Jha, M. (2020b). Eyes in the sky. *Rest of world*. Retrieved from: https://restofworld.org/2020/india-magh-mela/. Accessed 2 June 2021.

Joshi, N. (2020). Leveraging the benefits of AR and VR in the events industry. *Allerin*. Retrieved from: https://www.allerin.com/blog/leveraging-the-benefits-of-ar-and-vr-in-the-events-industry. Accessed 26 July 2021.

Makerobos. (n.d.). *Event management Chatbot: Uses cases, benefits and templates*. Retrieved from: https://www.makerobos.com/chatbot-templates/event-management-chatbot. Accessed 26 July 2021.

Maruti Techlabs. (n.d.). *How can Chatbots help make your event a huge success?* Retrieved from: https://marutitech.com/event-chatbot/. Accessed 27 July 2021.

Mathrubhumi. (2021). *Kalasamoohathinu Unarvekaan Mazha Mizhi Mega Streaming 28 Muthal*. Retrieved from: https://mathrubhumi.co.in/mazhamizhi-mega-streaming-from-august-28/. Accessed 30 July 2021.

Mint. (2021). *D4 Drone System, used for Republic Day Security, can prevent Jammu like attacks: DRDO*. Retrieved from: https://www.livemint.com/news/india/d4-drone-system-used-for-republic-day-security-can-prevent-jammu-like-attacks-drdo-11625205553835.html. Accessed 24 July 2021.

Renesas. (n.d.). *Introduction to wireless battery charging*. Retrieved from: https://www.renesas.com/us/en/products/power-management/wireless-power/introduction-to-wireless-battery-charging. Accessed 25 July 2021.

Saini, D. K. (2019). Event technology trends shaping the future of events in India. *Evenzhub*. Retrieved from: https://www.evenzhub.com/blog/event-technology-trends-shaping-the-future-of-events-in-india. Accessed 3 June 2021.

Security Today. (2016). *NEC India to deploy facial recognition in Andhra Pradesh festival celebrations*. Retrieved from: https://www.securitytoday.in/blog/nec-india-to-deploy-facial-recognition-in-andhra-pradesh-festival-celebrations/. Accessed 3 June 2021.

Shankar, A. (2019). An exhibition for VR makers. *Mint*. Retrieved from https://www.livemint.com/mint-lounge/features/an-exhibition-for-vr-makers-11574432002210.html. Accessed 15 Sept 2021.

Sharma, P. (2019). Role of technology in evolving the event management industry. *The CEO Magazine*. Retrieved from: https://www.theceo.in/technology/role-of-technology-in-evolving-the-event-management-industry. Accessed 5 June 2021.

Sheth, H. (2017). Event Planning 2.0: How technology has changed the face of event organizing. *TSNN*. Retrieved from https://www.tsnn.com/blog/event-planning-20-how-technology-has-changed-face-event-organizing. Accessed 15 June 2021.

Sightcorp. (n.d.). *Everything about emotion recognition*. Retrieved from: https://sightcorp.com/knowledge-base/emotion-recognition/. Accessed 3 June 2021.

Sourceforge. (n.d.). *Mobile event apps in India*. Retrieved from: https://sourceforge.net/software/mobile-event/india/. Accessed 25 July 2021.

Techopedia. (n.d.). *Mobile application*. Retrieved from: https://www.techopedia.com/definition/2953/mobile-application-mobile-app. Accessed 16 July 2020.

Telangana Today. (2021). *Telangana's AI-based crowd management tool selected for NASSCOM award*. Retrieved from: https://telanganatoday.com/telanganas-ai-based-crowd-management-tool-selected-for-nasscom-award. Accessed 15 Sept 2021.

Thales. (n.d.). *Facial recognition: top 7 trends (tech, vendors, markets, use cases & latest news)*. Retrieved from: https://www.thalesgroup.com/en/markets/digital-identity-and-security/government/biometrics/facial-recognition. Accessed 25 July 2021.

The Sherp. (2020). The live streaming of festivals is a global phenomenon that has reached India. *Sherpa Land*. Retrieved from: https://www.festivalsherpa.com/the-live-streaming-of-festivals-is-a-global-phenomenon-that-has-reached-india/. Accessed 20 July 2021.

The Week. (2021). *Govt working on real-time translation tool for Indian languages: MeitY*. Retrieved from: https://www.theweek.in/news/india/2021/06/19/govt-working-on-real-time-translation-tool-for-indian-languages-meity.html. Accessed 25 July 2021.

Thomas, A. (2020). 10 Stunning use cases of AR/VR for event management. *Datamatics*. Retrieved from: https://blog.datamatics.com/10-stunning-use-cases-of-ar-vr-for-event-management. Accessed 25 July 2021.

Tutorialspoint. (n.d.). *Artificial intelligence – Overview*. Retrieved from: https://www.tutorialspoint.com/artificial_intelligence/artificial_intelligence_overview.htm. Accessed 15 July 2021.

WhatIs.com. (n.d.-a). *Augmented reality*. Retrieved from: https://whatis.techtarget.com/definition/augmented-reality-AR. Accessed 14 July 2021.

WhatIs.com. (n.d.-b). *Virtual reality*. Retrieved from: https://whatis.techtarget.com/definition/virtual-reality. Accessed 12 July 2021.

Yolmo, Y. R. (2019). Top 7 Blockchain developments in India: 2019. *AnalyticsIndiaMag*. Retrieved from: https://analyticsindiamag.com/top-7-blockchain-developments-in-india-2019/. Accessed 25 July 2021.

Saroop Roy B. R., BTech, MTA is currently Assistant Professor in Travel and Tourism as well Co-ordinator of Centre for Responsible Tourism at Kerala Institute of Tourism and Travel Studies (KITTS). Before that, he had worked with Thenmala Ecotourism Promotion Society and was also the Kerala Co-ordinator of EQUATIONS, a Bangalore based research, campaign and advocacy organization on tourism policies and issues for 8 years. His expertise areas include tourism policy formulation, ecotourism, responsible tourism, rural tourism and community linked tourism projects. He participated in the SBSTTA 13 Meeting of the Convention on Biological Diversity held at Rome in February 2008. He also led the Indian delegation of Ecotourism practitioners to attend the South African and Indian Practitioners' Workshop to exchange lessons and best practices on using Nature Tourism as a force for Poverty Alleviation at Johannesburg in May 2008. He has played a significant role in the formulation and implementation of Responsible Tourism Initiative in the state of Kerala. On the policy front, he was the drafting committee member of the Kerala Tourism Policy 2012, Revamping of Homestay Classification Scheme for Kerala 2015 and preparation of Charter for Green Carpet initiative for Kerala Tourism 2016. He co-authored the book "An Introduction to business of tourism" published by SAGE India in May 2017 bringing out various dimensions of tourism with an academic perspective.

Chapter 13
Digital Development of Tourism Events in India

Sachin Gupta, Kiran Meena, and Priyanka Jingar

Introduction

Virtual Reality (VR) has captivated the interest of both the tourism business and researchers. Computerized and virtual innovation has progressed quickly and reasonably, in the field of the business as well as in the formation of tourism spots and the safeguarding cultural heritage. Mainstream tourism is developing into one that is socially zed, low carbon, and dependable. Furthermore, as vacationer inspiration and customer request become more altered and various, and traveler locations cannot address the issues of multi-evened out market requests, the tourism area is presently nearly change and update. Subsequently, it isn't normal that the principal tourism areas began to utilize this new and inventive innovation to draw in sightseers to visit. Past investigations found that PC created recreations of 3D pictures improve sightseers' craving to visit objections. Simultaneously, the investigation demonstrated the requirement for an easy to use and outwardly engaging VR application to guarantee sightseers' social expectations to visit the objective. Understanding the experience of travelers is basic for planning applications that work on the experience of sightseers who are keen on find out about the objective. These innovations can likewise be utilized to oversee assets, assess, plan, screen progressively, and settle on auxiliary choices around here. We can work on the environment, increment benefits, and make a happier with setting for sightseers. We can possibly work on the environment, increment administration, and establish a

S. Gupta (✉) · P. Jingar
Department of Business Administration, Mohanlal Sukhadia University,
Udaipur, Rajasthan, India
e-mail: drsachinbusinessadministration@mlsu.ac.in

K. Meena
Department of Banking and Business Economics, Mohanlal Sukhadia University,
Udaipur, Rajasthan, India

A. Hassan (ed.), *Technology Application in Tourism Fairs, Festivals and Events in Asia*, https://doi.org/10.1007/978-981-16-8070-0_13

charming environment for sightseers. At long last, the lovely places will defend the mode, suggesting that it will acquire new open doors in its own turn of events. At the point when virtual reality innovation is applied in the advertising show of computerized picturesque area, dynamic arranging, online virtual visit, etc., the impact is remarkable. The objectives of this study are: first, to study the contribution of virtual reality and its basic features; second, to examine the influence of virtual reality and virtual reality tours on traditional tourism; and finally, to study the benefits of adopting virtual reality platforms in tourism industry.

Virtual Reality and Its Features

VR is a PC produced intuitive environment that permits players to make a plenty of simulated encounters, both genuine and unbelievable (Hobson & Williams, 1995). Virtual reality applications, in which individuals can insight and find out about genuine objections without truly visiting them, are currently broadly examined in tourism contemplates; I as a method for decreasing the adverse consequence of tourism, (ii) as an advertising device to build the engaging quality of the objective, and (iii) as an expected danger to the movement business. As indicated by Cheong (1995), the tourism and travel industry will at long last understand its maximum capacity. Albeit large numbers of his forecasts have worked out, a few investigations in the writing have analyzed virtual reality and its utilization in tourism.

Williams and Hobson (1995) inform that for numerous years, software engineers, futurists, and researchers have estimated about virtual reality encounters. As of late, there has been a great deal of media energy and conversation with respect to the likely uses of virtual reality in the public arena. This article will feature a portion of the positive ramifications of VR in the tourism business, just as address a portion of the technology's practical inadequacies. Furthermore, some future advancement will be featured to set up when and regardless of whether VR is probably going to follow through on the guarantee it makes.

Huang (2011) believes that motivation behind Technology Acceptance Model's specialized acknowledgment factors of telepresence, saw value, and saw usability are significant perspectives related with both virtual vacationer encounters and social purposes. Besides, the discoveries show that mental necessities for independence, skill, and relatedness, as characterized by the Self-Determination Theory, are exceptionally identified with virtual vacationer encounters and conduct goals. As indicated by the discoveries of this investigation, tourism experts should address the mental requests of clients for skill, independence, and relatedness while building virtual objections. Moreover, tourism advertisers ought to consider the sorts of data gave just as the media used to introduce the data, by including recordings and music.

Tavakoli and Mura (2015) inspected this investigation about the encounters of voyagers in virtual traveler areas are yet for the most part obscure. This is particularly obvious when gendered characters and examples of conduct among Iranian ladies in virtual tourism locales are thought of. To address this hole, this paper explores the examples of conduct of Iranian female vacationers while going in Second Life. This investigation applies virtual ethnography, otherwise called ethnography. The discoveries uncover that in virtual traveler settings, members reject "subjected" sexual orientation biases about Muslim ladies' substantial portrayals in Iran. This paper investigates Iranian female vacationers' examples of conduct going in Second Life. Nonetheless, their gendered exhibitions likewise acknowledge other 'subjected' cliché portrayals of womanliness, for the most part emphasized by the media in numerous Western social orders. Generally, this paper gives a more inside and out comprehension of Iranian ladies' vacationer conduct in virtual traveler locations, a subject dismissed by tourism scholastics.

Zhu and Huang (2013) learn about tourism data is inadequate, intuitive encounters are missing, and smart administration is difficult to accomplish. Generally, a computerized picturesque area is a theoretical and imagined region upheld by 3D technology, mixed media technology, VR technology, etc. Applying VR to tourism is a phenomenal methodology that advances computerized picturesque spots, dynamic arranging, and online virtual visits. In light of a conversation of the present status of virtual reality application, the article makes a few ideas for future improvement of thoughts, stages, and frameworks.

Jayendran and Rejikumar (2018) reveal insight into the potential advantages of presenting VR into movement and tourism for further developed objective picture. The use of VR in movement and tourism has a significant influence in building a vacationer's psychological picture of a location. An individual's inclination to visit the saw area increments in the wake of survey a 360-degree video, which is since VR makes a virtual encounter, or all in all, a practical encounter of the genuine objective. VR fills in as an answer for objective availability for the old, people with versatility issues, and individuals experiencing alarm problems.

Liu (2019) intends to utilize virtual reality technology to construct a virtual tourism framework in Guangxi minority beautiful spots, investigate intuitive presentation technology in virtual frameworks, center around intelligent virtual showcase plan and creation strategies, further develop client experience, and empower guests to Experience the fun of virtual tourism in a vivid manner, and improve the worth and picture of ethnic minority grand spots, and upgrade the center intensity of the picturesque spot.

Pestek and Sarvan (2021) clarify how late patterns in VR have changed the manner in which the tourism and friendliness ventures convey their contributions and address the issues of sightseers. This paper distinguishes and examinations what VR technology has meant for the tourism and cordiality enterprises through three primary touchpoints: future tourism arranging and the executives, technology-based advertising of tourism objections, and VR potential in changing buyer prerequisites. These ideas were analyzed to distinguish changing powers and potential perspective changes that tourism and accommodation providers and advertisers ought to consider.

Beck et al. (2019) state that motivation behind this investigation is to examine VR tourism studies and present a total cutting edge survey. As the mechanical meaning of the term VR has advanced, it presently joins an assortment of VR frameworks with fluctuating abilities. The all-encompassing objective of this investigation is to add to a far reaching comprehension of VR research in tourism. The presentation of new VR technology suggests a qualification for different VR frameworks utilized in the tourism business. During the past 3 years, research has zeroed in on the utilization of head-mounted presentations, which mirrors the transient development of VR technology. Notwithstanding the VR framework, most investigations look at VR as a showcasing instrument for advancement and correspondence purposes during the pre-travel stage, zeroing in on conduct perspectives. Advances in technology will yield new freedoms and application opportunities for the tourism business.

El-Said and Aziz (2021) incorporated the Technology Acceptance Model (TAM) and Protective Action Decision Model (PADM) models were utilized to distinguish the elements that impact an individual's choice to utilize VTs as impermanent options during seasons of emergency. Information were gathered from 401 respondents who had taken part in somewhere around one of the VTs repeating Egyptian legacy destinations. The outcomes propose that the TAM and PADM models' forerunners are successful at anticipating clients' expectation to take on VTs, and that reception goal well affects the probability of visiting the genuine site. Moreover, useful ramifications are accommodated site administrators to think about while picking VTs as a special instrument or as an elective item during seasons of emergency.

Contribution of Virtual Reality and Virtual Reality Tours on the Tourism Industry

The use of VR in tourism plays an important role in forming a tourist's mental view of an event. After watching a 360-degree video, a tourist's willingness to visit the watched event increases, which is because VR provides a simulated reality, or in another sense, a true experience of the actual event. For the elderly, people with mobility issues, and people with panic problems, VR is a solution to event connectivity. Technology is a significant piece of the travel & tourism industry, assisting organizations with everyday activities, while likewise further developing the client experience. Cell phones, shrewd speakers and AI aides have all assisted with expanding the pertinence of voice search the extent that technology patterns in the movement business are concerned. Another significant type of movement tech is the capacity to acknowledge contactless payments. This will permit venture out organizations to deal with installments substantially more rapidly, remembering for circumstances where clients don't approach cash, or their credit or charge card. It can likewise improve the client experience since it is an efficient device. Utilizing VR visits, clients can encounter everything from virtual lodgings visits and eateries, to tourist spots, public stops or even explicit exercises. The intuitiveness and drenching can likewise assist with giving you an upper hand over rivals who are not yet benefiting as much as possible from these sorts of tech patterns. Most VR visits or 360 visits are currently viable with standard internet browsers and can help the individuals who are hesitant to make a trip because of COVID to see the advantages all the more plainly.

VR is a reproduction of this present reality, or once in a while a fictional universe. In the tourism, the simulated genuine world is of more significance. The feeling of sight is vital in tourism, where a ton of encounters depend in visual incitement (Gutiérrez-Maldonado et al., 2016). Thus, VR is by all accounts an incredible fit for the tourism business, an industry that can flaunt outwardly exceptionally amazing objections. The incredible capability of VR in tourism advancement has likewise been distinguished by different scientists (Sussmann & Vanhegan, 2000; Johnson-Glenberg, 2018). As Williams and Hobson (1995, p. 425) put it: "According to an advertising viewpoint, VR can possibly upset the advancement and selling of tourism". Numerous developments are at first expected for use in explicit ventures. Just later, after a specific timeframe and further advancement has passed; the technology is adjusted for use in different ventures, like tourism. In view of this deferred accessibility for the tourism business, numerous tourism scientists and experts in various touristic endeavors may "miss" a specific innovative turn of events. Consequently, they are uninformed of the potential that a specific improvement could have for them (Guttentag, 2010).

A few organizations in tourism have effectively gotten the VR viewpoint and have executed virtual visits. Notwithstanding, these, to the most part, comprise of sewed scene photographs of the premises, similar as what is shown in Google's Street-View administration. This is basically false VR on the grounds that there is possibly restricted, if by any means, capacity for the client to explore, and no likelihood to collaborate with the substance shown. These applications are viewed as VR-type applications, showing that there is an interest in the business to utilize VR-based instruments to advance items and administrations.

While connecting VR to tourism, a few groups may consider VR to be a device to visit lodgings, objections or attractions before genuinely heading out to a particular area. In spite of the fact that this is a choice and may be helpful in the arranging interaction of an excursion, VR can likewise be significantly more than that. Focusing on tourism, one space of consideration inside the venture where we are collaborating is virtual reality. Despite the kind of VR, regardless of whether inn or tourism objective, whether ski region, exhibition hall or amusement park, VR can possibly make extraordinary encounters and recollections for visitors and clients.

Customary promoting measures channels like paper advertisements, printed handouts or TV and radio adverts. These have the burden of just contribution restricted bits of knowledge into an objective; they can't pass on a legitimate picture of what the customer can expect of an objective. In the event that the material provided by the Destination Marketing Organization (DMO) to travel planners is low quality or lacking fundamental data, customers and even travel planners the same have issues settling on educated choices (World Tourism Organization, 2008). Sussmann and Vanhegan (2000) likewise express that conventional showcasing implies like leaflets or short video clasps can just pass on a short and shallow look at an offer. This ought to be stayed away from, on the grounds that customers who travel to an objective with an absence of information on the objective might be frustrated when their assumptions are not met, along these lines bringing about an adverse result in the client life cycle referenced beforehand (Cheong, 1995).

VR can show extra data about attractions and different focal points the client is taking a gander at. This will bring about more reasonable assumptions in the customers' mind, representing a benefit, since when the customer visits the objective, his assumptions can be satisfied at an agreeable level, further developing his vacation insight (Cheong, 1995; Sussmann & Vanhegan, 2000; Huang et al., 2016).

Customary advertising implies that when utilizing printed materials for limited time exercises, DMOs consistently should design with care about what necessities printing and its amount is required. They generally need to weigh between the danger of overprinting, which would prompt superfluous costs for capacity, and under printing, which would be extremely disadvantageous at exchange fairs or different occasions. Another significant factor to consider as indicated by them is the quality, size and thickness of paper utilized. In the event that an objective has some expertise in a long stretch source market for their advancement, the expenses of cargo and conveyance unquestionably must be viewed as cautiously to not transform into incredible liabilities.

Virtual Reality for Tourism Events

The Adopting of Virtual Reality for Tourism Events

VR innovation, by and large, use the VR headset to establish a simulated environment. This simulated environment gives an encounter of virtual reality. The client utilizing the VR innovation encounters the environment generally in a 3 D computerized world. Recently this innovation is utilized by the tourism and hospitality companies. These organizations show places of interest and in areas by utilizing 3 D recordings and videos. There is a significant gap exists between clients and hoteliers. The clients at a far distance are commonly uninformed of the vibe of the hotel, the nature of the rooms, and so forth Hoteliers portray their inn on the site; give not many pictures and recordings that would not give a total picture to the clients. The larger part of the time, this sort of approach brings about uncertainty to the clients. The clients, who are at a far distance, cannot survey the atmosphere and environment of the hotel, quality of the lodgings, offices in the inn, and so on this issue is effortlessly tended to by VR applications.

There are different Virtual reality applications utilized in the Tourism and Hotel Industry. Many of them are Virtual lodging tours, Virtual travel experiences, and Virtual Booking interfaces. Virtual lodging visits incorporate the show of the inn environment, its offices as 3D recordings. This makes the clients experience a real time encounter about the offices in the lodging. The clients might want to encounter the movement and investigation of the places of interest before they move to the places of interest. This is finished via looking through pertinent data on the Internet as client surveys and so forth with the coming of Virtual reality advancements, there was an extreme change in the accessibility of data to clients. This data causes the clients to choose about their movement ahead of time.

VR advances are wonderful passages to travel and investigate the inconspicuous areas in advance. Advertisers offer virtual excursion to the clients by banding together themselves with the lodgings, the travel industry spots (for example historical centers, zoo parks and any remaining sporting spots). In the new past, there are not many lodgings in particular Marriott inns, Atlantis Dubai Hotels who accomplice themselves with the advertisers to give the clients a virtual encounter. Virtual Booking Interface permits the clients continuous, simulated insight of strolling through a plane and choosing the seat in the airplane. Clients likewise select different administrations like taxi administration and make the installment. This simple exhibition of booking the flight tickets and different administrations as 3 D-recordings permit the clients to book without help from anyone else, without moving toward any consultancies. This assists the carriers with turning into the cutting edge retailers. Barely any organizations, for example, Navitaire airlines utilize the administrations of virtual booking interface.

The Creation of Virtual Theme Parks

Amusement parks are often obliged by their colossal land use and vicinity to enormous populaces, just as huge interests in current ride innovations. Notwithstanding, another age of more modest, more altered "virtual amusement parks" or area based diversion habitats that utilization virtual reality technology is arising. Virtual amusement parks are accounted for to be more modest and more affordable to make, permitting them to be underlying thickly populated regions and quick open up new business sectors. Sega, a Japanese business, made the principal "Sega World" indoor amusement park in Osaka and Yokohama in 1994. They choose to construct another 50 virtual amusement parks in Asia-Pacific by 1997. The park will have 32 virtual reality stages, a submerged test system and customary amusement park rides including an inside crazy ride and so forth stunning game rides are accessible.

Use of Virtual Reality as Sales and Promotional Tool of Tourism Events

VR can possibly change the publicizing and offer of tourism from a promoting outlook. Visit administrators and travel planners will actually want to give forthcoming explorers a virtual visit through their ideal excursion. As opposed to leaflets and movies, which are inactive instruments, VR gives an intuitive encounter. In any case, the benefit that travel planners might have might be momentary. Utilizing virtual reality for movement deals and advancement would not exclusively be expensive for movement organizations, yet it would likewise require the learning of an altogether new kind of technology.

Benefits of Adopting Virtual Reality Platforms in Tourism Industry

VR innovation, by and large, use the VR headset to establish a simulated environment. This simulated environment gives an encounter of virtual reality. The client utilizing the VR innovation encounters the environment generally in a 3D computerized world. Recently this innovation is utilized by the tourism and hospitality companies. These organizations show places of interest and in areas by utilizing 3D recordings and videos. There is a significant gap exists between clients and hoteliers. The clients at a far distance are commonly uninformed of the vibe of the hotel, the nature of the rooms, and so forth Hoteliers portray their inn on the site; give not many pictures and recordings that would not give a total picture to the clients. The larger part of the time, this sort of approach brings about uncertainty to the clients. The clients, who are at a far distance, can't survey the atmosphere and environment of the hotel, quality of the lodgings, offices in the inn, and so on this issue is effortlessly tended to by Virtual reality applications.

There are different Virtual reality applications utilized in the Tourism and Hotel Industry. Many of them are Virtual lodging tours, Virtual travel experiences, and Virtual Booking interfaces. Virtual lodging visits incorporate the show of the inn environment, its offices as 3D recordings. This makes the clients experience a real time encounter about the offices in the lodging. The clients might want to encounter the movement and investigation of the places of interest before they move to the places of interest. This is finished via looking through pertinent data on the Internet as client surveys and so forth with the coming of Virtual reality advancements, there was an extreme change in the accessibility of data to clients. This data causes the clients to choose about their movement ahead of time.

VR advances are wonderful passages to travel and investigate the inconspicuous areas in advance. Advertisers give the virtual excursion to the clients into the inns and the travel industry spots by banding together themselves with the lodgings, the travel industry spots (for example historical centers, zoo parks and any remaining sporting spots. In the new past, there are not many lodgings in particular Marriott inns, Atlantis Dubai Hotels who accomplice themselves with the advertisers to give the clients a virtual encounter. Virtual Booking Interface permits the clients continuous, simulated insight of strolling through a plane and choosing the seat in the airplane. Clients likewise select different administrations like taxi administration and make the installment. This simple exhibition of booking the flight tickets and different administrations as 3D recordings permit the clients to book without help from anyone else, without moving toward any consultancies. This assists the carriers with turning into the cutting edge retailers. Barely any organizations, for example, Navitaire airlines utilize the administrations of virtual booking interface.

Conclusion

Over the long haul, the presence of virtual reality in the movement business region is extending. As a result of the expanding utilization of creative development, it is entirely expected that the movement business will show up at tremendous heights later on. The section zeroed in on the high level improvement of the movement business and housing industry, the primary progression in the movement business actually due to the modernized advancements recorded as a hard copy. In the advancement driven events, the rising in the AI impression in the movement business is a respectable sign. This is because it exhibits that the business can use the farthest down the line advancement to additionally foster viability and value. The customers of the movement business will benefit due to the dealt with level of satisfaction. All the while, the movement business attempts will really need to have better order over the cycles. The business cycles will be robotized all things considered, and the business exercises and shows will be streamlined in nature.

The virtual environments spur purchasers to visit the objections. The virtual environments actually change the whole arrangement of the businesses and uncommonly engaging world. The virtual environment absolutely overwhelms the gaming scene, film, and mixed media related businesses. The buyers' apparent conduct is influenced primarily by virtual reality lately. Web-based media is likewise an imperative factor in the travel industry and friendliness industry's development. In a similar way, the variation of virtual reality is fortifying through utilizing online media stages. Web-based media is a critical supporter of social and legacy the travel industry advancement as of late. Expanded reality likewise primarily examined social and legacy the travel industry. Discoveries of the part identified with expanded reality and virtual reality comparably accentuate the travel industry and neighborliness industry's advancing variable through innovation. It exhibits the computerized innovations like web-based media, virtual reality, and increased truths are indispensable for future advanced the travel industry. Advanced the travel industry can work across borders, and community work among countries is as of now filling as of late.

References

Beck, J., Rainoldi, M., & Egger, R. (2019). Virtual reality in tourism: A state of- the-art review. *Tourism Review, 74*(3), 586–612.

Cheong, R. (1995). The virtual threat to travel and tourism. *Tourism Management, 16*(6), 417–422.

El-Said, O., & Aziz, H. (2021). Virtual tours a means to an end: An analysis of virtual tours' role in tourism recovery post COVID-19. *Journal of Travel Research*. https://doi.org/10.1177/0047287521997567

Gutiérrez-Maldonado, J., Wiederhold, B. K., & Riva, G. (2016). Future directions: How virtual reality can further improve the assessment and treatment of eating disorders and obesity. *Cyberpsychology, Behavior and Social Networking, 19*(2), 148–153.

Guttentag, D. A. (2010). Virtual reality: Applications and implications for tourism. *Tourism Management, 31*(5), 637–651.

Hobson, J. S. P., & Williams, A. P. (1995). Virtual reality: A new horizon for the tourism. *Journal of Vacation Marketing, 1*(2), 124–135.

Huang, Y.-c. (2011). *Virtual tourism: Identifying the factors that affect a tourist's experience and behavioral intentions in a 3D virtual world*. Retrieved from: https://tigerprints.clemson.edu/all_dissertations/775. Accessed 20 Aug 2021.

Huang, Y. C., Backman, K. F., Backman, S. J., & Chang, L. L. (2016). Exploring the implications of virtual reality technology in tourism marketing: An integrated research framework. *International Journal of Tourism Research, 18*(2), 116–128.

Jayendran, L., & Rejikumar, G. (2018). A study on the potential benefits of applying virtual reality in travel and tourism for better destination image. *International Journal of Pure and Applied Mathematics, 118*(5), 669–672.

Johnson-Glenberg, M. C. (2018). Immersive VR and education: Embodied design principles that include gesture and hand controls. *Frontiers in Robotics and AI*. https://doi.org/10.3389/frobt.2018.00081

Liu, G. (2019). Design of virtual display system for Guangxi minority scenic spots based on virtual reality. In *2019 International Conference on Intelligent Transportation, Big Data & Smart City (ICITBS)* (pp. 409–412). https://doi.org/10.1109/ICITBS.2019.00107.

Pestek, A., & Sarvan, M. (2021). Virtual reality and modern tourism. *Journal of Tourism Futures, 7*(2), 245–250.

Sussmann, S., & Vanhegan, H. (2000). Virtual reality and the tourism product substitution or complement?. In *ECIS 2000 Proceedings*. Retrieved from: http://aisel.aisnet.org/ecis2000/117. Accessed 20 Aug 2021.

Tavakoli, R., & Mura, P. (2015). 'Journeys in Second Life' - Iranian Muslim women's behavior in virtual tourist destinations. *Tourism Management, 46*, 398–407.

Williams, P., & Hobson, J. S. P. (1995). Virtual reality and tourism: Fact or fantasy? *Tourism Management, 16*(6), 423–427.

World Tourism Organization. (UNWTO). (2008). *Handbook on E-marketing for tourism destinations*. World Tourism Organization.

Zhu, Y., & Huang, Z. F. (2013). Application of virtual reality technology on digital scenic spot. *Applied Mechanics and Materials, 333–335*, 702–706.

Dr Sachin Gupta has a brilliant academic Record. He has earned his Doctorate degree (PhD) from Department of Business Administration, University of Rajasthan, Jaipur. He completed his Dissertation Research during his MPhil degree on the topic "NGOs Role in developing economic condition of Rajasthan State", from Department of Business Administration, University of Rajasthan, Jaipur. He completed his Post Graduate Degrees specialties- One in Business Administration (Masters of Commerce) from Department of Business Administration, University of Rajasthan, Jaipur and another in Finance (MBA-Finance) from Rajasthan Technical University, Kota. Dr Gupta cracked UGC NET Exam in Management twice. He obtained his B.Com degree from University of Rajasthan, Jaipur. Dr Gupta completed Executive Development Program in Advanced Financial Management from XLRI-Jamshedpur, did Post Graduate Certificate Program in Finance from IIM Indore, completed UGC Sponsored Diploma Program in IFRS from Ramanujan College, University of Delhi in association with IAA (All India Accounting Association), NCR Chapter.

Kiran Meena is a PhD Scholar at the Banking and Business Economic Department, Mohanlal Sukhadiya University, Udaipur. She is doing a B.Com degree in commerce from the Rajasthan University, Jaipur in 2016. She is doing an M.com degree in the Department of Commerce and Finance at the Central University of Rajasthan Bandar Sindri, Ajmer in 2018. Her research interest includes the area of Artificial Intelligence in Banking, Fintech management, the impact of technology on consumers.

Priyanka Jingar is pursuing Doctoral Degree at Department of Business Administration, Mohanlal Sukhadia University, Udaipur Rajasthan. She completed her MPhil from Mohanlal Sukhadia University, Udaipur. She has done her Post Graduation and graduation in Commerce from Maharshi Dayanand Saraswati University, Ajmer. She Qualified UGC- NET in Commerce. Her professional interest lies in the area of research. Her accolade includes Three Patent Published by Intellectual Property India, Ministry of Commerce & Industry, Government of India and a Book published by Lap Lambert Academic Publishing, Germany. She has several Research papers published with reputed journals and Chapters published with Taylor & Francis and Apple Academic Press. She has presented Research papers at National as well as International Conferences.

Chapter 14
Understanding and Improving Digital Tourism Events in Pakistan

Kalsoom B. and Mehtab Alam

Introduction

The events and festivals existed a long time ago as part of significant human demonstration of specific culture (Ryan et al., 2020). It contains public display, civic rituals and forms of celebrations that is recognized as a set of communal creativity and practices with certain spaces or the time (Laing, 2018). The events are arranged with formation of natural events underlying diverse nature of events that encourage exhibitions and trade (Arantes & Sousa, 2021; Jovicic, 2019; Opute et al., 2020; Dai et al., 2019). The pre-historic groups and ancient historians reflected forms of social activities those provided with arrangement of ancient games of Olympics. Event industry through the establishment of certain games highlighted the size and interaction of the key interest groups that influence the non-tourism proceeding (Kharouf et al., 2020). It reflects the target audience through marketing and attraction of tourists in an additional segment of destinations (Bargeman & Richards, 2020). Framework conditions for the digital tourism events brought about the trajectory of change in the opening of travelling opportunities for tourist through online interactions and meetings (Li et al., 2018a). Since the digital technologies are profoundly changing the traveling business and marketing, there's need for using the productivity-increasing technologies to be applied and incorporating the operation and structure of tourism events (Elgammal et al., 2020).

Digital cultural events, followed by the virtually integrated events, for technological utilities and digital events of tourism is a fluctuating experience (Purnomo et al., 2020). For arrangement of such events the consideration of harmonizing event

Kalsoom B.
Center for Policy Studies, COMSATS University Islamabad, Islamabad, Pakistan

M. Alam (✉)
University of Cyberjaya, Cyberjaya, Selangor, Malaysia

A. Hassan (ed.), *Technology Application in Tourism Fairs, Festivals and Events in Asia*, https://doi.org/10.1007/978-981-16-8070-0_14

policies, digital resource structuring and the digital compatibilities are all associated with the digital cultural events (Korstanje, 2020). The presentation of cultural tourism through endless persuasive technology is the advance form of embracing a digital culture (Philip & Williams, 2019; Glebova, 2022; Moreno-Gil & Coca-Stefaniak, 2020). It provides for breaking of conventional hierarchy where limited associates of tourism events are encouraged to opt for decisions (Jeong et al., 2020a; Escandon-Barbosa & Salas-Paramo, 2021). Culture of digital tourism events prove animated innovation that guide social networking, and enable socialization of tourism activities in line with uprising events images, conferences, calls and accessing the data anywhere in the concerned sites of tourism (Calvaresi et al., 2019). Physical gears are no longer acceptable for many intermediaries involving the services of online purchasing ticket, hotels and the booking for vacation on a specified tourist destinaiton (Torres-Bagur et al., 2020). Presentation of the event content and registration through virtual integration of online event platform empowers the features of tourism to go beyond traditional customization and promoting growth of tourism activities (Dowell et al., 2019).

Events that comprise of innovative, adoptive and customer-centric tools state the digital trends for transmission of overseeing tourism experiences (Robinson & Schänzel, 2019). Resilient event policies of such kinds are altering theory and practice of event of tourism that guide the technology driven solutions for sounding destinations through digital platforms (Navarrete, 2019; Malchrowicz-Mośko et al., 2019). Reliance of tourist through tourist's technology and hospitality automation is the service for designing the implementation of economic process, market structure and application of e-tourism for virtual tours (Bolan & Simone-Charteris, 2018). Digital structure for e-tourism and resourcing the compatibility of digital harmonization are essentially important and viable for improving the travel demographics of cultural, language and evocative online travelling experience (Li et al., 2018b; Jeong et al., 2020b). Though, the utilities of technologies for developing countries are limited, it is still relevant in accumulation of tourism persona through social media (Sakdiyakorn et al., 2021). Emotional support, strong ties and the form of bonding for events arrangement through media is the endorsement of cultural connections through digital media platforms (Yousaf et al., 2021). Given the relevancy of digital utilities and cultural advancement, for tourism events in Pakistan, the research focused on the key insights that follow objectives of investigating digital utilities for arranging tourism events and exploring the technological practices of digital events of tourism in Pakistan (Khan et al., 2021).

Literature Review

Previous studies (Asif et al., 2019) provide for event of tourism implying digital resources and functions of travelling with event arrangement (Ashfaq et al., 2021; Naqvi et al., 2018). An event is the representation of a completely different picture that practices unique human needs of entertaining and exchanging the cultural and social experience. Studies of Thomson et al. (2020) provided for tourism event as the application of creating large-scale conventions, concerts, cultural festivals, and focusing on the attraction of tourists to the specific tourist's destination (Elgammal et al., 2020). The digital incorporation of events for tourism related practices are denoting the virtual, cultural and intellectual needs of tourist to overcome the issues of travelling and destination (Marine-Roig et al., 2017). The tourism events are the contact of digital infrastructure with attraction of large-scale technologies that organize the special event and develop scenario for the tourism planning and specified tourism policies (Baggio, 2018). The engagement of tourist experiences with digital sites followed for gaining the popularity of culture. The event organization follows the professional integration of cultural events, and harmonized digital experiences. The performance of a big gap for many scholars and researchers is unavailability of resource structure of tourism where the stages of preparation and amalgamation of technologies is attracted for promoting tourism associated events (Arantes & Sousa, 2021). Productive and advance technological tools are highlighted rending digital solutions and devices that connect events with the tourism.

Digital utilities by Armbrecht et al. (2021) is for the modern events of tourism as effective and provide for the notable devices that carry provisions of unwavering emergence of carriers. It explores the maintenance of smartphones, screens and format of digital cramps providing the use of barcodes and other cloud accessibility to events. For researchers like Carlisle et al. (2021), the use of bar codes or the resource structure of the technology is the important development for convening event messages through the cloud accessible devices. Significance of relevant devices for the integration of cultural and event practices requires the use of adaptive policies to keep the record of visitors and underline progress of managing events (Sorooshian et al., 2021). For the improvement, technologies and synchronization of databases where networking during the events is specific for targeted event communication. Integrated gamification, self-studying during the tourists' events and use of social networks are explored for establishing tourism-based contacts, with meeting and conferences (Mariani et al., 2018). The virtual events and compatibility of technical products simplified corresponding events of tourism. Readiness to imply these systems is undoubtedly being comprehensive connection among tourism event participants, organizers and administrators. Deprivation of the live communication followed by the reliance on local technological systems and efficient use of competitive tourist influx is the rational practice of linking national heritage with cultural tourism event (Baggio & Sainaghi, 2011). Therefore,

H^1: Digital Cultural Events are positively associated with utilities of technological events

Tourism events are understandable and improved through digitalized compatibility testing of the instrument over each event of tourism (El-Gohary, 2012). Specific testing of software and hardware systems meet the behaviors of correctly using applications for operational efficiency of tourist experience. Extensive human labor and operations for reducing the tourism event costs, following issues of customer services view the strength of destination image (Rahmani et al., 2015). Tourism events are changing with emergence of technological competencies' that guide and focused on the models of business. For reaching new markets of tourism, events in developing countries may adopt the market demand of enforcing digital protocols through cashless and contactless transactions (Richards, 2010). Utilization of digital resources and attached culture of e-business explore the possibility of traveling with events arrangement and letting event manager on the critical phase of implying smart tools (Khalifa, 2020). Virtual integration and assistance for meeting the necessities of arranging technological interactions and personalized experience for targeting the market campaign towards events of tourism. Research from various countries implies the adoption of compatible technology in managing events is a potential boosting of tourism activities where increase productivity and efficiency is well performed and connected (Truong et al., 2020). Small and large events are equally disrupted by the leverages of digitalization which shares the event activities and inter-connect the very small functions of creating value over the unused destinations of tourism.

H^2: Technological utilities are negatively associated with digital cultural events

Events exhibitions with technological maintenance is the steadiness that provide multiple opportunities of valuation and expected sharing of innovative models (Baggio & Del Chiappa, 2016; Wang et al., 2019; Bulin, 2016; Gargano, 2021; Martini et al., 2017; Bauernfeind & Mitsche, 2008; Nishimura et al., 2007; Alseiari et al., 2019; Vystoupil et al., 2017; Eid & El-Gohary, 2014; Häfner et al., 2020). Uptake for different technologies and digital resourcing, harmonizing, applying cultural events are lagging behind for developing states of Asia. Reasons for the deficiencies is the unreliable events management and admissible connection with the tourism stakeholders. Increasing capabilities value chains of technological sites are determining to share, consume and exchanging the events through digital platforms (Arantes & Sousa, 2021; Jovicic, 2019; Opute et al., 2020; Dai et al., 2019). For non-living entities of tourism events certain breaches of digital modalities is limited due to infrastructure, digital technology, nature, virtual integrated communities, digital policies, producers and travelling customization. Heterogeneous and fragmented tourism events demonstrate the coverage of structured and non-structured resources of digital utilities. The information intensive unit of each tourism events are reliable, with constitution of majority tourism enterprise and collective contribution over the event arrangement (Wang et al., 2019; Bulin, 2016; Gargano, 2021; Martini et al., 2017). The exchange of the values among event managers for inclusive arrangement of global values for trending technological change in the tourism events that review the experience, services and goods for delivering customization and hyper-personalization.

H^3: Digital Resource Structure is negatively associated with the digital cultural events

For the tourism events, robotics and automation are key trends and applications driving physical and automated systems. Harmonizing events with digital tourism network is a flexible and informal forum for the design of events under compatibility of digital resource structure. The relevant private and public stakeholders bring equal opportunities through virtually integrated network and boosting the innovative event capabilities (Nishimura et al., 2007; Alseiari et al., 2019; Vystoupil et al., 2017; Eid & El-Gohary, 2014; Häfner et al., 2020). Resource structure and revolutionizing skills development of tourists' staff is based on technical input and facilitation of using the technology in an effective manner. Models of business trending over each event is adoptive for processing the change and referred digital understanding as based on assets for the knowledge, where inconsistency of digital-transition. Interoperability, technology, productivity, and increase decision-making enhances the data-driven management of events for supporting the digital gains (Ashfaq et al., 2021; Naqvi et al., 2018). Efficient sales over each tourism event, increase automation, reducing the costs of physical exchange policy have comprehensive impacts over digital utilities. Virtual integration and technology-based culture follow the pace of transformation with compatible scales facilitate demands of consumer. Digital supportive policies and culture of marginal costs over benefiting the events of tourism have shaped the transformation of tourism supporting businesses for improving the performance of events.

H^4: Virtually Integrated Events are positively associated with digital cultural events

Events values and digital ecosystem are integrated virtually with management support following deep impact over resource structure and integration of resource structure (Martini et al., 2017; Wang et al., 2019; Bulin, 2016; Gargano, 2021). Informative-intensive view of structure of technological utilities outperforms tourism, ease of booking and use of internet. Payment systems and online platforms for tourism events maintained average share of travelling and holidays indicating variation of extents and differing speeds (Alseiari et al., 2019; Vystoupil et al., 2017; Eid & El-Gohary, 2014). The intensity of digital assessment with supportive indicators embedded technology in purchasing of accommodation and services rendered to hotels. Concentration of digitization performs recreation and entertainment activities is leading untapped potential where the application and adoption of digital event solutions may outsource certain activities of business causing online advertising and sales for the concerned event (Alseiari et al., 2019; Vystoupil et al., 2017; Eid & El-Gohary, 2014; Nishimura et al., 2007). Business practices with event tourism may improve with the understanding of digital indicators leading productivity and markets gaining value from the awareness of producing visitor experience. Such initiatives for capacity of digital resources, experimentation and opportunities of capturing acceleration of tourism events adapting services and products for expansion and creation of technological utilities and culture (Fig. 14.1).

H^5: Digital Compatibility is the positively associated with Digital Cultural Events

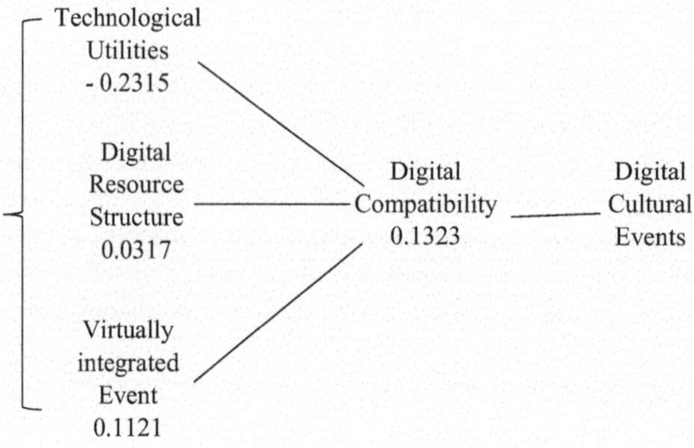

Fig. 14.1 Conceptual framework. (Source: authors' development)

Methodology

The events of tourism in Pakistan determinant with recreational potential and rich tourists' spots. Level of tourism is characterized by inadequate share of tourism events. For that matter, quantitative research is specified for the study and conducted the survey method to approach the participants (Bulin, 2016; Gargano, 2021; Martini et al., 2017; Bauernfeind & Mitsche, 2008; Nishimura et al., 2007; Alseiari et al., 2019; Vystoupil et al., 2017). The ability of quantitative study to complete statistical analysis of the events asked and responded over the questions. The creation of outstanding events focuses digital cultural events, virtually integrated events, technological utilities, digital tourism events, harmonization of event policies, digital resource structure, and digital compatibility (Baggio & Del Chiappa, 2016; Wang et al., 2019). Tourism events research for many authors is the facilitation and understanding of tourism events management, through the survey of concerned participants. Quantitative research continues to evolve from the context of managing events and exploring the relevant experiences. The participants incorporated the Pakistan Travel Mark as a leading event management organization. Key agencies, tourism media platforms, stakeholders, traveling agents, operators and concerned shareholders are approached through the online survey. Event managers, destination in charge, event coordinator for hospitality and tourism projects. It incorporates the lower staff and communication manager for the guidance and input clients attending events over the tourism engagement. Considering the deductive approach of the research, the investigation incorporated simple convenient sampling from Krejcie and Morgan (1970).

The sample size under the provided table of Krejcie and Morgan (1970), where questionnaire is distributed among 250 participants from the Pakistan Travel Mart, that comprises of key tourism event agencies, stakeholders, managers, planners of events and relevant individuals (Bauernfeind & Mitsche, 2008; Nishimura et al., 2007; Alseiari et al., 2019; Vystoupil et al., 2017; Eid & El-Gohary, 2014; Häfner et al., 2020). The table of sample size as provided with author is significant towards the efficiency of results and ascertaining relationship of each variable with the other, to comprehend event tourism analysis. The response over the distributed questionnaire is 209, signifying the highest ratio of answer from the distributed questionnaires. The statistical tools used while following the confidence level of 95% that drive marginal error over 0.0–0.5%, effectively influencing the data inclusion in the tables. The number of questions for each of the 36 questions distributed among variables of digital cultural events 06, virtually integrated events 05, technological utilities 05, digital tourism events 05, harmonization of event policies 05, digital resource structure 05, and digital compatibility 05. These questions explain the variation of events and needs of understanding, improving the digital utilities for tourist to visit Pakistan. Demographical explanation of the participants encourages the experience over events arrangement, training and education for events, sex, gender and age with marital status asked in the questionnaire.

For the events management in context of tourism each participant is response is significant for making sense of digital cultural events (Nishimura et al., 2007; Alseiari et al., 2019; Vystoupil et al., 2017; Eid & El-Gohary, 2014; Häfner et al., 2020; Bauernfeind & Mitsche, 2008). It direct for second part of questionnaire on the variable of virtually integrated events. It highlights the technological utilities where the next section of questionnaire focuses on the digital tourism events. It follows the role of harmonization of event policies section in the questionnaire that leads to the options on answering queries over digital resource structure, and digital compatibility of tourism events. The instrument of questionnaire under quantitative research is implied and it contain the options of five Likert scale that focuses on the issues of digital tourism event understanding and improvement. The conduct of the survey from participants during December 2020–May 2021 with the restrictive environment of physical distancing and maintaining safety measures for the pandemic. Questionnaire instrument after distributing among the participants provides for the tests of quantitative using the SPSS (Statistical Software Package). The tests implied reliability of Cronbach alpha, followed by the descriptive, ANOVA, regression and the correlation. These tests are applied in managing the data, checking suitability and using the tests to identify the formation of data is effective for reaching the proper outcome.

Discussion

For understanding and improving the event of tourism in Pakistan, the data analysis provided certain key construct incorporating the digital cultural events, virtually integrated events, technological utilities, digital tourism events, harmonization of event policies, digital resource structure, and digital compatibility. The distributed questionnaire response with the demographical explanation provides the ratio of 32% of female and 68% from the male members. Tourism event participant age stands at 42% with (24–35) years that follows the 28% as (35–45) and participants having age of (46–60) are calculated as 30% of the overall age demographics. It stands with events of tourism where events staff arrange an income of $500–1200 monthly which is significant for reflecting handsome activity of digital event over tourism and relevant sources. The outcomes of survey provided that 45.5% of workers are associated with digital technologies in one or other way. Remaining 14.5% are affiliated with computational use of understanding events of tourism. However, 40% of workers and employees need improvement for digitizing the tourism events.

For travelers, response online over the queries delays for 12–18 h, 1–2 days for critical issues and events response for arranging accommodation is 5 days' time in a week. There are online travelers reflected the ratios of 29.5% are satisfied with the compatibility of infrastructure of digital events. It reflects the use of 34.0% online event arrangement by the travelers and visitors wanted to know about the tourism culture and harmonization of polices leading to digital progress of events management in Pakistan. The rest of 36.5% of the online travelers and visitors are dissatisfied by the response and level of interaction of event staff, managers, and coordinator. The financial distribution of income and support from stakeholder to organize and understand events in Pakistan is at ratio of 30.5–42.5%. For the dealing digital event staff and stakeholders of Pakistan Travel Mart, manager is learning with 10% input from seniors, training staff for workers are contributing with 45% and improvement of digital events is 13% with 32% of workers and stakeholders contributing for arranging digital tourism events in Pakistan.

The questionnaire is tested with the data available in the SPSS provided that reliability value for the variables and overall results between 0.7 as minimum acceptability and 0.9 as highly significant showing the data as reliable with the instrument used in the study. For digital culture the value is at 0.7931, following the virtually integrated events where value is 0.832. The technological utilities stand at 0.812 with the value of digital resource structure as 0.9214. Further, digital compatibilities values are 0.831 as the significant where variables and data is reliable (Table 14.1).

For the research of digital tourism events, correlation provides for the relationship among values over each construct and variable. It contains numbers from +1

Table 14.1 Correlation table

Description	Digital cultural	Virtually integrated	Technological utilities	Digital resource structure	Digital compatibility
Digital cultural	1				
Virtually integrated	0.1121	1			
Technological utilities	−0.2315	0.2110	1		
Digital resource structure	0.0317	0.2431	−0.0316	1	
Digital compatibility	0.1323	0.1343	0.1632	−0.1231	1

Source: IBM SPSS Statistics v25

towards −1 and vice versa as determinant from the values of r. The available values leading with digital culture 1 towards virtually integrated events with 0.1121 where it is positive and constructive relation. It shows the application of virtual events for integration of resources is significant. Further relation of technological utilities at −0.2315 as weak and negative relation that provides the need for understanding and improving digital events in tourism. Values of 0.2110 provides for positive relation towards virtual integration of events. The digital resource structure reflects the values of 0.0317 with 0.2431 integration of events and negative towards the technological utilities at value of −0.0316. Further, digital compatibility 0.1323 with digital culture and 0.1343 value of the compatibility with virtual integration of events. It underlines the negative resource structure value of −0.1231 where technological utilities stand at 0.1632 showing the negative culture and useless structure that is not supporting the culture of digital event for tourism in Pakistan. The available values that contribute for the relationship of digital culture supporting events of tourism in Pakistan, however, there is significant lacking of technological utilities and structure which are the key contributors towards the event management and progress of using digital event of tourism in Pakistan (Table 14.2).

The provided values of the coefficient retreated with .0299471, with stander error of .0086147 and t-statistics of 3.14 is significant. However, the p-value of virtually integrated events is at 0.000 showing significant results with confidence interval of 95% and positive values of .0129323 and .0470316. For technological utilities the negative value of coefficient at −.05014132 with error of .0148323 that technological events of tourism lagging behind the specialized dealings of events. The values of the digital resource structure are negative at −.00992214 with error of .0074211 and t-statistics of −3.14 is reflecting the unviability and weak association of digital structure that counts the need for using structure to support the events of tourism while utilizing the digital resources.

However, p-value of 0.001 present significance of data that is used in the research with confidence interval 95% and values of .01126214 and .0064517 implying the digital compatibility for supporting the structure of digital events. Digital compatibility is provided with positive values but the non-accessibility to structure of the digital resources may cause issues for arranging the technological based events of tourism. Values of the digital compatibility are .3994357 having error of .0149234

Table 14.2 Tests and results

Digital cultural	Coefficient	Stander error	t-statistics	p-value	Confidence interval [95%]	
Virtually integrated	.0299471	.0086147	3.14	0.000	.0129323	.0470316
Technological utilities	−.5014132	.0148323	−33.22	0.000	−.5305511	−.4722213
Digital resource structure	−.0092214	.0074211	−3.41	0.001	.01126214	.0064517
Digital compatibility	.3994357	.0149234	26.11	0.000	.3702357	.4287471
Constant	.1997214	.0597318	3.19	0.002	.826157	.3167641

Source: IBM SPSS Statistics v25

Table 14.3 Regression model summary

Model summary[b]										
					Change statistics					
Model	*R*	R square	Adjusted R square	Std. error of the estimate	R Square change	F change	df1	df2	Sig. F change	Durbin-Watson
1	*.147*[a]	.068	.037	1.213	.061	1.901	4	154	.149	.914

Source: IBM SPSS Statistics v25

[a]Predictors: (Constant), Virtually Integrated Events, Technological Utilities, Digital Resource Structure, Digital Compatibility

[b]Dependent Variable: Digital Cultural

with t-statistics of 26.11. There is significance p-value of 0.000 with confidence interval .826157 and .3167641 as the positive sign towards the understanding and improving digital events in Pakistan. For that matter, values and tests run through SPSS provide for the lapses at one end and contribution of some variables to arrange the possible situation of event of tourism through digital sites (Table 14.3).

The model of the regression is the relevance of data fit for the research, that followed by various tests implying the provisions of r value at .037 as significant because of the margin given through the adjusted r square value of .037 and r square change at .061. Further the use of r at .147 with f change test value over 0.901 leads to determine that significance value of f change is .149 and .914 as the Durbin Watson. The nature and the character of relationship among the variables is highlighted through model summary while results show positive values of variables and data used in study. The significance of f change value provided the addition of the construct improved the prediction where coefficients have improved the values of model used in the study. The range of Durbin-Watson value is between 1.5 and 2.5 providing for the normal distribution of data and it shows the values are normal. The ANOVA value of P-0.000 shows the hypothesis implied in the study are true with variation of possible null hypothesis. Values greater than 0.05 shows no effect is observed.

Conclusion

Understanding and improving digital events in Pakistan carry certain lapses as well as positive aspects as tests of the results reflect each indicator. Provision of services, distribution of technological entities and implication of the quantitative results, prove through this research that pre-planning of digital events requires comprehensive documentation to meet the digital event of tourism. Digital cultural events, have the potential to compete with support from virtually integrated events, that apply technological utilities. Significance of digital tourism events, harmonization of event policies, is well-noted. However, the absence of digital resource structure, and digital compatibility view utilities of technology for event arrangement in an undesirable and feeble association. Ability to entertain and desire technological aspect of event development, progress and alignment with the density of event communication is regarded well. Still, there are long-remembering impression for tourists joining the digital events to know culture and heritage of a country, destination is failing to receive the message due to fail organization and formation of structure.

Digital event for tourism based on efficient structure is indicating immediate openness for innovative practices of technology with certain theoretical underpinning of intangible nature of events in context of tourism. Specifying the identifying the cases prevalent with practices of virtual meetings is common, however, in Pakistan there is requirement for accelerating the resumption of international travelling and arrival of tourist through managing event of tourism. Technology utilization with absorption of culture, heritage and destination image may provide a comprehensive look for event success and sustainability. Investing valuable assets and designing the efficient structure for virtual tours is based on information technology solutions. Much of the deliberations focused on analysis of various customer relationship followed with market structure of events, added virtual tourism and compatibility of policies that support the events of digitalization. Connection and navigations through satellite are effectively managing events on destinations or taking specific categories of internet connections through which events may be organized. For that matter the need to specify and limit the total digital utility through the innovative, adoptive and tourists' centric events causing digital trends in focusing on the emergence of technologies. The distribution, transmission, generation of digital cultural is linked with the virtually integrated events, and affect the technological utilities in a progressive conduct of arranging events.

Policy Implications

The research event implied for variety of options carrying engagement of integrated events of tourism to highlight the positive image and effective use of technologies. Portion of technologies allow the corporations to focus on arrangement of tourism events under single umbrella organization as in the form of Pakistan Travelling Mart. It constitutes for further amalgamation of such events arrangement and filling the gaps of structural technology changes followed with the utilization of digital technologies in a purposeful manner. The study implied for the use of tourism to bring major event planning and resourceful setting in different destinations of Pakistan. Highlighting such destinations provide for analysis of certain key construct incorporating the digital cultural events, virtually integrated events, technological utilities, and digital tourism events. The harmonization of event policies is recommended to be aligned with the digital resource structure, and digital compatibility where digital experience of technology focused on the processes that experience a positive aspect for the tourist to visit Pakistan physically.

The event of tourism and user interface over virtual engagement communicate for better method in the extension of this research as it is limited as certain areas and size of data to measure entire digital events in Pakistan. Implying the digital events for tourism follows social capital as a brand and it measure tourists trust in building effective structure of digital technologies. In future research, the authors may extend structural issues of event organization with accumulation of utilization of digital resources in a rightful manner for processing the values and motivation to international and domestic travelers to visit Pakistan as a place of destinations and diverse cultural heritage. Opportunities and offers for training the digital use of resources in tourism events may bring a plausible change in the negative picture, projected through malaise ideas and polices. Digital events are emerging signs of success for attracting the perceived negative concept, that may customize and modified to positive image building through range of successful events engaging intern atonal tourists for branding destinations.

References

Alseiari, H. A. S. M., Khalifa, G. S., & Bhaumick, A. (2019). Tourism destination competitiveness in UAE: The role of strategic leadership and strategic planning effectiveness. *International Journal of Recent Technology and Engineering (IJRTE), 8*(4), 860–865.

Arantes, L., & Sousa, B. (2021). Digital marketing and tourism: Case study applied to Barcelos. In *2021 16th Iberian Conference on Information Systems and Technologies (CISTI)*. IEEE. The 23rd–26th June 2021, pp. 1–6.

Armbrecht, J., Lundberg, E., Andersson, T. D., & Mykletun, R. J. (2021). 20 Years of Nordic event and festival research: A review and future research agenda. *Scandinavian Journal of Hospitality and Tourism, 21*(1), 49–59.

Ashfaq, J., Khan, A., Zulfiqar, U., & Ullah, M. (2021). Factors influencing destination image through social media in the pre-purchase period of tourism in Pakistan. *Sarhad Journal of Management Sciences, 7*(1), 103–122.

Asif, M., Zhiyong, D., Ullah, I., Nisar, M., Kalsoom, R., & Hussain, M. A. (2019). The study on China-Pakistan Economic Corridor (CPEC) as a language and culture changer in Pakistan. *Journal of the Punjab University Historical Society, 32*(1).

Baggio, R. (2018). Network analysis: Quantitative methods in tourism. In C. Cooper, S. Volo, W. C. Gartner, & N. Scott (Eds.), *The Sage handbook of tourism management* (pp. 150–170). Sage.

Baggio, R., & Del Chiappa, G. (2016). Complex tourism systems: A quantitative approach. In M. Uysal, Z. Schwartz, & E. Sirakaya-Turk (Eds.), *Management science in hospitality and tourism: Theory, practice and applications* (pp. 14–21). Apple Academic Press – CRC Press.

Baggio, R., & Sainaghi, R. (2011). Complex and chaotic tourism systems: Towards a quantitative approach. *International Journal of Contemporary Hospitality Management, 23*(6), 840–861.

Bargeman, B., & Richards, G. (2020). A new approach to understanding tourism practices. *Annals of Tourism Research, 84*, 102988.

Bauernfeind, U., & Mitsche, N. (2008). The application of the data envelopment analysis for tourism website evaluation. *Information Technology & Tourism, 10*(3), 245–257.

Bolan, P., & Simone-Charteris, M. (2018). 'Shining a digital light on the dark': Harnessing online media to improve the dark tourism experience. In R. P. Stone, R. Hartmann, T. Seaton, R. Sharpley, & L. White (Eds.), *The Palgrave handbook of dark tourism studies* (pp. 727–746). Palgrave Macmillan.

Bulin, D. (2016). Aspects of responsible tourism-a quantitative approach. *Amfiteatru Economic Journal, 18*(10), 781–797.

Calvaresi, D., Leis, M., Dubovitskaya, A., Schegg, R., & Schumacher, M. (2019). Trust in tourism via blockchain technology: Results from a systematic review. In J. Pesonen & J. Neidhardt (Eds.), *Information and communication technologies in tourism 2019* (pp. 304–317). Springer.
Carlisle, S., Ivanov, S., & Dijkmans, C. (2021). The digital skills divide: Evidence from the European tourism industry. *Journal of Tourism Futures*. https://doi.org/10.1108/JTF-07-2020-0114
Dai, T., Hein, C., & Zhang, T. (2019). Understanding how Amsterdam City tourism marketing addresses cruise tourists' motivations regarding culture. *Tourism Management Perspectives, 29*, 157–165.
Dowell, D., Garrod, B., & Turner, J. (2019). Understanding value creation and word-of-mouth behaviour at cultural events. *The Service Industries Journal, 39*(7–8), 498–518.
Eid, R., & El-Gohary, H. (2014). Testing and validating customer relationship management implementation constructs in Egyptian tourism organizations. *Journal of Travel & Tourism Marketing, 31*(3), 344–365.
Elgammal, I., Ferretti, M., Risitano, M., & Sorrentino, A. (2020). Does digital technology improve the visitor experience? A comparative study in the museum context. *International Journal of Tourism Policy, 10*(1), 47–67.
El-Gohary, H. (2012). Factors affecting E-Marketing adoption and implementation in tourism firms: An empirical investigation of Egyptian small tourism organisations. *Tourism Management, 33*(5), 1256–1269.
Escandon-Barbosa, D., & Salas-Paramo, J. (2021). Tourism amidst COVID-19: Consumer experience in luxury hotels booked through digital platforms. *Tourism Recreation Research*. https://doi.org/10.1080/02508281.2021.1940708
Gargano, R. (2021). Tourism competitiveness in Mediterranean countries: A quantitative approach. In F. Grasso & B. S. Sergi (Eds.), *Tourism in the Mediterranean Sea* (pp. 93–108). Emerald Publishing Limited.
Glebova, E. (2022). Sport tourism digital marketing. In D. Buhalis (Ed.), *Encyclopedia of tourism management and marketing* (pp. 1–4). Edward Elgar Publishing.
Häfner, F., Härting, R. C., & Kaim, R. (2020). Potentials of digital approaches in a tourism industry with changing customer needs–a quantitative study. In *2020 15th Conference on Computer Science and Information Systems (FedCSIS)*. IEEE. The 6th–9th September 2020, pp. 553–557.
Jeong, Y., Kim, E., & Kim, S. K. (2020a). Understanding active sport tourist behaviors in small-scale sports events: Stimulus-organism-response approach. *Sustainability, 12*(19), 8192.
Jeong, Y., Kim, S. K., & Yu, J. G. (2020b). Sustaining sporting destinations through improving tourists' mental and physical health in the tourism environment: The case of Korea. *International Journal of Environmental Research and Public Health, 17*(1), 122.
Jovicic, D. Z. (2019). From the traditional understanding of tourism destination to the smart tourism destination. *Current Issues in Tourism, 22*(3), 276–282.
Khalifa, G. S. (2020). Factors affecting tourism organization competitiveness: Implications for the Egyptian tourism industry. *African Journal of Hospitality, Tourism and Leisure, 9*(3), 116–130.
Khan, M. L., Ittefaq, M., Pantoja, Y. I. M., Raziq, M. M., & Malik, A. (2021). Public engagement model to analyze digital diplomacy on Twitter: A social media analytics framework. *International Journal of Communication, 15*, 1741–1769.
Kharouf, H., Biscaia, R., Garcia-Perez, A., & Hickman, E. (2020). Understanding online event experience: The importance of communication, engagement and interaction. *Journal of Business Research, 121*, 735–746.
Korstanje, M. E. (2020). Tourism as a vehicle towards recovery. In H. Séraphin, M. E. Korstanje, & V. G. B. Gowreesunkar (Eds.), *Post-disaster and post-conflict tourism* (pp. 19–36). Apple Academic Press – CRC Press.
Krejcie, R. V., & Morgan, D. W. (1970). Determining sample size for research activities. *Educational and Psychological Measurement, 30*(3), 607–610.
Laing, J. (2018). Festival and event tourism research: Current and future perspectives. *Tourism Management Perspectives, 25*, 165–168.

Li, J., Xu, L., Tang, L., Wang, S., & Li, L. (2018a). Big data in tourism research: A literature review. *Tourism Management, 68*, 301–323.

Li, J., Pearce, P. L., & Low, D. (2018b). Media representation of digital-free tourism: A critical discourse analysis. *Tourism Management, 69*, 317–329.

Malchrowicz-Mośko, E., Botiková, Z., & Poczta, J. (2019). "Because We Don't Want to Run in Smog": Problems with the sustainable management of sport event tourism in protected areas (A case study of national parks in Poland and Slovakia). *Sustainability, 11*(2), 325.

Mariani, M. M., Mura, M., & Di Felice, M. (2018). The determinants of Facebook social engagement for national tourism organizations' Facebook pages: A quantitative approach. *Journal of Destination Marketing & Management, 8*, 312–325.

Marine-Roig, E., Martin-Fuentes, E., & Daries-Ramon, N. (2017). User-generated social media events in tourism. *Sustainability, 9*(12), 2250.

Martini, U., Buffa, F., & Notaro, S. (2017). Community participation, natural resource management and the creation of innovative tourism products: Evidence from Italian networks of reserves in the Alps. *Sustainability, 9*(12), 2314.

Moreno-Gil, S., & Coca-Stefaniak, J. A. (2020). "Guest editorial": Overtourism and the sharing economy–tourism cities at a crossroads. *International Journal of Tourism Cities, 6*(1), 1–7.

Naqvi, M. H. A., Jiang, Y., Naqvi, M. H., Miao, M., Liang, C., & Mehmood, S. (2018). The effect of cultural heritage tourism on tourist word of mouth: The case of lok versa festival, Pakistan. *Sustainability, 10*(7), 2391.

Navarrete, T. (2019). Digital heritage tourism: Innovations in museums. *World Leisure Journal, 61*(3), 200–214.

Nishimura, S., Waryszak, R., & King, B. (2007). The use of guidebooks by Japanese overseas tourists: A quantitative approach. *Journal of Travel Research, 45*(3), 275–284.

Opute, A. P., Irene, B. N., & Iwu, C. G. (2020). Tourism service and digital technologies: A value creation perspective. *African Journal of Hospitality, Tourism and Leisure, 9*(2), 1–18.

Philip, L., & Williams, F. (2019). Remote rural home-based businesses and digital inequalities: Understanding needs and expectations in a digitally underserved community. *Journal of Rural Studies, 68*, 306–318.

Purnomo, A., Idris, I., & Kurniawan, B. (2020). Understanding local community in managing sustainable tourism at Baluran National Park-Indonesia. *GeoJournal of Tourism & Geosites, 29*(2), 508–520.

Rahmani, A., Fakhraee, A., Karami, S., & Kamari, Z. (2015). A quantitative approach to estimating carrying capacity in determining the ecological capability of urban tourism areas (case study: Eram Boulevard of Hamadan city). *Asia Pacific Journal of Tourism Research, 20*(7), 807–821.

Richards, G. (2010). The traditional quantitative approach. Surveying cultural tourists: Lessons from the ATLAS cultural tourism research project. In G. Richards & W. Munsters (Eds.), *Cultural tourism research methods* (pp. 13–32). CABI.

Robinson, V. M., & Schänzel, H. A. (2019). A tourism inflex: Generation Z travel experiences. *Journal of Tourism Futures, 5*(2), 127–141.

Ryan, W. G., Fenton, A., Ahmed, W., & Scarf, P. (2020). Recognizing events 4.0: The digital maturity of events. *International Journal of Event and Festival Management, 11*(1), 47–68.

Sakdiyakorn, M., Golubovskaya, M., & Solnet, D. (2021). Understanding generation Z through collective consciousness: Impacts for hospitality work and employment. *International Journal of Hospitality Management, 94*, 102822.

Sorooshian, S., Azizan, N. A., & Ismail, Y. (2021). Influence of readiness measures on planning tourism digital shift. *Academy of Strategic Management Journal, 20*, 1–6.

Thomson, A., Kennelly, M., & Toohey, K. (2020). A systematic quantitative literature review of empirical research on large-scale sport events' social legacies. *Leisure Studies, 39*(6), 859–876.

Torres-Bagur, M., Ribas, A., & Vila-Subirós, J. (2020). Understanding the key factors that influence efficient water-saving practices among tourists: A Mediterranean case study. *Water, 12*(8), 2083.

Truong, Q. H., Nguyen, A. T., Trinh, Q. A., Trinh, T. N. L., & Hens, L. (2020). Hierarchical variance analysis: A quantitative approach for relevant factor exploration and confirmation of perceived tourism impacts. *International Journal of Environmental Research and Public Health, 17*(8), 2786.

Vystoupil, J., Šauer, M., & Repík, O. (2017). Quantitative analysis of tourism potential in the Czech Republic. *Acta Universitatis Agriculturae et Silviculturae Mendelianae Brunensis, 65*(3), 1085–1098.

Wang, X., Sun, J., & Wen, H. (2019). Tourism seasonality, online user rating and hotel price: A quantitative approach based on the hedonic price model. *International Journal of Hospitality Management, 79*, 140–147.

Yousaf, S., Razzaq, A., & Fan, X. (2021). Understanding tourists' motivations to launch a boycott on social media: A case study of the# BoycottMurree campaign in Pakistan. *Journal of Vacation Marketing*, 1356766721993861.

Dr Kalsoom B. is Assistant Professor at Centre for Policy Studies, COMSATS University Pakistan. Key Competent Areas in Research are evidence based opportunity mapping & Spatial distribution of Public facilities, inequality, explicit planning of public facilities at regional and local level – Big data, Urban Sustainability, Foresight & Public Policy – Public Sector Governance, Public Management & Reforms – Governance, inter-provincial & Local Government – Public Services Ethics, Social Equity, Inequality – Crowdsourcing, Social Innovation in Public Sector – behavioral insights in Public Sector, Outsourcing, Collaborative Management. She has more than 15 years of extensive experience in teaching and research. Her recent research findings appear in various International, National Journals and conference proceedings on issues of public sector reforms, public Policy, governance, sustainable Management and foreign affairs of China in higher education.

Mehtab Alam is a PhD scholar from University of Cyberjaya Selangor, Malaysia. Tourism Destination Management is the primary subject of doctoral research that follows image branding of the country. It is based on Master's research on Soft Image and Image Development of Pakistan completed for degree of Master of Philosophy (MPhil). He obtained degree in Masters of Arts in English Language & Literature and also holds Bachelor of Education from Allama Iqbal Open University, Islamabad. In addition to this, he serves in various Public and Private Institutions those include the National Assembly, Parliament of Pakistan as Research Associate and Ministry of Foreign Affairs on assignments/tasks of image development. As a writing specialist he also served in Skyscrapers Pvt. Ltd. Over the years, his interest in research is developed and allows him to keep in touch with the advancement in research on various topics and subjects. These incorporated the field of Social, Communication, & Management Sciences, and issues of international concerns like Public Diplomacy, Foreign Policy. During the time, Alam interacted with highly qualified and experienced individuals from various backgrounds. Alam has several articles/chapters publications.

Chapter 15
Information Technology and Its Challenges Towards Sri Lankan Event Tourism: Event Managers' Perspective

A. M. D. B. Nawarathna, R. S. S. W. Arachchi, and W. M. A. H. Bandara

Introduction

The tourism industry has grown and diversified steadily throughout the past decades, being one of the world's fastest expanding economic sectors. Hence, United Nation World Tourism Organization (UNWTO) mentions that the tourism Industry approaches or exceeds oil exports, food items, and vehicle revenue. Especially, Meetings, Incentive, Conference and Exhibition related to business activities play a significant role in the global tourism industry (UNWTO, 2021). Thus, the event sector can stimulate economic growth and infrastructural development in a variety of nations. For instance, the Business Visits & Events Partnership (2021) mentions that the event industry in the United Kingdom (UK) generated 700,000 job opportunities for the community while generating £31billion income through the direct expense to the UK's economy in 2019. Weber and Ladkin (2003) state that today's Event tourism sector covers the world, with rising competition from established and growing markets, necessitating a growing need to address and address key strategic concerns to remain competitive. According to the International Congress and Convention Association (2016), concerns identified were advertising, infrastructure, human resources management, and service quality challenges.

As Tan et al. (2009) state, the application of information and communication technology tools, such as the computer network, software, and hardware necessary for internet connectivity, is called Information Technology (IT). In 1988, Poon

A. M. D. B. Nawarathna · W. M. A. H. Bandara
Department of Tourism Studies, Faculty of Management, Uva Wellassa University, Badulla, Sri Lanka

R. S. S. W. Arachchi (✉)
Department of Tourism Management, Faculty of Management Studies, Sabaragamuwa University, Belihuloya, Sri Lanka
e-mail: rangana@mgt.sab.ac.lk

A. Hassan (ed.), *Technology Application in Tourism Fairs, Festivals and Events in Asia*, https://doi.org/10.1007/978-981-16-8070-0_15

identified that IT is a particular utensil for the tourism sector. In the modern era, IT advancement continues rapidly, providing enormous potential and tangible benefits to suppliers, buyers, and intermediaries in each event tourism sector in the twenty-first century (Davidson et al., 2002). However, Mistilis and Dwyer (2000) state that the management of IT is critical in event tourism. Since it entails developing IT skills deemed necessary for the creation of comparative edge in event tourism. But tourism sector is hesitant to accept new technology since they mistakenly believe that technology cannot provide high service standards and a high-quality client experience. Hence, the most challenging hurdle in the modern event tourism market is IT. Furthermore, Maznevski and Chudoba (2000) and Papagiannopoulos et al. (2009) describe the influence of technical improvements in the event and hospitality sectors in great detail.

According to Sri Lanka Convention Bureau (SLCB), Sri Lanka is an attractive option for business activities. Since the island, located on the trade networks among east and west, provides excellent venues and facilities. Hence, the Event tourism sector in Sri Lanka dates back about four decades. Sri Lanka held the 5th Non-Alignment Conference in 1976, which 83 international leaders attended. According to Sri Lanka Tourism Development Authority's (SLTDA) statistical reports, Event tourist arrivals in Sri Lanka are steadily increasing every year. It accounted for 32% of overall arrivals in Sri Lanka (SLTDA, 2018).

Furthermore, according to the SLCB annual report 2019, the SLCB has made numerous efforts to advertise Sri Lanka as a destination for international event tourism, including several events, resources, and initiatives. For instance, Promotions and familiarisation tours in India, Germany, France, Australia, Spain and China. Further, as per the SLCB annual report, renovated Bandaranaike Memorial International Conference Hall (BMICH), Magam Ruhunupura International Convention Center (MRICC), Lotus Pond Theater, and leading hotels in Sri Lanka are performing a significant role in event Tourism.

However, SLTDA has noted some existing problems confronting the Sri Lankan event tourism industry: green events and maintaining information technology to establish responsible event tourism. The need for a large-scale convention Centre that can accommodate at least 5000 delegates and handle large-scale shows (SLTDA, 2020). In addition, SLCB (2019) mention that one of the most critical factors shaping the upcoming event tourism sector in Sri Lanka is IT. Hence IT can be highlighted as a challenge for Sri Lankan event tourism. As a result, there is a need for research to address the challenge of IT on Event Tourism. However, this sort of research has not been undertaken in the Sri Lankan context previously. Hence, there is a significant empirical gap that needed to be addressed with new knowledge. Thus, the primary purpose of this study is to investigate the effects of IT and its Challenges towards Sri Lankan Event Tourism with particular reference to the Event Managers' Perspectives.

Literature Review and Theoretical Framework

Event Tourism

Events have been identified as a critical and crucial aspect in the development process and the formulation of strategies. Hence, many event management specialists have presented various interpretations of the event. As per Lau (2015), events are gatherings of people who have a shared interest and come together in one location. The gathering's location must be planned ahead of time. In addition, the event site will have enough space and amenities to suit the demands of people who attend the meeting. Thus, Tassiopoulos (2005) mentioned that there are four goals of event tourism. Such as to create a favorable image for a destination; to expand the traditional tourist season; to spread tourist demand more evenly through an; to attract foreign and domestic visitors. According to Pendit (1999), event tourism comprises numerous meetings, incentives, conventions, and events, such as conferences, congresses, and trade exhibitions. This sector has been regarded as the most active and vital tourist industry sector, focused on business rather than leisure (Hamid et al., 2013).

When considering Sri Lankan Context, Figures of SLTDA statistics, event travelers accounted for 32% of overall tourist arrivals in 2019. Furthermore, about130,000 event travelers visit Sri Lanka each year, and the country generates 160 million dollars from event tourism. As per SLTDA's Tourism Strategic Plan (2017–2020), Currently Sri Lanka has several event tourism promoting agencies, Such as the Sri Lanka Convention Bureau (SLCB), Sri Lanka Tourism Promotion Bureau (SLTPB), Sri Lanka Association of Professional Conferences Exhibitions and Event Organizers (SLAPCEO), Professional Conference Organizers (PCO), Professional Exhibition Organizers (PEO), Event Management Companies (EMC), Destination Management Companies (DMC).

Information Technology

According to Werthner and Ricci (2004), IT refers to computer technology solutions that aid an organization's performance, operations, and strategy. Therefore, the purpose of information technology is to increase the productiveness of an organization. In addition, Sarosa and Zowghi (2003) suggested that The information technology component will comprise hardware (scanners, printers, computers, etc.) and software (operating systems, programming languages, office applications). However, Onn and Sorooshian (2013) described that IT encompasses any software program and necessary equipment, computer production, computer design, digital data exchange, and enterprise resource planning that positively influence organizational productivity.

IT and Event Tourism

According to Wöber (2003), the topic of IT is critical in the everyday actions of managers of the tourism industry. Since IT aid in providing the required information and organizing and performing sales marketing and distribution tactics. Thus, from a management standpoint, IT plays a crucial role in decision-making throughout all the sectors in the tourism industry. In addition, the accuracy of information about accessibility, amenities, attractions, and venue activities, boost visitor satisfaction. Travelers now have more information and options to choose their destination. Hence, the tourism sector needs to adopt IT to satisfy tourists and thrive in the long run (Buhalis & O'Connor, 2005). According to Akpan and Shanker (2017), IT establishes a new competitive global market by reaching a global audience, obtaining quick market information, conducting electronic commercial transactions, enhancing financial efficiency, and opening new markets.

Goldblatt (1990) describes that writers are meticulously chronicling previous literature on the effect and advancement in IT in tourism industries. Accordingly, Mistilis and Dwyer (2000) state that Modern society and event tourism have evolved to rely significantly on the information and services offered by IT. Mainly, IT provides both physical and intangible rewards that help business organizations generate money and achieve consumer demand outcomes. Thus, it involves developing IT skills deemed necessary for providing quality service and achieving competitive advantages in event tourism. As Flowers and Gregson (2012) describe that IT has been increasingly crucial in event management. Since more events incorporate IT to influence many event components, the value that IT can contribute to event management will boost pleasure, potential value, attachment to the event, image, income, reduce time, expense, and cash. Every stage of the event planning and implementation process may be aided by IT. Further, Draper et al. (2018) have listed five significant IT benefits in the event sector. Such as increasing audience participation, defining performance, reliable communication, obtaining valuable information and influencing.

The COVID-19 epidemic, on the other hand, decimated event tourism. As a result, event planners were forced to rely on IT systems to serve in these circumstances. Hence, Ranasinghe et al. (2020) transformed the event tourism industry into a hybrid and virtual format with the COVID-19 pandemic. A small group of people may be physically present in the hybrid concept while the rest of the audience watches the event from afar with modern online platforms. In addition, a majority of conferences and meetings have been designed as online events in the virtual method, and this format has been perhaps the best option in these difficult circumstances.

Challenges on Implementing IT in Event Tourism

According to Buhalis and Zoge (2007), traditionally, tourism managers have seen IT and their service as two separate and incompatible concepts, and IT has relegated itself to a supporting position. Hence, despite the growing importance of IT, many executives are still hesitant. Otherwise outright opposed, to incorporate IT into their high-level decision-making processes in tourism. However, according to Ziyadin et al. (2019), events grow in size every year. In this, Event managers have power over events in the early phases. However, there would be a significant gap in the planning phase between the actual performances on large sites due to the lack of incorporation with IT. As a result, IT should be introduced into those firms where it still hasn't been employed to solve this challenge. Furthermore, as well as those where it has already been employed more specialized IT should be introduced.

Concept Indicator Model

The conceptual Indicator model demonstrates the research methodology in graphic and textual formats to help the reader comprehend the research strategy. The conceptual indication for this study is illustrated as follows (Fig. 15.1).

Methodology

This study focuses on primary data gathered through structured interviews with event firms registered with the Sri Lanka Association of Professional Conference, Exhibition and Event Organisers (SLAPCEO) in Sri Lanka. The purposive sampling method was employed for this research, and 15 event firms in Sri Lanka who were members of SLAPCEO were interviewed for this study as the sample size. This study typically consists of 15 critical questions, with additional sub-questions. The interview guide was established, and it included basic questions about the company's history and questions concerning the influence of IT on event tourism. The general questions were designed to elicit background information about the firm and motives for arranging an event.

The main objective of the research was to identify the challenges of the adaptation of information technology in Sri Lankan event tourism. Further, the last

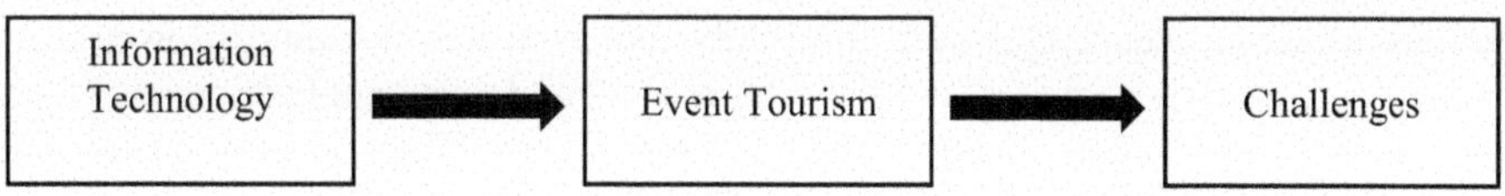

Fig. 15.1 Conceptual indicator model. (Source: Developed by the researchers, 2020)

question aimed to determine the respondents' opinion about promoting IT in a better and more efficient way to increase the number of events in the Sri Lankan context.

Respondents in the interviews were contacted via telephone, e-mail, in-person, or a mixture of all three. All 15 interviews were conducted via phone and person, and the length of each interview ranged from 20 to 30 min. After the interview, each discussion was documented and re-examined by the researcher. Each transcript was submitted to the appropriate responder to validate that the data collected was correct. The revised transcripts were then analyzed.

To increase the validity and reliability of the research, the researcher used different methods as member checking and debriefing of the research. First, the researcher shared a copy of the interview transcript with each participant to check the validity of the work. It has been done to ensure participants review what they said, add more information if they want to, and edit what they said. Further, the researcher shared the interview transcripts, methodology and final report with the impartial peers and feedback to enhance reliability and ensure validity.

The researcher chose content analysis as the data analysis approach in the study. It is the most generally used approach of qualitative data analysis that focuses on detecting patterned importance throughout a dataset. It seeks to find meaningful patterns from the data that give a solution to the research topic.

Results and Discussions

Technology is becoming increasingly crucial in the event industry that is increasingly affecting various aspects of the event. Adopting information technologies in the event industry can provide the quality service and create the competitive advantage. Currently, in Sri Lanka, event management organizations organize events that are not fully equipped with modern technology.

According to the findings, 73% of organizations used at least a few information technologies in their event operations. The central focus of the study was to identify the challenges of the adaptation of information technology in Sri Lankan event tourism. Analyzing the respondents' interview responses, the researcher identified five challenges: cost, resistance to change, tech-savvy employees, lack of government support, and security issues.

The first challenge that occurs in adapting information technology is cost. It becomes the delaying factor of any organization which contemplates the IT initiatives in the event industry. Various technologies make it easier for event planners to organize more significant events. Event planners have been used information technology to some extent. However, this study indicated that it requires lower-cost technological infrastructures, and they did not integrate new applications of IT is due to cost. The previous studies found that adapting new IT applications can create a 20% increase in event attendance and reduce the cost by up to 30%.

In Sri Lanka, you can do whatever the requirement. But then we have to pay a huge some. Because of that, people do not want to go ahead with this (Participant 07: Personal communication, 2020).

This barrier is complex because it can manifest in different ways. The challenges are allocating resources toward information technologies measures and acquiring the necessary resources in finances, staffing and time. Finally, the third challenge is identifying cost-effective solutions when information technology, alternatives, and practices are costlier than conventional ones. Therefore, overcoming the above impressions and adapting IT in the event industry is a challenge.

The second challenge of adapting IT in the Sri Lankan event industry is resistance to change. Some event managers want to carry out something manually, and their minds cannot be changed. On the other hand, most people want to live and make choices without sacrificing the world to meet their present needs. And also, customers want to carry out something physically, and their mentality of filling something on paper cannot be changed. Their mentality and perception of being difficult to achieve are critical elements that influence the effectiveness of the event industry.

Some people need the documentation process. We have something through the internet, but people are not bothered because most people are not familiar with the computers and internet (Participant 11: Personal Communication, 2020).

According to the research, 50% of the respondents indicated the lack of tech-savvy employees as the challenge to adopt IT for the Sri Lankan event industry. The spread of new information technologies affects the development of the event industry. Therefore, getting the right skills in the workforce's hands is vital for the industry's future expansion. And also, employees tend to work with manual methods, and they do not bother new knowledge about IT in the event industry.

The local government plays an essential role in promoting IT adaptation in the event industry. They are responsible for providing the required facilities, bringing significant economic benefits to the event industry. A lack of government support is another challenge associated with adopting IT for the events industry. According to Mistilis and Dwyer (2000), the government is an essential interested party in event tourism with a significant role in training, accreditation standards and marketing. The least discussed, and the referenced barrier was still a significant challenge and a critical need. Findings cited 41% of the respondents mentioned the need for improved government support, adopt IT, and policy-making to the Sri Lankan event industry.

Government should not do anything for the events industry to initiate IT, and they do not create policies regarding information technologies. They are not bothered (Participant 08: Personal Communication, 2020).

The fifth challenge of adapting IT in the Sri Lankan event industry is security issues. According to the research, the researcher has identified two issues as online security issues and venue security issues. Online security is a complex system designed to prevent the theft of credit cards or as simple as using antivirus software to protect the device against malware and viruses. The researcher has identified the three technologies from the respondents to secure online security at their event operations as payment gateways, IT team and firewall implementation.

> *Yes. We have an IT team to monitor the online security and protect the cyber-attacks and fraudulent transactions* (Participant 03: Personal Communication, 2020).

And also, venue security is essential to organize a successful event. The event organizers must ensure that the site or venue of the event is safe for workers and visitors. According to the research, the researcher has identified three technologies from the respondents to secure venue security at their event operations as security officers, CCTV cameras and metal detectors. By providing security and protection, showing potential business partners, investors and associates that genuinely care for their well-being as well as their interests. Further, it is essential to document and review venue security and event security management operations, contact and activities for future venue security.

Conclusion and Implications

Festivals, meetings, conferences, exhibitions, incentives, sports, and a variety of other events are all growing in popularity, and they contribute significantly to business and leisure tourism (Nawarathna, 2017). Adaptation of Information technology in the event industry has been recognized as a vital and vital factor in the development process and the decision-making of business strategies. And also, event planners forced to adopt new technologies during this COVID-19 pandemic. The findings revealed that most of the event organizations currently used information technologies to some extent in their event operations. Further, Sri Lanka cannot organize fully technologized events due to challenges associated with adopting information technologies. Therefore, integrating the technology with event planning tools through websites, apps, social media marketing, and other IT initiatives provides exciting opportunities to make each event memorable.

The study addressed the empirical gap in the industry that caters to organizations' modern trend in highlighting their IT applications in the highly sophisticated market. However, this research is more significant in revealing the more real evidence in diversified event organization, irrespective of size, scale or scope, that are interviews with the accurate flow of inputs and outcomes. Further, this research reveals responsibility can be seen as a fundamental aspect of events and a mechanism to the organization's longevity within a changing event environment, leading to management competency and an aspect of organizational excellence.

The findings elucidate practical managerial implications for event organizers, government and residential communities where events are held (Nawarathna & Arachchi, 2021). Maximize the awareness of attendees and organizers, invest in the latest event technologies, government involvement in policy framework and private, public partnership create long-term opportunities and improve the quality in Sri Lankan event industry.

Future Research Directions

This research was conducted to study information technology and its challenges towards Sri Lankan event tourism; event managers' Perspective. Future researchers can conduct the same research on a broader basis based on all the event stakeholders as attendees, responsible government organizations, local community, and private event management organizations. Since the study was conducted in the qualitative method, it cannot provide significant proof regarding the concept indicator model. Future research can conduct this study in a quantitative analysis method to validate the concept of the indicator model.

References

Akpan, I. J., & Shanker, M. (2017). The confirmed realities and myths about the benefits and costs of 3D visualisation and virtual reality in discrete event modelling and simulation: A descriptive meta-analysis of evidence from research and practice. *Computers & Industrial Engineering, 112*, 197–211.

Buhalis, D., & O'Connor, P. (2005). Information communication technology revolutionising tourism. *Tourism Recreation Research, 30*(3), 7–16.

Buhalis, D., & Zoge, M. (2007). The strategic impact of the Internet on the tourism industry. In M. Sigala, L. Mich, & J. Murphy (Eds.), *Information and communication technologies in tourism 2007* (pp. 481–492). Springer.

Business Visits & Events Partnership. (2021). *Home*. Retrieved from: https://www.businessvisitsandeventspartnership.com/. Accessed 11 June 2021.

Davidson, R., Alford, P., & Seaton, T. (2002). The European meetings, incentives, conferences, and exhibitions (MICE) sectors are used to use the information and communications technology. *Journal of Convention & Exhibition Management, 4*(2), 17–36.

Draper, J., Thomas, L., & Fenich, G. G. (2018). Event management research over the past 12 years: What are the current trends in research methods, data collection, data analysis procedures, and event types? *Journal of Convention & Event Tourism, 19*(1), 3–24.

Flowers, A. A., & Gregson, K. (2012). Decision-making factors in selecting virtual worlds for events: Advocacy, computer efficacy, perceived risks, and collaborative benefits. *Event Management, 16*(4), 319–334.

Goldblatt, J. J. (1990). *Special events: The art and science of celebration*. Van Nostrand Reinhold International.

Hamid, M., Ismail, N., Fuza, Z. I. M., Ahmad, K. N., & Awang, K. W. (2013). Sustainable tourism development practices of mice venue providers in the east coast region, peninsula Malaysia. *Journal of Tourism, Hospitality & Culinary Arts, 5*(2), 19–29.

Lau, C. (2015). Meetings, Incentives, Conventions and Exhibitions (MICE). *Journal of Convention and Event Tourism, 2*(4), 150–163.

Maznevski, M. L., & Chudoba, K. M. (2000). Bridging space over time: Global virtual team dynamics and effectiveness. *Organisation Science. INFORMS, 11*(5), 473–492.

Mistilis, N., & Dwyer, L. (2000). Information technology and service standards in MICE tourism. *Journal of Convention & Exhibition Management, 2*(1), 55–65.

Nawarathna, A. M. D. B. (2017). An analysis of the push and pull motives for choosing Sri Lanka as the wedding tourism destination: With special reference to Southern Province. *International Journal of Engineering and Management Research, 7*(5), 256–261.

Nawarathna, A. M. D. B., & Arachchi, R. S. S. W. (2021). A study on sustainable event management practices in Sri Lanka; event Managers' perspective. *Tourism and Sustainable Development Review Journal, 2*(1), 49–64.
Onn, C. W., & Sorooshian, S. (2013). Mini literature analysis on information technology definition. *Information and Knowledge Management, 3*(2), 139–140.
Papagiannopoulos, P., Xenikos, D. G., & Vouddas, P. (2009). Event management and group communications: The case of the 2004 Olympic Games in Athens. *Event Management, 13*(2), 103–116.
Pendit, N. S. (1999). *Wisata Konvensi.* Potensi Gede Bisnis Besar PT Gramedia Pustaka Utama.
Poon, A. (1988). Tourism and information technologies. *Annals of Tourism Research, 15*(4), 531–549.
Ranasinghe, R., Damunupola, A., Wijesundara, S., Karunarathna, C., Nawarathna, D., Gamage, S., & Idroos, A. A. (2020). *Tourism after corona: Impacts of COVID 19 pandemic and way forward for tourism, hotel and mice industry in Sri Lanka.* Retrieved from: https://ssrn.com/abstract=3587170. Accessed 10 June 2021.
Sarosa, S., & Zowghi, D. (2003). Strategy for adopting information technology for SMEs: Experience in adopting e-mail within an Indonesian furniture company. *Electronic Journal of Information Systems Evaluation, 6*(2), 165–176.
Sri Lanka Convention Bureau. (SLCB). (2019). *Annual Report.* SLCB.
Sri Lanka Tourism Development Authority. (SLTDA). (2018). *Tourism Strategic Plan (2017–2020).* SLTDA.
Sri Lanka Tourism Development Authority. (SLTDA). (2020). *Sri Lanka Tourism, year in review.* SLTDA.
Tan, K. S., Chong, S. C., Lin, B., & Eze, U. C. (2009). Internet-based ICT adoption: Evidence from Malaysian SMEs. *Industrial Management & Data Systems, 109*(2), 224–244.
Tassiopoulos, D. (2005). *Event management: A professional and developmental approach.* Juta and Company Ltd.
United Nation World Tourism Organization (UNWTO). (2021). *Home.* Retrieved from: https://www.unwto.org/. Accessed 9 June 2021.
Weber, K., & Ladkin, A. (2003). The convention industry in Australia and the United Kingdom: Key issues and competitive forces. *Journal of Travel Research, 42*(2), 125–132.
Werthner, H., & Ricci, F. (2004). E-commerce and tourism. *Communications of the ACM, 47*(12), 101–105.
Wöber, K. W. (2003). Information supply in tourism management by marketing decision support systems. *Tourism Management, 24*(3), 241–255.
Ziyadin, S., Koryagina, E., Grigoryan, T., Tovma, N., & Ismail, G. Z. (2019). Specificity of using information technologies in the digital transformation of event tourism. *International Journal of Civil Engineering and Technology, 10*(1), 998–1010.

A. M. D. B. Nawarathna is a graduate in Hospitality, Tourism and Events Management (BBM) from the Uva Wellassa University of Sri Lanka and currently reading for an MBA in Tourism from the Sabaragamuwa University of Sri Lanka. He is a full-time faculty member of the Department of Tourism Studies. He also is a member of different committees and projects in the university and has coordinated several events organized in the university. His research interests are in Event Management, MICE tourism, Travel Agency and Tour Operations Management, Tourism Planning and Management, Destination Management and Marketing etc.

Dr R. S. S. W. Arachchi graduated from the Sabaragamuwa University of Sri Lanka before receiving his Master degree from Colombo University, Sri Lanka and a PhD degree from Management and Science University in Malaysia. He is currently working as the Head of, Department of Tourism Management and a Senior Lecturer of Faculty of Management Studies, Sabaragamuwa University of Sri Lanka, Belihuloya. His primary teaching and research areas are eco-tourism, community-based tourism, sustainable tourism development, responsible tourism, environment and tourism resources and homestay tourism in Asian Countries. He has published nearly 20 articles in well-recognised journals both in Sri Lanka and overseas. Dr Arachchi is an editorial board member of the Journal of Tourism and Hospitality Management, International Journal of Education Humanities and Social Science, and Global Association for Humanities and Social Science Research.

W. M. A. H. Bandara is a graduate of Tourism and Hospitality Management (BSc) from the Rajarata University of Sri Lanka. He is an academic member of the Department of Tourism Studies of the Uva Wellassa University of Sri Lanka. In addition, He represents several projects undergoing in the university. His research interests are Hospitality Management and Tourist Behaviors.

Chapter 16
Technology-Based Tourism Development Mechanism in China with Events, Festivals and Fairs

Syed Far Abid Hossain, Kazi Mohiuddin, Faiza Tanaz Ahsan, Hasan Mahmud, Arslan Amzad, and Shazia Irshad

Introduction

Tourism is a key strategy for economic development that is viewed as one of the world's largest sectors that can shape economic health (Samala et al., 2020). Tourism contributes to the economy by generating employment, earning foreign currencies, improving local infrastructure, and branding the country's image. Due to geographical location and cultural heritage, China is endowed with a lot of tourism opportunities. Leveraging these opportunities, China is attracting tourists from different parts of the world. Tourism is now considered the fastest-growing industry in China (Zhao & Liu, 2020). Every year, the government invests many amounts in developing tourism infrastructure and modernization of facilities. Government promotes tourism by relaxing visa requirements, improving private and public transportation, conserving natural resources, modernizing and developing hotels and accommodations, and improving customer service. But due to the omnipresence of technology, these traditional measures are no longer alone sufficient to attract new visitors and retain existing travelers. Like any other sector, tourism is also largely influenced by

S. F. A. Hossain (✉)
College of Business Administration, IUBAT-International University of Business Agriculture and Technology, Dhaka, Bangladesh

K. Mohiuddin
Shanghai Maritime University, Shanghai, China

F. T. Ahsan
North South University, Dhaka, Bangladesh

H. Mahmud
LUT School of Engineering Science, LUT University, Lappeenranta, Finland

A. Amzad · S. Irshad
Xi'an Jiaotong University, Xi'an, China

A. Hassan (ed.), *Technology Application in Tourism Fairs, Festivals and Events in Asia*, https://doi.org/10.1007/978-981-16-8070-0_16

technology (Tussyadiah, 2020). Research is plenty on tourism development through the traditional way but is scantly using the technology. Technology has much to offer in tourism sectors ranging from ticket booking to trip plans to optimize the tour experience. Technology makes easer many travel-related activities. Nowadays, excursionists and tour marketers use various technology-based solutions for enhancing the travel experience. Few of them to mention are AI-based facial recognition for hassle-free check-ins, virtual reality (VR) for virtual hotel tours, travel experience, and booking interfaces, voice-based or text-based chatbots for enhanced customer experience, robots for taking care of front desk activities, room services and luggage check-ins, Google maps with a visual positioning system (VPS) for a live view of the position and landmarks, natural language processing based language translator as a local guide, and various optimization tools based on Maximum Likelihood Algorithm (Samala et al., 2020; Zeng et al., 2019).

Being a technological frontier, China has already paid attention to implement a technology-based mechanism for the development of the tourism sector. According to a report by Honghao (2019), Director of China Tourism Academy, China is focusing on improving the tourism industry, adopting digital technologies such as 5G, blockchain technology, artificial intelligence, Internet of Things, big data, cloud computing, VR/AR, modular construction, big health, 3D printing, and self-driving cars, etc. However, it is not clear how China attracts tourists with these technology applications in this contemporary era, how customers perceive the application of these technologies in the tourism industry, and how this trend promotes the quality of life (TQOL) in China. In this context, this chapter will explore the role of technology-based tourism development mechanisms in China with the help of tourism events, festivals, and fairs. The chapter specifically focuses on tourism Apps development that may affect tourism events, festivals, and fairs and improve tourism and quality of life (TQOL).

Various Tourism Spots in China

China emerged in the global tourism market as a latecomer. China was closed to the outside world until its ‘open-door policy in the 1970s. Later in 1986, China started considering tourism a tool for economic development, investing heavily in infrastructure and facilities development. This initiative led the country to attract a huge number of inbound tourists. Geographically and culturally, China is abundant with many tourism amenities, such as scenic landscapes, cultural traditions, and historical places. China’s tourism spots can be classified into the following categories: world heritage sites (cultural, natural, mixed), theme parks, museums, and temple fairs.

China, jointly with Italy, is the home of the largest number of World Heritage Sites (55) declared by the United Nations Educational, Scientific, and Cultural Organization (UNESCO) (UNESCO, 2021). Such heritages significantly impact the development of tourism of a country (Yang et al., 2010). According to UNESCO, heritages are classified into a cultural, natural, and mixed (cultural and natural) heritage (Yang et al., 2010). China has 37 social, 14 natural, and 4 diverse world heritage sites. Some of the most visited world heritage sites are The Great Wall, Forbidden City, Mukden Palace, Terra-Cotta Warrior, Sichuan Giant Panda Sanctuaries, Mount Huangshan, Potala Palace, etc. Theme parks and amusement parks are also attractive tourist attractions. China is the home of some top-performing theme parks. Some notable theme parks are the Chimelong Ocean Kingdom, Ocean Park in Hong Kong, Flying Kiss Theme Park in Wulong, Disneyland in Hong Kong, Songcheng Park in Hangzhou, Overseas China Town (OCT), Changzhou Dinosaur Park, Window of the World in Shenzhen, Guangzhou Changlong Holiday Resort, Beijing Happy Valley, Shenzhen Happy Valley, and Splendid China Folk Village. In addition to the above, China is also famous for its world-class modern museums. China's National Museum is the world's third most visited museum. Other museums include Hong Kong Science Museum, China Geological Museum, Shaanxi History Museum, Shanghai Museum, etc. They are an excellent source of knowledge about China and its history. China is also known for its temple gatherings, which taken place on the eve of the Chinese New Year Celebration. Visitors assume the chance to enjoy traditional folk performances, handicraft works, and local snacks at a temple fair. Some notable temple gatherings are Ditan Park Temple Fair, Longtan Park Temple Fair, Dongyue Temple gatherings, and Grand View Garden Temple Fair.

Based on the attraction regarding safety, cleanliness, sanitation, shopping, and transportation, the Chinese authority has been using a rating system to categorize the tourism spots. These are 1A, 2A, 3A, 4A, and 5A (the highest level). Visiting 5A and 4A spots are more worthwhile than 3A, 2A, and 1A. Few examples of 5A spots are Overseas Chinese Town in Shenzhen, The Great Wall, The Oriental Pearl TV Tower, and Shanghai Wildlife Park. Naturally, China is endowed with a lot of scenic spots. In addition, China has a rich cultural heritage. Together all of China has many great tourist spots to offer to tourists.

Tourism, Events, Fairs and Festivals in China

Tourism fairs and festivals are exhibitions that cater to the travel and tourism sector. To obtain information and establish national and international travel arrangements, visitors may communicate with tour operators, travel agencies, hotels, and airlines. Tourism is important to countries because of its economic and social advantages, and they invest heavily to gain a portion of this market. Therefore, fairs provide effective marketing and promotional prospects of tourism values and fascinate both leisure and business tourists to the destinations. They also enable governments to promote themselves internationally, contributing to intercultural exchange, and aid in the development of domestic and international tourism. Popular tourism exhibitions have a favorable impact on the destination's image while also providing economic and social advantages to the host city (Emir & Zencir, 2018).

China is a pioneer in organizing tourism fairs and festivals for its visitors from all over the world. Being one of the world's largest economies and a culturally diversified country, China has various traditional and cultural festivals to delight its visitors. While festivals are an important part of Chinese culture, most Chinese festivals in 2021 are centered on the lunar calendar and are rather intriguing. Several tourism festivals are organized in different regions to provide; thus, they offer a formal and large channel for tourists to get a lot of information about tour resources and local culture in one place. These traditional holidays will truly immerse you in Chinese culture. The Qingming (Tomb Sweeping) Festival, Lantern Festival, the Winter Solstice the Double Seven Festival are all yearly festivities. The festivals included here were chosen for their traditional Chinese flavor. Because they are observed according to the Chinese calendar, the dates frequently change from year to year.

In the context of events, festivals and fairs play a significant part in people's lives. It has substantial importance to countries because of its economic and social advantages, and they invest heavily to gain a portion of this market (Emir & Zencir, 2018). It is a fast-expanding global industry and an essential component of the service economy. Tourism has become a very powerful sector for growing local gross domestic product (GDP) in many countries (Zhang et al., 2021). The estimated domestic market value has surged tenfold in the last decade, reported by the China National Tourism Bureau and Shenwan Hongyuan Securities. Domestic tourism listed firms are quickly expanding in China and internationally as domestic tourists grow (Geng et al., 2021). In many ethnic areas of China, tourism in ethnic communities has become an important conduit for poverty alleviation and rural vitalization. Over the last several years, a significant amount of social capital has been spent in ethnic villages to grow tourism under the auspices of a government effort, resulting in various governance structures in ethnic village tourism (Qianlang et al., 2021). China is a wonderful tourist destination that conducts various tourism events, such as fairs, and supports various activities to increase interest in the resources they provide. Tourist fairs make a substantial contribution to all tourism stakeholders. Tourism fairs and festivals serve both entertainment and culture and economic and social development by allowing cultural value exchange in China.

Technology-Based Tourism Development in China

China is a fantastic example of a country that is developing its tourist industry through technology. Technology has aided in reducing expenses, improving operational efficiency, customer experience, and enhancing services. Improved bookings, guest service systems, and communication may benefit both companies and customers. By the way, tourism enterprises have intensely benefited from electronic labor to exchange expensive human endeavors. Technology-based tourism not only saves money on labor but also assists China in avoiding customer service concerns. The advanced technology and big data statistics have progressively contributed to the tourism industry in recent years, inspiring more meaningful uses in the new tourist management model (Liu, 2021).

All facets of our daily lives and surroundings, including the effective operations of commercial enterprises and government/non-government organizations, are enhanced by modern technology. The level of living and quality of life and services will undoubtedly increase as travel technologies are utilized and implanted in the Chinese government's planned developmental aims to grow and modernize the Chinese tourism sector. The overall income generated by China's travel and tourism business is presently estimated to be around 3 trillion Yuan. As the industry grows, the country is well-positioned to overtake the United States of America as the world's second-largest tourist economy. The China National Tourism Administration announced a 5-year tourism strategy for China. China's tourism industry has grown to the point that the number of visitors visiting the country each year has exceeded 6.7 billion (Hou, 2021).

Digital platforms can help visitors and local agents develop collaborative networks, emphasizing the value of discussion and the potential for continuous innovation (Tymoshchuk et al., 2021). These technologies are being used by an increasing number of Chinese tourism destinations, including theme parks, museums, galleries, and even zoos, to attract tourists. The Palace Museum is a pioneer in the application of these tools. The world-famous museum, also known as the Forbidden City, put on a multimedia show to bring its mute historical artifacts to life. Visitors felt like they were going around the iconic site during ancient times while wearing a VR helmet in a moving seat. In Changsha, Hunan Province, the intelligent robot tour guide was put to action in a picturesque location. Tourists are shown around by the robot, which also provides a quick overview of the gorgeous area. It could entertain guests by singing and dancing. Some hotels have also implemented cutting-edge technology, such as artificial intelligence and intelligent robotics. Guests may control the lighting, drapes, and television in their hotel room by speaking to a smart speaker. They can also inquire about dinner hours, order room service, and reserve a taxi.

They are capable of entertaining clients with singing and stories while keeping them company. Ctrip, Qunar, Fliggy, and Tuniu, among China's online travel agencies, have helped modernize travel by providing one-stop booking services for hotels, plane tickets, train tickets, and bookings to visit picturesque locations (Sang, 2018). In practically every aspect of China's tourist sector, cutting-edge technology has been creatively employed. Tourists may enjoy interesting, immersive, and engaging visual experiences thanks to technologies like artificial intelligence (AI), augmented reality (AR), and virtual reality (VR).

Artificial Intelligence (AI) in the Chinese Tourism Industry

Artificial Intelligence (AI) is the imitation of natural human intelligence used to perform different human and other animals like image processing, language recognition, and processing, object detection with the help of built-in expert systems. Most definitions of artificial intelligence focus on it as a subfield of computer science or in terms of how machines can mimic human intelligence (Tussyadiah, 2020). The advent of technology is compelling AI in every field, with tourism is no exception. China has several tourist attractions spots that can contribute a lot to its economy. The advent of technology based on AI is taking the world to new dimensions. Unlike the previous days when there were no road maps available, it was difficult for the tourist to explore locations in the visited area. But now China is providing its tourists with several apps like Google Maps, Baidu Maps, Autonavi Navigations is another good example of a car's GPS supplied by China. All these apps offer content in the English language to facilitate foreign customers in a better way. Traveling agencies in China are mingling AI and human services to provide their tourists the best of what they want. Things ranging from customized trips, online booking, online payment systems, keeping the record of travelers (keep records of customers in their internal control system so they can manage their next visit more appropriately by remembering their key requirements). Mioji is another tool used by travel agencies in China where tourist can plan their trip. Like they have to add their current city, the city they want to visit, their duration of the journey, the number of people, and their preferences regarding trip, they can get details of nearby hotels and attractions by simply entering these particulars and transportation in nanoseconds. In the past language was a barrier which was impeding tourist to travel.

Still, now artificial intelligence makes it possible to communicate which local Chinese are not aware of the English language. With artificial machine devices like WT2 (smart earphone translator), travelers can communicate with them and obliterate the language gap. Recently, China has developed facial security check-in at their airports. This is again another step towards improving the quality of life provided to the tourist. They don't have to wait in long lines and ques. for security check now. The Chinese government is enforcing AI importance everywhere. In China, the instrumental value of artificial intelligence and "intelligence" as an important direction of social governance in destinations have been highly valued by governments at all levels (Davenport et al., 2020). There are in-site installed voice guides in different locations working as tour guides, inform the tourist about vintage history, etc. Digital marketing through AI is enabling the agencies to do promotions by incorporating more details.

Use of Robotics in the Tourism Industry

Robots become a force to be reckoned with as piloted advancements. Their utilization is getting progressively far-reaching, and they are viewed as arising innovations in the travel industry. In China, to tackle bulky heck in and out for travelers, the Hangzhou government began from the front work area of the hotel. It dispatched a "hotel registration in 30 s" administration. Vacationers can look at in and check rapidly without anyone else with just three stages, sweep and analyze ID cards, discover, arrange and affirm registration, make room cards. Both the registration experience of travelers and the productivity of administration are improved. With specialized help, the robot persistently adjusts and assesses application situations. In 2018, Alibaba's first automated lodging, "Fly Zoo Hotel", arrived in Hangzhou (Lo et al., 2019). Albeit full-robot lodgings like this may, in any case, be uncommon today, hotels throughout the planet have carried out smart robotization in some client-situated activities, for example, programmed registration, individual virtual colleagues, and feast conveyance robots. For sightseers, robots in travel scenes can give more definite and helpful administrations, yet robots themselves are likewise a fascination.

Robots have steadily gotten comfortable to individuals going in China, as the nation is making a beeline for its travel industry 3.0. A wide scope of bleeding-edge advances has been inventively applied in pretty much all aspects of China's travel industry. A developing number of China's vacation destinations, including amusement parks, exhibition halls, displays, and even zoos, apply those innovations to draw in guests. In April, an intelligent local robot escort was put to use in a beautiful region in Changsha, Hunan Area. The robot named Yoyo showed sightseers around and gave a short presentation for the picturesque spot. It could sing and move to delight guests. Also, because of voice acknowledgment programming, it could even answer inquiries from travelers. Cutting-edge transportation administrations give vacationers in China a lovelier and effective travel insight. The overhauled high-velocity rail line network empowers vacationers to go to distant spots for less cash and time.

What is more, the facial acknowledgment frameworks utilized at an expanding number of air terminals and rail line stations help accelerate security screening and ticket checks for voyagers. When discussing the disintegration and danger of artificial consciousness to individuals, a few analysts will utilize the expression "human-driven" to depict the deficiency of natural worth. Characteristic worth doesn't fill any need, so its impact on moral decisions is significant. Albeit the ebb and flow level of innovation is still at the phase of feeble artificial brainpower and just stays at straightforward specialized help, it is important to keep a levelheaded mentality and get ready for research on robot morals-related frameworks. Automation does not simply mean replacing the workforce with machines but integrating machines into an autonomous system that can complete a process without workforce assistance (Liu, 2019).

Mobile Apps in Chinese Tourism Industry

Tourism Mobile apps assist with travel planning, ticket bookings, cab booking, accommodation bookings, route mapping, etc. Apps are transforming the way we travel, and China is at the forefront of this trend. Tourists may make their trip to China a lot simpler by downloading these apps ahead of time. In recent years, diverse information and communication technologies have increased the smart city idea of the tourist business (ICT). The amount of travel sale services via mobile

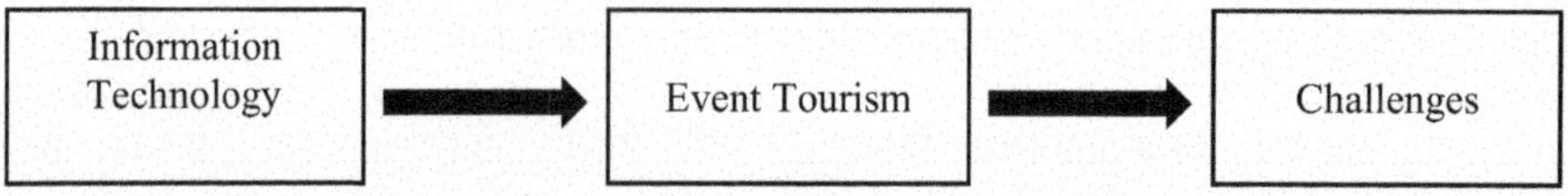

Fig. 16.1 Various applications (Apps). (Source: Authors' own creation)

internet is growing rapidly in the tourist business. Researchers and travel industry experts are already noticing disparities in usability between travel websites and travel apps as the mobile channel becomes more prevalent in the tourism business (Xu et al., 2014). The online travel industry is getting increasingly competitive, and customer demand is diversifying. As a result, online travel marketing strategy and product line diversifying have grown popular (Jia et al., 2016). Digital platforms can help visitors and local agents develop collaborative networks, emphasizing the value of discussion and the potential for constant improvements. As a result, digital media may assist travelers in co-creating their most valuable experiences with other visitors, influencing the success of a tourism location (Tymoshchuk et al., 2021). End-users can arrange their trip route using a variety of commercially available smartphone applications. On the other hand, existing mobile applications have various flaws, such as not being adaptive to every city's sustainable infrastructure. Apps can sometimes be limited in trip booking, with just static trip task scheduling and no dynamic geo-location-based job rescheduling (Krishnamurthi et al., 2021).

Apps are transforming the way we travel in China. Before traveling to China, download these apps to make things a lot simpler. It can help to overcome linguistic barriers and acquire a tourist's bearings more quickly (Fig. 16.1).

In China, using a VPN is required since Google apps cannot be used without one. Baidu Convert is one of the greatest language-learning applications available. It will save you time if you cannot understand menus, signs or cannot translate the name of a tourist destination into Chinese. WeChat is China's most popular messaging app, offering features such as text, voice, and video chatting, photo and video sharing, food ordering, travel and hotel booking, and more. WeChat pay/Alipay is the most popular smartphone-based app which connects with bank card help to make very easy payment. If any tourist finds themselves in a situation where there are no cabs available, DiDi is the app to use. It's essentially the Chinese equivalent of Uber and can be a good alternative when the metro is closed and the buses aren't running. The most popular map app is Baidu Maps, which is more accurate, up-to-date, and dependable for navigating China. The applications will show you the best routes to get you to your destination. Mobike is a bike-sharing software that allows users to navigate the city without being trapped in traffic. Meituan is a well-known online food delivery service. Ordering different dishes may be handy, quick, and typically less expensive than visiting a restaurant because restaurants frequently provide fantastic discounts and coupons to their clients. Taobao is China's largest and most dependable e-commerce website, where practically any goods may be purchased. It's linked to Alipay, which makes it simple to make a payment quickly and effortlessly. Since the travel and tourist business is developing, there is a significant need for apps in this sector. The perfect smartphone app should please end-users while also growing China's tourism industry. It provides substantial benefits and adjustments for both company owners and visitors in China.

Social Networking Sites (SNSs) to Support Tourism Fairs in China

Social networking sites (SNSs) are user communication-based platforms that can play an essential role in exchanging electronic word-of-mouth (eWOM) between travelers (Luo & Zhong, 2015). Individuals are using different digital platforms as a social network, which might include a strategic choice of information and self-related photographs, in which they create their social network profile. However, the influence of SNS use and online social support in tourism has been examined scarcely in previous studies. SNS helps satisfy tourism's social demands by creating a mobile and networked community that is a complicated link to face-to-face connection, media communication, co-existence, digital closeness, physical and virtual mobility. It has been established that smart technology and SNSs have been used in tourism at every stage, from seeing a dream to recollection (Morrison, 2019). The findings of (Shen et al., 2020) study show that SNSs influence and effective communication mediums due to the function and features. The number of people using SNSs demonstrates the popularity of SNSs. In 2018, the number of SM users globally estimated at 2.65 billion, with a predicted growth of nearly 3.1 billion in 2021 (Statista, 2019). SNSs have altered tourism decision-making.

To organize any event in China, each tourism firm must realize that Western social media apps such as Twitter, Facebook, YouTube, etc., are forbidden. However, several alternative SNS's are accessible for SM marketing in China, including WeChat, QQ, Weibo, Douyin, Tencent video, Youku, Xiaohongshu, Douban, Meituan, Toutiao, Figgy (Feizhu), Qunar, and a few dating apps, including Tantan, MoMo, Tinder, etc. People in China spend considerable time on social media platforms, especially WeChat, in their daily lives. In comparison to western texting apps, WeChat provides far better features. It has over 1 billion active monthly users. It is part of moments offers an opportunity to all the companies, including tourism organizations, to post an ad in moments section similar to Facebook newsfeed. It is a useful tool to support tourism fairs. Sina Weibo was introduced in 2009 as a microblogging network in China. Videos, photos, and gifs can be uploaded here. They may also follow a person to read and share posts. Chinese people use Weibo to get information, to share life, and the opportunities to interact with others. These things have made Weibo an unbelievably successful social media marketing channel in China to support tourism fairs.

QQ is Tencent's another instant messaging software that offers numerous services to users and advertisers, similar to WeChat. Tencent Video is likewise an excellent platform to reach a wide range of people like YouTube. The target audience can be reached by using several methods via Tencent Video and related advertising tools. Xiao-Hong-Shu is another e-commerce site with a variety of elements often seen in social media. Douban is also an SNS focusing on self-expression and a way of living. Users commonly talk about books, films, music, and events on it. This app may be particularly powerful in utilizing word-of-mouth marketing which is important for tourism fairs. Toutiao is a news dissemination application that aims to provide information of all types with complicated algorithms. In the Chinese landscape of social media, Douyin has become a force to consider. With 400 active users per month and an average time of 52 min per user per day makes it a wonderful SNS to support tourism fairs. Tourism fairs have also been supported by dating apps like Momo, Tantan, Tinder, etc. These are instant messaging apps while traveling apps like C-trip, Figgy, Qunar, etc., are also useful advertising tools.

Tourism and Quality of Life (TQOL) in China

Tourism is one of the largest industries. Tourism has a significant positive impact on the well-being of both tourists and hosts. The research found that people derive overall life satisfaction from non-leisure and leisure life (Neal et al., 2004). Traveling/tourism provides an excellent source of leisure life satisfaction. Traveling helps tourists experience a new culture, environment, and people and provides a memorable experience, thus increasing overall life satisfaction. On the other hand, tourism also affects the host's economy through generating employments, attracting foreign investments, mobilizing business activity, increasing property value, and developing infrastructure (Liang & Hui, 2016), thereby influencing the quality of life (QOL) of residents (Uysal et al., 2016). Quality of life entails a set of subjective and objective features that shape people's lives (Stiglitz et al., 2009). Particular dimensions represent the people's assessments of their lives in general or in specific categories such as family, work, and financial circumstances (Stiglitz et al., 2009). It is a cognitive exercise measured by emotional contents such as feelings and perceptions about life (Ridderstaat et al., 2014). On the other hand, objective features are characterized by the instrumental value of the conditions and resources—tourism chips in both dimensions of QOL. Residents get opportunities to meet new people from a new culture, share knowledge and experience, and have more contacts, who may become good friends (Liang & Hui, 2016). Furthermore, cultural tourism based on heritage, historical place, nature, local artifacts uphold the image and identity of the destinations (Xie, 2006). This experience helps develop emotional well-being and enhances the QOL of residents (Neal et al., 2004). Finally, the economic benefits of tourism are well established in the literature. Tourism fuels the growth of the local economy, creating employment, generating revenue, and contributing to GDP (Li, 2002; Uysal et al., 2016).

China is one of the biggest tourist destinations. Following the Travel & Tourism Competitiveness Report 2017 of the World Economic Forum (WEF) of 2017, China ranked 13th in 2019 rose from 71th in 2007 (Zhao & Liu, 2020). According to

United Nations World Tourism Organization (UNWTO) (2020), China is the fourth most visited country globally for the last couple of years. As many as 66 million international tourists visited China in 2019 and spent nearly 39 billion USD, the 11th highest tourists' expenditure in the world (United Nations World Tourism Organization, 2019). The number of domestic tourists is also remarkable. According to the Ministry of Culture and Tourism (2019), there were 6 billion registered domestic trips that indicate a greater impact of domestic tourism on the Chinese economy. The country's tourism industry earned $935 billion in revenue in 2019. According to CGTN (2020), the tourism sector contributed to 11.05% of total GDP and accounted for 10.31% of employment of the whole country. China earns substantial foreign currency from tourism and is benefitted from other tourism-related economic activities. Many idle places and historical sites have become vibrant and moneymakers. Tourists mostly consume services from transportation, shopping, accommodation, catering, entertainment, and sightseeing, thus contributing to the development of sectors related to these services (Liu & Li, 2019). Tourism plays a key role in developing the quality of life of tourists (domestic and inbound) and hosts. Tourism provides leisure life satisfaction to the tourists by offering a memorable experience. It contributes to developing the quality of life in China through increased economic activities and promoting emotional well-being.

Tourism Development Through Smart Cities in China

A smart city is an urban area that uses different types of electronic data collection sensors to supply information used to manage assets and resources efficiently. Urban communities throughout the planet progressively perceive the smart travel industry city idea and related procedures as methods for upgrading manageable conditions. Especially for urban communities confronting inhabitants' negative insights towards the travel industry, shrewd the travel industry city engages a city to meet people's high expectations by making urban spaces that inhabitants and guests can partake in together. China takes advantage of this trend-setting innovation in the remarkable expansion of its data, which consistently cooperates with the biologic world and empowers its travel business to anticipate travelers' social advantages. A brilliant city to centers around friendly and social life and it might improve metropolitan social cooperation.

Given that the travel industry in China is profoundly identified with social marvels, the idea of 'savvy' has been adjusted to the setting of the travel industry. As another popular expression in China, 'savvy innovation' has been altogether changing how guests consider perusing sites. PDAs and versatile applications are abused for settling on different travel choices, like transportation, convenience, and exercises accessible at an ideal travel industry objective. The travel industry exercises in China dependent on shrewd innovation, savvy the travel industry, is a social marvel. The brilliant travel industry experience addresses innovative intervened encounters that consolidate personalization, setting mindfulness and ongoing observing.

China's travel industry is a social and financial wonder that includes the development of individuals to places outside their typical climate for individual or expert purposes, and the principal reason for voyaging is for social inundation. At the start of the 1990s, the travel industry got one of the elective types of the travel industry, gone against to mass the travel industry. Travelers show a proactive approach and make encounters effectively while voyaging. Subsequently, travelers' heterogeneous inclinations have drawn consideration from objections, and the objective urban communities try to give tweaked top-notch travel encounters. Since the Web impacts the circulation of the travel industry data and deals, the travel industry is looking for fruitful e-the travel industry systems by creating sites, electronic trade, and related administration. The adoption of these e-businesses is flourishing China's tourism industry. China can develop its tourism by forcefully pushing the urban communities to the new travel industry plan forward with the huge advancement of intelligent innovation. It demonstrates another travel industry supported by incorporated endeavors to gather and examine information extricated from different sources in the mix with the utilization of cutting-edge data advances to change go encounters to make them more enhanced, productive, and supportable. In this regard, innovation can change how guests settle on various voyages, like transportation, convenience, and exercises accessible at an ideal travel industry objective.

Recent Challenge in the Tourism Industry in China Due to COVID-19

All industries have suffered insecurity since the COVID-19 breakout in late 2019, but this situation is most noticeable in the tourism and hotel industries. Being an industry dependent on human mobility and close cooperation, the hospitality and tourism business is the co-creator and key recipient of the pandemic and related consequences (AIEST, 2020). The crises nearly caused the shutdown of all industries, including restaurants, hotels, airlines, and drastically changing everything in China and the whole world. Especially in the hotel and tourism sector, almost all companies lost earnings. In his report under COVID-19 on the Chinese tourist market in early 2020, Koe (2020) indicates that cafes and lodging are empty since they do not have customers. As travelers are absent, the transportation companies face major disasters. International tourist operations are also declining substantially in China, damaging the tourism industry and travel companies involved in outbound tourism suffering losses. Tourist attractions, restaurants, and sports events were suspended for all international and domestic travelers and till now just partially opened, resulting in a huge loss to this industry.

The COVID-19 wreaked havoc on China's hotel industry first and foremost. As the first nation afflicted by health crises in China, its hotel business has faced severe hurdles (Zhang et al., 2020). For survival, many hoteliers, especially the owners of individual hotels, have transferred their property or suspended the business operation temporarily. Mounting evidence indicated that the pandemic is appeared to be on the verge. Tourists became cautious and began to cancel nonessential trip arrangements. For example, Dossen revealed that it got more than 40% of hotel cancellation messages (Wu, 2020). Zhang's research on 498 hotels found that OCC's total percentage has plunged by 89% during 14 and 28 January and, afterward, stayed about 10% until the end of February. In important cities, the hotel profits per accessible accommodation declined by over 85% yearly during February (Zhang et al., 2020). The hotel industry has been forced to experience its worst outcomes in recent history due to border closures, lockdown measures, and travel restrictions. Economic losses and hotel shutdowns seem to be too prevalent and caused insecurity which most had not encountered before.

On 28 March 2020, China shortly blocked its borders with foreign nationals to prevent a 2nd COVID-19 outbreak. Since then, many countries took the same actions and restricted foreign flights. Several countries, including Europe and the USA, have suspended trading and tourism with China due to concerns about the viral spread (Al Jazeera, 2020). Owing to the cancellation of flights from and to China due to COVID-19, the airline sector is badly affected (World Economic Forum, 2020). The Cruise lines also have had minimal passengers since the pandemic began (The New York Times, 2020).

Though the local airline has started its domestic flight operations, international flights are still limited, affecting its business. The traveling restrictions include COVID Test, and Quarantine problems still affect tourism as travelers have to quarantine their expenses. Days of quarantine are also different, leading to the cancellation of many tours. This pandemic has caused the tourism sector to increase and remain competitive in innovativeness. Relaxing limitations on traveling and vaccination delivery aren't sufficient for travelers to return. Trust, safety, and comfort are the challenges that will be important to revitalize tourism, and contribution to generating peace by the hospitality industry will be vital.

Conclusion and Strategies to Overcome the Situation

The tourism sector has to establish a framework to understand and execute disaster management strategy in the context of COVID-19 and for the increasingly catastrophe-driven world (Chan et al., 2019). A complete view of a pandemic scenario and ideas for handling it in each step is provided in several frameworks. Based on wider research in the tourist and hospitality industry regarding disaster management, Faulkner (2001, p. 144) suggested a standard method broadly used to examine and develop tourist disaster management. This framework identifies three components – the catastrophe phase, the reaction aspects, and the key aspects of the disaster management techniques. Effective business catastrophe management techniques may reduce risk and speed up the recovery.

Human Resource Strategy

High-performing personnel must be regarded as great assets for the company during catastrophes; those people help businesses to regain their performance once the pandemic has been calmed down. Businesses should thus endeavor to retain this personnel during the pandemic. It also creates an opportunity to teach employees online, assist the growth of employees, encourage employees to take a yearly vacation, and minimize working days and hours (McKercher & TRINET, 2020).

Communication and Leadership Strategy

With a proper assessment of pandemics, tourist firms should prepare themselves for various circumstances and efficiently adapt to anti-pandemic policies. In communication, the COVID-19 scenario demands innovation that can make information sharing and decision-making smooth and responsive (Moorhouse, 2020). Companies must build an effective remote control and management system for deploying personnel and holding meetings using a mixture of online office technology, tools, and platforms.

Strategies for Service Delivery

In times of crisis, the restoration of consumer trust is necessary. Interestingly, numerous hotel and airline brands started free cancellation and reservation support during the pandemic, offered prolonged membership in the loyalty program, and boosted membership perks to reassure clients (Sun et al., 2020). The tourism industry must adopt stringent hygiene and sanitation steps to ensure the housing environment for customers, including comprehensive disinfection, food hygiene control, the distribution of masks, the delivery of online medical services, health checkups, and closure of laundries, gymnasiums, and other public places and amenities. Enhanced smart contactless service delivery (including self-check-in, face-scanning, remote control, voice control, robots for room service, and zero-second check-out).

CSR Strategies

The significant growth in the number of infected people during the pandemic's peak caused a lack of hospital wards. Hospital expansions were required to confine and treat infected individuals. For the returned non-local employees, some hotels in large cities were designated as quarantine stations. Meanwhile, because of the global spread of COVID-19, arriving tourists were subjected to the same quarantine requirements. Hotels near major international airports acted as quarantine stations for coming travelers to prevent infection transmission. This is a great strategy to keep up the business in hard times.

Financial Strategy

The Tourism industry should promote the events through proper channels to revive their operations and cover their financial losses. Moreover, tourist companies should pay attention to regulations published by relevant government organizations and solicit monetary support from them (Sun et al., 2020). In addition, the municipal council must subsidize businesses, especially hotels that acted as hospital expansions, medical personnel lodging, and quarantine stations during the pandemic.

Standard Operating Procedure for Crisis Management

To improve effective crisis management skills to combat the pandemic, the tourism industry needs to build a responsive and efficient Standard Operating Procedure. McKinsey and the company (2020) noted a renewed confidence in Chinese customers and a peculiar increased interest in environmentally friendly products through numerous stages of the COVID-19 pandemic.

References

AIEST. (2020). *The response of tourism businesses vis-à-vis the economic ramifications of SARS-CoV-2*. Retrieved from: https://www.aiest.org/fileadmin/ablage/dokumente/Covid-Reports/Report_20200418_Business_Response.pdf. Accessed: the 1st July 2021.

Al Jazeera. (2020). *China coronavirus death toll jumps to 563*. Retrieved from: https://www.aljazeera.com/news/2020/02/china-coronavirus-outbreak-latest-updates-200206001100173.html. Accessed the 1st July 2021.

Chan, C., Nozu, K., & Cheung, T. O. (2019). Tourism and natural disaster management process: Perception of tourism stakeholders in the case of Kumamoto earthquake in Japan. *Current Issues in Tourism, 23*(15), 1864–1885.

CGTN. (2020). *China's total tourism revenue in 2019 reached 6.63 trillion yuan, up 11% year on year (CGTN)*. Retrieved from: https://news.cgtn.com/news/2020-03-11/China-s-total-tourism-revenue-in-2019-reached-6-63-trillion-yuan-OLDZXQvVXG/index.html. Accessed the 1st July, 2021.

Davenport, T., Guha, A., Grewal, D., & Bressgott, T. (2020). How artificial intelligence will change the future of marketing. *Journal of the Academy of Marketing Science, 48*(1), 24–42.

Emir, O., & Zencir, B. (2018). Opinions of business representatives on international tourism fairs: A qualitative research study. *Journal of Business Research - Turk, 10*(4), 1100–1109.

Faulkner, B. (2001). Towards a framework for tourism disaster management. *Tourism Management, 22*(2), 135–147.
Geng, S., Liu, S., & Liao, X. (2021). Operating performance of tourism listed companies in China: The perspective of economic value added. *SAGE Open, 11*(1), 215824402098106.
Honghao, Y. (2019). 2019 Report on the development of China's tourism and accommodation industry. .
Hou, T. (2021). Virtual tourism simulation system based on VR technology. *Journal of Physics: Conference Series, 1881*(3), 032084.
Jia, Z., Li, D., & He, F. (2016). Analysis and reviews on tourism and travel mobile apps of China. *Proceedings of the International Conference on Electronics, Mechanics, Culture and Medicine*. https://doi.org/10.2991/emcm-15.2016.13
Koe, T. (2020). Coronavirus impact: Nutrition retailers hit by drastic cut in China tourist numbers. *Nutra Ingredients Asia*. Retrieved from: https://www.nutraingredients-asia.com/Article/2020/02/03/Coronavirus-impact-Nutrition-retailers-hit-by-drastic-cut-in-China-tourist-numbers. Accessed: the 1st July 2021.
Krishnamurthi, R., Gopinathan, D., & Kumar, A. (2021). Trip-I-Plan: A mobile application for task scheduling in smart city's sustainable infrastructure. *Digital Cities Roadmap*. https://doi.org/10.1002/9781119792079.ch11
Li, Y. (2002). The impact of tourism in China on local communities. *Asian Studies Review, 26*, 471–486.
Liang, Z. X., & Hui, T. K. (2016). Residents' quality of life and attitudes toward tourism development in China. *Tourism Management, 57*, 56–67.
Liu, X. Y. (2019). Cognitive logic and technical path of artificial intelligence reshaping government service process. *E-government, 11*, 104–111.
Liu, Y. (2021). The application of big data in the intelligent tourism management mode is explored. *Journal of Physics: Conference Series, 1881*(3), 032080.
Liu, Y., & Li, W. (2019). The comprehensive economic contribution of China's tourism trade export. *Modern Economy, 10*, 1095–1106.
Lo, Y. L., Li, Y., & Chan, K. C. (2019). Contemporary innovation in China. *The Chinese Economy, 52*(5), 387–399.
Luo, Q., & Zhong, D. (2015). Using social network analysis to explain communication characteristics of travel-related electronic word-of-mouth on social networking sites. *Tourism Management, 46*, 274–282.
McKercher, B., & TRINET. (2020). A tourism crisis recovery checklist. *Travel and Tourism Research Association (TTRA)*. Retrieved from: https://ttra.com/wp-content/uploads/2020/04/TRINET-COVID-19-Recovery.pdf. Accessed: the 1st July 2021.
McKinsey and the company. (2020). *Understanding Chinese consumers*. Retrieved from: https://www.mckinsey.com/~/media/mckinsey/featured%20insights/china/china%20still%20the%20worlds%20growth%20engine%20after%20covid%2019/mckinsey%20china%20consumer%20report%202021.pdf. Accessed: the 9th July 2021.
Ministry of Culture and Tourism. (2019). *China reports surge in domestic tourism*. Retrieved from: http://english.www.gov.cn/statecouncil/ministries/202006/21/content_WS5eef34d5c6d0a6946639c7e1.html. Accessed: the 1st July 2021.
Moorhouse, B. L. (2020). Adaptations to a face-to-face initial teacher education course 'forced' online due to the COVID-19 pandemic. *Journal of Education for Teaching, 46*(4), 609–611.
Morrison, A. M. (2019). *Marketing and managing tourism destinations* (2nd ed.). Routledge.
Neal, J. D., Sirgy, M. J., & Uysal, M. (2004). Measuring the effect of tourism services on travelers? Quality of life: Further validation. *Social Indicators Research, 69*, 243–277.
Qianlang, S., Huanhuan, W., & Mengxue, L. (2021). Comparative study on the sustainable livelihood of ethnic tourism villages in China. In *Proceedings of the 2021 6th International Conference on Social Sciences and Economic Development (ICSSED 2021)*. https://doi.org/10.2991/assehr.k.210407.174.

Ridderstaat, J., Croes, R., & Nijkamp, P. (2014). The tourism development-quality of life nexus in a small island destination. *Journal of Travel Research, 55*(1), 79–94.

Samala, N., Katkam, B. S., Bellamkonda, R. S., & Rodriguez, R. V. (2020). Impact of AI and robotics in the tourism sector: A critical insight. *Journal of Tourism Futures*. https://doi.org/10.1108/JTF-07-2019-0065

Sang, Y. (2018). How is technology transforming Chinese tourism? *Chinese Plus*. Retrieved from: https://chinaplus.cri.cn/news/china/9/20180927/188985.html. Accessed: the 1st July 2021.

Shen, S., Sotiriadis, M., & Zhou, Q. (2020). Could smart tourists be sustainable and responsible as well? The contribution of social networking sites to improving their sustainable and responsible behavior. *Sustainability, 12*(4), 1470.

Statista. (2019). *Number of social media users worldwide*. Retrieved from: https://www.statista.com/statistics/278414/number-of-worldwide-social-network-users. Accessed: the 1st July 2021.

Stiglitz, J. E., Sen, A., & Fitoussi, J. P. (2009). *Report by the Commission on the Measurement of Economic Performance and Social Progress*. Retrieved from: https://ec.europa.eu/eurostat/documents/8131721/8131772/Stiglitz-Sen-Fitoussi-Commission-report.pdf. Accessed: the 1st July 2021.

Sun, Z., Li, R., He, P., & Shi, Y. (2020). Perspective crisis: The impact of COVID-19 on hotel industry and its response. *Meadin.com*. Retrieved from: https://www.meadin.com/yj/212678.html. Accessed: the 1st July 2021.

The New York Times. (2020). In coronavirus, $45-Billion cruise industry faces a big challenge. *The New York Times*. Retrieved from. https://www.nytimes.com/2020/02/12/travel/coronavirus-cruises-travel.html. Accessed: the 1st July 2021.

Tussyadiah, I. (2020). A review of research into automation in tourism: Launching the annals of tourism research curated collection on artificial intelligence and robotics in tourism. *Annals of Tourism Research, 81*, 102883.

Tymoshchuk, O., Almeida, A. M., Pedro, L., Antunes, M. J., Ramos, F., Oliveira, E., & Carvalho, D. (2021). Digital technologies as tools to promote tourism and territorial development. *Impact of New Media in Tourism*. https://doi.org/10.4018/978-1-7998-7095-1.ch016

United Nations Educational, Scientific and Cultural Organization (UNESCO). (2021). *China - UNESCO World Heritage Centre*. Retrieved from: http://whc.unesco.org/en/statesparties/cn. Accessed: the 1st July 2021.

United Nations World Tourism Organization (UNWTO). (2019). *International tourism highlights*. Retrieved from: https://www.e-unwto.org/doi/pdf/10.18111/9789284421152. Accessed: the 1st July 2021.

United Nations World Tourism Organization (UNWTO). (2020). UNWTO World Tourism Barometer and Statistical Annex, December 2020. *UNWTO World Tourism Barometer, 18*, 1–36.

Uysal, M., Sirgy, M. J., Woo, E., & Kim, H. L. (2016). Quality of life (QOL) and well-being research in tourism. *Tourism Management, 53*, 244–261.

World Economic Forum (WEF). (2017). *The travel & tourism competitiveness report 2017*. Retrieved from: https://www.weforum.org/reports/the-travel-tourism-competitiveness-report-2017. Accessed: the 1st July 2021.

World Economic Forum (WEF). (2020). *Trapped tourists: How is the coronavirus affecting travel?* Retrieved from: https://www.weforum.org/agenda/2020/02/the-coronavirus-will-hit-the-tourism-and-travel-sector-hard. Accessed: the 1st July 2021.

Wu, W. (2020). After the epidemic prevention and war, the hotel group's scale and brand export. In *China Cultural Tourism and Hospitality Industry Digital Summit, 2020*. Retrieved from: http://www.tripvivid.com/articles/26594. Accessed: the 1st July 2021.

Xie, P. F. (2006). Developing industrial heritage tourism: A case study of the proposed jeep museum in Toledo, Ohio. *Tourism Management, 27*, 1321–1330.

Xu, R., Liu, L., & Panneerselvam, J. (2014). User experience evaluation of Chinese travel app software. In *2014 IEEE international conference on computer and information technology, 2014* (pp. 610–615). https://doi.org/10.1109/CIT.2014.109

Yang, C. H., Lin, H. L., & Han, C. C. (2010). Analysis of international tourist arrivals in China: The role of World Heritage Sites. *Tourism Management, 31*, 827–837.

Zeng, Y., Lu, E., Sun, Y., & Tian, R. (2019). Responsible facial recognition and beyond. Retrieved from: arXiv preprint arXiv:1909.12935 Accessed: the 1st July 2021.

Zhang, C., Cui, R., Xu, C., & Wang J. (2020). COVID-19 and the Chinese hotel sector. *HVS*. Retrieved from: https://www.hvs.com/article/8723-COVID-19-and-the-Chinese-Hotel-Sector. Accessed: the 1st July 2021.

Zhang, H., Song, H., Wen, L., & Liu, C. (2021). Forecasting tourism recovery amid COVID-19. *Annals of Tourism Research, 87*, 103149.

Zhao, Y., & Liu, B. (2020). The evolution and new trends of China's tourism industry. *National Accounting Review, 2*, 337–353.

Syed Far Abid Hossain is a PhD researcher at the School of Management of Xi'an Jiaotong University, China. He graduated from the University of Sunderland, London, UK with an MBA degree. He is a Senior Lecturer and MBA program Coordinator at the College of Business Administration, International University of Business Agriculture and Technology (IUBAT), Dhaka, Bangladesh. He has published numerous papers in indexed and reputed international journals. Also, he has achieved the 'Outstanding International Student Award', Dean's scholarship and Full bright scholarship from the Chinese government. His research interests include women entrepreneurship, innovation, mobile-phone usage, academic performance, social networking, and TPACK. He has participated numerous reputed international conferences in the USA, China, and Bangladesh. He is appointed as a reviewer by famous publishers as Elsevier, Emerald, Wiley, SAGE, Springer, Frontiers, Taylor and Francis and IGI Global.

Kazi Mohiuddin is a PhD Researcher at Shanghai Maritime University, China. He has published numerous papers in well-indexed reputed international journals. His research interests are: tourism management, maritime management, Artificial Intelligence in tourism and so on.

Faiza Tanaz Ahsan is a Research Assistant at North South University. Her research interests are: Women Empowerment, Edge-cutting technology in business, Artificial Intelligence, bank lending practices, credit management, cloud computing, Budget taxation and public investment, trade and globalization and so on.

Hasan Mahmud is a Doctoral student at LUT School of Engineering Science, LUT University, Finland. He has 12 years of industry experience in information technology, banking, and FMCG. His research interests are in information technology and business management.

Arslan Amzad is a Graduate Student at Xi'an Jiaotong University, China. His research interests are: tourism management, technology management in organization, job engagement in tourism and so on.

Shazia Irshad is a Graduate Student at Xi'an Jiaotong University, China. Her research interests are: innovation, smart tourism, machine learning, AI in tourism and so on.

Chapter 17
Approaching the Paradox: Loving and Hating Technology Applications of Indonesia's Cultural Events

Putu Diah Sastri Pitanatri and Luh Yusni Wiarti

Introduction

In destinations that rely on culture as the primary capital for tourism growth, expert criticism of tourist sites is frequently a point of contention. The fundamental issue is the contradiction between cultural tourism and traditional tourist (Picard, 2006). Especially in an increasingly connected world that is becoming "borderless" in order to foster the development of a "global culture", one must progressively scrutinize the culture's authenticity. Are indigenous cultures still cultures that are consumed by visitors visiting the destination? Or is it simply a "staged culture" that is purposely staged in order to profit economically from tourist activities?

As the tourist discussion proceeds, academics' concerns about authenticity have increased (Hamilton-Smith, 1987; Maoz, 2018; Unger et al., 2016). Indeed, as technology and modes of mobility advance, the movement of tourists between and among places becomes increasingly unstoppable. It's worth debating if the culture formed by tourism is truly based on cultural exchange, or on a shared transnational culture that transcends national or regional culture; or on what is referred to as a "global tourist culture" (Urry, 1990). Cultural tourism is undoubtedly a fascinating subject to debate from a variety of scholarly perspectives.

Apart from the question of authenticity, global tourism actors are presently compelled by technology to maximize cultural potential. For instance, K-Pop concerts in Korea are staged virtually. Another area that will require additional research in the future is the application of new technology to cultural tourism experiences. While Virtual Reality (a computer-generated three-dimensional environment) and

P. D. S. Pitanatri (✉)
School of Postgraduate of Tourism Studies, Gadjah Mada University, Yogyakarta, Indonesia

L. Y. Wiarti
Tourism Studies, Bali Tourism Polytechnic, Bali, Indonesia

A. Hassan (ed.), *Technology Application in Tourism Fairs, Festivals and Events in Asia*, https://doi.org/10.1007/978-981-16-8070-0_17

Augmented Reality (the projection of computer-generated visuals onto a real-world view) are not new concepts, their full potential is only now becoming apparent (Wiltshier & Clarke, 2017). This was demonstrated during the COVID-19 pandemic, which necessitated concert performances being conducted virtually (Milman et al., 2020; Richards, 2020a). Previously, the model augmented reality had gone viral when the Pokémon Go game was utilized to pique young people's interest in museums and as a promotional tool for increasing product sales (Bonus et al., 2018; Juhász & Hochmair, 2017; Rauschnabel et al., 2017). As a sort of "flow" of cultural conceptions, so-called trans-modern culture.

Globalization, on the other hand, is impossible without advancements in information and communication technologies. Its spread was likewise rapid and widespread, not just within wealthy countries with strong economic growth rates, but also across the borders of developing and poor countries with low economic growth rates. Globalization has become an inescapable phenomenon in this scenario (Pitanatri, 2020; Sheller, 2014). It must be recognized that the developed countries are the primary participants in the current globalization process. Local values inside a country are "exported" to be distributed globally as global values. These governments are able to accomplish this relatively easily because they regulate the flow of information and communication technology across nation-state borders.

In comparison, emerging countries are unable to disseminate their indigenous values due to their low competitive strength. As a result, developing countries frequently become bystanders to the admission and development of developed country ideals that are regarded as global values within their borders. In Indonesia, it is clear that the younger generation adores K-Pop culture, anime, and even Indian-style soap operas. Indonesians also have a preference for foreign cuisine. *Pasta, sushi,* and *gelato*, for example, are symbols of this nation's youthful generation's socioeconomic status.

For Indonesia, the pervasiveness of Western values brought about by globalization poses a danger to indigenous culture, which shapes the region's image. Regional arts such as *ludruk, ketoprak, wayang, gamelan,* and traditional dances face grave risks as pop culture develops. Western dishes that are becoming increasingly popular with the public due to their perceived modernity (Pitanatri, 2016). The traditional culture that values *tepo seliro* (tolerance and hospitality), and elder respect is also under siege by promiscuity and individualistic views brought forth by globalization.

In such a context, failures to adapt to globalization can result in the erasure of indigenous culture. Errors in devising measures to preserve indigenous culture can also result in indigenous culture being increasingly abandoned by individuals who have become increasingly infatuated with the culture engendered by globalization. This is unquestionably a significant threat to the survival of Indonesia's indigenous identity and culture. However, the response to globalization can also be limited by adopting technology in an effort to preserve culture (Meiliana et al., 2017). Maintaining the existence of local culture in the midst of the currents of globalization, can be done with strategies and steps to strengthen and maintain local cultural identity.

In Indonesia, significant progress has been made in the application of technology to event organization. When communication and information technology growth was slower than it is now, the values of Indonesian cultural identity were still strongly cherished by the community. However, when foreign identity values enter people's homes more easily and swiftly through information transformation, the values of Indonesian cultural identity appear to be diminishing.

Thus, preserving culture entails managing, documenting, and sharing information and knowledge in accordance with the noble values of local and national culture, leveraging the complexity of information and communication technology. Thus, each ethnic group's culture can be passed on to future generations in order to preserve culture as an identity themselves and demonstrate the Indonesian nation's existence to the rest of the world.

According to Jansson (2020), digitization is the process of converting media from printed, audio, and video formats to digital formats. Digitization is further developed with the concept of converting tacit knowledge, which is personal, specific, and generally difficult to formalize and communicate to other parties, into explicit knowledge, which is knowledge that has been formulated and presented in written form, such as regulations or books of literature. This explicit knowledge is manifested through the use of information and communication technologies in the form of well-managed papers. The documented knowledge can be used as educational material for future generations, ensuring that the process of information and knowledge transmission runs smoothly.

Considering the preceding context, the literature in this chapter reflects the dynamic growth of cultural tourism in Indonesia. With a touch of technology, events, festivals, and cultural fairs actually create space and opportunity for culture to remain sustainable. By examining two cultural tourist hotspots in Indonesia, namely Bali and Yogyakarta, this article intends to provide a critical assessment of the development of culture, which is presently inextricably linked to technology.

This chapter is divided into three sections to aid readers in comprehending the flow. The first section will discuss the growth of cultural tourism in Indonesia. This part examines historical studies from an academic standpoint, as well as the debates that have arisen as a result of the emergence of cultural tourism. The second section of this chapter will discuss how cultural events with a technological twist are a source of love and hate in Indonesia. Numerous cultural activities will be presented in Bali and Yogyakarta to hone the technique. Both culture tourism and technology have the potential to be mutually useful and complementary. Finally, the third section of this chapter will address the breadth of cultural tourism via the lens of technology, which can be utilized as a springboard for scholarly discussion. As a result of the debates, in the section "Creative Economy: A New Spectrum of Cultural Tourism in Indonesia", new horizons for cultural tourism research is elaborated. Indonesia is an excellent case study for increasing tourist literacy and a model for the development of cultural tourism in the Southeast Asian region.

Traditional to Digital: The Development of Events and Cultural Tourism in Indonesia

Since the 1960s, tourism has developed an empirical "scientific attraction" among experts in Indonesia, as both positive and negative aspects became apparent (Mahagangga et al., 2018). Understanding the negative repercussions of tourism from an early age aids in destination development planning. It will be more difficult to overcome the bad consequences of the past, particularly with regard to highly abstract cultural norms.

Academically, the impact of tourism on experts is perpetually dichotomous. There are some who favor tourism on the basis of getting numerous benefits, those who oppose tourism on the belief that tourism destroys culture, the environment, and nature, and those who fall somewhere in between the two opposing camps (Pitana & Gayatri, 2005).

Cultural tourism emerged as a theme in the 1970s, with an emphasis on authenticity, which is a critical component of cultural tourism. On the one hand, authenticity serves as a distinguishing feature of cultural identity, but on the other side, it becomes a source of contention when authenticity is "sold" as a tourism commodity.

In the 1920s and 1930s, the colonial administration promoted and controlled Indonesian tourism. Following independence, the central government launched the promotion of international tourism with the assistance of foreign advisors. Making Bali a "storefront" for Indonesia through the central government's designation of the island of a thousand temples as a "living museum" as a milestone in the development of Indonesian cultural tourism following the independence (Picard, 2006). Bali serves as a model and catalyst for domestic tourism development due to its artistic and cultural legacy, which is Indonesia's most valuable asset. Bali is Indonesia's representative island, where Balinese music, dances, paintings, and sculptures are the works of Indonesia's most internationally acclaimed artists.

Putra and Paturusi (2017) provide three stages of development (milestones) for cultural tourism in Bali. The first stage was in 1902, where Balinese saw the arrival of Bali's "first tourist", H. Van Kol, a Dutch parliamentarian who traveled to Indonesia, especially Bali, on his own dime (not on official business) and for pleasure. The impressions of his written travels became the basis for the Dutch colonial government's ethical political policies aimed at preserving culture and promoting Bali as a tourism destination. Second, at 1910, when the idea of establishing a Balinese museum in Denpasar was born. The museum's purpose was to store Balinese cultural objects in order to prevent them from being lost to time. Apart from its cultural preservation mission, this museum, which opened to the public in 1932, also serves as a tourist attraction. Third, at 1994 where the establishment of the Official Tourist Bureau, a subsidiary of KPM (*Koninklijke Paketvaart-Maatschappij*), in Bali as a branch of the company's Batavia office.

Bali tourism's metamorphosis began with the development of cultural tourism, which was then employed as a strategy and branding as an artistic and cultural destination with economic potential for the development of Balinese art and culture.

Marine tourism, culinary tours, adventure, spiritual tourism, village tourism, and MICE (Meeting, Incentive, Convention, and Exhibition) are all defining characteristics of tourism in Bali (Putra & Paturusi, 2017).

The growing technology and its impact on contemporary life demonstrate that we are at the start of the Fourth Industrial Revolution, a new period in which digitalization is reshaping and impacting society in novel and frequently unexpected ways. It is critical to evaluate the nature of the shift we are experiencing and how we can ensure that this revolution benefits everyone, collectively and individually.

In cultural tourism, digitization is not a novel concept. Tourism was one of the first industries on a global scale to automate commercial operations, establishing the precedent for online flight and hotel bookings. As information and communication technology (ICT) becomes a worldwide phenomenon, tourism has been a consistent early adopter of new technologies and platforms.

On the Dutch Museum's YouTube account, several collections of paintings are exhibited. The museum's information is offered to inform the development of Rijks Studio, a website that allows people to curate digital collections online (https://www.rijksmuseum.nl/en/rijksstudio). These two digital platforms are used to facilitate encounters, conversations, and discoveries within the walls of museums, all of which are a component of "engaging/emerging" the client. Social media has evolved into a new mode of communication for acquiring digital clients. Online communication has primarily been used to duplicate traditional modes of communication, market items, and even sell exhibition tickets.

Today, the use of technology in the form of mobile applications to disseminate cultural heritage data in a destination in order to give integrated information on "tangible" and "intangible" cultural assets to the general public, including potential tourists, is prevalent. Every public and private institution in Europe digitizes physical cultural assets, including not only photos but also audio/video recordings of performing arts, sports, folkloristic practices, monuments, and landscapes. Intangible cultural assets, such as oral histories, food and beverage traditions, and regional customs, to ensure that no component is lost in the digital age (Anike & Chinelo, 2017).

The utilization of technology in tourism locations for the purpose of providing and sharing information. Provide information about efforts to interpret attractions in destinations, travel timetables, and maps of tourist sites, lodging, and other amenities. There is a two-way flow of information, notably from stakeholders as providers to visitors as users.

Tourism events can be regarded as the process of organizing, developing, and promoting tourism destinations with the goal of establishing a self-image and acting as an attraction (Çelik & Çetinkaya, 2013). Events play a critical role in the tourist sector because they have the ability to generate significant economic, social, cultural, and environmental advantages for a location (Buultjens & Cairncross, 2015). Tourism events can also help promote a destination's image, as they can work as an attractant, increasing the interest of both foreign and domestic tourists in a given location (Damster, 2005).

From the beginning to the present, tourism events have shown to be beneficial in terms of increasing or maintaining its attraction. For example, Barcelona has grown in global recognition and experienced a favorable economic impact as a result of hosting the 1992 Olympics (Casellas et al., 2012; Proni, 2008) as has Sydney as a result of holding the 2000 Olympics. Socioeconomic status that is unique and more acknowledged (Waitt, 2003). Cities such as Pasadena, which is well-known for its Flower Festival events, and Spain which is well-known for its Matador Festival, are excellent instances of how events are gaining prominence.

With the nomination of Indonesia to host major international MICE events, such the Asian-African Conference 18–24 April 1955 at Gedung Merdeka, Bandung and the X 1992 Non-Aligned Summit in Jakarta, Indonesia, events, particularly MICE, have become a component of Indonesia's tourism development. As a result of the epidemic, technology is now compulsory to assist MICE organizations in conducting events.

In Indonesia, the function of events in generating or enhancing visitor attraction is becoming increasingly apparent. Indonesia has a huge event-based tourism potential, as each region and province in the country has its own distinct blend of cultures and characteristics, as well as a million attractions. The majority of them have been integrated into national, regional, and international attractions or tourist destinations as festivals and events. Hosting events in these tourism destinations has a long tradition and a distinct regional feel. As proven by its sustained expansion year after year, the ensuing impact is similarly favourable.

In 2017, 98 national tourism events were organized around the archipelago by 34 provinces to promote Indonesian tourism (Go Explore Indonesia, 2017) with each region hosting a key event. The Ministry of Tourism released a list of the top 100 tourist attractions in Indonesia in 2018, which is presented in the 2019 "100 Calendar of Events" (CNN Indonesia, 2018). To further emphasize, in 2019, the Ministry of Tourism and Creative Economy promote MICE as part of its efforts to improve the quality of destinations and to increase the number of events held in order to attract more foreign tourists to Indonesia (Kompas.com, 2020).

On the one hand, the event seeks to attract a large number of people simultaneously, but on the other side, the Covid-19 Pandemic necessitates the opposite. A critical strategy that has been demonstrated to significantly slow the spread of the virus is social distance. Though this technique was extensively (though not totally) adopted in the United States in March and April 2020, it is uncertain how long (and to what extent) it will be required to remain in place; some estimates predict that it may be required, to varied degrees, through the year 2022 (Kissler et al., 2020).

The COVID-19 epidemic teaches us about a variety of new habits, including those that apply to the world of events. A survey conducted by online media PCMA Convene Magazine on event planners and suppliers regarding the impact of the corona virus revealed several key findings from respondents (1230 people). According the research, 87 percent of respondents had cancelled an event, while the remaining respondents were debating whether to cancel or postpone a future event. Seven out of ten responders have shifted their face-to-face event to a virtual platform in part or entirely, and many see this as a long-term solution during the epidemic, continuing alongside in-person gatherings going forward (Russell, 2020). Indeed, the coronavirus epidemic may have allayed a long-held fear that virtual events might suffocate future in-person ones; less than a quarter indicated that this is a source of concern for them; 62% indicated that this is not a source of concern for them. 63% of respondents indicated that they are using this time to build additional skills, with the majority focusing on developing virtual event tactics and becoming more familiar with online platforms. Due to COVID-19, several events that were postponed and virtual events were held.

Going virtual is an alternative as the current state of the event industry is strongly reliant on technology. Understanding that physical humans cannot coexist during the pandemic, transferring them to digital encounters becomes even more vital (Russell & Stenning, 2020). While the human experience will never be precisely replicated in a digital environment, the concepts used to build an in-person experience can be applied to digital experiences: what causes delight, what develops memory, and how we interact and engage with people are all crucial (Hartman, 2021; Richards, 2020b; United Nations World Tourism Organization, 2020).

One of the implementations of virtual events in Indonesia is the Djakarta Warehouse Project Virtual (DWP-V). The event is a 2-day virtual event experience held on 19 & 20 December 2020. This event is an example of an annual music event that is transformed digitally. Live performance was transformed into virtual due to limitations during the COVID-19 pandemic. This virtual festivals streams exclusively at djakartawarehouse.com and it is free of charge. However, audience need to have a V-Pass with the unique login information to gain further access within djakartawarehouse.com on the event date (Djakarta Warehouse, 2020).

DWP Virtual presents performances from a number of EDM musicians from a number of countries ranging from Martin Garrix, Vini Vici, Armin van Buuren, Brennan Heart, Curbi, Valentino Khan, What So Not and many others. A number of domestic musicians also enlivened this virtual electronic music festival such as Ramengvrl, Hezkia, Ridwan G, Trillions, Devarra, Fun on A Weekend, Joyo, Patricia Schuldtz, Preachja Crew; Wow and Yasmin. Including collaborations from Dipha Barus with Yura Yunita, Nadin Amizah, Kallula, and Indies, as well as performances from Asty and Billy Tanner (CNN Indonesia, 2020).

According to the same article by CNN Indonesia (2020), DWP-V event necessitated a 1.5-month procedure for stage preparation and other staging requirements – which was one response to the pandemic's crisis conditions. Numerous things were attempted for the first time at the DWP event in Jakarta in previous years. On the one hand, offline events have been cancelled since March 2020, but because DWP is a long-awaited event, the organizers choose to hold it virtually and for free to demonstrate to its fans and supporters that the event is still exists.

The DWP-V event demonstrates that something new is being discovered in the world of events. In a crisis and limitation situation, technology that was previously a secondary consideration becomes a main consideration. Additionally, technology creates a new experience for both organizers and fans. It delivers a holistic experience for technology providers, pre-event, during the event, and post-event. During the pre-event phase, technology assists organizers in marketing their events, particularly in the transmission of all information via various digital channels, including the organizer's official website. Technology is also critical in preparing the virtual festivities that will take place on "D" Day. During the event phase, technology enables organizers to execute the highest-performing events possible for their fans. At the post event stage, technology once again demonstrates its value in communicating organizers' sentiments of thanks to consumers, including sponsors, and engagement activities with customers regarding upcoming events (Fig. 17.1).

Although packaged virtually, the appearance of DWP-V was accomplished in three distinct steps. As depicted in the image, there are three major stages: Stadium, Cosmic Station, and Neon Jungle, all of which can be freely selected by the audience at home. Each stage features a unique concept created employing cutting-edge technology, including the use of green screen technology to simulate the look of a "real" stage, similar to prior DWP events. This is done in order to preserve participants' experience from prior years' DWP events. Additionally, DWP has prepared a special stage for foreign performers to record their performances prior to being broadcast live at the DWP Virtual event, which will take place in Ibiza, Sydney, Bangkok, and China. Along with a new visual idea, some DWP-V services have been virtualized, including interactive chat, virtual maps, and special item sales that can be ordered through the DWP mall (Fig. 17.2).

The image above depicts the DWP-V 2020 event's features, which demonstrate that the organizers' revenue options were not through ticket sales, but rather through sponsorship and retail sales. It was accommodated in the DWP-V mall component, as well as food and beverage sales listed at the "Eat & Beats Feature".

From the perspective of their fans and audience, technology creates new experiences through digital and virtual experiences. Prior to the presentation, the audience

Fig. 17.1 Virtual stage DWP-V 2020. (Source: Djakarta Warehouse, 2020)

Fig. 17.2 Stage and merchandise features in DWP-V 2020. (Source: Djakarta Warehouse, 2020)

enjoyed a digital experience in which they could acquire information about the DWP-V. Additionally, fans also earn experience from registering for the site, which is organized by numerous groups of potential viewers. To obtain a V-Pass, fans must register by providing information from their identification card. Registration is necessary for fans who are Indonesian citizens and reside in Indonesia. If the KTP (Indonesian ID Card) number is not recognized during the registration procedure, it is possible that the ID Card number has not been entered into the database maintained by the Department of Population and Civil Registration in Indonesia. In this scenario, the organizer suggests that fans must register for DWPV through another ID Card. Citizens of Indonesia who reside outside the country may register using their Passport credentials. Only viewers aged 18 years and older are permitted to access DWP-V, which can be viewed in a single internet browser on a single device during the event and is available only in Indonesia and Asia. Additionally, numerous access points are obtained in order to optimize the viewing experience for viewers at home, including the usage of HDMI cables, Airplay, and Chromecast, as detailed on the organizers' official website.

The Djakarta Warehouse Project organizer also prefers to organize future events in a combination, namely events held offline (face to face) as well as virtual (CNN Indonesia, 2020). Several of the obstacles associated with planning virtual events include the need for a robust internet network capable of handling anticipated high traffic, participant registration mistakes, and IT limits that impede the event's seamless execution. Additionally, the virtual organization now generates less than optimal revenue due to its reliance on sponsors and online goods sales. In 2020, revenues from the sale of products at the DWP mall were donated to pandemic-related activists.

For events in Bali with a greater emphasis on art, culture, and entertainment, technology is rapidly being incorporated. Certain art performances, like performances in general, are staged according to staging standards, which include the use of decorations, sound systems, and lighting. The event is co-designed with the purpose of drawing a small audience through tight health regulations and live hybrid online broadcasts via the YouTube channel.

A total of 12 performances have been scheduled from June 2021 to May 22, 2022. The show aims to reintroduce artists to the art world, entertain the community through art, financially support artists during periods of inactivity or performance, restore Bali's tourism conditions, and encourage people to learn about traditional arts in Bali, as well as assisting traditional artists in performing at regional, national, and international festivals. The performance features traditional dance and gamelan performances from a variety of traditions at a variety of locations throughout Bali, as well as training sessions and talk shows with the performers themselves. These performances are collaboration between artist, Ministry of Tourism and Creative Economy and Antida Event Organizer (Table 17.1).

Technology is utilized to create marketing campaigns for performances, which are disseminated via a variety of social media platforms. Throughout the event,

Table 17.1 Performances program 2021–2022

No	Date	Program	Place/Venue
1	20th June, 2021	Tradition Art of Okokan & Tektekan	Jatiluwih Tourism Village, Tabanan Bali
2	25th July, 2021	Tradition Art of Sanghyang Dedari and Sanghyang Jaran Dances	Arma Museum, Ubud Bali
3	21st August, 2021	Tradition Art of Rejang Renteng dance and Gamelan Gambang	Kertagosa, Klungkung Bali
4	26th September, 2021	Tradition Art of Gong Kebyar, Rare Kual, Luh Menek- Singaraja	Gedong Kirtya Buleleng Bali
5	31st October, 2021	Tradition Art of Genggong Kutus, Selonding Sraya	Pura Desa Batuan, Gianyar Bali
6	28th November, 2021	Contemporary Composer- Tradition music of Gamelan Werdhi Cwaram	Art Center Denpasar Bali
7	19th December, 2021	Tradition Art of Genjek and Gebug Ende	Taman Ujung Karangasem Bali
8	23th January, 2022	Tradition Art of Jegog Suar Agung	Jembrana, Negara, Bali
9	20th February, 2022	Tradition Art of Legong Peliatan Oka Dalem Dance	Peliatan Ubud Bali
10	20th March, 2022	Tradition Art of Wong Puppet – Batuan	Br. Puaya Sukawati Bali
11	24th April, 2022	Art of Barong Brutuk Trunyan	Trunyan Kintamani Bali
12	22nd May, 2022	Contemporary Composer of Ceraken Subandi	Denpasar Bali

Source: Antida Music Production (2021)

technology is employed to enhance the stage act, which is streamed live via YouTube. Additionally, in the post-event time, digital technology is utilized to communicate the results of completed activities, such as notice of the audience's level of enthusiasm as measured by views, likes, and subscribes obtained during live streaming (Fig. 17.3).

Another stage performance was also launched in Ubud Bali in partnership with the Ministry of Education and Culture of the Republic of Indonesia as a way of employing technology to add a touch of stage art/performing art to Bali. The traditional dance performance with Dayu Ani on the Online Art Stage was streamed live on the YouTube channel, which received 800 live views and a total of 2700 spectators. This demonstrates one of the applications of digital technology, which is used to broadcast staging activities with the purpose of increasing the event's rating and disseminating information about the performance's existence (Fig. 17.4).

Other performances are virtually shown in the image below, which were held to reintroduce the art of performing throughout the pandemic, utilizing technology to provide an enjoyable experience for the audience. To encourage audience involvement, the organizers allow viewers to comment on the live chat section on the YouTube channel where this presentation is being streamed live. This event drew 600 live viewers and a total of 1600 online watchers (Fig. 17.5).

In the aftermath of the Covid-19 pandemic, technology permits the hosting of virtual events. In the future, more events will be able to be held virtually for a fee,

Fig. 17.3 Traditional art of Okokan & Tektekan. (Source: Antida Music Production, 2021)

Fig. 17.4 Online streaming art stage performance. (Source: Antida Music Production, 2021)

Fig. 17.5 Performance of Panggung Bertutur Sebab Cinta. (Source: Antida Music Production, 2021)

giving economic benefits not just to organizers but also to destinations similar to those associated with face-to-face events. Consumers may find a fresh solution to numerous travel limits through technology that provides digital/virtual experiences. Event organizers can capitalize on market potential despite other travel constraints through the use of technology.

Love-Hate Paradox Between Cultural Events and Technology

Case Study: Bali

Global tourists have identified Ubud-Bali as an international tourism destination since the 1920s. Tjokorda Gede Agung Sukawati, the King of Ubud, was one of the initiators. He had a visionary concept of opening Bali to the world while retaining its "self" and "culture". To support this cause, he recruited Walter Spies to Ubud to educate the populace about the need of preserving traditional arts. Walter Spies also polished the art of Balinese dancing, in partnership with Ubud artists, into the Kecak Dance, which has remained popular to this day. This is where the residents of Ubud Bali begin their search for events that have the potential to increase the attractiveness of cultural tourism on a national and international scale. Several of Ubud's internationally well-known events include, the Bali Spirit Festival, Ubud Food Festival, Ubud Village Jazz Festival, the Ubud and Beyond Festival, Maybank Marathon was held in 2012; the Ubud Writers and Readers Festival and the Ubud and Beyond Festival.

Today, covid-19 influenced a variety of events in Bali, most notably in the Ubud district. All national and international events in 2020 were cancelled. Numerous Ubud's international event organizers have been unable to confirm the rescheduling of the event. Virtual events, according to some event organizers, are incompatible with the Ubud event model. Bali Spirit Festival's essence is in the interaction of actors and viewers, which needs participants' immediate presence. If the Ubud Village Jazz Festival is held virtually due to a lack of participation, the organizers will be unable to reap financial rewards.

As an example, the Bali Spirit Festival. The festival, which once drew over 1000 domestic and international participants and 700 volunteers-needs community involvement and has a fairly strong multiplier effect on the destination. Local inhabitants' provision of bamboo as a decorative material, transportation, indigenous cuisine, and participant housing all limit virtual participation (Fig. 17.6).

Another example is Ubud Village Jazz Festival. Previously the event was held to create economic viability at villages in Bali. Apart from the participant experience with the concept of "village", which cannot be replicated online, the multiplier impact of face-to-face activities is difficult to produce. When the Festival is held virtually, the organizers have not reaped financial rewards (Fig. 17.7).

Differ to Bali Spirit Festival and The Ubud Village Jazz Festival; The Ubud Writers Festival, which serves as a forum for writers, makes it extremely easy to hold the event electronically. Throughout 2020, the event is held virtually as a means of engagement between the organizer and its clients, announcing the event's presence (Fig. 17.8).

The event virtually despises the wellness notion due to "lack of experience" and "lack of multiplier effects". Thematic events benefit from virtual technology, which enables them to be implemented at a "lower cost". Meanwhile, traditional art performances as an effort to realize flexibility and fluidity enter the new carriage called

Fig. 17.6 Bali Spirit Festival. (Source: BaliSpirit Festival, 2019)

Fig. 17.7 Ubud Village Jazz Festival. (Source: Ubud Village Jazz Festival, 2019)

the virtual stage, it may be considered that they have not found a bright spot. The nature of the virtual stage seems to contradict the style and character of traditional art.

The virtual world has always longed for something filmic, while the nature of traditional art leads to contemplative intimacy. The way to enjoy it is not only seen but also heard, felt, and internalized. The fact that traditional arts are able to grow and live in a unique world where crowds, eye contact with the audience directly,

Fig. 17.8 Ubud Writers Festival during the pandemic. (Source: Ubud Writers & Readers Festival, 2021)

body touch, are the benchmarks for success. Therefore, when it comes to appearing virtually, there are clear references and roadmaps on how the ideal mechanism when art and tradition must metamorphose into hashtags and trends.

Case Study: Sendratari Ramayana at Prambanan

The Ramayana sendratari (ballet) performance took place against the backdrop of Prambanan, Indonesia's most ancient Hindu temple. The combination of traditional nuances and cutting-edge technology such as lighting, audio-visual, and laser technology ensures that this dance remains contemporary. The issue therefore becomes one of dilution of indigenous culture's core. If one may imagine, it is just this type of behavior that contributes to traditional culture's appeal and exclusivity. For both local and international tourists, it has become a cliché that a trip to Yogyakarta would be incomplete without watching a Ramayana ballet performance.

As illustrated in Fig. 17.9, the performance is also recognized as the top attraction in the Sleman Yogyakarta region on TripAdvisor. The performance adds to the temple's charm. Tourists can now visit Prambanan Temple at night, rather than only during the day.

Nonetheless, the dispute over the validity of culture, which modernisation degraded through this performance, continues to this day. However, there are

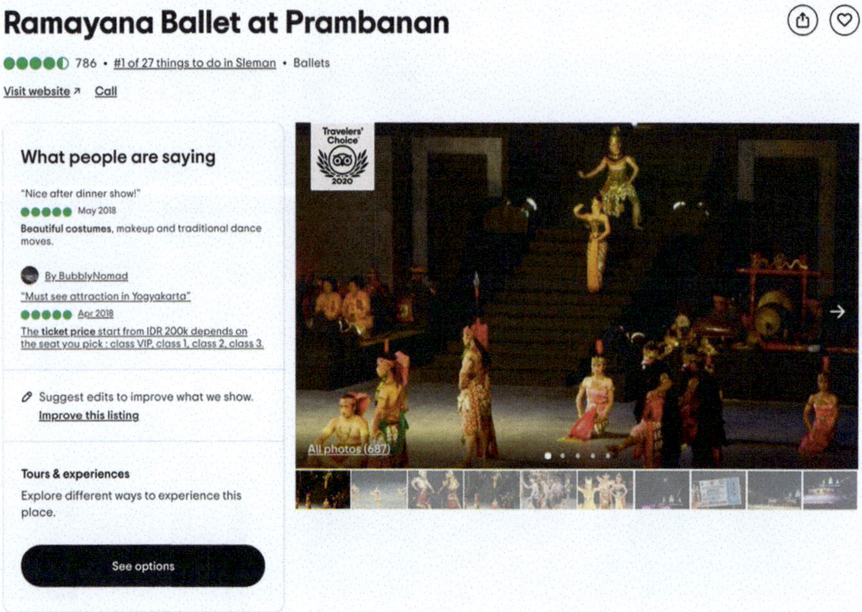

Fig. 17.9 *Sendratari* Ramayana (ballet) on TripAdvisor. (Source: TripAdvisor, 2021)

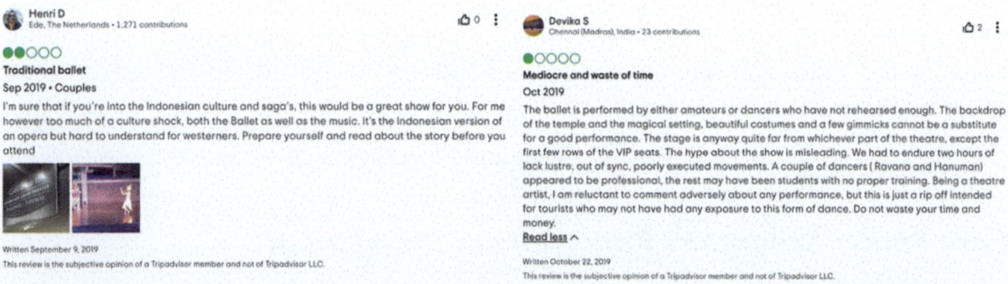

Fig. 17.10 Negative reviews from tourist on TripAdvisor. (Source: TripAdvisor, 2021)

individuals who are opposed to the alteration and those who believe that it truly degrades the indigenous culture. One of the debates that erupted was the title of this theatrical show, is called "Ballet". Tourists accustomed to western-style theatrics will undoubtedly be perplexed by Indonesian-style ballet Additionally, the presentation is in Javanese, a language that many people are unfamiliar with. This has also raised concerns among foreign visitors to Yogyakarta who witnessed the show. Ballet, is quite dissimilar to typical Indonesian *sendratari* by western standards, as illustrated by the following two TripAdvisor reviews (Fig. 17.10).

On the other hand, the performance is adored by tourists with a variety of cultural objectives and adventure seekers. The paradox of enjoying and despising technology cannot be perceived only through one lens. A humanist would interpret this as a sign of cultural deterioration. Art has evolved from a necessity of the soul to a necessity of life in the modern day. Art that interacts with modern instruments such as computers, cameras, and video recorders undergoes significant transformation. Art has lost its purity as an expression of the soul that created it and now has a million meanings and 'interests.' With the advancement of technology, art has become an instantaneous activity that anybody can participate in. In contrast to the Vincent Willem van Gogh or Leonardo da Vinci eras, art was regarded as sacrosanct during the time and could only be created by those with the status of artists. This is in direct opposition to the founding idea of art in classical civilization, which is based on *memesis,* or mimicking the original item rather than depicting it.

In comparison, a tourism business technopreneur may view this as an opportunity to grow the younger generation's tourist market share. On the one hand, tourists who enjoy attractions will view it as an exciting spectacle that will become a must-see during their visit to Yogyakarta. As a result, the concept of combining technology and ancient cultural understanding is truly unique and intriguing. Technological advancements are made and must be accommodated. On the other hand, traditional cultural wealth is preserved as a timeless intellectual legacy.

Creative Economy: A New Spectrum of Cultural Tourism in Indonesia

As seen by the two case studies above, the dynamics of cultural tourism has enable scholars to delve deeper into new spheres of culture. Cultural tourism can no longer be viewed as a symbol of stagnant petrification, which the younger generation has finally abandoned as "outdated" and unappealing. The development of creative cultural tourism really contributes to the preservation of indigenous culture that was previously unpopular with the younger generation.

The rising use of technology in cultural tourism, as well as the blending of real-world and virtual experiences, will undoubtedly be another major field of research in the future. Especially in the aftermath of the COVID-19 pandemic, which fundamentally altered how visitors and operators interact with virtual technology. Of course, it will be fascinating to study how this can result in broader sociocultural shifts.

In an academic study through a search on the Scopus database, it was found that cultural tourism publications have been around since 1967. The emergence of cultural tourism is a social phenomenon and as an object of academic study can be traced back to the post-World War 2 surge in recreational travel (Richards, 2018). In Europe, travel helps increase intercultural understanding and rebuild economies devastated by the second world war. In the 1960s and 1970s, European incomes and consumption continued to increase; as well as international travel, and cultural consumption.

Pitanatri (2021) asserts that the theme of cultural tourism research in Indonesia will continue to evolve following its epoch. If cultural tourism was once defined by its consumer and economic elements, the creative economy has emerged as a research issue for many national and international experts, as illustrated in the Fig. 17.11.

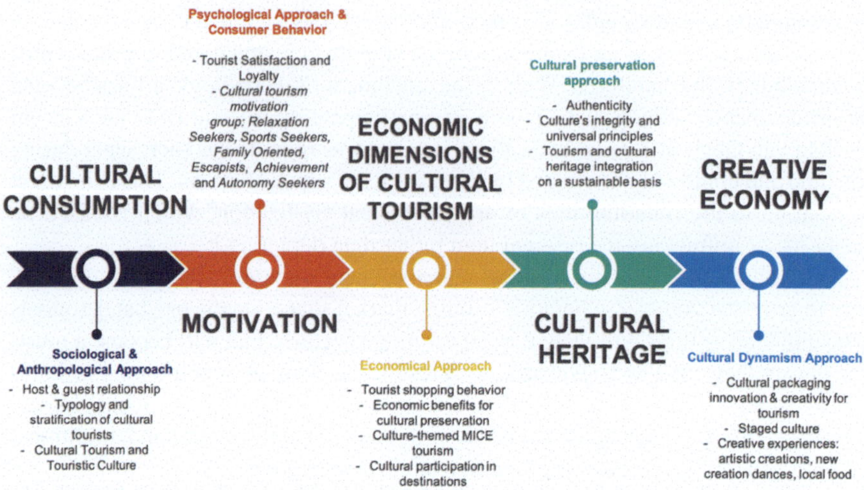

Fig. 17.11 Academic research on cultural tourism. (Source: Pitanatri, 2021)

Cultural tourism studies have evolved to include topics such as cultural consumption, tourist motivation, economic elements, cultural legacy, and creative economy. The term 'creative economy' is one of several concepts used to emphasize the importance of the creative process and science in a holistic economy (LY, 2018; Richards, 2018). Contemporary cultural tourism's development gives a chance to strengthen the link between tourism and the creative economy.

As the globe undergoes a 4T revolution (technology, telecommunications, transportation, and tourism), regional boundaries will become increasingly blurred, eventually ending in the development of a global village. As Picard (2006) notes, this situation exemplifies the deterioration of cultural assets. True evidence of deterioration of cultural heritage can be found in a multitude of ways, including clothes, language, and information technology. Thus, cultural tourism and technology should complement one another rather than compete. Globalization and artistic innovation are unquestionably beneficial to the development of cultural tourism. The way the creative concept is conceptualized will govern the manufacturing and marketing concepts (Pappas, 2019; Pitanatri, 2018; Xie et al., 2019). The concept's and its target market's path will be determined by the power of invention and originality. Additionally, when cultural activities are linked to technology, a phenomenon called creative digital transformation develops. Digital transformation is a process and strategy for radically altering how businesses operate and serve their consumers via the use of digital technologies.

Discussions

Cultural digitalization is a concept that refers to the process of leveraging information and communication technology to enhance usability in the field of culture, most notably in terms of managing, documenting, and disseminating information and knowledge concerning cultural aspects. Cultural digitalization is advancing at a rapid pace, encompassing religious systems, social systems or social organizations, knowledge systems, languages, arts, livelihood or economic systems, and living equipment or technological systems.

Cultural arts education must be capable of providing students with an appreciative and creative experience while also contributing to their future knowledge, attitudes, and skill development. To achieve these objectives, inventive and creative efforts are required in developing methodologies or learning models adapted to certain characteristics. Therefore, the development and application of information technology-based culture, or digitalization, is vital for resurrecting cultural values such as self-identity and the Indonesian nation's richness.

Combining art and technology through "live streaming" is the only way to express ideas and ideas for performing arts. It is also emphasized that the Pandemic did kill business in the world of events and performing arts, but did not kill ideas and ideas related to art. Traditions must be maintained, preserved and documented, to be able to bridge the younger generation who tend to be more familiar with technology. Therefore, technology can be used to bridge the documentation which is expected to accommodate this traditional art to be communicated to local communities including tourists.

Conclusion

The dynamics of cultural tourism, aided by technology, create unlimited possibilities for scholars and researchers to delve deeper into new spheres of culture. The future will also see the growth of methodologies from other disciplines, one of which is on the creative economy spectrum.

However, one of the most major issues is defining the term 'culture' and its interpretation, which has substantial ramifications for the definition of cultural tourism. Cultural tourism can no longer be viewed as a symbol of stagnant petrification, which the younger generation has finally abandoned as "outdated" and unappealing. Although it is frequently a paradox, the integration of technology with tourism in cultural events is an excellent way to preserve the destination's culture.

The rising use of technology in cultural tourism, as well as the blending of real-world and virtual experiences, will undoubtedly be another major field of research in the future. The aftermath of the Covid-19 pandemic, will drastically alter how visitors and operators interact with technology. Naturally, it will be fascinating to observe how this can result in broader socio-cultural shifts.

References

Anike, L. O., & Chinelo, M. I. (2017, February). Digitization in research for cultural, commercial and scientific development. *American Based Research Journal, 6*(2), 1–5.

Antida Music Production. (2021). *Performance program 2021–2022*. Retrieved from: https://www.antidamusic.com/#. Accessed: 24 July 2021.

BaliSpirit Festival. (2019). *Bali Spirit Festival Gallery 2019*. Retrieved from: https://www.balispiritfestival.com/gallery/2019. Accessed: 21 July 2021.

Bonus, J. A., Peebles, A., Mares, M. L., & Sarmiento, I. G. (2018). Look on the bright side (of media effects): Pokémon Go as a catalyst for positive life experiences. *Media Psychology, 21*(2), 263–287.

Buultjens, J., & Cairncross, G. (2015). Event tourism in remote areas: An examination of the Birdsville races. *Journal of Place Management and Development, 8*(1), 69–84.

Casellas, A., Dot-Jutgla, E., & Pallares-Barbera, M. (2012). Artists, cultural gentrification and public policy. *Urbani Izziv, 23*, S104–S114.

Çelik, S., & Çetinkaya, M. Y. (2013). Festivals in event tourism: The case of International Izmir Art Festival. *International Journal of Contemporary Economics & Administrative Sciences, 3*, 1–21.

CNN Indonesia. (2018). *Kemenpar Rilis 100 calendar of events 2019*. Retrieved from: https://www.cnnindonesia.com/gaya-hidup/20181222191805-269-355751/kemenpar-rilis-100-calendar-of-events-2019. Accessed: 28 July 2021.

CNN Indonesia. (2020). *DWP 2020 Gelar Konser Virtual Gratis*. Retrieved from: https://www.cnnindonesia.com/hiburan/20201203175410-227-577697/dwp-2020-gelar-konser-virtual-gratis. Accessed: 28 July 2021.

Damster, G. (2005). *Event management: A professional and developmental approach*. Juta and Company Ltd.

Djakarta Warehouse. (2020). *Djakarta Warehouse Project Virtual 19 & 20 Dec 2020*. Retrieved from: DjakartaWarehouse.com. Accessed: 28 July 2021.

Go Explore Indonesia. (2017). *Even dan Festival 2017*. Retrieved from: https://goexplore.id/. Accessed: 28 July 2021.

Hamilton-Smith, E. (1987). Four kinds of tourism? *Annals of Tourism Research, 14*(3), 332–344.

Hartman, S. (2021). Adaptive tourism areas in times of change. *Annals of Tourism Research, 87*(C), 102987.

Jansson, A. (2020). The transmedia tourist: A theory of how digitalization reinforces the de-differentiation of tourism and social life. *Tourist Studies, 20*(4), 391–408.

Juhász, L., & Hochmair, H. H. (2017). Where to catch 'em all?–a geographic analysis of Pokémon Go locations. *Geo-Spatial Information Science, 20*(3), 241–251.

Kissler, S. M., Tedijanto, C., Goldstein, E., Grad, Y. H., & Lipsitch, M. (2020). Projecting the transmission dynamics of SARS-CoV-2 through the postpandemic period. *Science, 368*(6493), 860–868.

Kompas.com (2020). *Garap Pariwisata, Wishnutama Bakal Kembangkan Event Wisata*. Retrieved from: https://travel.kompas.com/read/2019/10/23/153020127/garap-pariwisata-wishnutama-bakal-kembangkan-event-wisata?page=all. Accessed: 28 July 2021.

LY, M. B. (2018). An application of Butler's (1980) Tourist area life cycle to Saly (Senegal). *International Journal for Innovation Education and Research, 6*(1), 47–56.

Mahagangga, I. G. A., Suryawan, I., Anom, I., & Kusumanegara, I. (2018). *Evolusi Pariwisata di Indonesia*. Cakra Media Utama.

Maoz, D. (2018). The mutual gaze. *Annals of Tourism Research, 49*(1), 283–299.

Meiliana, I. D., Hidayat, M. R., Amalina, N. V., & Suryani, D. (2017). Mobile smart travelling application for Indonesia tourism. *Procedia Computer Science, 116*, 556–563.

Milman, A., Tasci, A. D. A., & Wei, W. (2020). Crowded and popular: The two sides of the coin affecting theme-park experience, satisfaction, and loyalty. *Journal of Destination Marketing and Management, 18*, 100468.

Pappas, N. (2019). The complexity of consumer experience formulation in the sharing economy. *International Journal of Hospitality Management, 77*, 415–424.

Picard, M. (2006). *Bali: Pariwisata Budaya dan Budaya Pariwisata*. Kepustakaan Populer Gramedia.

Pitana, I. G., & Gayatri, G. G. (2005). *Sosiologi Pariwisata*. Yogyakarta.

Pitanatri, P. D. S. (2016). *Inovasi dalam kompetisi: Usaha kuliner lokal menciptakan keunggulan kompetitif di Ubud*. 3(1), 1–14.

Pitanatri, P. D. S. (2020). Wisatawan Flashpacker: Sebuah Pendekatan Epistemologis. *Jurnal Master Pariwisata (JUMPA), 7*(1), 77.

Pitanatri, P. D. S. (2018). Esensi dalam Eksistensi : Sharing economy pada pengelolaan homestay tiga destinasi prioritas di Indonesia. In I. W. Mertha & P. D. S. Pitanatri (Eds.), *Homestay: Mozaik Pariwisata Berbasis Kerakyatan* (pp. 37–52). Pusat Penelitian dan Pengabdian Kepada Masyarakat Sekolah Tinggi Pariwisata Nusa Dua Bali.

Pitanatri, P. D. S. (2021). Pariwisata budaya dalam tinjauan riset: dulu, kini, dan nanti. In I. B. P. Puja, P. A. Aryasih, & I. G. N. A. Suprastayasa (Eds.), *Kamala: Kepariwisataan Berbasis Masyarakat, Budaya, dan Berkelanjutan* (pp. 71–89). Pusat Penelitian dan Pengabdian Kepada Masyarakat Sekolah Tinggi Pariwisata Nusa Dua Bali.

Proni, M. W. (2008). A reinvenção dos jogos olímpicos: um projeto de marketing. *Esporte e Sociedade, 3*(9), 1–35.

Putra, I. N. D., & Paturusi, S. A. (2017). *Metamorfosis pariwisata Bali: tantangan membangun pariwisata berkelanjutan*. Denpasar-Bali.

Rauschnabel, P. A., Rossmann, A., & tom Dieck, M. C. (2017). An adoption framework for mobile augmented reality games: The case of Pokémon Go. *Computers in Human Behavior, 76*, 276–286.

Richards, G. (2020a, June). Tourism in challenging times: Resilience or creativity? *Tourism Today*, 1–8.

Richards, G. (2020b). Designing creative places: The role of creative tourism. *Annals of Tourism Research, 85*, 102922.

Richards, G. (2018). Cultural tourism: A review of recent research and trends. *Journal of Hospitality and Tourism Management, 36*, 12–21.

Russell, M. (2020). COVID-19 impact on events research: Top-line results for planners. *PCMA*. Retrieved from: https://www.pcma.org/covid-19-impact-events-industry-planners-survey-results/. Accessed: 28 July 2021.

Russell, W., & Stenning, A. (2020). Beyond active travel: Children, play and community on streets during and after the coronavirus lockdown. *Cities & Health*. https://doi.org/10.1080/23748834.2020.1795386

Sheller, M. (2014). The new mobilities paradigm for a live sociology. *Current Sociology, 62*(6), 789–811.

TripAdvisor. (2021). *Ramayana Ballet at Prambanan*. Retrieved from: https://bit.ly/2UXScWd. Accessed: 28 July 2021.

Ubud Village Jazz Festival. (2019). *Ubud Village Jazz Festival*. Retrieved from: https://ubudvillagejazzfestival.com/about-uvjf. Accessed: 12 July 2021.

Ubud Writers & Readers Festival (UWRF). (2021). *Ubud Writers Festival 2021*. Retrieved from: https://www.ubudwritersfestival.com/about/. Accessed: 28 July 2021.

Unger, O., Uriely, N., & Fuchs, G. (2016). The business travel experience. *Annals of Tourism Research, 61*, 142–156.

Urry, J. (1990). The "Consumption" of tourism. *Sociology, 24*(1), 23–35.

Waitt, G. (2003). Social impacts of the Sydney Olympics. *Annals of Tourism Research, 30*(1), 194–215.

Wiltshier, P., & Clarke, A. (2017). Virtual cultural tourism: Six pillars of VCT using co-creation, value exchange and exchange value. *Tourism and Hospitality Research, 17*(4), 372–383.

United Nations World Tourism Organization (UNWTO). (2020). *Covid - 19 related travel restrictions a global review for tourism* (pp. 1–11). UNWTO.

Xie, K., Mao (Eddie), Z., & Wu, J. (2019). Learning from peers: The effect of sales history disclosure on peer-to-peer short-term rental purchases. *International Journal of Hospitality Management, 76*, 173–183.

Putu Diah Sastri Pitanatri is a Lecturer in the Hospitality Department, Bali Tourism Polytechnic, Indonesia. She is currently a PhD candidate of Tourism Studies at Gadjah Mada University. She is active in various national and international associations. She edited the book "Homestay: A Mosaics of Community Based Tourism" with I Wayan Mertha in 2018, and "Culinary Tourism: A New Attributes of Ubud" with Nyoman Darma Putra in 2016. She also wrote a book chapter titled "Big Data Research for Tourism Marketing in Indonesia" in 2021. She has received 1st place in Win Way Award (2018) as the best Lecturer under the Ministry of Tourism Republic of Indonesia and best paper presenter in OBOROT-Palembang 1st International Conference in 2018. Till date, she is also an active reviewer in national journals and speaker in many international conferences. Her research interests are on big data, tourist studies and sustainable tourism management.

Luh Yusni Wiarti is a Lecturer and the Head of Tourism Department, Bali Tourism Polytechnic, Indonesia. She has got her PhD of Tourism Studies at Udayana University. Previously, she obtained her Masters in Double Degree Program at Udayana University (Tourism Studies) and University of Angers (Tourism and Hospitality Management), France. She actively participated in various national and international conferences such as ITSA Conference in Greenwich University London, INTACT Conference in Gajahmada University Indonesia, ISOT and IHTC Conference in UPI Bandung Indonesia, the ASEAN Tourism Research Conference at James Cooks University in Singapore, the International Conference on Tourism and Business (ICTB) in Lucerne University of Applied Science and Art in Switzerland, etc. She won the best paper awarded at INTACT conference (2016). She has written book chapters titled "Homestay's opportunity and challenge: A study case of local participation in KEK Mandalika Lombok, Indonesia" in 2016, and "Gastronomy in Ubud: Opportunity and challenge of the legendary cultural destination model in Bali" in 2021. Her current research focuses on events and festivals, as well as topics in destination management.

Chapter 18
Technology Application in Tourism Fairs, Festivals and Events in Malaysia

Sharina Osman

Introduction

Technology advancement revolutionized tourism industry and bring new business opportunities globally. Tourism-related organizations and Internet corporations are joining together to take advantage of the new market opportunities provided by technological advancements. Through the development of the Internet and the World Wide Web (WWW), information technology has introduced e-commerce. These changes have influenced consumer and industry behavior in the travel and tourism industries. Consumers now have additional vacation and financial planning options. 95% of web users have looked for travel-related information on the internet. 93% used destination websites to seek travel-related information, while nearly half used e-mail (Lehto et al., 2006). Search engines were used by nearly three-quarters of the tourists before making online tourism product and service purchases. Technology gadgets such as Virtual Reality (VR), mobile phones, smart phones, and hand-held devices, in addition to the Internet, have improved customers' tourism and travel experiences. Access to information on travel, destinations, bookings, payments, hotels, attractions, and map quality are all areas that may have been improved.

Information and communication technologies (ICT) have had an influence on the travel and tourism industry for at least 50 years, especially in the areas of automation and distribution channel networking. To obtain a competitive edge, local travel agencies who used the international distribution system altered their marketing techniques and diversified their offerings. Increased visibility in additional market sectors has resulted from a better understanding and use of the technology by local tourism businesses. By registering with online travel operators, online advertising

S. Osman (✉)
Tourism Section, Business School, Universiti Kuala Lumpur (UniKL),
Kuala Lumpur, Malaysia
e-mail: sharina@unikl.edu.my

A. Hassan (ed.), *Technology Application in Tourism Fairs, Festivals and Events in Asia*, https://doi.org/10.1007/978-981-16-8070-0_18

agencies, and inter-organizational reservation systems, businesses have expanded their services globally. In addition, several firms have developed websites that allow consumers to plan, book, and pay for their holidays online. Businesses that have not taken advantage of the Internet will be unable to compete, so they must seize the chance now before being out-competed by those who have. The development and advancement of the travel and tourist business has been aided by information technology. Improved information accessibility, increased competitiveness, and a broader global market of consumers and enterprises are all long-term benefits of technology.

Events, Fair and Festival in Malaysia

Malaysia is known for its values of multiculturalism and diversity in terms of races, ethnic groups and religions. Because of this, nearly every month, there are multiple events and festival being celebrated. Even, the West and East Malaysia often have their separate cultural celebrations either religious or cultural in nature. Those in the East Malaysia are still celebrating traditions which rooted from the ancient pagan especially by the indigenous tribes. They preserve their traditions with the festival today though it is no longer involved the ancient rituals. In addition, today, in Malaysia, celebrating festivals and organizing events has become one of the tourism product that attract tourist from all over the world to experience even the religious ones. In fact, many of the events are highly anticipated by tourists and often receive favorably wide participations.

Le Tour De Langkawi

The Tour de Langkawi is an annual countrywide multiple stage bicycle race of Malaysia. This event is held every year. The tournament is named after the Langkawi archipelago, which hosted the first ever edition. Since 1996, the race has been contested every year in February. It normally consists of 10-day segments (stages) spread out over 10 days, although in recent years, it has been reduced to eight stages. The route changes every year but the Genting Highlands climb, the most difficult in the tour always stays in the list. The International Cycling Union (UCI) has recognized the Tour de Langkawi as a 2.HC road race on the UCI Asia Tour calendar. In 2020, the race will be a part of the new UCI ProSeries. To the finish, all stages are timed. The aggregate time for each of the completed stages is added together, and the rider with the lowest total time wins the race and wears the yellow jersey. In this tour, some other competitions are also held. These are: the sprinters' points classification, the climbers' mountains classification, the Asian riders' classification, the competing teams' classification, and the competing Asian teams' classification etc.

Thaipusam Festival

After Deepavali, Thaipusam is the Hindus' most important festival, which usually takes place in January or February. It is a day of penance and thanksgiving dedicated to Kartikeya or Lord Murugan, known as the Hindhu God of war. Thaipusam is acknowledged as public holiday in many Malaysian states. Devotees pay tribute to Lord Subramaniam by piercing their tongues, faces, and bodies with hooks and sharp skewers weighted down with oranges on this day of penance and thankfulness. Thaipusam is not for the faint of heart, and Western visitors are often taken aback by the public self-flagellation. The Batu Caves attracts around 10,000 visitors each year during this festival and it is highly anticipated by tourists for the unforgettable and fantastic experience. Hindus assemble at the Sri Mahamariaman Temple on Jalan Tun HS Lee (Petaling Street neighborhood/Chinatown) the night before Thaipusam. The devotees will then set out at midnight on a 15-km (roughly 8-h) walk to the Batu Caves, where they would arrive the following morning. The long journey culminates in a 272-step climb to the cave entrance. During Thaipusam, it is estimated that over a million people visit the Batu Caves, so get prepared for a busy, occasionally claustrophobic and chaotic experience. It can get very packed at times that people are shoving the masses forward. All are welcome to climb the stairs to the actual caves, including devotees, visitors, and locals. At countless altars, Hindus will pray and express gratitude. The area surrounding the Batu Caves is densely packed with small kiosks selling food, snacks, drinks, and a variety of religious artefacts. Walking past the small booths, as well as the stairs leading up to the caves, can be a bit congested. Foreigners can get up close and personal with the activity since there is a special dedicated area where members of the press (as well as tourists) are permitted to take close-up photographs and videos of the Hindus waiting to begin their ascent up the stairwell leading to the Batu Caves.

Pasir Gudang World Kite Festival

The Pasir Gudang World Kite Festival is one of the most prestigious annual events in Johor and has been included in the Johor State Tourism Calendar as well as the Malaysian Tourism Calendar. The Pasir Gudang Municipal Council (MPPG) has been running it successfully since 1995, with the help of Tourism Johor and the Johor Tourist Association. It is an exhibition with booths showing kites, as well as kite-making workshops and demonstrations, held in Pasir Gudang, Johor. The festival has continuously been a venue for visitors from more than 40 nations to exhibit beautiful themes and aesthetic value in traditional Malaysian handcrafted art produced from natural materials such as bamboo and paper. Food and refreshment stalls, as well as souvenirs and collectibles kiosks, will be available on the fairgrounds. Visitors can also visit the famed Pasir Gudang Kite Museum and watch traditional local performances while attending the kite festival.

The Regatta Lepa Semporna

The Regatta Lepa water festival, which has been hosted annually in Semporna since 1994, is one of Malaysia's and Borneo's oldest and most distinctive events of its kind. The event pays tribute to the itinerant Bajau Laut (Sea Gypsies), whose cultural legacy has been passing down for centuries. The Floating Parade of Lepa Boats is the festival's highlight. Each Lepa is associated with a distinct family or community in the region. The Celebes Sea is dotted with a flotilla of colorful boats that compete for the title of "Most Beautiful Lepa". The Lepa Tug of War, small dugout boat (the Kelleh-kelleh) race, and the duck capturing competition are among the festival's main highlights are some marine sporting events. Late into the night, Semporna's stadium hosts Igal-igal dance performances, "Lepa Car" competition, fireworks display, karaoke, beauty pageants and cultural music shows.

MyBalloonFiesta

MyBalloonFiesta or formerly known as the "Putrajaya International Hot Air Balloon Fiesta" is a Kuala Lumpur-based annual event. This is a widely popular event enjoyed by thousands of tourists and takes place in Desa Parkcity. This event is a type of reminiscent of the Adirondack Hot Air Balloon Festival of New York. It is a fantastic opportunity to grab some fantastic travel photos. Visitors may expect to see a lot of both bright teardrop and oddly shaped balloons. Hot air balloons styled like Darth Vader's head, a cat, Doraemon, a clown, a pink elephant's head, and even a big peacock are included in the event. Tourists are required to purchase the MyBalloon Club pass to ride the tethered balloon during the festival, which also includes a gourmet buffet breakfast, live jazz concerts, and a private viewing area where to watch the balloons inflate. Those who are not keen on the thought of riding in a hot air balloon need not worry, as this 4-day event offers a range of entertainment options. Aside from hot air balloon rides, there are cultural performances, food trucks balloon workshops, and a nighttime firework to create interests to attend this event.

Technologies Supporting Tourism Industry

The past decade trends have seen the advancement of technologies and its complex features in assisting day to day activities. The use of technologies in the tourism industry has a profound impact in reshaping tourist experience while creating an unforgettable and impactful means to the attendees. Today, the tourism sector is expected to be heavily reliant on technology, both in terms of specialized interactive systems (e.g., self-check in kiosks, travel management systems, in-room entertainment) and overall purposive systems for the delivery or augmentation of services (Lehaney et al., 1999). (e.g., mobile phone Apps, agency websites) (Slivar et al., 2019; Wani et al., 2017). With the introduction of technologies as the Internet of Things (IoT), Artificial Intelligence (AI), Blockchain, Big Data Analytics, Location-based Services, or Virtual Reality (VR) and Augmented Reality (AR), and their application in tourism (Peceny et al., 2019), a large portion of the production and delivery of tourism products and services is now fully automated or digitalized (Peceny et al., 2019). (Ivanov, 2020).

The Use of Mobile Technology

Consumers are able to access information and conduct transactions across national, cultural, and linguistic boundaries as a result of increased knowledge of globalisation and the availability of technology (Laudon & Laudon, 2006). Mobile technology refers to a portable device that is used mostly in cellular communication and a few other relevant areas. This employs a framework that permits multiple transmitters for delivering data on a single channel simultaneously. With the tremendous rise in smartphone ownership, consumers can now utilise their phones as erudite tour guides. Smartphones could provide travellers with advance assistance to overcome problems in their trips, such as ride direction, destination suggestions, payment, and even language barriers, by using built-in software development. According to TripAdvisor, 45% of consumers utilise their smartphone for all aspects of their vacation planning. Thus, emphasising the need for corporate services and communications to exploit the implications of mobile devices.

The usage of mobile technology aids the planning and travelling for tourist while boosting their overall experience. Tourist can travel with ease as the planning process are within the reach of their hands and made easily accessible, where the bookings of accommodation, transportation and maps are all embedded and integrated in mobile devices. Furthermore, the application of mobile devices in tourism can be a pushing factor for tourists to travel since they no longer need to face the unnecessary hustle of queuing up to buy tickets at the counter or having to ask around for directions by simply accessing their mobile devices. Tourists may be more motivated to travel and will be able to fully enjoy their travels without any worries to get the full experience of the destination and travels. In addition to that, tourists are able to compare and discover new places with the aid of mobile technology, by means they can also avoid places with bad reviews by conducting research beforehand. Therefore, mobile technology allows users to optimise the best possible travelling experience while.

Social Media Influences

Social media is a device-based technology that allows individuals to establish virtual communities and networks to exchange ideas, views, and information. By design, social media is Internet-based, allowing individuals to easily exchange material via electronic methods. The material includes personal information, files, videos, and photos. On a computer, tablet, or smartphone, users engage with social media via web-based software or applications. People tend to be snapping more photographs on their trips as a by-product of the impact of social media platforms such as Instagram, Facebook, and Snapchat. As a result, it appears that the urge to share travel experiences is a common need among travellers. The company may also profit from social media shares since it serves as word of mouth and expands the consumer base. Furthermore, mobile-enabled technology will alter how firms interact with online travellers. Unfortunately, the concept of sharing is gradually erasing the boundaries of privacy and personal identity, as personal data might be used for shameful behaviour. Users may, without a doubt, use social media traits to create false news, fake reviews, or false experiences in order to damage competitors in a dirty PR scam. As a consequence, technology in the tourist industry will be driven by the development of social media and the trend of having an online presence. Companies would also need to plan for complications with online communication. On the other hand, it could also be the means for free marketing when tourists share their interesting experience visiting a place. It can be the determinants of success for the business when they are able to generate positive reviews and recommendations from the visiting customers as the customers would most likely to share with their friends and families of their great experience and persuade them to visit the place. In essence, businesses must incorporate this sharing culture into their daily operations, as unfavourable evaluations will undoubtedly reflect negatively on the company's brand image and reputation.

Augmented Reality (AR) and Virtual Reality (VR)

Augmented reality (AR) is a technologically enhanced version of reality that is generated using digital visual components, audio, or other sensory stimuli. Virtual reality (VR) is a type of computer-generated simulation where the user applies electronic equipment such as customized eyeglasses with a screen or sensors-equipped gloves to interact with an artificial three-dimensional world. In this replicated artificial environment, the user can have a realistic-feeling experience. Both AR and VR have made inroads into the tourism industry, and the reality is that they are growing in popularity as a result of the various advantages they provide. More corporations are using technology to show consumers a stateroom on a space ship or send them to the Taj Mahal for a few seconds. We may now "teleport" to the most remote parts of the globe without ever leaving our sofa. EVEREST VR, for example, allows users to see the top of the world without having to climb to the top. Alternatively, tourists can wonder around the Colosseum and visit the Machu Picchu in the same day, taking in the sights and sounds.

The Internet of Things (IoT)

The Internet of Things (IoT) is said to have the potential for revolutionizing the tourism business. In Iot, sensors that are connected to the Internet integrate into products such as bags, cars, buildings, and others. In addition, travellers can benefit from non-document travel identification. Consequently, it alters how people prepare for their vacation. Instead of having to give multiple documents such as a hotel confirmation, a user ID, a passport, and so on, visitors now have additional options for proving their identity. Therefore, recognition technology in the tourism business provides great benefits to users, such as increased security, non-transferability of services, and a high level of efficiency.

Big Data

Data on visitors may be used by tourism sector operators to select the best target for marketing campaigns. Also, they analyze their database for looking at the amount spent with the reason for the trip, and the country of origin. Thy then cross-check such data and information with public data from government sources for creating the most appropriate customer profile and for obtaining a higher success rate. This allows for better segmentation of their campaigns, which increases their efficacy and reduces the amount of money spent on campaigns.

Blockchain

Blockchain is a highly potential technology to reshaped the way we live. Despite the fact that it is most of time associated with finance, it appears to have an influence on travel as well. Although there has not been much testing, it may be beneficial for authenticating people at airports, ensuring openness in tourist evaluations, and for making payments secure and simple.

The Application of Technologies in Events, Fair and Festival

Before the wider acceptance of event technologies, the participation and hosting of events were fairly laborious processes. The list of attendees was printed and used on-site for confirming the registration. However, they could not be updated in real time and became outdated. Promotional method was constrained to certain channels and, as a consequence, to only the audience they could reach. In addition, determining the effectiveness of the event's many components were also practically impossible. Event technology strives to alleviate most, if not all, of these issues, allowing the planner and guests to have a more streamlined experience with more systematic data collection and fewer miscommunications. According to event technology trends, there is a comprehensive stack of technical applications that may be incorporated to provide a customized experience for each and every guest. From the minute they first see an advertising until they leave the event, technology is impacting every element of the tourist experience. Learning from industry players involved in organising and managing major events and festivals in Malaysia, the adoption of event technologies such as mobile devices, social media, AR, VR, IoT, Big Data and Blockchain are applicable to serve various functions while enhancing the overall process and the outcome of the events, fair and festival in Malaysia.

Capturing and Increasing Engagements

One of the most difficult aspects of organising or participating in an event is capturing the audiences' attention. Unfortunately, content consumers at present time are bombarded with information from all sides, so consumers have learned to filter out the majority of it unless the organiser can pique their interest. When used effectively, technology can have a positive impact on audience interaction and engagement throughout an event. Attendees can, for example, respond in real time to questions posed by speakers using live response systems. This led to the creation of increased engagement when attendees are able to interact with the organiser and promotion of events become easier as organisers can deploy the use of technologies to capture and attract participant to join the event by promoting enthusiastic anticipation in prospects attendees. Moreover, technologies also allow organisers to reach wider scope of audience through the use of social media platforms and exploit the "sharing culture" on platforms such as Instagram, Facebook, Snapchat and many more. It facilitates in getting a quicker response and word of mouth to promote the events and relatively cheaper than the traditional marketing techniques of printing posters and banners. Organisers can share their event's digital poster online for better awareness and engagements. In addition, there are also paid advertisements that assist in targeting specific users that fits the criteria and the ads of the event will appear on their profile which can be a useful tool for organisers to incorporate in their marketing and promotion. Organisers can get feedbacks from attendees post events by conducting an online survey or by having attendees to rate the event.

Measuring Success

In defining success for an event, objectives needed to be established beforehand. Regardless of the metric, technology can facilitate in measuring it without the need for laborious calculations. In some cases, the user may only need to log into a dashboard to get the answer, or they may need to spend time setting up the application technology before to the event and then checking in once it is over. Technology at events will make the process of measuring and monitoring success easier and faster, whether it is event surveys, or tracking social media activity, or simply the overall number of attendees. Moreover, the results produced will be accurate and free from human error hence organisers can measure the success of the events efficiently and identify areas of improvements for future events. Tools such as survey monkey, google forms are platforms that can be implemented to get attendees feedbacks and recommendations for the purpose of measuring the successfulness of the event. By conducting it digitally, attendees can even answer the surveys at the comfort of their own home and time by using their own devices, which would help in getting more response than manual feedback methods.

Effective Communications

Advertised times do not indicate when an event will begin or end. Instead, weeks or even months of planning are generally committed leading up to the event, with time left over to wrap and tie things up. By developing and distributing unique, relevant, and timely material, event technology allows users to engage with guests far before and far after the event is planned for taking place. Users could, for example, employ email marketing campaigns to help generate enthusiasm for the event by proposing sessions that a certain attendee might enjoy on the basis of their previous click-through and the interactions with content of the organizers. Organisers might use a smartphone app to remind participants of session start times or to allow them to plan one-on-one time with specific vendors during the event. Following up on an event, users may conduct a survey to provide further information about the sessions that a participant attended. Keeping in touch with the event's attendee is also crucial for the full experience of the participants, hence creative messages and development of valuable video content are essential.

Data Collection

In addition to assessing success, organizer must be able to obtain insights and relevant data for fueling the analysis that actually determine the success. For example, data can also assist organiser in improving specific areas of the event and ensuring their guests have the greatest possible experience each time they interact with their company. Technological applications that can be applied for such specific capacities. These include tracking the organizers' the open rates of email communications, the use of radio-frequency identification (RFID) technology for measuring traffic flow within the event space, and to launch questionnaires and surveys in the event app on smart devices. These users can become able for easily and rapidly altering the data for spotting trends and for utilizing them for informing real-time or future decisions when collected digitally.

Making an Impression

Aside from the practical uses and benefits, the application of technologies in events an also aid organizers captivate attendees with memorable experiences. Attendees can remember the technical applications like sound, lighting, and even AR and VR experiences, making them eager to return to the next event. When it comes to Mixed Reality (MR) technologies, going for the biggest "wow" factor can sometimes result in the most memorable experiences. Also, this is important not to overlook the finer elements that could be improved with much easier and simpler technologies. Utilizing sound and light to define the event's ambiance and rhythm, highlight a certain product or location, or give the venue a dramatic impact are all simple ways to improve the guest's experience.

Conclusion

The range of innumerable festivals and events in Malaysia reflect its vibrant cultural diversity and religious tolerance of its citizen. Taking place throughout the year, it attracts locals and tourists to attend the event. With the advancement and innovation of technologies, fairs, festival and event management had reshaped the tourist experience in a more interesting and memorable manner. The technology had also impact the tourist experience and ways of travel through the use of various event technologies such as Mobile devices, Social media, AR and VR, IoT, Big Data and Blockchain. All these are integrated in most events for different purposes including capturing and increasing engagements, measuring success, consistent communications, data collection and making an impression. Therefore, the application of technologies in managing events, fairs and festival are part and parcel in determining the successfulness of the events, fairs and festival in Malaysia.

References

Ivanov, S. (2020). The impact of automation on tourism and hospitality jobs. *Information Technology & Tourism, 22*(2), 205–215.

Laudon, K. C., & Laudon, J. P. (2006). *Management information systems*. Pearson Education Limited.

Lehaney, B., Clarke, S., Kimberlee, V., & Spencer-Matthews, S. (1999). The human side of information systems development: A case of an intervention at a British visitor attraction. *Journal of Organizational and End User Computing (JOEUC), 11*(4), 33–39.

Lehto, X. Y., Kim, D.-Y., & Morrison, A. M. (2006). The effect of prior destination experience on online information search behavior. *Tourism and Hospitality Research, 6*(2), 160–178.

Peceny, U. S., Urbančič, J., Mokorel, S., Kuralt, V., & Ilijaš, T. (2019). Tourism 4.0: Challenges in marketing a paradigm shift. In consumer behavior and marketing. *IntechOpen*. https://doi.org/10.5772/intechopen.84762

Slivar, I., Stankov, U., & Pavluković, V. (2019). Case study: Delegated distribution: Hotels should be warned! An example from Croatia. *Transnationational Marketing Journal, 7*(2), 245–256.

Wani, M., Raghavan, V., Abraham, D., & Kleist, V. (2017). Beyond utilitarian factors: User experience and travel company website successes. *Information Systems Frontiers, 19*(4), 769–785.

Dr Sharina Osman is a Senior Lecturer and the Head of Tourism Section at Universiti Kuala Lumpur Business School. She earned her PhD degree in Management from the University of Exeter, United Kingdom; Master of Human Resource Management and BBA (Marketing) Honors from Universiti Utara Malaysia (UUM), Malaysia. Her research interest is around Identity, Image and Reputation, Organizational Culture, Gastronomic Tourism, Tourism Destination Identity & Branding, Participative Tourism Development and e-Tourism. Dr. Osman carries with her an array of experience which ranges of Human Resource to Marketing (Image & Branding) and Tourism. She is also the Editorial Advisory Board for several journals and was also invited as Keynote or Guest Speakers in seminars. Being in an academic and in the research environment, coupled with vast hands-on experience in the industry, she helps organizations improve the competency and capacity building through trainings and consultations. She believes in interacting and experiential learning and has conducted many workshops to different sectors, including government and private at both local and international.

Chapter 19
Exploring the Role of Technology Application in Tourism Events, Festivals and Fairs in the United Arab Emirates: Strategies in the Post Pandemic Period

Syed Far Abid Hossain, Faiza Tanaz Ahsan, Armana Hakim Nadi, Musanna Ahmed, and Hafsa Neyamah

Introduction

Amidst the global pandemic where all the nations and its commercial sectors faced difficulties in sustaining the activities, however the airlines and tourism sectors have particularly been affected severely. Prior to COVID-19, tourism had been a significant sector for the world economy contributing up to 10% of Global GDP and over 300 million jobs approximately (Behsudi, 2020). Tourist countries of the Middle east for instance Jordan, Dubai, the United Arab Emirates (UAE), Oman and Qatar to name a few have experienced negative impact of the crisis that other countries. Fear of job loss, industry shock and GDP shrink are amongst the few disasters these country are prone to. Not only recreational tours, but also the significant pilgrimage of the Muslims known as Hajj which takes place at Saudi Arabia was suspended in Medina and Mecca. The border restrictions have further caused tourism companies to shut down or minimize their operations and as a result in the UAE, the Meetings, Incentives, Conferences & Exhibitions (MICE) industry faced a sharp decline in demand leading to noticeable material losses. It is important the MICE industry strives for an optimal solution which involves a competitive survival strategy for the most impactful sector of the country.

The country's largest airlines companies along with the hospitality sectors have received enormous support from the government fund, support schemes and capital packages. Oman, for instance came up with a campaign named 'Within Oman' where the citizens are encouraged to patronize the local airlines, hotels and resorts.

S. F. A. Hossain (✉) · A. H. Nadi · M. Ahmed · H. Neyamah
College of Business Administration, IUBAT-International University of Business Agriculture and Technology, Dhaka, Bangladesh

F. T. Ahsan
North South University, Dhaka, Bangladesh

A. Hassan (ed.), *Technology Application in Tourism Fairs, Festivals and Events in Asia*, https://doi.org/10.1007/978-981-16-8070-0_19

The campaign aims to focus on the natural assets of Oman and draw in more travelers by offering them lucrative traveling facilities. Hafidh Al-Hadhrami, the Asset Manager of Tourism Group of Oman stated that the tourism push can highlight Oman's exclusive heritage, natural, and cultural experiences and will allow local people to appreciate the country's diversity (Arab News, 2020). Besides singularly working on mending the travel industry, Middle eastern countries are collaborating with European countries to ensure a rise in tourism; for instance, UAE and Bahrain signed an agreement with UK under the corridor travel list, which encourages travelers to move across the borders to visit the middle eastern countries while maintaining a safety quarantine period along with inbound travel until a mutual visa waiver. This will increase the volume of incoming tourists an d the country will see a boost in the economic recovery. The Middle eastern countries have also decided to lift travel bans from South Africa and neighbouring nations.

Prospect of Tourism Events in the UAE

In the Middle East, the UAE has risen as the wealthiest country (Staff Reporter, 2021). As the financial condition is so superior for this country, there are lots of investments are coming in diverse divisions. Investment in event tourism has been considered as one of the successful ones as tourists are as of now fascinated by the luxurious offerings of the UAE. There are many Arabian celebrations and events which have brought a lot of visitors to the UAE. These events are also bringing financial growth and it is to talk about that tourism is supplanting oil as a source of income for the country (IvyPanda, 2019). As the information of Dubai Online (2020), Dubai has the aim to serve at least 25 million tourists in a year by 2025; it also mentioned that due to the Coronavirus pandemic, they have already lost 5.09% visitors compared to 2019; and that the number of tourists dropped to 5.51 million because of the COVID-19 (Arab News (2021), Kazoba et al. (2016), Travel Triangle (2021), Zaidan (2017)

Implementing event tourism might bring lots of positive results in the sector of tourism and the hospitality industry. There might be an upward slopping curve which may represent the positive impact on a nation's expenditure, creation of employment, labor supply, and also the standard of living (Skoultsos & Tsartas, 2009). If the UAE invests more in event tourism, it will be able to create a good standard of living. There will be scopes for more employment. The rate of unemployed people will be decreased if they ensure event tourism efficiently. There are some other benefits such as they can spread the positivity of their places. These events will enlighten people to explore more in the UAE.

As this pandemic is going on, we already mentioned that a tougher period is going in the sector of tourism. The UAE should conduct some virtual events to gain more interest in their tourism sector or to keep the interest as before. There are many events are going on worldwide such as global investment, real estate, and residency shows (10 times, n.d.). The UAE is one of the most exciting places for travelers and it can be taken as an opportunity to utilize this chance to gain more investors.

Internet is one of the powerful tools to reach people. Digital marketing can be used to reach out to many places. Due to COVID-19, many online platforms have been created to offer people to participate and see gather experience. The UAE is one of the destinations which serves rich culture, ancient monuments and also a great taste of art. Event tourism can be implemented successfully by representing their arts, and their architecture by online platforms. The museums and art galleries such as Louvre Abu Dhabi, Sharjah Art Museum, Salsali Private Museum, etc. can be brought online events in this pandemic period. If tourists think about offline events, there are many to offer for the tourists. Many upcoming events are inviting visitors to taste the culture and luxury such as wedding events, real estate events, and IT events, etc. There is a huge prospect in the UAE if we ensure sustainable event tourism.

Various Tourism Spots in the UAE

Burj Khalifa

The Burj Khalifa is a famous skyscraper in Dubai, United Arab Emirates, and the tallest building in the world. Its height is 2722 feet and the ceiling can reach a height of 2717 feet. It was developed for Dubai and aims to gain international recognition. It was originally called Burj Dubai. But it was later renamed to commemorate Khalifa bin Zayed Al Nahyan, President of the United Arab Emirates and ruler of Abu Dhabi. The building was designated as the tallest building in the world and broke many height records. The Burj Khalifa is a mixed-use development project intended to become a large-scale central building. It has a total of 9 hotels, including the famous Dubai Downtown Address Hotel, 30,000 houses, at least 19 residential skyscrapers, 7.4 acres of parkland, 12 hectares of Burj Khalifa man-made lake, and Dubai Mall (Thrillophilia, 2021).

Sheikh Zayed Grand Mosque

The Sheikh Zayed Grand Mosque is a huge modern mosque with implausible beauty. This mosque uses the advantages of contemporary design and ancient craftsmanship to integrate ancient and modern techniques and styles provides a new interpretation of Islamic architecture. It must be visited during any Abu Dhabi itinerary. Its interior uses a lot of gold, mosaics, and glass products, and a lot of marble, and its dazzling white stone exterior forms a sharp contrast under the blue UAE sky (Lee, 2021).

Global Village

Global Village is the region's first family destination for culture, entertainment, and shopping. This is a unique and comprehensive destination where tourists can enjoy the best shopping, food, and entertainment experience in the world, and provide guests with the largest and most diverse activities, performances, and activities in the region. It provides many interesting surprises for millions of outdoor guests (TripAdvisor, 2021).

Hajar Mountains

This is one of the ideal destinations for nature lovers to visit the United Arab Emirates. The mesmerizing scenery along the winding roller-coaster road is the highlight of this place (Fleet Management, 2019).

Yas Island

This is an exciting man-made island that can be an ideal destination for all types of tourists. It would be a good idea to relax or take a fascinating cruise at Yas Water World Park. Tourists can even choose to drive a Ferrari on a personal test track (Fleet Management, 2019).

Jebel Hafeet

A must-see for anyone looking to take in a great view, the Jebel Hafeet summit is a favorite day trip from Abu Dhabi and Al Ain. It is the second-highest peak in the United Arab Emirates, and sweeping desert vistas welcome visitors to the top of the winding road. Following the winding road up the mountain road, tourists are rewarded with a view across the vast desert of the region located in Al Ain (Edarabia, 2021).

Al-Bidyah Mosque

The most seasoned mosque in the United Arab Emirates, Al Bidyah is one of the UAE and the Emirate of Fujairah's most significant chronological destinations. The whole surrounding zone is an archaeological site, where a lot of ceramic and metal artifacts have been excavated. It is positioned 35 km from Fujairah (Edarabia, 2021).

Various Tourism Fairs and Festivals in the UAE

Over the past decade, local governments have conducted several festivals and events across the United Arab Emirates (UAE) and have been able to increase the number of tourists in the region (iExplore, n.d.). There are some festivals and fairs such as Awafi Festivals, Shopping Festivals, Dubai Shopping Festival, Festival of Taste, Dubai International Film Festival, Art Dubai, and Dubai International Jazz Festival, etc. have been very successful to attract and amuse the tourists.

UAE Awafi Festival

The Awafi festival is a yearly celebration of the traditions and old times of the union, mainly in the shape of outdoor games, making it one of the few conventional celebrations of the United Arab Emirates. It is held in summer every year. This festival is mainly popular for its numerous competitions to scale the tallest dunes in various kinds of improvised vehicles. There are also a lot of cultural occasions during the event, including a Heritage Village show, the main bazaar as well as coffee shops and art shows. This is one of the most excellent celebrations of the United Arab Emirates and is a must-visit place for tourists.

Dubai Shopping Festival

Shopaholic's Paradise-this event lasts over a month and is one of the best seasonal festivals in the UAE. This is a time when all shopping outlets in Dubai are lowering their prices. In other words, the city streets are full of shoppers in the world. Besides, there is the magic and energy of concerts, fireworks, and other entertainment, which is one of the best events and festivals in Dubai, UAE.

Festival of Taste

The Dubai Festival of taste is an annual food festival held in the famous Dubai Media City. The festival attracts food lovers who gather from the world's renowned first-class chefs to prepare delicious dishes for a variety of events. It also explains cooking tips and how to make it look good. Of course, the Festival of Taste is one of the most popular food fairs in Dubai (Crystal Travel, n.d.).

Dubai International Film Festival

At this popular UAE festival, travelers can see Hollywood's biggest stars descending into the city to screen and discuss their latest projects with fans and commentators. But the biggest attraction of many festivals is the platform they offer to local filmmakers. After all, it is one of the leading promoters of Arab films, which has screened more than 5000 locally produced films since 2004. This is a place where tourists can see and experience other sights of the world (Expatica, n.d.).

Art Dubai

The annual Art Dubai fair has been an important platform for artists from all through the region. It has been involved in drawing on the city's mysterious blend of artists', curators' and art lovers, thereby creating a mounting stage for imaginative artwork and native aptitude. The flourishing contemporary art displays and achievement of Art Dubai, construct Dubai a crucial visit for any art connoisseur (Dubai.com, n.d.).

Dubai International Jazz Festival

The city of Dubai is host to this astonishing jazz festival that is featured by some of the top stars in the jazz and pop music industry. The carnival is one of the only jazz celebrations to occur in the territory. Legends such as Duran, John Legend, and Ricky Martin have often featured in this annual celebration of melody in the region.

The Role of the Technology in the Tourism Industry

Tourism and technology, both are considered as one of the most enormous and effective industries (Benckendorff et al., 2019) and the combination of these two fields by taking the tourism industry and implementing technology in it would generate an outcome that can lead to the content for the tourists when they go or decides to go for a visit to any place which eventually leads to a satisfactory feedback from the tourists. Technology has made our lives easier in a number of ways, and it also plays a significant role in the tourism industry, ranging from the fundamental to the strategic activities as well as enabling them to cope up with the competition within the industry (Khatri, 2019). The application of technology enables the tourism industry to operate in an efficient manner which eventually satisfies and brings light and positivity to the tourism industry. As tourism connects people all around the world which seemed impossible even decades ago (Benckendorff et al., 2019), the integration of technology has empowered and highly facilitated the bridge for connecting all around the world and can also gather and visualize details of the destination even before traveling there and this strategy also brings a great accomplishment for the industry as people are getting more attracted and if they travel, that would generate profit for the tourism industry as well. The advancement of innovative ideas and practices has also heated up the competition within the industry. However, technology has helped to match this competitive environment, so that the tourism industry does not fall behind and catch up with the changing disturbance and unsettlement that comes up as days come by, making the use of technology in the tourism industry unavoidable (Khatri, 2019). As we know that the tourism industry is one of the driving forces of the economy of UAE, the extensive adoption and implementation of up to date technologies in this industry during recent years has constitutionally reconstructed the method in which the exchange of services are being provided and received (Kansakar et al., 2019).

Technologies might bring change the situation and environment which is encountered by people and their ability to come up with inferential services which is a key aspect (Dias et al., 2021) that transmutes the tourism industry into a more comprehensible, approachable, and highly enjoyable. Moreover, the evolution of technology over the years has strongly contributed to the tourism industry by attracting people from various countries which has led to the number of tourists increasing as time passes by Eichelberger et al. (2020). Thus, the application of technology has a notable role in the tourism industry as technology in this field has transformed into a strategic armament for the tourism products, pinpointing and recognition, demonstration and finally the circulation, which highly aids to gain competitive advantage which is highly beneficial for the industry that can lead to success easily (Bayram, 2020).

The Role of Apps in Promoting the Tourism Industry

In recent years, the tourism industry has been implementing the use of apps in order to promote and support their business which would help them to run with higher efficiency and accuracy. When access to mobile apps was not available, people used to follow the traditional method of purchasing by visiting an agent. Even a lot of people were left unaware of the process of dealing with the tourism industry. Thus, mobile apps have made these tasks easier and reliable as the role of apps in promoting the tourism industry is quite vital and significant in the development and promotion of the tourism industry. The role of apps in the tourist sector is that they help to connect the dots inside a community, making it a more livable, desirable, viable, and sustainable place to live (Wayne, 2016). The use of apps in tourism facilitates portability, social integration, and territorial marketing by implying that the use of technologies are extending the tourist and hospitality sector in such a manner that tourism, as well as its products and services, would be available to those who are temporarily or permanently paralyzed (Dias et al., 2021) and gives individuals who are unable to access many of their dreams, such as traveling, hope, and tourism technology make these aspirations a reality. The tourist sector has adopted technology-enhanced tourism, such as Airbnb, which is regarded as one of the industry's most astonishing and spectacular developments (Guttentag, 2019).

Airbnb is a smartphone application that allows users to access and book without having to leave their homes, but still adhering to pandemic laws, such as maintaining social distance and keeping passengers safe. This smartphone application assists its travelers by studying and scrutinizing a large data set encompassing all of the online evaluations and comments available on the internet. It was also revealed that customers try to appraise their experience based on the previous hotel stays collected from online references (Cheng & Jin, 2019). Through technology-enhanced tourist sites and applications, tourists may also assess and predetermine their selections and through certain mobile applications, these tasks are made possible even without physically being present at that place. Bookings.com is a wonderful source of information for academic researchers in the hospitality business since it can collect millions of hotel evaluations in a timely, cost-effective, and appropriate manner (Mellinas et al., 2015). Because of the rapid innovation and growth of technology, E-commerce has stretched out tremendously, resulting in a concentration of online travel ticket buying platforms (Noronha & Rao, 2017). Trip.com which was formerly known as Ctrip.com International, Ltd. is also one of the popular platforms in the tourism industry which was established in June 1999 in Shanghai. For starters, the company offers independent business and leisure visitors countrywide travel-related internets services such as hotel booking, air ticketing, and package tours. Moreover, Trip.com (n.d.) promises to provide more than 1.2 million hotels in 200 countries and regions, as well as over 2 million airline itineraries linking more than 5000 cities and claims to provide booking services for flights, hotels, trains, car rentals, airport transfers, excursions, and attraction tickets which highly facilitates the customers with their operation and also demonstrates the importance and significance of the role of apps in promoting the tourism industry.

The Role of Artificial Intelligence (AI) to Attract Tourism Events

The application of artificial intelligence, also known as AI, in the tourism industry and their events have made arrangements related to traveling and hospitality in those events less complicated and untroubled. Artificial intelligence in the tourism industry and its events created enhancement and exceeded the abilities of the traditional search engines and also the interaction between people which also followed the traditional method (Kazak et al., 2020). The application of artificial intelligence has been encouraged due to its bright sides, and is already being used in some travel services which aided to examine enormous amounts of data and grasp the experiences of tourists and as well as themselves which also helps to attain customer orders (Kazak et al., 2020). In recent years, it has been witnessed that in tourism events like the hotel, food, and beverages, meeting, and as well as in agreements artificial intelligence has been applied (Yang et al., 2020) and facilitates the industry by attracting customers with the use of these innovative and creative technologies. A glance into AI in the year 2020 demonstrates that it has already established itself in a range of settings, including business, tourism, and everyday life, despite the fact that they are often disregarded (Tussyadiah & Miller, 2019).The citation "Tussyadiah & Miller, 2018" has been changed to "Tussyadiah and Miller 2019" to match the author name/date in the reference list. Please check if the change is fine.

Artificial intelligence systems, for instance, are used and applied to improve customer service, monitor and report specific company objectives, and assist customers with queries and transactions (Ivanov et al., 2017). Thus, artificial intelligence can be useful in a number of ways to our customers in tourism events as they assist customers with queries and transactions which helps our customers to get the solution to their questions, as well as it attracts them. Moreover, artificial intelligence aids the tourism events to be run with higher efficiency and accuracy as the responses and feedback are being generated by technology which would give the correct output to the problems being faced. Robots are becoming more widespread in the hotel industry as a result of rapid technological advancements. While much of the study has concentrated on the technical and functional elements of robotics, little is known about how robots might be productive members of human service teams. This study analyzes the effects of integrating non-human service providers into human teams with an emphasis on trust, using a human-robot interaction (HRI) perspective (Simon et al., 2020). Artificial intelligence in tourism facilitates portability, social integration, and territorial marketing by implying that smart cities are expanding the tourist and hospitality business in such a manner that tourism, as well as its products and services, would be available to those who are temporarily or permanently paralyzed (Dias et al., 2021). Additionally, AI can boost efficiency and productivity by lowering service costs, improving safety and security, assisting decision-making processes, and assisting visitors in navigating through smart places (Wang et al., 2020). The use of intelligent automation in travel and tourism is likely to grow in the future, thanks to advances in artificial intelligence (AI) and associated technologies (Tussyadiah, 2020).

The Role of Social Networking Sites (SNSs) to Attract Tourism Festivals

The arrival of COVID-19 brought an episode of disaster and loss of lives (Browning et al., 2021). It disrupted several industries to various degree, especially industries that rely on movement and transportation. Tourism and hospitality industry suffered a lot from measure like travel restrictions and newly found health concerns. Here we focus on the tourism sector in the UAE, the challenges it faces and the potential solutions especially those utilizing technology to address those issues for seeking more variety (Hossain et al., 2019a, b) in tourism. It also aims to see the variety of ways technology can facilitate in tourism events, festivals and fairs that will take place in UAE that will take place in the post-pandemic period of COVID-19. To begin with, the hospitality industry of the country needs to bloom and heal from the impact of the COVID-19. For this, there needs to be taken actions as recommended by UAE Hospitality industry. As was said in the paper, authority bodies and financial institutions need to come together to help hoteliers make it through as well as recommending several reforms with the help of SNSs (Hossain, 2019). As for the crisis, support of financial and government institutions are critical to the sectors prosperity in the long run. The Think Tank recommends that the Government provide a 25–50% subsidy for staff accommodation costs and employment permits, as well as waive visa charges as a result of the high, ongoing accommodation and food costs associated with staff who are unable to return home. Taxes, including municipality and VAT should also be temporarily waived, and non-essential security enhancements should also be suspended. Alongside these, it also suggests establishing an official industry body to provide a forum, as well as a unified voice for the benefit of the hospitality sector. This provides for a representative body on the part of all hoteliers in the UAE.

Sustainability can be achieved through the introduction of a national and/or emirate-level hotel development strategy that will manage the supply of hotels and other hospitality assets in a holistic way, while ensuring appropriate assets are available to meet future needs. Another problem is the rise of multiple empty units in hotels due to low demand and social distancing measures, which is draining efficiency. Concerning this, the report advices to introduce a legislation that enables a more flexible use of space, such as combined residential and business units or multipurpose conference facilities during the pandemic. Safety and stability is vital for employees in the industry. This can be ensured by making employment insurance mandatory for all employees to ensure their welfare in case of an emergency. Technology has great promises for boosting tourism as well as revolutionizing it. Such as by introducing new ways to attract tourists by using solutions such as alternative tourism experience through virtual reality and video sightseeing that lets users look around various locations around the world to choose a destination for their vacations, all at the comfort of their home. Online streaming platforms and projects like Google Arts and Culture have let many hidden gems and tourist favorites resurface to newer tourists. That is just the beginning. Besides this, we have services such as trip planners with tools that are quick, customizable as per customer needs and provides great efficiency. Examples of such services includes things that packs an all-in-one solution for exploring destinations, highlighting the experiences that can be found and dynamic maps for easy navigation. Now for some clarification, it needs to be stated technologies implemented to aid tourism during COVID-19 is going to stay post-COVID-19 despite the reasons that the technologies are built around health precautions and alter society in unexpected ways (Nurunnabi et al., 2020).

Technologies such as digital ID's, though originally used for carrying vaccine and immunity passports, are being talked about by government and leaders to implement it for a variety of purposes. As well as this, contactless check-in improves convenience by reducing the amount of time tourists spent in the processes, apart from preventing shared checkpoints which promotes hygiene. Finding accommodation has now been made easier by services which act as a one-stop solution to enable property owners to manage all their rental deals across Booking, Expedia, Airbnb all in one place. Last but not least, technologies are not only meant to be used by customers and travelers, as some like the Internet of Things (IoT) and Big Data helps authorities generate insights from real-time data analytics to develop digital strategies and initiatives to make traveling more efficient and secure.

Various Challenges in the Contemporary Era

Tourism industry has faced numerous challenges in the past, and more profoundly during the COVID-19 pandemic (Baloch et al., 2021). The challenges range from sustainability issues, sanitary measures to those of financial nature. To begin with, The COVID-19 pandemic has brought its own problems (Khoshaim et al., 2020). One such problem includes guaranteeing sanitary measures. This extra demand poses a significant challenge for the tourism sector. Investment of capital is crucial in accomplishing this. Due to loosening of travel restrictions and resumption of mass tourism, the situation needs to be dealt with differently. Materialization of services through different websites and platforms can be a first step, as it will improve convenience and ease in planning trips, booking hotels and tickets etc. Retention of employees has always been a challenge, but the situation worsened during the pandemic. About one-third of workers quit their jobs after working for just 6 months. About 45% of workers manage to hold on for an average of 2 years. Besides securing their positions with backups and insurances in cases of emergency and aids from government funding, referral-based hires is a significant method in finding employees with talent and skill. This also applies to tourist guides, who mostly work independently. They often rely on tips from tourist besides regular work. Hotels are one of the worst-affected sectors amid the COVID-19 pandemic. Due to low occupancy rates, investors are reluctant to invest. Research by McKinsey Institute suggests that recovery to pre-COVID-19 levels (Khoshaim et al., 2020) could take until 2023—or later. Besides challenges posed by the pandemics itself, the industry has faced restrains irrespective of the era, such challenges need to be addressed to propel it to new heights. One such challenge comes in the form of taxes. Governments of the world set heavy taxes in these industry. This is observable through several expenses for tourism. This makes it hard for both tourists and companies. Governments should understand that tourists already contribute to the local economy, so making extra profits from heavy taxes would be detrimental to the welfare of the industry. Travel marketing facer many uncertainties and obstacles along the way. Lack of thought and creativity made a lot of advertising seem false, inadequate and exaggerated.

Marketers can counter this by altering advertisements to be personalized, up to date and relevant. One of the major challenge in tourism is security. The increase in number of people traveling domestically and internationally in the past had bumped up demand for accommodation services, which translates to extra pressure for hotel managers to keep things running smoothly. According to some statistics, maintenance costs constitute 60% of a hotels' total operating costs. Considering these, COVID-19 acts as an additional burden (Khoshaim et al., 2020), given that hotel occupancy rates have hit unprecedented low. Last, but not least, Security is a top concern and one of the most difficult obstacle in the industry. This is primarily due to lack of synergy and connect with law enforcement departments and the ever-changing nature of crime. A designated security infrastructure and personnel trained in tourist protection is vital for the industry.

How to Overcome the Challenges with Tourism Events, Festivals and Fairs?

Events, festivals and fairs especially those of mega scales such as the Olympic Games and The FIFA World Cup accumulate a massive number of tourists. Although tourists can visit the respective city at any time, a limited-time event or events that are only held once in a few years can urge visitors to claim those opportunities and visit the place during the event. For example, although many people are interested in visiting China (Nurunnabi et al., 2020), the time around Chinese New Year celebrations provide unique opportunity to part in the joy by visiting fairs and festivals. This opens tourist to many opportunities to indulge and interact with local communities. For attracting such tourists, event organizers need to use tools like social networks to attract like-minded and interested tourists (Nurunnabi et al., 2020). But use of such social networks should be narrowed to building audience in a few, rather than spreading too thin across a large number of networks. As there is a wide range of platforms to choose from, the best success can be found by surveying the foreign attendees at these events and how they came to know about the event. As we mentioned earlier, about how online streaming platforms helps get vacation destinations exposure and brings in tourists. Besides that, online platforms like social media helps in marketing, by persuading tourists about particular fairs and festivals to a broad audience, thus accumulating a larger audience in the aftermath.

Social media provides many other factors that drive and influence visitors to festival tourism especially in the outbreak situation (Khoshaim et al., 2020). First of all, it lets you to reap the rewards of building social capital and feelings of participation among all. It helps shape consumer behavior, by playing a role in influencing decisions. WOM or as 'word of mouth' as we know is the discussion between customers about products. Online communication has now paved the road for what can now be referred as electronic word of mouth or eWOM. This eWOM takes place through social media, traveler's blogs, and online forums. Despite all, it has its own downsides such as the inability to assess credibility and negative comments from previous visitors de-motivating future travelers. Event organizers can check for this feedback to further improve their service.

Possibility of UAE in Developing Smart Cities in the Future

It is the dream of every country to achieve the status of owning smart cities. It comes hand in hand with giving out better access to the internet and technological tools to citizens of the country. Now technology is being injected more directly into the lives of residents. Smartphones have become the keys to the city, providing millions of people with real-time information on transport, traffic, health services, safety warnings, and community news (McKinsey & Company, 2018).

Creating a smart city does not end at installing a few technological infrastructure and interfaces here and there and ending the responsibility rather training the management and sending out user guidance to the citizens is essentially important. It is in the Vision 2021 target of the UAE to achieve a fully operating smart city and for this, strategies have been laid out since 2014 to include solar power and energy panels to make sure of covering the efficiency and environment targets while achieving a digitized city form. Based on the technological advancements of the fourth industrial revolution, the city is planning to equip federal level strategies that involve the use of block chain, AI, renewable energy and smart mobility to achieve their target (Mansoori, 2020).

The vision behind this mega project was announced by the ministry with the motto written on their Smart Dubai website as stated that due to Dubai's rapid growth in many economic sectors, conventional processes have to be modified on a regular basis to maintain efficiency and speed. Government efficiency has become increasingly important, particularly in consumer and government-to-government (G2G) services (Smart Dubai, n.d.).

The initiatives are: AI Lab, Happiness Agenda, Blockchain, Paperless, AI Principles and Ethics, Data First, Startup Support, Smart Cities Global Network (Smart Dubai, n.d.). In fact, as of 2021, the citizens are enjoying the benefits of dwelling in a smart city. Starting from sufficient internet coverage throughout the city to paperless government services, all have been implemented following a proper framework. To ensure inclusiveness of the smart city the government joined forces with the big houses of the private sectors such as MasterCard, Mena to create a dynamic, sustainable strategic growth. (Holtmeier, 2021). Representatives of Smart Dubai have declared the following statements regarding their progress so far: living in a sustainable city is considered the most interesting innovation in future smart cities by 53% of UAE respondents, and this percentage rises with respondents' age (Holtmeier, 2021). The movement and strategies are going to be further discussed in the Abu Dhabi Smart City Summit which is to be held on 23–24 November, 2021. In this summit the government has requested all the industrial leaders to come forward with their ideas, funding and edge-cutting innovative tools in order to sustain the ongoing plan of building the smart cities.

Post Pandemic Period: A Challenge in the Tourism Industry and Possible Solutions

Modern days tourism is unreachable without the use of internet. Whether it is about looking at the top tourist nations, checking out reviews of hotels or the best tourist spots, the internet is human's initial guide to discover tourism and hospitality spots. Each significant issue in the tourist and hospitality sector has its own website, which may be divided into four categories: holding corporate identification websites, chains, concession and membership websites, individual site and facility websites, and portals and vitals websites. It is clear that the tourist agency's increasing competition in the global tourism industry necessitates extraordinary efforts and investments in promotion, resources, expertise, and quality in order to achieve satisfying growth (Batinić, 2013). Thus, the travel agencies and hospitality must incorporate the latest technologies and have the expertise to make best use of it so they stay ahead of the vast completion that exists globally. Firstly, it is believed that the markets should be researched, for instance in the middle eastern nations use of strata devices and public experiences helped countries like Oman and Jeddah attract the traditional, cultural and resource filled aspects of tourism. In the case of middle Eastern countries, mobile devices are used in every aspect of planning a trip starting from booking trips online, buying tickets and looking up hotels. This gives vast opportunities to the hospitality firms, airlines and tour planners who utilize the devices to connect with their customers and plan their business ideas in accordance to customers' feedback. As the generations are being more tech savvy, there has been a noticeable shift in traveling pattern using which the key players of the traveling and hospitality industry are creating business schemes using which they can attain competitive advantage (Chebib, 2014). It was reported that tourism and hospitality market in UAE is driven by a rise in visitors to both Dubai and Abu Dhabi. It is one of the most active markets in UAE in general and Dubai in particular which welcomes millions of visitors per year and it is expected to grow to 25 million visitors for EXPO2022 (Hashmi & Rose, 2021). Thus it can be forecasted that although the travel industry's performance was disrupted due to global pandemic, but by acquiring the policies mentioned and use of technology to analyses the market conditions, traveling patters could be brought back to normal very soon.

Conclusion

This chapter attempts to explore the role of technology application in tourism events, festivals, and fairs in the The United Arab Emirates (UAE) during the post-pandemic period of COVID-19. The chapter specifically explore a different perspective on various technical Apps based with the latest technology that may affect tourism events, festivals, and fairs. Future work may focus on the global role of upgraded-technology Apps in sustainable tourism development in collaboration with tourism festivals, events, and fairs around the globe. The study identifies how tourists are affected by technology application. It also reveals an increased tourism development opportunity with the collaboration of tourism festivals, events, and fairs. Advancement of the latest technology application in tourism industry is recommended for sustainable tourism development. The perspective of the UAE is brought into the discussion only. Future research may explore the same phenomenon globally with big data or multi-country analysis. The chapter reveals that technology application in tourism festivals, events, and fairs can ensure sustainable tourism development in the UAE, especially in the post-pandemic period of COVID-19.

References

10 times. (n.d.). *Virtual events- Travel and tourism*. Retrieved from https://10times.com/online/travel-tourism. Accessed: 23 Aug 2021.

Arab News. (2020). *Oman pushes local tourism in new campaign amid COVID-19 pandemic*. Retrieved from: https://www.arabnews.com/node/1731216/middle-east. Accessed: 13 Apr 2021.

Arab News. (2021). *Egypt to produce up to 80 million Sinovac vaccine doses annually*. Retrieved from: https://www.arabnews.com/node/1839786/middle-east. Accessed: 24 May 2021.

Baloch, G. M., Sundarasen, S., Chinna, K., Nurunnabi, M., Kamaludin, K., Khoshaim, H. B., & AlSukayt, A. (2021). COVID-19: Exploring impacts of the pandemic and lockdown on mental health of Pakistani students. *PeerJ, 9*, e10612.

Batinić, I. (2013). The role and importance of the internet in contemporary tourism in travel agencies business. *International Journal of Cognitive Research in Science, Engineering and Education, 1*(2), 119–122.

Bayram, G. E. (2020). Impact of information technology on tourism. In A. Hassan & A. Sharma (Eds.), *The emerald handbook of ICT in tourism and hospitality* (pp. 243–258). Emerald Publishing Limited.

Behsudi, A. (2020). Wish you were here. *International Monetary Fund*. Retrieved from: https://www.imf.org/external/pubs/ft/fandd/2020/12/impact-of-the-pandemic-on-tourism-behsudi.htm. Accessed: 21 May 2021.

Benckendorff, P. J., Xiang, Z., & Sheldon, P. J. (2019). *Tourism information technology*. CABI.

Browning, M. H. E. M., Larson, L. R., Sharaievska, I., Rigolon, A., McAnirlin, O., Mullenbach, L., Cloutier, S., Vu, T. M., Thomsen, J., Reigner, N., Metcalf, E. C., D'Antonio, A., Helbich, M., Bratman, G. N., & Alvarez, H. O. (2021). Psychological impacts from COVID-19 among university students: Risk factors across seven states in the United States. *PLoS One, 16*(1), e0245327.

Chebib, K. (2014). The impact of technology on travel and tourism in the Middle East. *Euromonitor International*. Retrieved from: https://blog.euromonitor.com/the-impact-of-technology-on-travel-and-tourism-in-the-middle-east/. Accessed: 28 May 2021.

Cheng, M., & Jin, X. (2019). What do Airbnb users care about? An analysis of online review comments. *International Journal of Hospitality Management, 76*, 58–70.

Crystal Travel. (n.d.). *25 most beautiful fairs, events, and festivals of Dubai*. Retrieved from: https://blog.crystaltravel.co.uk/fairs-and-festivals-in-dubai/. Accessed: 15 Apr 2021.

Dias, A., Santinha, G., Rodrigues, M., Queirós, A., Rodrigues, C., & Rocha, N. P. (2021). Smart cities and accessible tourism: A systematic review. In C. Eusébio, L. Teixeira, & M. J. Carneiro (Eds.), *ICT tools and applications for accessible tourism* (pp. 96–114). IGI Global.

Dubai Online. (2020). *Dubai tourism statistics*. Retrieved from: https://www.dubai-online.com/essential/tourism-statistics/. Accessed: 28 May 2021.

Dubai.com. (n.d.). *Dubai festivals*. Retrieved from: https://www.dubai.com/v/festivals/. Accessed: 13 May 2021.

Edarabia. (2021). *Top 15 places to visit in the UAE*. Retrieved from: https://www.edarabia.com/top-places-visit-uae/#8. Accessed: 30 April 2021.

Eichelberger, S., Peters, M., Pikkemaat, B., & Chan, C. S. (2020). Entrepreneurial ecosystems in smart cities for tourism development: From stakeholder perceptions to regional tourism policy implications. *Journal of Hospitality and Tourism Management, 45*, 319–329.

Expatica. (n.d.). *The best festivals in the UAE you have to visit*. Retrieved from: https://www.expatica.com/ae/lifestyle/holidays/best-festivals-in-the-united-arab-emirates-70925/. Accessed: 13 May 2021.

Fleet Management. (2019). *30 must-visit tourist attractions in UAE*. Retrieved from: https://www.fleetmanagement.ae/blog/tourist-attractions-uae/. Accessed: 12 May 2021.

Guttentag, D. (2019). Progress on Airbnb: A literature review. *Journal of Hospitality and Tourism Technology, 10*(4), 814–844.

Hashmi, S. & Rose, I. (2021). Rescheduled Dubai Expo hopes to attract 25 million visits. *BBC News*. Retrieved from: https://www.bbc.co.uk/news/business-56682427. Accessed: 24 May 2021.

Holtmeier, L. (2021). Majority of UAE residents happy to live in a smart city – study. *Arabian Business*. Retrieved from: https://www.arabianbusiness.com/technology/459455-majority-of-uae-residents-say-they-are-happy-to-live-in-smart-city-study-finds. Accessed: 30 Apr 2021.

Hossain, S. F. A. (2019). Social networking and its role in media entrepreneurship: Evaluating the use of mobile phones in the context of online shopping–a review. *Journal of Media Management and Entrepreneurship (JMME), 1*(1), 73–86.

Hossain, S. F. A., Nurunnabi, M., Hussain, K., & Saha, S. K. (2019a). Effects of variety-seeking intention by mobile phone usage on university students' academic performance. *Cogent Education, 6*(1), 1574692.

Hossain, S. F. A., Shan, X., & Nurunnabi, M. (2019b). Is M-learning a challenge?: Students attitudes toward the sustainable learning and performance. *International Journal of e-Collaboration (IJeC), 15*(1), 21–37.

iExplore. (n.d.). *United Arab Emirates holidays and festivals. Adventure Tourism - Experiential Travel Guides*. Retrieved from: https://www.iexplore.com/articles/travel-guides/middle-east/united-arab-emirates/festivals-and-events. Accessed: 30 Apr 2021.

Ivanov, S. H., Webster, C., & Berezina, K. (2017). Adoption of robots and service automation by tourism and hospitality companies. *Revista Turismo & Desenvolvimento, 27*(28), 1501–1517.

IvyPanda. (2019). *Tourism and economy in UAE*. Retrieved from: https://ivypanda.com/essays/tourism-and-economy-in-uae/. Accessed: 28 May 2021.

Kansakar, P., Munir, A., & Shabani, N. (2019). Technology in the hospitality industry: Prospects and challenges. *IEEE Consumer Electronics Magazine, 8*(3), 60–65.

Kazak, A. N., Chetyrbok, P. V., & Oleinikov, N. N. (2020). Artificial intelligence in the tourism sphere. *IOP Conference Series: Earth and Environmental Science, 421*, 042020.

Kazoba, H., Massawe, M. D., & Msavange, M. (2016). Assessment on the role of internet in tourism marketing: A case study of Mwanza City-Tanzania. *Journal of Tourism, Hospitality and Sports, 22*, 22–44.

Khatri, I. (2019). Information technology in tourism & hospitality industry: A review of ten years' publications. *Journal of Tourism and Hospitality Education, 9*, 74–87.

Khoshaim, H. B., Al-Sukayt, A., Chinna, K., Nurunnabi, M., Sundarasen, S., Kamaludin, K., Baloch, G. M., & Hossain, S. F. A. (2020). Anxiety level of university students during COVID-19 in Saudi Arabia. *Frontiers in Psychiatry, 11*, 1397.

Lee, J. (2021). 14 top-rated tourist attractions in the United Arab Emirates. *PlanetWare*. Retrieved from: https://www.planetware.com/tourist-attractions/united-arab-emirates-uae.htm. Accessed: 30 Apr 2021.

Mansoori, H. A. (2020). From dream to reality: Developing sustainable smart cities in the United Arab Emirates. *MyITU*. Retrieved from: https://www.itu.int/en/myitu/News/2020/08/24/06/46/Smart-sustainable-city-development-UAE. Accessed: 1 July 2021.

McKinsey & Company. (2018). *Smart cities: Digital solutions for a more livable future*. Retrieved from: https://www.mckinsey.com/business-functions/operations/our-insights/smart-cities-digital-solutions-for-a-more-livable-future. Accessed: 30 Apr 2021.

Mellinas, J. P., María-Dolores, S. M. M., & García, J. J. B. (2015). Booking. com: The unexpected scoring system. *Tourism Management, 49*, 72–74.

Noronha, A. K., & Rao, P. S. (2017). Effect of website quality on customer satisfaction and purchase intention in online travel ticket booking websites. *Management, 7*(5), 168–173.

Nurunnabi, M., Hossain, S. F. A., Chinna, K., Sundarasen, S., Khoshaim, H. B., Kamaludin, K., Gul Baloch, M., Sukayt, A., & Shan, X. (2020). Coping strategies of students for anxiety during the COVID-19 pandemic in China: A cross-sectional study. *F1000 Research, 9*, 1115.

Simon, O., Neuhofer, B., & Egger, R. (2020). Human-robot interaction: Conceptualising trust in frontline teams through LEGO® serious play®. *Tourism Management Perspectives, 35*, 100692.

Skoultsos, S. G., & Tsartas, P. (2009). Event tourism: Statement and questions. *Tourismos: An International Multidisciplinary Journal of Tourism, 4*(4), 293–310.

Smart Dubai. (n.d.). *Our initiatives*. Retrieved from: https://www.smartdubai.ae/initiatives. Accessed: 23 Aug 2021.

Staff Reporter. (2021). UAE wealthiest nation in Middle East, Dubai richest city. *Khaleej Times*. Retrieved from: https://khaleejtimes.com/business/uae-wealthiest-nation-in-middle-east-dubai-richest-city. Accessed: 28 May 2021.

Thrillophilia. (2021). *Burj Khalifa, UAE: How to reach, best time & tips*. Retrieved from: https://www.thrillophilia.com/attractions/burj-khalifa. Accessed: 30 Apr 2021.

Travel Triangle. (2021). *10 festivals in UAE that you must attend to experience the joy of the Arabic culture*. Retrieved from: https://traveltriangle.com/blog/festivals-in-united-arab-emirates/. Accessed: 30 Apr 2021.

TripAdvisor. (2021). *Global village (Dubai) - 2021 all you need to know BEFORE you go*. Retrieved from: https://www.tripadvisor.com/Attraction_Review-g295424-d2038312-Reviews-Global_Village-Dubai_Emirate_of_Dubai.html. Accessed: 30 Apr 2021.

Trip.com. (n.d.). *Hotels*. Retrieved from: https://uk.trip.com/hotels/. Accessed: 30 Apr 2021.

Tussyadiah, I. (2020). A review of research into automation in tourism: Launching the annals of tourism research curated collection on artificial intelligence and robotics in tourism. *Annals of Tourism Research, 81*, 102883.

Tussyadiah, I., & Miller, G. (2019). Perceived impacts of artificial intelligence and responses to positive behaviour change intervention. In J. Pesonen & J. Neidhardt (Eds.), *Information and communication technologies in tourism 2019* (pp. 359–370). Springer.

Wang, J., Xie, C., Huang, Q., & Morrison, A. M. (2020). Smart tourism destination experiences: The mediating impact of arousal levels. *Tourism Management Perspectives, 35*, 100707.

Wayne, S. (2016). How smart cities are leading the way to smart tourism. *Hotel Management*. Retrieved from: https://www.hotelmanagement.net/tech/how-smart-cities-are-leading-way-to-smart-tourism. Accessed: 19 Apr 2021.

Yang, L., Henthorne, T. L., & George, B. (2020). Artificial intelligence and robotics technology in the hospitality industry: Current applications and future trends. In B. George & J. Paul (Eds.), *Digital transformation in business and society* (pp. 211–228). Palgrave Macmillan.

Zaidan, E. (2017). Analysis of ICT usage patterns, benefits and barriers in tourism SMEs in the Middle Eastern countries: The case of Dubai in UAE. *Journal of Vacation Marketing, 23*(3), 248–263.

Syed Far Abid Hossain is a PhD researcher at the School of Management of Xi'an Jiaotong University, China. He graduated from the University of Sunderland, London, UK with an MBA degree. He is a Senior Lecturer and MBA Program Coordinator in the College of Business Administration, International University of Business Agriculture and Technology (IUBAT), Dhaka, Bangladesh. He has published numerous papers in indexed and reputed international journals. Also, he has achieved the 'Outstanding International Student Award', Dean's scholarship and Full bright scholarship from the Chinese government. His research interests include women entrepreneurship, innovation, mobile-phone usage, academic performance, social networking, and TPACK. He has participated numerous reputed international conferences in the USA, China, and Bangladesh. He is appointed as a reviewer by famous publishers as Elsevier, Emerald, Wiley, SAGE, Springer, Frontiers, Taylor and Francis and IGI Global.

Faiza Tanaz Ahsan is a Research Assistant at North South University. Her research interests are: Women Empowerment, Edge-cutting technology in business, Artificial Intelligence, bank lending practices, credit management, cloud computing, Budget taxation and public investment, trade and globalization and so on.

Armana Hakim Nadi is a Research Assistant at IUBAT-International University of Business Agriculture and Technology, Dhaka, Bangladesh. She enjoys writing and relishes to play with words. As she pursues her career in teaching, it helped her to engage with the topic involving education and have a superior perspective over it.

Musanna Ahmed is a Research Assistant at IUBAT-International University of Business Agriculture and Technology, Dhaka, Bangladesh. His research interest is in Tourism and Hospitality Science and Word of Mouth marketing.

Hafsa Neyamah is a Research Assistant at IUBAT-International University of Business Agriculture and Technology, Dhaka, Bangladesh. Her research interests are tourism and economic development.

Chapter 20
Technology Usage in Tourism Events, Fairs and Festivals: Samples from Turkey

Aslı Özge Özgen Çiğdemli and Ceren Avcı

Introduction

The diversification of tourism activities and the differentiation of tourism products that may affect tourists' choice of destination are of great importance in terms of the intense competition environment of the tourism sector. The identification of a tourism destination with a specific event is an important element in product differentiation. In this context, event tourism is used as an element of tourism attraction included in systematically developing, planning, and organizing, marketing and promoting events (Tassiopoulus, 2005). The focus of a visit within the framework of event tourism is to attend an event; however, people participating in the event also get the opportunity to benefit from tourism activities. The demand for event tourism leads to the organization of numerous and varied events around the world, in the fields of socio-cultural activities, commerce, politics, arts, entertainment and sports.

Depending on the content, events are classified as "cultural celebrations (festivals, carnivals, commemorations, religious events); political and state (summits, royal occasions, political events, VIP visits); arts and entertainment (concerts, award ceremonies); business and trade (meetings, conventions, consumer and trade shows, fairs, markets); educational and scientific (conferences, seminars, clinics); sport (amateur/professional, spectator/participant); recreational (sport or games for fun) or private events (weddings, parties, socials)" (Getz, 2008: 404). Events are also categorized according to their size as local, regional, large or mega events (Allen et al., 2002; Getz, 2008). Again, Getz (2008) stated that mega events create both high value and high tourism demand. These classifications differ in terms of the

A. Ö. Ö. Çiğdemli (✉)
Department of Recreation Management, Osmaniye Korkut Ata University, Osmaniye, Turkey

C. Avcı
Department of Gastronomy and Culinary Arts, Şırnak University, Sirnak, Turkey

A. Hassan (ed.), *Technology Application in Tourism Fairs, Festivals and Events in Asia*, https://doi.org/10.1007/978-981-16-8070-0_20

visitor and participant audience and the period of the event, and offer the opportunity to evaluate leisure time, experience cultural and social diversity, meet special needs, achieve self-realization, or just relax.

In the 2000s, a new classification, described as "digital events", was added to the agenda, led by technological transformation. Digital events have started to be classified as technology-based events (events designed according to technological tools and technology integral to the event), online events (online virtual fairs, virtual world events, events and meetings on virtual media) and transmediatic events (Seyfi, 2017). Although the participants are passive state in events based on technological tools, it can be said that the activities are interesting and creative. There are few human-induced risks in these events, and they can be used as a viral marketing tool. On the other hand, events designed around technology tools are more passive in events where technology is designed within the event. Online events, led by virtual environments such as Second Life, RuneSpace, or World of Warcraft, create a sense of reality, and are frequently used by companies, corporate brands and event marketers to create virtual worlds and virtual fairs, while meetings with online communication tools can be held without time and place restrictions (Seyfi, 2017).

Considering that events offer opportunities to both the host society and the participants, the technology revolution, which has changed the social and business world, has gained a new dynamism through scientific developments and innovation. The widespread use of smart phones and the increasing speed and coverage area of the internet have been game-changers for hotels, travel businesses, transportation businesses, entertainment businesses, food and beverage businesses, as well as tourism. Technology-based systems are used for virtual and smart events, in the realization of events, smart ticketing systems, sensors and monitoring devices, participant tracking, security and health applications, and the organization of employees. In this section, the importance of the events for tourism, the technological applications used in events, festivals and fairs and the situation of technology usage for events organized in Turkey are discussed.

The Importance of Events, Fairs and Festivals for Tourism

Within the scope of diversification of the touristic product, events as a tourism tool are considered as social and cultural activities to have a good time (McDonnell et al., 1998). Ritchie and Crouch (2005); they may be large activities that are one-off, occurring or repeated within a certain period and are developed to increase short- or long-term awareness and income of a tourist destination. In an attempt to ensure the personal development of individuals and to get away from their daily routines in their spare time (Jago & Shaw, 1998), religious celebrations may be organized, depending on the cultural characteristics of the society, or informal meetings organized for reasons such as birth and death (Ferdinand & Shaw, 2012). Local products or meals, music, movies, cultural and artistic diversions, as well as sporting and scientific activities are included in the events.

Although such events originally emerged for reasons such as holy days, entertainment, celebrations and competitions other than tourism, they have always been an attraction for tourists, from small local fairs to global mega-events such as the Olympics. In this context, the main features that distinguish special events from other attractions that are open to the public are that they have a celebratory purpose and a theme, are organized once a year or less, have a pre-planned opening and closing date, contain one or more activities, and are arranged in a touristic area (Getz, 1989).

Events create a tourism potential in terms of the region where they are held. The image of the event increases the recognition of the touristic region, possibly countering any negative image that exists, offers a different product from competing destinations (Baum et al., 2013), and attracts investors to the destination. Events also spread tourism demand to different regions at different times, increase the amount that visitors spend, and improves the infrastructure in the touristic region, thus increasing trade and optimizing the demand for other attractions and activities in the destination (Getz, 1997). In this context, event tourism, which is considered an alternative type of tourism, is defined as "planning, developing and marketing events as primary or secondary tourist attractions in order to maximize the number of tourists participating in the events" (Getz, 1997: 16).

In addition to the economic and strategic management functions of events for tourism, they have positive social, cultural, psychological, physical and personal effects, such as increasing visitor numbers and expenditure (Stokes, 2006), developing social pride, ensuring the preservation and reinforcement of traditions and values (Henderson et al., 2010), heightening awareness of the destination outside the local community (Ritchie, 1984), or increasing active and passive sports participation of the local people where the event is held (Jago et al., 2010). In this context, event studies, which were previously handled as economic and marketing issues, have begun to be examined from the point of view of the motivations, attitudes, behaviors and satisfaction of the participants (Lockstone-Binney & Ong, 2019).

Apart from positive effects, events can also have negative impacts on the destination and local people. Ritchie (1984) compiled the negative effects of events as an increase in prices in the region where the event is held (economic); environmental destruction and crowding (physical); the commodification of socio-cultural values and the alienation of the event from its original values (social and cultural); misunderstandings in host-guest relations; defensive behaviors (psychological); political exploitation of local people due to daily political concerns; negative image formation as a result of organizing inappropriate activities due to inadequate planning; and the negative attitude of current initiatives towards events (commercial and tourism).

As long as events are planned and organized by professional managers, they are considered as a means of creating demand and a core attraction in terms of tourism (Ritchie & Crouch, 2005), and they have played an active role in tourism development. In this regard, while serving to promote and contribute to image development, events are important recreational and socio-cultural activities for the local people. In addition, events can be essential elements in diversifying and maintaining touristic products, raising awareness in the destination, and improving the infrastructure and environmental quality of the destination.

Technology Applications in Event Tourism

The use of technology is an inherent element in the organization of events. Event organizers use high-tech equipment, such as overhead projectors, slide projectors and video players, while creating multimedia productions utilizing recording, sound and image editing for presentations during the event. Digital transformation of computer-sourced text, graphics and videos, microphones, speakers, video conferencing, soundtracks and visuals are important technical aspects of the events (Shone & Parry, 2010). Internet and technological transformations affect event planning and management in terms of participation, awareness, marketing, evaluation, communication and networking.

In the management and marketing of events, it is possible to obtain information that is easily understood, quickly accessible and sufficient for the global demands of potential participants and stakeholders. Influencing and measuring the behavior of attendees and visitors, in particular, can be realized with smart, technology-based applications. From the start of these digital possibilities, cloud computing technologies are used at every stage for the acquisition, storage and analysis of big data before, during and after the event. In this framework, information about the type of event, event packages, employees and customers is stored in a way that creates an entire database of event details (Pinjari & Nur, 2020). In addition, web-based access creates many advantages in terms of providing temporal and spatial flexibility in the content of events. Bronson (2018) stated that mobile applications, social media, e-mail marketing, marketing automation (marketing clouds, automated systems that track the time potential event participants spend on the site, smart advertising strategies, software systems used for event campaigns, etc.) and live broadcasting used in events, facilitate the interaction and communication of the planners, and provide cost and time control for event planning. Better feedback can be obtained from the participants through social media that offers a broad database and creates a quality network, making social media a good promotion tool in the marketing of events. PhotoShooter, Instawall, Tweetwall, SocialBreak and Messenger are some examples of technology that can be used for events, festivals and fairs.

The most important of the technological innovations used to manage event events are mobile applications. Mobile applications provide advantages for event participants for information that can be stored in the cloud without the need for a physical ticket, preventing waste of time in ticket sales at the box office, and ensuring participation in events with QR codes and barcodes. Various games can be organized during the event with QR codes, and videos, visual and unlimited content can be provided to contribute to the event. Mobile applications also contribute to the modernization of events with gamification in the organization, finding the participant's location, benefiting from navigation, individualized information and welcome opportunities, social networks, demo videos during the event, and targeted messages for participant and group segments (Ziyadin et al., 2019). Artificial intelligence is another important technology tool used in the planning, implementation and management of events, fairs and festivals. Artificial intelligence, especially when evaluated in terms of tourism events, provides an opportunity to develop events by properly utilizing the data of the participants, with its capabilities to distinguish voices and faces, facilitate service delivery and improve visitor experience (Neuhofer et al., 2020). For example, booking.com plans to use machine learning and artificial intelligence tools to provide personalized recommendations to visitors about events. Marriott Hotels also offer event reservations and sales with virtual reality tours (Statista, 2020).

The Internet of Things (IoT) is another technology tool used in the management of events. IoT serves to meet the needs of event participants, such as determining their location in the event area, identifying crowded areas, finding a parking place, accessing health services, monitoring points of traffic density and crowding, choosing the most suitable public transportation vehicle, or choosing an appropriate restaurant or entertainment venue. From the event manager's point of view, it is critical to minimize damage to the ecosystem by means of environmental sensors, to provide security measures with CCTV and monitoring monitors, to direct employees to points where there is a concentration of participants, and to determine the participant profile (Threlfall, 2017; Wise & Heidari, 2019). The effective use of technology is among the basic elements that determine the performance of both the event itself and the event managers, organizers and planners (Shone & Parry, 2010) (Table 20.1).

Other special digital tools used for events are Tweetwall, QR code, Digital Signage, PhotoShooter, Video Mapping, wearable devices, digital screens, radio frequency systems, digital scene design, Adobe Connect Pro (Seyfi, 2017). Tweetwall is a web-based application developed specifically for use in events. By means of the application, tweets sent using certain hashtags and keywords during the event can be reflected on the screens with live moderation and a visualization specially designed for the event. The system is used through TweetWallPro.com, and in addition to features such as keyword and list definition, design customization, filtering and management, detailed reporting, there is also sponsor promotion during the tweet stream. The system can be shaped according to needs, in which different activities such as voting and competition can be carried out (SosyalMedya.Co, 2021). Unlimited information, messages, videos and all kinds of content can be provided in limited areas with QR code (Onat, 2016). Since QR code can be used together with different applications, such as social media and Google Maps, it provides ease of use in events. Digital Signage is a digital format which combines hardware, software and content, displayed on the screen, including text, images, animations, videos and sound. This application can easily attract the attention of many participants (Moipro, 2020). PhotoShooter is an application that allows photos of the participants with their logos, taken by teams that shoot mobile during the event, to be integrated with other digital communication tools (Seyfi, 2017). Video mapping is a projection technique used to give an impression of the natural texture

Table 20.1 Technology usage in tourism events, festivals and fairs

	Technology tool				
Event Stages	Mobile apps	Web	Social media	Artificial intelligence (AI)	Internet of Things
Before Events (Pre stage)	Digital ticketing, Data Collection via Software	e-mail info	Instagram Stories	Chatbots	Social mobile crowd sensing
		Tourism and/or destination and/ or event's own websites	Blogs	Microchips	Smartphone tracking monitors
				Virtual Assistants	CCTV cameras
				Self-Driving Transport	Local weather forecasts
				Biometric Recognition	Environmental sensors
				VR tours/AR	
During Events	Digital admission, Data Collection via Software		Instagram	Access Gates	CCTV Monitoring attached with AI
	Social networks		Instawall	Wearable Devices	
			Tweetwall Stories with snapshots	Holograms	
			GIFs	Cameras	
			Blogs	Service Robots	
			Photographs accompanied by tweets	Drones	
			Video-Demos	Live Broadcast	
After Events (Post stage)	Data Collection via Software,	Podcasts	Blogs	Virtual Maps	
			Storytelling	Info Boards	

Source: AI column adapted from Neuhofer et al., 2020; other columns adapted by authors

of objects on a screen, using spatially augmented virtual reality. These objects can be buildings, small indoor objects or landscapes, even theater stages (Haberler.Com, 2020). Adobe Connect Pro provides virtual meeting and training software that allows individuals in different locations to communicate with each other over the Internet (Seyfi, 2017).

Another application developed for events is a mobile guide. Mobile guides aim to provide assistance during a trip by providing tourists with information about the attractions in their location, and services related to transportation to facilitate mobility, using a smart area infrastructure (Smirnov et al., 2013). In addition to providing great advantages in terms of increasing visitor demand and experience, informing, reducing waiting time, efficient planning of personnel, and providing security, digital technologies also provide an immediate response to health problems at events. Thanks to the monitoring of the heartbeat and blood sugar of visitors and employees with mobile applications and alarm systems, health measures can be taken with immediate interventions (Threlfall, 2017).

Technology Applications in Events, Fairs and Festivals Organized in Turkey

Social media and other technological platforms are an indispensable marketing area for events. Technology applications create great advantages in terms of time and cost, reaching a large number of audiences in a short time, without space and time restrictions. They can attract mass attention, using hashtags to bring together people who share the same interests, and providing a flexible operation area for event planners. Social media-based basic digital applications in the management of events are QR codes, RFID (Radio Frequency Identification), Tweetwall (Twitter social networking wall), 360 Selfie, Presstagram, SocialBreak and Social Goal applications (Akay, 2014).

In terms of festivals, events and fairs, Turkey has a wide coverage locally, nationally and internationally (see details also on Turkish Culture Portal, 2021). Culture, films, arts, sports, opera and ballet, competition, music, gastronomy, entertainment and many other events are organized in Turkey. Music festivals that include both local and international jazz, pop, classical, rock, dance, hip hop, Latin or traditional tunes are held. Many sports activities, such as football, basketball, volleyball, tennis, athletics and other outdoor sports, are held throughout the year. Film and theater festivals, ballet performances and book fairs are held, too (GoTurkeyTourism, n.d.). Various activities are organized by the Istanbul Foundation for Culture and Arts (İKSV) within the scope of film, music, design, and theater, and İKSV organizes film screenings called the "FilmEkimi" every year (Istanbul Foundation for Culture and Arts, 2021a). There are also tasting workshops, member presentations or talks as part of İKSV events (Istanbul Foundation for Culture and Arts, 2021b).

On examining the Turkey Festival List, it is seen that music and film festivals take place intensively. In addition to these, there are events and festivals on literature, puppetry and theater. Within the scope of tourism, activities carried out based on provinces and bearing the names of provinces stand out. In addition to these, it is seen that digitalization comes to the fore in tourism festivals, as elsewhere around the World. Within the scope of digital tourism festivals, digital training programs and workshops are organized for those operating in the sector (Web Marketing Festival, n.d.).

Just as the Walt Disney Company has started a Magic Band application that can be connected to the infrastructure of the theme park to reduce waiting times, there is a smart wristband application in Turkey, especially at music festivals. In this way, participant safety can be ensured (GlobalData, 2018). Another application similar to the MagicBand application is the Beacons technology, which is also a geolocation system. In this way, it can be determined which areas of the event location are visited the most or which stands are attracting more attention. The participation of robots in events as translators is another example of technology in use (Erdem, 2021). Spatial-temporal behaviors of participants can also be analyzed with smartphone applications at various festivals (Bustard et al., 2019). All these applications are still in the preparatory stage in Turkey. However, one widely used application is the smart card application; this enables event participants to spend by loading the desired amount into the card.

Mobile event applications are another type of application that finds widespread use in Turkey. With these, participants can be kept up to date with current event information by instant notifications and can obtain event tickets through the application. IKSV Mobil and MyBilet are examples of these applications (Erdem, 2021). Along with these, the Biletix application, which has the largest event database in Turkey, enables people to purchase tickets, obtain information about events within 15 km of their location via GPS, learn about events in the venue, get 360-degree views of popular event venues, and choose the most ideal place in the event area.

It is also possible to choose events by their features. Mobile applications that provide information about culture, art, cinema, theater, exhibition, stand-up and sports events and competitions include Culture & Art Events, Istanbul. net.tr, Arts & Culture, Ankara Event, TimeOut Istanbul, What Do I Do, Fair, Congress, and Conference.

Another application is KonserV, which has developed a reward system in return for the tickets purchased and the events attended, and awards points for reading barcodes through the camera of the mobile app (Öncel, 2016). There is a Cultural Activities application in the "Mobile Applications" section of the e-Ankara Applications tab of the Metropolitan Municipality in Ankara, the capital of Turkey, with links to theater, children's theater, cinema, concerts, festivals & events, seminar conversation, sports, exhibitions, tourism, travel guides, event calendars, centers, while announcements of upcoming events can be seen at the top. The application also allows users to create a calendar for any clicked museum trip and to share it with other users (Duran & Uygur, 2019). With the myAntalya application, implemented by the Antalya Metropolitan Municipality, NFC tags and QR codes, enable the tourist to determine activities in the area (Çelik & Topsakal, 2017).

At the opening of the EXPO 2016 Fair held in Antalya on the national Children's Festival holiday, 23 April, the formation of the cosmos was shown and, with the help of another projection, reflected on the Children's Island. The seasons around the world and in Turkey were shown with kaleidoscopes and a video mapping show (HaberTurk, 2016). Hosting many different summits and meetings after the fair, the EXPO 2016 Area continues to contribute to tourism and benefits from social media opportunities (Expo Antalya, 2021).

Another example of a technology-based event is the Crystal Apple Creativity Festival. The festival is considered one of the biggest events in Turkey, and is used by marketing and advertising experts. Transmediatic storytelling and multiple intelligence applications were used in the festival, where new technological tools were tested. Other examples are Cosmic Yoga Women's Meeting and the Virtual Fair Application, which creates the opportunity to talk live with the students by giving the university's own promotional materials to the participants from the digital environment (Seyfi, 2017).

Conclusion

The COVID-19 pandemic has brought forward the digitalization process that is expected to take place in the coming years, and in a sense, has given us an idea of what the world of the future will be like (Çelikiz, 2021). Before the pandemic, festivals and events used mobile applications, digital ticketing, collected information through software, organized websites or podcasts about events, exchanged views or shared experiences through social media; in short, mobile technologies and web technologies. Now it seems that Artificial Intelligence may be added to the list.

The IoT refers to the set of Internet-connected devices and is intended to further expand digital data. The use of artificial intelligence, on the other hand, is focused on developing insights and predicting behavior, using algorithms in big data. In this process, brands that have digital channels quickly understand customer preferences in depth, prepare digital experiences and improve the services they offer with the feedback they receive (Cançelik, 2021). Since 2010, augmented reality applications, virtual reality, mobile applications, cloud computing and online data storage, wearable technologies, social media, Google analytics, review sites, collaborative online environments, web 2.0 technology and chatbots have also been available in the tourism sector (Ülkü & Göktaş, 2021). In festivals or events, location-based technologies, mobile applications, smart ticket applications and social media applications are frequently used. It is expected that Turkey is also preparing for technology applications such as artificial intelligence and the IoT and will include these applications in service marketing in the future, despite various concerns.

Considering the events and festivals organized by Turkey, smart ticket applications and location-based applications can determine where there are dense accumulations of participants, how much time is spent in which area, and can enable organization and spatial arrangements to be made accordingly. Especially in province-based events and festivals, people and events that have an important place in that province can be presented to participants with a hologram application. Robot technology can be used in all areas where service delivery takes place, as well as before and during the organization of events and festivals. With wearable smart devices, it may be possible for potential participants to pre-evaluate an event or festival event to ensure their participation.

The main issue in the use of all these technologies is to get them into the hands of individuals and institutions, with the awareness and skills to ensure that the technologies are used in accordance with their purpose. If this does not happen, history will repeat itself and it will become very possible for the world to be plunged into a dystopia under the control of technology, just like the atomic bombing disasters in Hiroshima and Nagasaki.

References

Akay, R. A. (2014). Etkinlik yönetimi uygulamalarında yaratıcı rekabet ve sosyal medyanın entegrasyonu. *The Turkish Online Journal of Design, Art and Communication, 4*(4), 55–70.

Allen, J., O'Toole, W., Harris, R., & McDonnell, I. (2002). *Festival and special event management*. Wiley.

Baum, T., Lockstone-Binney, L., & Robertson, M. (2013). Event studies: Finding fool's gold at the rainbow's end? *International Journal of Event and Festival Management, 4*(3), 179–185.

Bronson, R. (2018). The role of technologies in event management. *StartUsmagazine*. Retrieved from: https://magazine.startus.cc/role-technologies-event-management/. Accessed: 27 July 2021.

Bustard, J. R. T., Bolan, P., Devine, A., & Hutchinson, K. (2019). The emerging smart event experience: An interpretative phenomenological analysis. *Tourism Review, 74*(1), 116–128.

Cançelik, M. (2021). Dijital Dönemde İşletmecilik. In A. Ülkü (Ed.), *içinde Dijital Dönüşüm ve Turizm* (pp. 1–25). Nobel Yayınları.

Çelik, P., & Topsakal, Y. (2017). Akıllı Turizm Destinasyonları: Antalya Destinasyonunun Akıllı Turizm Uygulamalarının İncelenmesi. *Journal of Travel and Hospitality Management, 14*(3), 149–166.

Çelikiz, E. (2021). Dijital Çağa Geçiş ve İletişim. In A. Ülkü (Ed.), *içinde Dijital Dönüşüm ve Turizm* (pp. 26–50). Nobel Yayınları.

Duran, G., & Uygur, S. M. (2019). Akıllı Turizm Destinasyonları Kapsamında Ankara'nın Akıllı Turizm Uygulamalarına Yönelik Bir Araştırma. *3. Uluslararası Turizmin Geleceği Kongresi: İnovasyon, Girişimcilik ve Sürdürebilirlik Kongresi*.

Erdem, A. (2021). Akıllı Etkinlikler ve Turizm. In A. Ülkü (Ed.), *içinde Dijital Dönüşüm ve Turizm* (pp. 217–230). Nobel Yayınları.

Expo Antalya. (2021). Retrieved from: https://www.facebook.com/expo2016?ref=br_rs. Accessed: 25 Aug 2021.

Ferdinand, N., & Shaw, S. J. (2012). Events in our changing world. In N. Ferdinand & P. Kitchin (Eds.), *Events management: An international approach* (pp. 5–22). Sage Publications.

Getz, D. (1989). Special events. *Tourism Management, 10*(2), 125–137.

Getz, D. (1997). *Event management & event tourism*. Cognizant Communication Corporation.

Getz, D. (2008). Event tourism: Definition, evolution, and research. *Tourism Management, 29*(3), 403–428.

GlobalData. (2018). *Top 6 technology trends to watch out for in the travel and tourism industry in 2018*. Retrieved from: https://www.globaldata.com/top-6-technology-trends-watch-travel-tourism-industry-2018/. Accessed: 8 July 2021.

GoTurkeyTourism. (n.d.). *Events and festivals*. Retrieved from: https://www.goturkeytourism.com/tr/festivaller-etkinlikler/turkiyede-festivaller-ve-etkinlikler.html#. Accessed: 7 July 2021.

Haberler.Com. (2020). *Mapping ne demek? Mapping gösterisi nedir, ne anlama gelir?* Retrieved from: https://www.haberler.com/mapping-ne-demek-mapping-gosterisi-nedir-ne-13651742-haberi/. Accessed: 26 Aug 2021.

HaberTürk. (2016). *Expo 2016 Çocuk Adası'na Video Mapping Uygulaması*. Retrieved from: https://www.haberturk.com/yerel-haberler/haber/43792921-expo-2016-cocuk-adasina-video-mapping-uygulamasi. Accessed: 25 Aug 2021.

Henderson, J. C., Foo, K., Lim, H., & Yip, S. (2010). Sports events and tourism: The Singapore Formula One Grand Prix. *International Journal of Event and Festival Management, 1*(1), 60–73.

Istanbul Foundation for Culture and Arts. (İKSV). (2021a). *Istanbul film festival*. Retrieved from: https://film.iksv.org/tr. Accessed: 7 July 2021.

Istanbul Foundation for Culture and Arts. (İKSV). (2021b). *Lale Kart*. Retrieved from: https://lalekart.iksv.org/. Accessed: 7 July 2021.

Jago, K. L., & Shaw, N. R. (1998). Special events: A conceptual and definitional framework. *Festival Management and Event Tourism, 5*(1), 21–32.

Jago, L., Dwyer, L., Lipman, G., Lill, D., & Vorster, S. (2010). Optimising the potential of mega-events: An overview. *International Journal of Event and Festival Management, 1*(3), 220–237.

Lockstone-Binney, L., & Ong, F. (2019). Event studies: Progression and future in the field. In S. Beeton & A. Morrison (Eds.), *The study of food, tourism, hospitality, and events, tourism, hospitality & event management* (pp. 37–46). Springer Nature.

McDonnell, I., Johnny, A., & O'Toole, W. (1998). *Festival and special event management*. Wiley.

Moipro. (2020). *Digital Signage Nedir?* Retrieved from: https://www.moipro.com.tr/digital-signage-nedir/. Accessed: 26 Aug 2021.

Neuhofer, B., Magnus, B., & Celuch, K. (2020). The impact of artificial intelligence on event experiences: A scenario technique approach. *Electronic Markets*. https://doi.org/10.1007/s12525-020-00433-4

Onat, K. (2016). *QR Kod nedir? Ne İşe Yarar? Nerelerde Kullanılır? Nasıl Taranır?* Retrieved from: https://www.webtekno.com/mobil/qr-code-tarama-h11806.html. Accessed: 26 Aug 2021.

Öncel, Ü. (2016). *Şehirdeki etkinlikleri kaçırmamanız için en iyi 12 etkinlik uygulaması (10 May 2016)*. Retrieved from: https://webrazzi.com/2016/05/10/en-iyi-etkinlik-uygulamalari/. Accessed: 25 Aug 2021.

Pinjari, K., & Nur, K. (2020). Smart event management system. *International Journal of Computer Science Trends and Technology, 4*(2), 161–164.

Ritchie, J. R. B. (1984). Assessing the impact of hallmark events: Conceptual and research issues. *Journal of Travel Research, 23*(1), 2–11.

Ritchie, J. R., & Crouch, G. I. (2005). *The competitive destination, a sustainable tourism perspective*. CABI Publishing.

Seyfi, M. (2017). *Halkla İlişkiler Perspektifinde Dijital Etkinlik Yönetimi*. Der Yayınları.

Shone, A., & Parry, B. (2010). *Successful Event Management* (3rd ed.). Cengage Learning.

Smirnov, A., Kashevnik, A., Balandin, S. I., & Laizane, S. (2013). Intelligent mobile tourist guide. In S. Balandin, S. Andreev, & Y. Koucheryavy (Eds.), *Internet of things, smart spaces, and next generation networking* (pp. 94–106). Springer.

SosyalMedya.Co. (2021). *TweetWall Pro: Etkinlikler İçin Twitter Duvarı Uygulaması*. Retrieved from: https://sosyalmedya.co/tweetwall-pro/. Accessed: 26 Aug 2021.

Statista. (2020). *Travel & tourism report: Mobility market outlook, October 2020*. Retrieved from: https://www.statista.com/global-consumer-survey. Accessed: 31 July 2021.

Stokes, R. (2006). Network-based strategy making for events tourism. *European Journal of Marketing, 40*(5/6), 682–695.

Tassiopoulus, D. (2005). *Event management: A professional and developmental approach*. Juta Academic.

Threlfall, R. (2017). Smart infrastructure: Managing major events. *KPMG*. Retrieved from: https://assets.kpmg/content/dam/kpmg/uk/pdf/2017/11/smart-infrastructure-managing-major-events.pdf. Accessed: 20 July 2021.

Türkiye Kültür Portalı. (2021). *Etkinlikler*. Retrieved from: https://www.kulturportali.gov.tr/turkiye/genel/etkinlik?ilId=0&keyword=&kategori=0&nitelik=0&gorsel=0&nearest=0&etiket=&hariciEtiket=&sayfa=2&lat=0&lang=0&baslangic=&bitis=&clickDate. Accessed: 26 Aug 2021.

Ülkü, A., & Göktaş, L. S. (2021). Dijital Dönüşüm ve Turizm. In A. Ülkü (Ed.), *içinde Dijital Dönüşüm ve Turizm* (pp. 51–71). Nobel Yayınları.

Web Marketing Festival. (n.d.). *Digital tourism festival*. Retrieved from: https://www.webmarketingfestival.com/digital-tourism-festival/. Accessed: 7 July 2021.

Wise, N., & Heidari, H. (2019). Developing smart tourism destinations with the internet of things. In M. Sigala, R. Rahimi, & M. Thelwall (Eds.), *Big data and innovation in tourism, travel, and hospitality–managerial approaches, techniques, and applications* (pp. 21–29). Springer.

Ziyadin, S., Koryagina, E., Grigoryan, T., Tovma, N., & Ismail, G. Z. (2019). Specificity of using information technologies in the digital transformation of event tourism. *International Journal of Civil Engineering and Technology, 10*(1), 998–1010.

Assistant Professor Dr Aslı Özge Özgen Çiğdemli graduated from Ankara University, Faculty of Political Sciences, Department of International Relations in 2006. She completed a master's degree in International Relations in 2010 at Süleyman Demirel University, with a thesis entitled "European Union Immigration Policies: Historical Background and Current Situation". She took her PhD degree in 2016 from the Hospitality and Tourism Management Department of the Akdeniz University; her dissertation was "The Comparative Analysis of Mediterranean Cruise Destinations: The Position of Turkey". Dr. Çiğdemli worked at the Republic of Turkey Ministry of Trade between 2009 and 2017 and became a faculty member at the School of Tourism and Hotel Management of Şırnak University in 2017, before moving to Osmaniye Korkut Ata University in 2020 as an Assistant Professor in the Faculty of Kadirli Applied Science, Department of Recreation Management. Her main areas of interest are tourism marketing, tourism policy, cruise tourism, technology and tourism, consumer behavior in tourism and strategic marketing management.

Dr Ceren Avcı completed her undergraduate education at Balıkesir University School of Tourism and Hotel Management in 2008. She completed her Master's thesis and doctorate dissertation on Tourism Management at Gazi University, Faculty of Tourism. The subject of her Master's thesis is occupational health and safety and the subject of the doctoral dissertation is on spatial relations in urban tourism. Avcı worked as a research assistant at Gazi University, Faculty of Tourism, Department of Tourism Management between 2012 and 2017. She has been working at Şırnak University since 2017. Her academic interests are tourism philosophy, tourism psychology and tourism sociology.

Part VII
Challenges and Solutions

Chapter 21
Tourism Events, Festivals and Digital Technology Applications in Asia: Socio-Cultural Drawbacks and Ways to Overcome

H. M. Kamrul Hassan and Mohammed Shahedul Quader

Introduction

In the last few decades, it has been observed that rapid and huge development in digital technology brought revolutionary changes in the tourism and hospitality industry around the globe. In fact, such change has created a new widespread dimension. As a result, the tourism sector has been established as the greatest category of products and services sold continuously through the online system (Abou-Shouk et al., 2013; Navío-Marco et al., 2018). Researchers in tourism and hospitality management firmly acknowledged that introducing advanced digital technology into the tourism industry brought diverse effects in the economy (Buhalis, 2020; Navío-Marco et al., 2018; Willis et al., 2017). However, very few sporadic researchers have identified the negative consequences of the widespread digital technology within tourism sectors (Li et al., 2018; Tribe & Mkono, 2017). To meet the demand in the competitive markets, industry players envision a completely different scenario by focusing on sophisticated digital technology-driven tourism services to expand the business (Xu et al., 2019). Nowadays, international tourists expect a complete integrated tourism package that helps them understand the required facilities and support systems available at a destination. Ensuring such an integrated service system requires a solid understanding and synthesising of various tourists' behavior trends and expectations along with digital technology-based infrastructure, reengineering management system and synergetic stakeholder collaboration (McCabe et al., 2012).

Recently, the overall tourist numbers around the globe have grown significantly, where Asia and the Pacific have played a pivotal role in the recent rise of tourism, accounting for 37% of worldwide travel expenditures and about one-fourth of global arrivals (UNWTO/GTREC, 2018). Asia Pacific destinations represent nearly half of

H. M. K. Hassan (✉) · M. S. Quader
Department of Marketing, University of Chittagong, Chattogram, Bangladesh
e-mail: kamrul@cu.ac.bd

A. Hassan (ed.), *Technology Application in Tourism Fairs, Festivals and Events in Asia*, https://doi.org/10.1007/978-981-16-8070-0_21

the world's top 10 most visited cities, according to the MasterCard Global Destination Cities Index (GDCI) 2019 (MasterCard, 2019). A variety of tourism-based cultural festivals and events are held across the Asian continent annually, which provide tourists with a greater understanding of the country's unique cultural and spiritual heritage. Each nation in Asia has its own distinctive, unique festivals and events that reveal a lot about the traditional rituals, beliefs and values that give a wonderful glimpse into the peoples' customs and standard of living. Currently, in Asia, diverse forms of digital technology have been embraced by different cultures, resulting in people adopting innovative tourism approaches (Ooi, 2019). Technological advancement becomes a crucial factor in global cultural convergence, pushing both Eastern and Western cultures towards a single set of values and gradually declining the cultural distinctions between the East and West (Salehan et al., 2018). With the advent of digital technologies, conventional modes of travel planning and experience have altered and reshaped digital tourism and how tourist companies manage their businesses (Liang et al., 2017; Navío-Marco et al., 2018). Due to its increased dominance in the e-commerce industry, Chinese e-tourism is regarded as one of the largest marketplaces in the world (McKinsey & Company, 2016). Hence, Asian tourism events and festivals are changing their way of organizing, commemoration, and exposure with the evolution of digital technology.

In the recent situation, significant attention has been grown in tourism focusing culture as creativity has been integrated as a critical instrument for implementing innovative ideas (Long & Morpeth, 2016). The growing convergence of tourism and creativity has been facilitated by the quest for alternative tourism development models. Incorporating creativity with tourism has become a popular diversification strategy, particularly cultural tourism (Getz, 2008). A variety of events, festivals, restorations of ancient structures that encourage visitors to interact with attractions through joining tourism fairs, events, and exhibitions have become mainstream in the current tourism industry (Richards, 2020). Despite growing recognition, in tourism, like in other disciplines of study, acknowledging the possible negative implications of digital technology is still not well established (Egger et al., 2020).

Moreover, a significant increase in discussions of digital-free tourism has occurred over the previous decade, both in traditional media as well as in the digital space that includes webpages and social media (Li et al., 2018; Pawłowska-Legwand & Matoga, 2020). Exploring the detrimental effects of digital technology on tourism events and festivals has led to consider this study. To address this gap, this qualitative research investigated the drawbacks of digital technology on tourism events and festivals and addressed the possible alternative solutions to overcome new challenges related to digital technology. The findings will provide valuable insights into the negative aspects of digital technology in tourist events and festivals and set the groundwork for further research. Practitioners can understand how digital technology might have a negative influence on the growth of tourism events and festivals.

In this research, a qualitative approach was employed to undertake well-defined research objectives and to gain a detailed knowledge of the drawbacks of digital technology on tourist events and festivals. By nature, this represents a desk-based research study, and from the perspective of research philosophy, the Interpretivism method has been given emphasis (Goldkuhl, 2012). Quantitative data is challenging to handle in a systematic manner due to the extent and imperative nature of the issue (Rahman, 2016). Due to the inherent complexity of the topic, this research study considered academic literature instead of conducting primary research. Therefore, the most cost-effective way to synthesize and analyze findings from various data analyses is through desk-based research (Lewis-Beck et al., 2012). For systematic analysis and comprehensive interpretation on the critical concerns regarding drawbacks of digital technologies in tourism events and festivals, relevant valid secondary data have been considered. Using Scopus and Google Scholar, the methodology reviewed the pertinent papers that are identified and accessible. Additional reference materials were retrieved by accessing the following library resources: Springer, ScienceDirect, Emerald, and JSTOR. Search queries covered words including "impact of digital technology" along with related keywords like "tourist events" and "tourism festivals". Findings and analyses from the included studies were summarized, organized and depicted in a presentable manner.

Tourism Events

Tourism events have been recognized as an efficient way for destinations to boost up tourism (Getz, 2008; Kelly & Fairley, 2018). Tourism events are often focused on a single topic or issue and activities geared for professionals, researchers, trade organizations, and related specific groups that are deliberately formed to increase the popularity of a tourist place (Lu et al., 2020). Tourism has been significantly driven by events that play a pivotal role in the development and marketing strategies of the majority of locations. The functions and effects of organized events in tourism are widely documented, and their relevance for destination competitiveness is evolving. Even though the majority of events have significant tourism potential, small get-togethers, weddings, reunions, celebrations etc. events are also generating tourists' attention in the current situation more than ever before. In addition, most major cities now have top-notch convention and exhibition spaces for events like meetings, trade shows, conferences, conventions, and exhibitions, alongside related marketing and event organizing teams to promote tourism (Getz, 2008).

Additionally, the sports event-focused tourism market has expanded rapidly since 2000. FIFA World Cup and the Olympics both serve as mega-sports events that foster sport tourism development (Pranić et al., 2012). Qatar, one of the world's smallest countries by population, is anticipating a record 1.5 million tourists for the 2022 FIFA World Cup (The Peninsula, 2018). More active kinds of sport involvement that necessitate travel include the confluence of sport management and sport studies with tourism. Innovative marketing opportunities have been developed by integrating digital technology, which has made it possible for tourist organizations to reach many consumers through tourism events. This sort of digital technology could be accessed by anybody across the globe or even continents away, providing them with the ability to have a complete and authentic presentation of such tourism events. While events are obviously more exciting to visitors and the media, they create many extrinsic and intrinsic adverse effects that are frequently overlooked when event planners evaluate their accomplishments (Dawson & Jöns, 2018; Perić, 2018). Focusing on the alternative ways to mitigate negative consequences towards the society and environment has become a priority issue for tourism events since the local community members perceive these events as tourist developments that affect both the social and natural environment as dangerous (Ahmed, 2017; Collins et al., 2009).

Tourism Festivals

Festivals have become a very important part of the tourism industry. Cultural heritage, sociology, anthropology and historical events influence festivals in the socio-economic atmosphere around the world (Dychkovskyy & Ivanov, 2020). Festivals can be considered mono or multi-ethnic, or maybe even mono or multi-artistic in nature, serving as a means to identify inhabitants in a community by reflecting its vibe or a proclamation about its presence and traditions (Saleh & Ryan, 1993). The meaningful concept of the festival is explained and evaluated by various academic scholars from multidimensional contexts (Jang & Feng, 2007; Girish & Chen, 2017; Vajirakachorn & Chongwatpol, 2017). The festival is considered a mass theatrical festive situation that demonstrates certain types of achievements at different public places in representing the value of the cultural phenomenon of various regions or ethnic groups (Dychkovskyy & Ivanov, 2020). In addition, festivals can spark powerful and enriching experiences for tourism by facilitating the joint celebration of a single event with like-minded individuals for a limited time. Through festivals, tourism may inspire communities to maintain or restore cultural traditions and related opportunities that increase awareness and acceptance of other cultures, customs, and ways of living. The use of digital and online media may substantially raise the value of the festival business by supporting to preserve, growth, and promote festivals (Zou et al., 2021). As people from various ethnic and socio-economic backgrounds attend festivals, each generation may have distinct responses to the digital technology that may lead to a unique effect on attitudes, preferences, and cultural practices (Llopis-Amorós et al., 2019). People are well aware that festivals may significantly influence society, community, business, and culture, especially when they are created to attract more tourists and help improve the overall image of a place (Jang & Feng, 2007; Vajirakachorn & Chongwatpol, 2017). Despite its positive effects and the tremendous favorable economic impact, many issues significantly affect the local community in a negative way, including excessive noise, mob violence, theft, traffic, parking troubles, trash, and environmental pollution (Moisescu et al., 2019).

Strategic Framework

Diverse functions of events and festivals business take place within a very dynamic environment. Forces in the macro-environment (political, economic, socio-cultural and technological) are mostly uncontrollable, and actors in the industry environment are more or less controllable constituents. At different stages of the industry life cycle, rapid and incremental changes have been observed in the tourism business environment. A certain degree of uncertainties and risks exist within the digital business environment. The use of digital technology by festival and events companies, to some extent, create barriers towards achieving their business mission. Three stages are required to be followed by policymakers. Firstly, the industry policymakers need to redesign a pragmatic strategic framework by applying management functions (planning, organizing, leading and controlling) basically to identify problems caused by digital technology. Secondly, formulate and implement realistic strategies to convert the threats into innovative opportunities and sustainable business competitiveness. Finally, all kinds of supportive and positive endeavors need to be ensured to move the business forward towards achieving mission and vision by meeting the expectations of diverse stakeholders and enhancing socio-economic contributions. In addition, after the execution of these three stages, managers might require to adopt business process reengineering to adapt with continuous growth and development of digital technology to move festival tourism business forward towards a sustainable position (see Fig. 21.1). Managers should manage the digital situation in a professional manner because a professional elite can go a long way towards initiating economic growth in the ever-changing digital era.

Socio-Cultural Drawbacks of Digital Technology in Tourism Events and Festivals

Increasing Mental Anxiety, Distress and Other Psychological Effects

A detrimental influence on travel companions' interactions occurs with the usage of limitless digital communication, which prevents personal conversations between fellow travelers (Dickinson et al., 2016). The widespread use of digital technology, particularly mobile devices and social media, has resulted in increased anxiety, tension, psychological disorders, lack of sleep, and a reduction in interpersonal connections (Beyens et al., 2016). When travelers are trying to find some rest and relaxation through participating in tourism events and festivals, the option to continually connect to business-related concerns via digital technology can cause the blurring of work and leisure time, which eventually lead to mental fatigue and hinder their healing (Egger et al., 2020). Even if any tourist is attached to digital technology while experiencing tourism events and festivals, they are diverted from fully

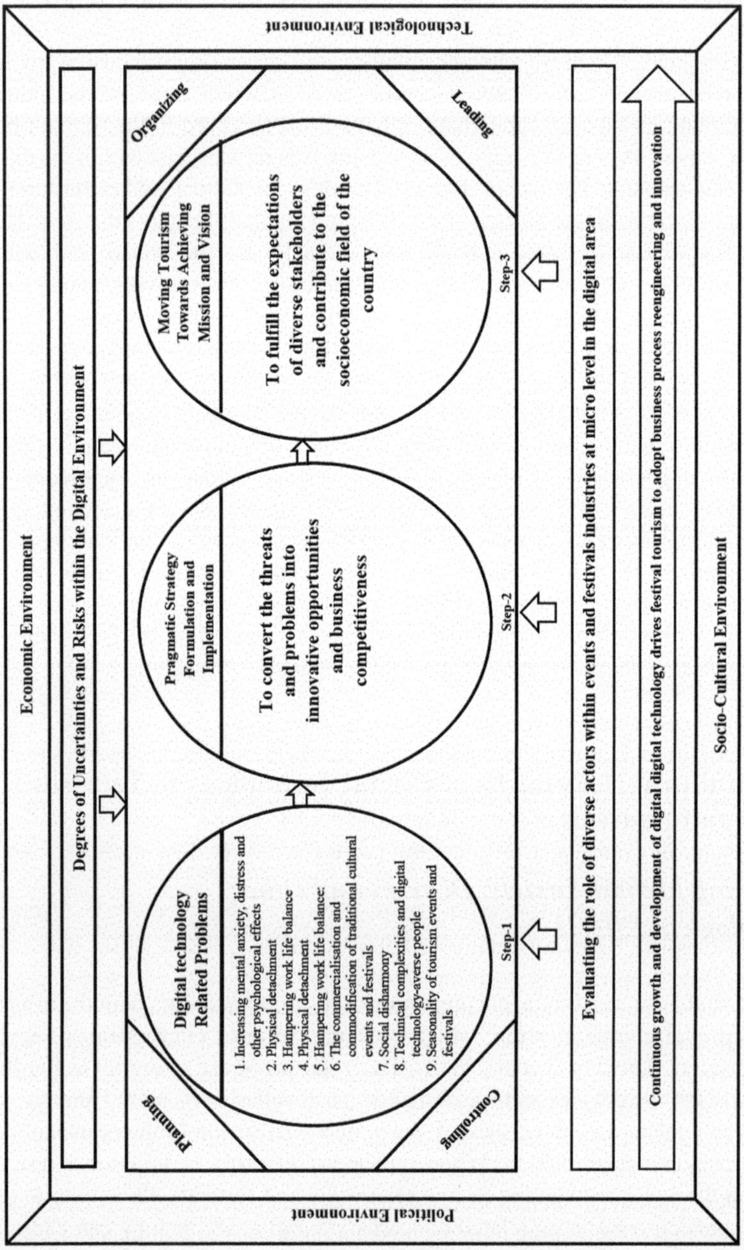

Fig. 21.1 A strategic framework. (Source: authors' own work)

experiencing the real-world situations, sights, sounds, social contacts, the thrill of adventure, and relationships with people in the events and festivals, which can lead to psychological and physical stress for visitors. As people increasingly use devices such as smartphones and computers to access the web, their behavior might become controlled by these devices. It eventually causes them to get drawn into a swirl of online activity that can lead to obsessive or habitual behavior and even addiction (Harwood et al., 2014).

Physical Detachment

With the advent of digital technology, tourism events and festivals in Asian countries may substantially impact travelers' interactions with others. Every participant who participates in an event is always looking to establish new relationships or solidify existing ones. It is well-known that people travel because they seek new relationships (Moscardo, 2017). When the participants were not using digital technology, connecting with new contacts was perceived as "much convenient" and "more genuine". As a whole, people are eager to experience natural human interaction while travelling, although digital technology hinders this (Egger et al., 2020; Tribe & Mkono, 2017). Another reason why travelers face social alienation with strangers is their technological dependence and the embarrassment they feel when trying to converse with someone they have not discovered on any social media or the Internet platforms. It reveals that the use of digital technologies has the potential to decrease ties between travelers and locals (Ayeh, 2018; Egger et al., 2020).

Hampering Work-Life Balance

Travelers can pre-design their future courses of action for a specific period of suitable time if they have access to authentic information about the events. Therefore, it is true that valid information to individual tourists has a momentous impact on multiple aspects of the tourist decision making particularly when selecting a tourist destination to visit (Xiang et al., 2015). However, work stress or pressure is created through the digital platform using the media, or social networks will have a detrimental influence on the choice to proceed with plans to attend any tourism activities such as events and festivals (Villacé-Molinero et al., 2021). Recently, tourists have grown more concerned about the risks posed by the Internet and mobile phones because of their availability, but at the same time, tourists are seeking opportunities for recreational purposes that do not involve using these digital technologies. As a result, people subconsciously de-emphasize their use of the Internet and mobile phones and implement a new trend called unplugged tourism (Pawłowska-Legwand & Matoga, 2020). Furthermore, a detrimental influence on travel companions' interactions occurs with the usage of limitless digital communication, which prevents personal conversations between fellow travelers (Dickinson et al., 2016). Shields (2011) brought a concept to wanderlust, meaning inherent urge and motivation of travelers to get to the unfamiliar and to explore the unknown, which might be stifled and hindered if digital technology impedes that desire.

The Commercialization and Commodification of Traditional Cultural Events and Festivals

Due to the rapid usage of digital technology, tourism events and festivals from host communities inevitably exposed to different cultures may be commodified and commercialized, resulting in pseudo-cultural events with no cultural value for the local people or the tourist. Because of technological advancements, culture may be sacrificed for the sake of commercialization, resulting in the creation of greater economic value at the cost of cultural value (Pratheep, 2017). An increasing number of examples of negative consequences of digital technology on socio-cultural values and activities of the host community may be found while organizing tourism events and festivals, and such kinds of activities are expanding at a greater rate (Moisescu et al., 2019). Even though the use of digital technology in marketing tourism events and festivals will boost cultural globalization while strengthening cross-cultural relationships, it may also have some negative consequences, including cultural invasion and the propagation of violent culture. A growing number of developing countries in Asia are concerned about digital technology that might undermine their traditional cultural identities, practices, and languages. As a result, the origins of traditional cultures in Asia are at risk of losing their original values for continuing to exist, which may raise the possibility that traditional culture will be overtaken by new cultural phenomena affected by digital technology and spread out.

Social Disharmony

A lot of challenges exist for using digital technologies in the way of strengthening social harmony (Davison, 2017). The long-term viability of tourism might be threatened by being influenced by any negative consequences of tourism events and festivals, generating discomfort, displeasure, or anger within the local community. Cultural sharing through social media can trigger hostility and loss of cultural pride, which could create negative impacts and social imbalance in society. Tourists' disrespect for religious festivals and their propensity to wear sacred objects as decorations or clothing may undermine religious importance and negatively impact the local population's socio-cultural and spiritual wellbeing of the local population (Dilsaver, 1979). Similarly, the propagation of dangerous activities (such as illegal gambling, violence, delinquency, and crime, increasing drug and alcohol usage, as well as rising prostitution and sex tourism) might be a negative consequence of increasing tourism events and festivals. Again, technological advancements can create imbalance and disparity between the tourists and the local community. As locals may feel uncomfortable owing to visitors and may begin to believe that tourists have the quality of life, which might have a significant impact on the mind-set of local community, even though tourists fail to recognize it (Ferreira et al., 2021).

Technical Complexities and Technology-Averse People

Although events and festival organizers can apply digital technology to facilitate their tourism services and processes; however, it should bear in mind that there are a good number of constraints associated with the use of such technology (Gössling, 2021; Munar et al., 2013). Digital technology can be extremely challenging to negotiate; with so many inexperienced users and a relatively small tour operating company, finding one's way around the Internet is initially very time-consuming and frustrating. It is even also time-consuming for experienced users, particularly in high demand situations when many other users are already involved in the system. Furthermore, most developing countries of Asia have a huge number of ageing people with very limited technical knowledge who are likely to form the largest segment of the tourism market. Aged people are usually risk-averse and less comfortable with digital technology (Hanson, 2011). In addition, older people have a deep-seated phobia of dealing with anything unknown; additionally, digital technology usage necessitates a considerable level of risk-taking, especially for novice users (Jones & Bednar, 2020). As a result, they are unlikely to be habituated to digital technology.

Privacy and Security Concerns

In the events and festival business, information exchange is crucial since there is a huge flow of information in both directions, including tourists, tour operators, auxiliary service providers and tourism organizations (Dhaware & Vanwari, 2020). To plan a trip, tourists must process substantial amounts of information related to the various steps involved, from pre-trip planning (including booking the trip, applying for visas, and payment) to on-trip experiences (enjoying local destinations, validating judgments) and writing about the experience to post-trip evaluation by sharing knowledge in online media (blogs, social networking sites etc.) (Tussyadiah et al., 2019). As a consequence, significant security issues are raised because of physical and digital privacy concerns. In order to generate appropriate suggestions for visiting desired tourism events and festivals, accessing sensitive user information, including date of birth, gender, location, interests, mobility concerns, prior visits, etc. is necessary for the online platform (Abu-Ghosh et al., 2018; Dhaware &

Vanwari, 2020). However, participants tend to be more ready to share personal information if they believe they will be provided tailored suggestions that may even lead to negative results such as privacy concerns. Furthermore, many users use digital platforms (i.e., social networks including Facebook, YouTube, Instagram), and sharing their travel experiences, both while they are travelling and after their return. Providing these sorts of information typically involves disclosing personal information of family members and friends travelling together or being visited, which eventually poses new privacy concerns.

Seasonality of Tourism Events and Festivals

Tourism appears to be the most effective economic sector in creating income and employment in less developing, peripheral nations with limited growth potential, mostly in Asia. In these countries, tourism events with the combination of digital technology help expand visitors' options for entertainment and consumption while it may also expand the economic sustainability of festivals and events since more tourists come through and patronize such activities utilizing online social media (Wang et al., 2019). However, even though tourism enhanced the rate of economic growth, as soon as digital technology integrates with it, it contributed to economic inequality. With the advancement of digital technologies, tourism starts representing the best use of natural and human resources in this case, so long as resources are efficiently allocated. Although events and festivals are well shared through websites, blogs and online social networks, tourism events and festivals are annually arranged, affecting local community living (Suess et al., 2021). Locals have stated that this leads to job seasonality and sales related to tourism's activities and employment, which is mostly low-paid and rising housing rents in those areas, which may affect those who do not own their own homes. Moreover, compared to other industries, as the tourism industry requires employees with low skills and expertise, most companies offer a poor compensation scale and fewer opportunities for promotion.

Recommended Policies to Overcome Challenges

Digital Free Tourism

It is evident that when digital devices were restricted, face-to-face communication increased, and this led to an increase in freedom and personal confidence for tourists since they had to rely on themselves throughout their trips that directly led to an increase in their personal interactions. To lessen tourists' web dependency, anxiety, tension as well as obsession, researchers have proposed paying attention to real-life rather than virtual life through introducing the idea of "digital-free tourism", also termed as "technology dead-zones" (Egger et al., 2020; Pearce & Gretzel, 2012). The idea of travelling without being attached to digital technology has supported to get the real essence of tourism; through ensuring better work-life balance, enhancing wellbeing, and reinvigorating people's attention to the real world (Egger et al., 2020). Furthermore, digital-free tourism can be offered as a vacation feature that can attract tourists, more specifically younger generations, who want to decrease or limit their technology usage instead of eliminating it (Egger et al., 2020). Again, digital free tourism can be designed for different trip durations, holiday types, activities, and places that can benefit consumers who are unable to put down their technology for short periods (Egger et al., 2020; Li et al., 2020).

Increased Government Intervention

Digital technology has evolved into a platform for social change. Through the use of social media, individuals may communicate and share experiences, views, and opinions, therefore shaping the public's attitudes and behaviors on a wide range of social issues and topics (Alabau-Montoya & Ruiz-Molina, 2020; Buhalis, 2020; Lanouar & Goaied, 2019). Recently, as social media platforms such as Twitter, Facebook, Weibo have grown in popularity, the government has discovered a new approach to ensure social harmony and stability. Forcing governments to impose tougher penalties on individuals who misuse digital technology and mobilizing global support for this cause appears to be significant. Therefore, it is crucial to ensure that national governments continue to be engaged in the fight against social imbalance and discord to develop the tourist sector. Digital technology should contribute to the prevention and control of terrorism, violence, delinquency, and crime and the improvement of the tourist industry, which plays a critical role in the growth of Asian economies. Nowadays, nevertheless, the comprehensive use of digital technology and social media in crisis communication has outperformed the use of traditional media since they generate better results and wide-ranging public engagement (Hays et al., 2013). Hence, it should be utilized to achieve a better outcome for the society. Furthermore, the government should have a well-accepted regulatory policy aim to manage the industry situation. It will not be wise to ignore digital technology; rather, critical problems need to be addressed and come up with realistic solutions within a well-accepted legal environment (Hojeghan & Esfangareh, 2011). Therefore, government should move forward to enhance the growth and development of digital technology by taking positive initiatives, which includes (a) pragmatic planning to solve critical constraints related to digital technology; (b) introducing new regulatory structure; (c) investing IT infrastructure for enhancing capacity; (d) ensuring technical training to generate skilled human resource in IT; and (e) tourism sector and evaluating the digital environment periodically.

Understanding the Culture and Environment

The accelerating growth of digital technologies in the tourism industry throughout Asia in particular and the overall globe, in general, is evident. Festival and event management businesses are already started using Internet digital technology to promote their services. With the aid of digital technologies, travel businesses may engage in interactive marketing campaigns that force them to redefine their approach to research, planning, and developing effective campaigns. Further consideration will enable event organizers to see that the conventional mass marketing system has come to an end and that the future techno-driven condition will be described by mass customization and person-to-person marketing in significant numbers. Despite the fact that digital technology provides a novel means of marketing tourism events and festivals, policymakers in the tourism industry should primarily understand the culture of the target market and embrace the informal code of conduct that these customers or clients observe. It is not appropriate to consider managers of tourist event business firms to be a panacea capable of performing quick miracles inside an organization's system. The event managers are required to investigate networking, exhibition formats, costs of digital technologies, and the company's aims before developing the data highway. As technological advancement progresses, its potential benefits, reach, and application will become more and more apparent.

Strong Privacy Safeguards

Information privacy is indeed a major issue for consumers while they are accessing the web since they have no control over who has access to their personal information, and when this is found breached, hence information privacy is severely jeopardized. Consumers always had security concerns, but as the Internet has developed, those concerns have been revised to match changing consumer demands. Privacy concerns are a significant issue that companies encounter in the contemporary digital world because of the growing number of threats, hazards, and possibilities posed by technological innovations (Anic et al., 2019). Privacy concerns may be shaped and changed with the growth of digital technology. It is inevitable that the Internet's privacy regulations be updated and regulated regularly to handle issues such as potential attacks, information privacy issues, and data protection behavior. Tourism-related companies should implement strong privacy safeguards to prevent unauthorized entities from accessing their customers' personal information (Afolabi et al., 2020). The tourism industry may exert more control regarding privacy issues through primarily sharing users' personal data with legitimate entities and solely with the user's permission. Finally, the contribution and cooperation of government authorities are vital in offering technical and managerial assistance that may encourage travelers to enhance their use of digital technologies.

Convergence with Digital Technology

The introduction, integration and interrelationship of innovative digital technologies and travel tourism may positively influence the preservation and maintenance of events and festivals belonging to cultural heritage (Hoon & Hong, 2019). Unfortunately, key stakeholders in the tourism sector have already embraced digital technology for the purpose of business development; the cultural industry has still yet to recognize the significance of technological applications for the preservation of cultural heritage. Because of the integration of resource conservation and management, digital technologies have the potential to bring up new opportunities for cultural and tourism authorities (Lee et al., 2020). One of the most popular ways is integrating digital technologies with heritage tourism, a tourism strategy that begins by exploiting the convergence points between the cultural sector and the tourism business (Ardissono et al., 2012; Hoon & Hong, 2019). Digital technologies-based heritage tourism development might be an acceptable answer for the government and other institutional authorities to act in order to be relevant to the cultural and tourism sectors. Again, it is also necessary to ensure that the involvement of the local community should be oriented with technological means, which may balance between accessing the cultural heritage as well as safeguarding it.

Conclusion

Digital technology has created huge innovative marketing opportunities in the Asian tourism sector, especially for tourism events and festivals. Festivals and events are not only limited to the town or community; instead, they spread worldwide and bring communities closer. Rather than a local event, it turns into a global celebration. Digital technology has enhanced the ability of tourism companies in different countries of Asia to reach the masses in the phenomenon, which can be accessed by people, countries or even continents away and reach all these people effectively and efficiently. Again, the developmental trend and convergence of digital technologies have created a new paradigm and opportunities for policymakers to focus on restructuring the tourism events and festival industry worldwide. A perfect synergy between macro-level strategy and micro-level strategy is required to be assured by addressing to mitigate the negative effects of digital technology in the travel and tourism industry in particular and socio-economy in general. It is noticeable that the private sector dominates the tourism market at the micro-level. The government should have a well-accepted policy aim to manage the industry situation. It will not be wise to ignore digital technology; rather, critical problems need to be addressed and come up with a realistic solution within a well-accepted legal environment. The policy agendas must incorporate: (a) improving legal framework, (b) government itself being involved in online procurement and (c) promoting e- transformation in tourism and related industries and (d) government should move forward to enhance the growth and development of digital technology by taking positive initiatives which include: pragmatic planning to solve critical constraints related to digital technology, introducing new regulatory structure, investing IT infrastructure for enhancing capacity, ensuring technical training to generate skilled human resource in IT and tourism sector and evaluating the digital environment periodically. Policymakers should bear in mind that tourism is an information-sensitive industry, and information digital technology is driving the economic wheel of various industries in developing countries. In this era of digitalization of industries, all levels of stakeholders need to have a clear picture of the techno-driven business situation along with the implications and impact of digital technology development. Pragmatic policies could change the constraints into innovative opportunities to develop and manage sustainable competitive tourism events and festivals. Conducting a review of existing events and festival-based research indicates that some aspects are needed to be highlighted in the industry:

- A mutually supportive strategic synergy between the community's cultural practices and technological innovations.
- A combined strategy of policy and digital technology initiatives along with synergistic benefits.
- Coordination impacts of policies and digital technologies deployed in reaction to an event.
- Issues caused by ineffective policy and technological strategies further aggravated.
- Good correlation between market performance and innovative offerings for the entire industry.
- A well-connected and managed value chain is needed for gaining sustainable socio-cultural benefit.

In a world of increasing competition, there is no doubt that tourism events and festival businesses must struggle to achieve distinctive performance by optimizing technology resources and enhancing efficiencies. It is practically true that no technologies, once applied, remain unmodified or are being replaced through the rest of the business. Therefore, it is required to improve the usage of digital technology in a timely manner along with the appropriate assistance. Again, consumers should be motivated to embrace the good aspects of new digital technologies and be warned about the bad implications of technology if discovered. On the other hand, policymakers dealing with tourism events and festivals have been directed to lead and manage travel and tourism companies to cope with the demand for growth, increasing competition and the complexity, risks and frenetic pace of change of the digital era.

References

Abou-Shouk, M., Lim, W. M., & Megicks, P. (2013). E-commerce and small tourism businesses in developing countries: Drivers versus boundaries of adoption. *Tourism Planning and Development, 10*(3), 249–266.

Abu-Ghosh, D. H., Al-Dmour, H., Alalwan, A. A., & Al-Dmour, R. H. (2018). Factors affecting Jordanian consumers' attitudes towards Facebook advertising: Case study of tourism. In Y. Dwivedi et al. (Eds.), *Emerging markets from a multidisciplinary perspective. Advances in theory and practice of emerging markets* (pp. 285–302). Springer.

Afolabi, O. O., Ozturen, A., & Ilkan, M. (2020). Effects of privacy concern, risk, and information control in a smart tourism destination. *Economic Research-Ekonomska Istrazivanja*. https://doi.org/10.1080/1331677X.2020.1867215

Ahmed, T. S. A. A. (2017). A triple bottom line analysis of the impacts of the Hail International Rally in Saudi Arabia. *Managing Sport and Leisure, 22*(4), 276–309.

Alabau-Montoya, J., & Ruiz-Molina, M. E. (2020). Enhancing visitor experience with war heritage tourism through information and communication technologies: Evidence from Spanish Civil War museums and sites. *Journal of Heritage Tourism, 15*(5), 500–510.

Anic, I. D., Budak, J., Rajh, E., Recher, V., Skare, V., & Skrinjaric, B. (2019). Extended model of online privacy concern: What drives consumers' decisions? *Online Information Review, 43*(5), 799–817.

Ardissono, L., Kuflik, T., & Petrelli, D. (2012). Personalization in cultural heritage: The road travelled and the one ahead. *User Modeling and User-Adapted Interaction, 22*(1–2), 73–99.

Ayeh, J. K. (2018). Distracted gaze: Problematic use of mobile technologies in vacation contexts. *Tourism Management Perspectives, 26*, 31–38.

Beyens, I., Frison, E., & Eggermont, S. (2016). "I don't want to miss a thing": Adolescents' fear of missing out and its relationship to adolescents' social needs, Facebook use, and Facebook related stress. *Computers in Human Behavior, 64*, 1–8.

Buhalis, D. (2020). Technology in tourism-from information communication technologies to eTourism and smart tourism towards ambient intelligence tourism: A perspective article. *Tourism Review, 75*(1), 267–272.

Collins, A., Jones, C., & Munday, M. (2009). Assessing the environmental impacts of mega sporting events: Two options? *Tourism Management, 30*(6), 828–837.

Davison, R. M. (2017). Facilitating social harmony through ICTs. In *14th International Conference on Social Implications of Computers in Developing Countries (ICT4D)*, Yogyakarta, pp. 3–9.

Dawson, J., & Jöns, H. (2018). Unravelling legacy: A triadic actor-network theory approach to understanding the outcomes of mega events. *Journal of Sport and Tourism, 22*(1), 43–65.

Dhaware, M., & Vanwari, P. (2020). A tourism and travel recommendation system based on user-location vector. *Lecture Notes in Electrical Engineering, 601*, 1429–1437.

Dickinson, J. E., Hibbert, J. F., & Filimonau, V. (2016). Mobile technology and the tourist experience: (Dis)connection at the campsite. *Tourism Management, 57*, 193–201.

Dilsaver, L. M. (1979). Some notes on the negative impacts of international tourism. *Yearbook, Association of Pacific Coast Geographers, 41*, 109–118.

Dychkovskyy, S., & Ivanov, S. (2020). Festival tourism as part of international tourism and a factor in the development of cultural tourism. *Informacijos Mokslai, 89*, 73–82.

Egger, I., Lei, S. I., & Wassler, P. (2020). Digital free tourism – An exploratory study of tourist motivations. *Tourism Management, 79*, 104098.

Ferreira, F. A., Castro, C., & Gomes, A. S. (2021). Positive and negative social-cultural, economic and environmental impacts of tourism on residents. *Smart Innovation, Systems and Technologies, 208*, 288–298.

Getz, D. (2008). Event tourism: Definition, evolution, and research. *Tourism Management, 29*(3), 403–428.

Girish, V. G., & Chen, C. F. (2017). Authenticity, experience, and loyalty in the festival context: Evidence from the San Fermin festival, Spain. *Current Issues in Tourism, 20*(15), 1551–1556.

Goldkuhl, G. (2012). Pragmatism vs interpretivism in qualitative information systems research. *European Journal of Information Systems, 21*(2), 135–146.
Gössling, S. (2021). Tourism, technology and ICT: A critical review of affordances and concessions. *Journal of Sustainable Tourism, 29*(5), 733–750.
Hanson, V. L. (2011). Technology skill and age: What will be the same 20 years from now? *Universal Access in the Information Society, 10*(4), 443–452.
Harwood, J., Dooley, J. J., Scott, A. J., & Joiner, R. (2014). Constantly connected - The effects of smart-devices on mental health. *Computers in Human Behavior, 34*, 267–272.
Hays, S., Page, S. J., & Buhalis, D. (2013). Social media as a destination marketing tool: Its use by national tourism organisations. *Current Issues in Tourism, 16*(3), 211–239.
Hojeghan, S. B., & Esfangareh, A. N. (2011). Digital economy and tourism impacts, influences and challenges. *Procedia - Social and Behavioral Sciences, 19*, 308–316.
Hoon, Y. J., & Hong, S. (2019). Three-dimensional digital documentation of cultural heritage site based on the convergence of terrestrial laser scanning and unmanned aerial vehicle photogrammetry. *ISPRS International Journal of Geo-Information, 8*(2), 53.
Jang, S. C., & Feng, R. (2007). Temporal destination revisit intention: The effects of novelty seeking and satisfaction. *Tourism Management, 28*(2), 580–590.
Jones, J., & Bednar, P. (2020). Are the elderly averse to technology? *Lecture Notes in Information Systems and Organisation, 33*, 25–32.
Kelly, D. M., & Fairley, S. (2018). What about the event? How do tourism leveraging strategies affect small-scale events? *Tourism Management, 64*, 335–345.
Lanouar, C., & Goaied, M. (2019). Tourism, terrorism and political violence in Tunisia: Evidence from Markov-switching models. *Tourism Management, 70*, 404–418.
Lee, P., Hunter, W. C., & Chung, N. (2020). Smart tourism city: Developments and transformations. *Sustainability, 12*(10), 3958.
Lewis-Beck, M., Bryman, A., & Futing Liao, T. (Eds.). (2012). The SAGE Encyclopedia of social science research methods. In *The SAGE Encyclopedia of social science research methods*. SAGE Publications.
Li, J., Pearce, P. L., & Low, D. (2018). Media representation of digital-free tourism: A critical discourse analysis. *Tourism Management, 69*, 317–329.
Li, J., Pearce, P. L., & Oktadiana, H. (2020). Can digital-free tourism build character strengths? *Annals of Tourism Research, 85*, 103037.
Liang, S., Schuckert, M., Law, R., & Guo, X. (2017). A retrospective analysis and future trends of E-tourism research in China. *Journal of China Tourism Research, 13*(3), 231–256.
Llopis-Amorós, M. P., Gil-Saura, I., Ruiz-Molina, M. E., & Fuentes-Blasco, M. (2019). Social media communications and festival brand equity: Millennials vs Centennials. *Journal of Hospitality and Tourism Management, 40*, 134–144.
Long, P., & Morpeth, N. D. (Eds.). (2016). *Tourism and the creative industries: Theories, policies and practice*. Routledge.
Lu, S., Zhu, W., & Wei, J. (2020). Assessing the impacts of tourism events on city development in China: A perspective of event system. *Current Issues in Tourism, 23*(12), 1528–1541.
MasterCard. (2019). *Global Destination Cities Index 2019*. Retrieved from: https://newsroom.mastercard.com/wp-content/uploads/2019/09/GDCI-Global-Report-FINAL1.pdf. Accessed: 1 Aug 2021.
McCabe, S., Sharples, M., & Foster, C. (2012). Stakeholder engagement in the design of scenarios of technology-enhanced tourism services. *Tourism Management Perspectives, 4*, 36–44.
McKinsey & Company. (2016). *How savvy, social shoppers are transforming Chinese e-commerce*. Retrieved from: https://www.mckinsey.com/industries/retail/our-insights/how-savvy-social-shoppers-are-transforming-chinese-e-commerce#. Accessed: 1 Aug 2021.
Moisescu, O. I., Gică, O. A., Coroş, M. M., & Yallop, A. C. (2019). The UNTOLD story: Event tourism's negative impact on residents' community life and wellbeing. *Worldwide Hospitality and Tourism Themes, 11*(5), 492–505.

Moscardo, G. (2017). Exploring mindfulness and stories in tourist experiences. *International Journal of Culture, Tourism, and Hospitality Research, 11*(2), 111–124.

Munar, A. M., Gyimóthy, S., & Cai, L. (2013). Tourism social media: A new research agenda. *Tourism Social Science Series, 18*, 1–15.

Navío-Marco, J., Ruiz-Gómez, L. M., & Sevilla-Sevilla, C. (2018). Progress in information technology and tourism management: 30 years on and 20 years after the internet - Revisiting Buhalis & Law's landmark study about eTourism. *Tourism Management, 69*, 460–470.

Ooi, C. S. (2019). Asian tourists and cultural complexity: Implications for practice and the Asianisation of tourism scholarship. *Tourism Management Perspectives, 31*, 14–23.

Pawłowska-Legwand, A., & Matoga, L. (2020). Disconnect from the digital world to reconnect with the real life: An analysis of the potential for development of unplugged tourism on the example of Poland. *Tourism Planning and Development*. https://doi.org/10.1080/21568316.2020.1842487

Pearce, P. L., & Gretzel, U. (2012). Tourism in technology dead zones: Documenting experiential dimensions. *International Journal of Tourism Sciences, 12*(2), 1–20.

Perić, M. (2018). Estimating the perceived socio-economic impacts of hosting large-scale sport tourism events. *Social Sciences, 7*, 176.

Pranić, L., Petrić, L., & Cetinić, L. (2012). Host population perceptions of the social impacts of sport tourism events in transition countries: Evidence from Croatia. *International Journal of Event and Festival Management, 3*(3), 236–256.

Pratheep, P. (2017). The impact of tourism on Indian culture. *KnE Social Sciences, 1*(3), 429–437.

Rahman, M. S. (2016). The advantages and disadvantages of using qualitative and quantitative approaches and methods in language "testing and assessment" research: A literature review. *Journal of Education and Learning, 6*(1), 102–112.

Richards, G. (2020). Designing creative places: The role of creative tourism. *Annals of Tourism Research, 85*, 102922.

Saleh, F., & Ryan, C. (1993). Jazz and knitwear. Factors that attract tourists to festivals. *Tourism Management, 14*(4), 289–297.

Salehan, M., Kim, D. J., & Lee, J. N. (2018). Are there any relationships between technology and cultural values? A country-level trend study of the association between information communication technology and cultural values. *Information and Management, 55*(6), 725–745.

Shields, P. O. (2011). A case for wanderlust: Travel behaviors of college students. *Journal of Travel and Tourism Marketing, 28*(4), 369–387.

Suess, C., Woosnam, K., Mody, M., Dogru, T., & Sirakaya Turk, E. (2021). Understanding how residents' emotional solidarity with Airbnb visitors influences perceptions of their impact on a community: The moderating role of prior experience staying at an Airbnb. *Journal of Travel Research, 60*(5), 1039–1060.

The Peninsula. (2018). *Qatar expects 1.5 million visitors as 2022 countdown begins - The Peninsula Qatar*. Retrieved from: https://www.thepeninsulaqatar.com/article/21/11/2018/Qatar-expects-1.5-million-visitors-as-2022-countdown-begins. Accessed: 1 Aug 2021.

Tribe, J., & Mkono, M. (2017). Not such smart tourism? The concept of e-lienation. *Annals of Tourism Research, 66*, 105–115.

Tussyadiah, I., Li, S., & Miller, G. (2019). Privacy protection in tourism: Where we are and where we should be heading for. In J. Pesonen & J. Neidhardt (Eds.), *Information and communication technologies in tourism 2019* (pp. 278–290). Springer.

UNWTO/GTREC. (2018). *UNWTO/GTERC Annual report on tourism trends - 2018 edition*. Retrieved from: https://www.unwto.org/global/publication/unwtogterc-asia-tourism-trends-2018-edition-0. Accessed: 1 Aug 2021.

Vajirakachorn, T., & Chongwatpol, J. (2017). Application of business intelligence in the tourism industry: A case study of a local food festival in Thailand. *Tourism Management Perspectives, 23*, 75–86.

Villacé-Molinero, T., Fernández-Muñoz, J. J., Orea-Giner, A., & Fuentes-Moraleda, L. (2021). Understanding the new post-COVID-19 risk scenario: Outlooks and challenges for a new era of tourism. *Tourism Management, 86*, 104324.

Wang, X., Sun, J., & Wen, H. (2019). Tourism seasonality, online user rating and hotel price: A quantitative approach based on the hedonic price model. *International Journal of Hospitality Management, 79*, 140–147.

Willis, C., Ladkin, A., Jain, J., & Clayton, W. (2017). Present whilst absent: Home and the business tourist gaze. *Annals of Tourism Research, 63*, 48–59.

Xiang, Z., Magnini, V. P., & Fesenmaier, D. R. (2015). Information technology and consumer behavior in travel and tourism: Insights from travel planning using the internet. *Journal of Retailing and Consumer Services, 22*, 244–249.

Xu, F., Huang, S., & Li, S. (2019). Time, money, or convenience: What determines Chinese consumers' continuance usage intention and behavior of using tourism mobile apps? *International Journal of Culture, Tourism, and Hospitality Research, 13*(3), 288–302.

Zou, Y., Meng, F., Bi, J., & Zhang, Q. (2021). Evaluating sustainability of cultural festival tourism: From the perspective of ecological niche. *Journal of Hospitality and Tourism Management, 48*, 191–199.

H. M. Kamrul Hassan is currently serving as Associate Professor in the Department of Marketing, University of Chittagong. His principal interest in teaching is developing, using, and disseminating comprehensive, affordable teaching and business education platform to be maximized both by faculty members and students. He studied and lectured in sustainable marketing, entrepreneurship and tourism. He has published articles in national and international journals. Moreover, he attended many local and international conferences in Bangladesh, India, Malaysia, Thailand, Indonesia, Czech Republic and Spain. In addition, he regularly writes on concurrent business issues at the academic and non-academic journal and newspapers.

Mohammed Shahedul Quader studied and lectured International Business, Marketing, and Shipping Management, working as Associate Professor in the Department of Marketing, University of Chittagong. He has completed his MBA in Shipping Management, MA in Marketing Management from University of Leicester and Middlesex University, UK respectively. He has more than 15 years of teaching experience in the field of international business. Besides teaching, he has published more than 30 articles in reputed local and international journals and joined international conferences local and abroad.

Chapter 22
Post COVID-19 Pandemic: The New Technology and Tourism Revolution for Tourism Fairs, Festivals and Events

Noraihan Mohamad

Introduction

The growth of Information and Communication Technology (ICT) has changed individual's behaviour and businesses' communication. The advancement of ICT has facilitated global interconnectivity where people has change their communication form, information searching and decision making (Ngernyuang & Wu, 2020). Most of business in various sector has also had utilize ICT to communicating their product and service, creating new opportunities, gaining competitive advantages, and sustaining business continuity including sub-sector in tourism industry such as accommodation, transportation, tourist destination, tourist attraction as well as festivals and events (Sigala et al., 2012). In this digital era, the collaboration between ICT and tourism sectors are seen as the perfect match in many ways. ICT plays a significant role in transforming tourism globally which it gradually creates a new paradigm, modify the structure of industry and rising an entire range of chances. By providing means for enlarging, managing, and allocating offerings global, ICTs allows peoples to identify, alter and purchase tourism products, as well as support the globalization of the business (Khan & Hossain, 2018).

Previously, events widely revealed as "animators of destination attractiveness" that are a "core element of the destination system" and a "bridge between the market for visitor attractions created by tourists and the use of events to fill the gap" (Getz & Page, 2016, p. 593). The most important thing is event industry has been acknowledged as one of the significant key predictors in nation's economic growth (Lyck et al., 2012). Nowadays, the festivals and events are well-known as a vigorous and fast-growing sector with strong alliances with tourism. The high value of

N. Mohamad (✉)
School of Hospitality and Creative Arts, Management and Science University, Selangor, Malaysia
e-mail: noraihan_mohamad@msu.edu.my

A. Hassan (ed.), *Technology Application in Tourism Fairs, Festivals and Events in Asia*, https://doi.org/10.1007/978-981-16-8070-0_22

alignment between festivals and events with tourism industry is obvious, where many cities, region and country has prioritized many tourism-oriented festivals and events by hosting business events (such as convention, trade shows, and conference), leisure travel events (such as cultural festivals, food festivals, heritage events) and sports tourism events as their long-term strategies for various development. Despite of that, tourism-oriented festivals and events also has utilized by many countries around the globe to attract both domestic and international tourist to a particular tourism destination, as well as to align the role of government agencies, industry players, tour operators, service providers and mass media during hosting the event (Mapjabil et al., 2015; Eck, 2018).

Malaysia is a Southern-East, multi-ethnic and multi-cultural country that consisting of 13 states and three federal territories. Based on that, there are minorities live in the country, including Bumiputras Sabah and Sarawak, Chinese and Indians. Due to this diversity, Malaysia popular as a colourful country that rich with multiple cultural, arts and traditions. It can be seen through the celebration of many festivals and events which always reflects to the cultural diversity and religious tolerance. For instance, celebration of Eid ad Fitr, Eid al Adha, Chinese New Year, Wesak Day, Gawai Festival, Thaipusam, Deepavali, Christmas, Mooncake Festival, Harvest Festival and National Day, where all Malaysian citizen with different background will celebrate the events together and respects each other.

Alongside arts and cultural festivals, Malaysia also has emphases sport-oriented tourism events as its great potential for various developments in the country has accepted (Mapjabil et al., 2015). In spite of Malaysia is a small and developing country with limited resources, Malaysia manages to utilize these resources to build the necessary infrastructures and all the supporting services like the Bukit Jalil Sport Complex, Putrajaya Water Sport Complex, Sepang International Circuit, and National Velodrome (Mapjabil, 2014). As venue for hosting many international sport events such as Commonwealth Games, Formula One Championship Car Racing (F1) Series, Moto GP Racing Series, and SEA Games, it was proven that Malaysia's sports facilities are among the best in the world. In addition, apart from serving as the venue for sport events, most of these facilities also being utilized as a training arena for aspiring enthusiasts and international athletes to come to Malaysia.

Despite of that, it is essential to understand that in hosting an events, media coverage is vital to promote the country as a whole, not just to highlight the capability of the country as a host, but to enhance its image as unique tourism destination. Previously, as a host for sports-oriented tourism events, Malaysia has been succeeded in securing the coverage of many televisions broadcasting networks both local and international. In result, Malaysia became well-known around the world. Truly, mass media plays a significant role in carrying out promotion and marketing duty for every phases of events: before, during and after the events.

As festivals and events take place for both physical and virtual spaces, ICT plays an important to create new style to analyse the relationship between the organizers and audiences. An electronic media based on Internet become the most influential communication platform between organizer and the audience. Internet platform allow the organizer to share all important information related to an event which needed by the audiences, while audiences easy to access the information by using a mobile application (Brown et al., 2020). Basically, all the information about local festivals and events can be accessed via government website such as Ministry of Tourism, Arts and Culture (MOTAC) and Tourism Malaysia, organizer's website, official website for the event, and numerous social media platform.

Over the years, before the COVID-19 pandemic, the event industry has been recognized and predictable to become a significant professional field in tourism industry, and therefore it has been widely discussed in the academic literature. As such, the integration of ICT into the management of tourism festivals and events has been increased, particularly to facilitate attendees' experiences (Van Winkle & Bueddefeld, 2020). Additionally, during the pandemic COVID-19 disaster, the employment of ICT in event industry is crucial because of the safety protocol such as social distancing and other safety measures obligatory by the authority bodies will ominously affect the operation of many events (Dillette & Ponting, 2020). Therefore, the aim of this chapter is to explore the role of ICT during crises situation and how it's integration throughout the tourism festivals and events industry will enhance the businesses revolutions and opportunities. The discussion focuses on the uniqueness of ICT to tourism festivals and events particularly during COVID-19 pandemic.

Tourism Industry in Malaysia

Tourism was categorized as one of the fasters industries that significantly contribute to the economic growth, socio-economic development as well as heritage and environment preservation and conservation. The tourism industry involves food and beverages services (F&B), business tours and holidays tours (domestic or international), transportation services (including railway services, cruise/ship services, public transport services), lodging services (including hostels, homestays, motels, guest house, hotels), as well as entertainment or destination spots (including shopping malls, theme parks, amusement, museums, beaches, forests, cinemas, rural tourism). Based on that, tourism industry generates opportunities for foreign exchange income through tourist's spending and activities which includes traveling cost, food cost, hotel staying, lodging service cost, and tourist's destination cost (Zubair & Shamsudin, 2021). Since the tourism industry is a huge industry, it provides more opportunities for career growth too either on a professional level or operational level.

The tourism industry is Malaysia's fifth largest industry after the oil, gas and energy industry, financial services industry, retail and wholesale industry, and palm oil industry (Tourism Malaysia, 2020a). It is also recognized as one of the significant contributors to the nation's foreign trade income (The Star, 2019; Karim, 2020). In the first quarter of 2019, the Malaysian tourism industry charted an upward trend when tourist expenditure reached RM21.4 billion compared with RM18.3 billion in 2018 (Tourism Malaysia, 2020b). Additionally, the per capita expenditure also recorded an increase to a total of RM3,300 in 2019 when compared with RM2,813.1 in 2018. In 2019, Malaysia also recorded a 2.7% spike in tourist arrivals (i.e., from 6,520,218 tourists in the previous year to 6,696,230 tourists) (Tourism Malaysia, 2020b). According to the records of 26,100,784 international tourists, most of them hailed from 10 different countries, namely, Vietnam (400,346), Philippines (421,908), Japan (424,694), South Korea (673,065), India (735,309), Brunei (1,216,123), Thailand (1,884,306), China (3,114,257), Indonesia (3,623,277), and Singapore (10,163,882) (Tourism Malaysia, 2020b). Nevertheless, international tourist influxes have dropped 83.4% from 26,100,784 in 2019 to 4,443,722 in 2020. Due to that, Malaysia's tourist receipts in 2020 is RM12.69 billion which decline 85.3% from RM86.14 billion in 2019. In addition, the tourists' average per capita spending also declines 11.3% from RM3,300 in 2019 to RM2,928 in 2020. Data release by Tourism Malaysia (2021) indicated that the top 10 international tourist influxes to Malaysia in 2020 were from Vietnam (64,184), Australia (72,680), Japan (74,383), South Korea (119,750), Brunei (136,020), India (155,883), Thailand (394,413), China (405,149),

Indonesia (711,723), and Singapore (1,545,255). Based on the data recorded by Pacific Asia Travel Association (PATA) and National Tourism Organization (NTO), not only Malaysia facing with a reduction in the volume of international tourist influxes, but other ASEAN countries also show a significant decrease, including Indonesia (−75%), Vietnam (−78.7%), Thailand (−83.2%), and Singapore (−85.7%).

Undeniably, the great plunge in international tourist influxes is associated with the shutting of Malaysian borders and travel restriction since 18 March 2020 because of the outbreak disease of COVID-19 pandemic and the Movement Control Order (MCO) by government of Malaysia. The effect of COVID-19 to Malaysia's tourism industry obviously demonstrated by the sub-sector of tourism such as accommodation, transportation, entertainment, food and beverage, as well as festivals and events. It can be said that tourism industry is the worst industry impacted by the COVID-19 which have to closed until some period of time which still not yet clear. Worst case scenario is when the campaign of Visiting Malaysia 2020 has announced cancel by the ministry which leaves the industry's expectation to contribute to at least RM 103.6 billion to Malaysia's Gross National Income with targeting 36 million of tourist arrival in 2020 (Tourism Malaysia, 2020c).

The airline or aviation industry is the nastiest as it was completely prohibition on travelling and all the airlines' services are shut. Due to the government's enforcement to control the spread of the virus, all flights have been cancelled and forcing millions of travellers to reschedule their travel plans (Campiranon, 2021). Clearly, the airline industry completely being the hardest hit industry cause by COVID-19 even risk of bankruptcy due to the travel restriction and flight cancellations (Yahya, 2020; IATA, 2021; Mazareanu, 2021). As reported by Minister of Transport, due to travel restriction, flights for 4.316 million passengers have been cancelled, 1.211 million passengers have been postponed, while 628,000 passengers request for refunds (RM399 million). Consequently, Malaysia's airline industry has to bear about RM13 billion loss which from Malindo Airways Sdn Bhd, AirAsia Group

Bhd, Malaysia Airlines Bhd, and also airport operators (Yahya, 2020). This three-airline industry are in worst situation, even they don't have enough capital to sustain in the industry. The best option they have is arranged their employees for forced retirement and unpaid leaves for an unknown period (Zubair & Shamsudin, 2021).

The COVID-19 also created massive impact on the hotel industry. As one of the significant boosters to the nation's economy, hotel industry was fall and decreasing Malaysia Gross Domestic Product (GDP). From January to March 2020 the COVID-19 cases continued increase. Consequently, tourists were withdrawing 170,084 hotel rooms booking and thus resulting fall of revenue of RM68,190,364 (Karim, 2020). Besides, due to the enforcement of travel restrictions and MCO, tourists have no better options but cancel the hotel rooms booking. Consequently, this unhealthy situation has forced many hotels in Malaysia to downsize their organization by sending their employees for unpaid leaves and termination, as well as salaries cutting.

For the past 20 years, event industry has been acknowledged by Malaysian government as a substance for nation's economic and social well-being. This can be confirmed by the fact that the development of a strategic plan initiated by the Malaysian government known as Business Events Roadmap with the main purpose is to make Malaysia as the main hub in hosting business events by 2020 not only within the Asia Pacific region but internationally also (MyCEB, 2016). Beside to highlight Malaysia as a significant business event destination, this strategic plan is included to increase international tourist arrivals for business events from 5% to 8% by 2020 (MyCEB, 2019). Even though the event industry has seen hard hit by the pandemic COVID-19, the goals of the strategic plan has remained with some revision and improvement in particularly by employment of element of ICT (Bernama, 2021a; Onan, 2020).

Tourism Fairs, Festivals and Events

The growth of tourism industry has been increase tourists' demands and expectations for unique experiences especially from festivals and events (Ngernyuang & Wu, 2020). Festivals are recognized as unique opportunities for an individual to explore and experience a specific culture, arts and tradition for a specific community. Earlier, Falassi (1987) defines festivals as an organized and unique event that encountered human cultures at a certain area where specific economic activities are conducted. Moreover, festival also defined as "an organized set of special events with a specific cultural man-made theme on a specific day or period, normally at a specific place, gathering people in mutual and direct contact to participate in the festival theme" (Lyck et al., 2012, p. 11). Based on the concepts, therefore festival tourism refers to tourist's travelling activities to a specific destination for gaining a festival experience.

Event, on the other hand widely used "to explain about specific rituals, presentations, performance and celebrations that are consciously planned and created to mark special occasion and/or to achieve particular social, cultural or corporate goals and objectives" (Bowdin, 2006, p. 14). Thus, event management referred to "the applied field of study and area of professional practice devoted to the design, production, and management of planned events, encompassing festivals and other celebrations, entertainment, recreation, political and state, scientific, sport and arts events, those in the domain of business and corporate affairs (including meetings, conventions, fairs and exhibitions), and those in the private domain (weddings and parties, and social events for affinity groups)" (Getz, 2008, p. 404).

Even though, festivals have been recognized as part of cultural activity throughout history, to understand about its management and administration is remaining important. According to Goeldner and Ritchie (2012), festivals and events is a dominant tool to attract tourists to a destination and to create an awareness and impression towards an area. Similar to that, Eck (2018) state that festival and event industry is a significant component of tourism since it's provides a unique opportunity to the host to develop a city, region or nation by reposition the destination. By hosting festivals or events, it significantly will contribute to the economic strengthen of the area and local community by involving media exposure, Internet marketing, partnership strategies, and infrastructure stimulation. Besides, festivals and events are always used as a medium to highlight the uniqueness of local community, promote the destination and local industries, encourage demand by motivating visitors to visit the destination, and thus increase revenue (Van Winkle & Bueddefeld, 2020).

As festivals and events are recognized as a significant sharing platform, local society have big opportunities to stimulate an awareness of their culture and heritage among tourists including to highlight the local society's lifestyle, nature-based attraction and historical features of an area. In spite of that, it will be encouraging local cultural activities, highlighting cultural institutions of an area, as well as encourage cohesion and a sense of local identity among residents (Suryawardani & Wiranatha, 2016). Due to these potentials, the diversity and quantity of festivals and events have been swiftly rising. On the same note, most of events' host including Destination Marketing Organizations (DMOs) are committed to host a festival or event properly and made it as a main attraction to make the destination stand out from competitors.

However, essential to understand that high numbers of tourist arrival at the destination will create negative impact such as environmental pollution, conflict with the local people, traffic jam and vandalism. Therefore, sustainability concept in event management plays an important role to balance between human activities, resources use, service quality, investment and amenities, and environmental impact (Jones, 2010; Nasir et al., 2019). As festivals and events involve huge gathering of people, it has an impact on the surrounding air, land and water. Thus, to reduce the impact, there many aspects need to be consider by the organizer including waste management, energy production, water management and transport. The most important thing is to make events' audience and stakeholders understand about the important of actual changes, as well as to their ongoing attitudes and behaviour.

As tourism industry significantly grow and returns high revenues to the economy of Malaysia including the local people, tourism festivals and events sector has been highlighted as one of the key drivers (Nasir et al., 2019). Generally, tourism festivals and events organized can be categorized as follow:

Sport Tourism Event

Sport tourism events are highlighted as one of the important tourism products which significantly contribute to the national income through the involvement of domestic and international participants, as well as large numbers of the world tourist arrivals (Mapjabil et al., 2015). Malaysia successfully had hosted the prestigious sport-oriented tourism events such as Commonwealth Games, Le Tour de Langkawi, Formula One Championship Car Racing (F1) Series, Moto GP Racing Series, SEA Games, Monsoon Cup, Langkawi Ironman Trathlon, and Royal Langkawi International Regatta. Statistic reveals that a single series of Formula One race event was contributed more than RM1.39 million to the national economy. Due to the high potential provided by sport tourism events, many countries around the globe have emphasis it as an important strategy for the country's development.

Business Events

The term of business events is widely being applied in industry and academic field. Referring to Edwards, Foley and Schlenker (2011, p. 3), "business events come in various forms which include meetings, conferences, conventions, incentive programs, exhibitions, seminars, summits and training". Recently, Yong and Ghazali (2018) posit that business event is an augmentation from the standard segment of Meetings, Incentives, Conferences and Exhibitions (MICE). On the same note, Wagen and White (2018) described that business events are usually commercially driven, and in many countries, it remains marked as MICE. As business events remains as the rapidest contributor to the tourism industry, community, and economy, some of Asian nations such as Japan, South Korea, Singapore and China are also highlight business event as a high-value sector and willing to provide significant resources to support its development.

Local Festivals

Local festivals include a variety of events that utilize cultural elements of local people within the city or region as a product to attract tourists for a certain period of time (Kim & Chang, 2020). Local festivals are meaningful to the local people, tourists as well as city and region based on numerous of its contributions. Local festivals can be categorized as food festival, film festival, music festival and other arts and cultural festival which have become an important platform for local people to introduce their local traditions, intangible heritage, uniqueness of ethnic, and natural panorama to the tourists (McKercher et al., 2006; Zhu & Kim, 2020). By discovering these elements, they are able to highlight the uniqueness of local cultural and tradition, and thus utilize them as remarkably sustainable tourism resources. Local festivals also contribute to enhancing tourists' demand to travel to the city. During their visit, tourists will pay a cost for using various conveniences such as accommodation, transportation, food and beverage, souvenir and other amenities. Additionally, local festivals also be a factor to increasing city's image and strengthening the identity as a tourist destination like for recreation purposes (Liu & Chen, 2007). Hence, local festivals have always acknowledged as a way of enriching the local tourism industry, developing destination image and energizing the economy.

The Pandemic COVID-19 and Malaysia's Event Industry

Undoubtedly, the tourism and hospitality industry were significant as one of the main contributors to the nation's revenue until 2019. However, a pandemic of SARS-CoV-2 (Severe Acute Respiratory Syndrome Corona Virus 2) unexpectedly emerged in Wuhan, China and spread to the city so rapidly before it completely spread worldwide. On 12 March 2020, the World Health Organization (WHO) officially declared the pandemic as COVID-19 (Corona Virus Disease 2019) which caused by novel of SARS-CoV-2 (Zubair & Shamsudin, 2021). Then, as people's movement and traveling around the world, it also spread the COVID-19 around the globe. Due to the rapid increase of the COVID-19 cases, many countries were announced their lockdown enforcement. Since early of 2020, many countries around the world have shutdowns all borders and restrict all travel and tourism activities (Dillette & Ponting, 2020). This enforcement as a response to the COVID-19 that has continued to spread across the world badly.

The COVID-19 has affected all industries, economies, livelihoods, public services and opportunities on all continents (UNWTO, 2021). The tourism industry is one of the industries that significantly destroyed with the impact of the pandemic and the economic downfall (Baldwin, 2020). Previously, global tourism industry has been experienced affected with variety of crises such as terrorism, natural disaster (e.g., earthquakes, tsunami, landslide, typhoon) and outbreak diseases (e.g., SARS, H1N1, Ebola, MERS-CoV), but it has been resistant in recovery (UNWTO, 2021). Even though, COVID-19 has caused many impacts and has given rise to profound and long-term changes (Dillette & Ponting, 2020). In addition, after a year the COVID-19 issues are remain ongoing and still have no clue when businesses industries in the world will back to normal. This situation indirectly drives to the downfall on global tourism industry including Malaysia (Karim, 2020; Zubair & Shamsudin, 2021).

Before the COVID-19 spread worldwide, the event industry had made a significant impression on the global economy and were a draw for business and leisure travellers alike. The disruption of the outbreak towards global economy will undeniably leave a long-term impact particularly to the event industry (Dillette & Ponting, 2020). Compared with other sectors of the tourism industry, the events industry required longer period of time to recover. As nature of events, it is requiring the organizer to attract attendees as many as they can. However, to reduce risks caused by COVID-19 people are advised to avoid any crowd or gatherings and always monitor their social distancing (Hao et al., 2020; Sigala, 2020). Accordingly, all kinds of events which involve large groups of people have been restricted, including festivals, meetings, concerts, conferences, sports, weddings, reunion, and family gatherings (Gössling et al., 2020). In results, all events around the globe have been postponed and cancelled, and therefore affected the stakeholders' businesses such as sponsors and suppliers.

According to Malaysia Convention and Exhibition Bureau (MyCEB) (n.d.), in 2020, 60 events had been cancelled while 28 events has postponed due to the implementation of MCO which include (1) convention such as Corporate Directors Conference Asia-Pacific 2020, World Congress of the World Federation of Hemophilia, Asian Oceanian Congress of Radiology 2020, Radiodays Asia 2020, and World Congress on Information Technology 2020; and (2) exhibition such as ASEAN Senior Care & Wellness Expo, Malaysia 2020, Malaysia Travel & Tourism Mart (MTTM) 2020, The Fit Malaysia 2020, Penang Signature Gold (PSG) 2020 and Malaysia China Trade Expo (MCTE) 2020 by Notice of Events Postponement & Cancellation -2021 (Malaysia Convention and Exhibition Bureau (MyCEB), n.d.). Additionally, there are also several cultural and traditions festivals that affected with the restriction such as Hari Raya Celebration, Thaipusam, Gawai Festival and National Day.

According to the survey conducted by the Malaysian Association of Convention and Exhibition Organizer and Suppliers (MACEOS), since the first MCO in 2020, business events industry's revenue losses are reach to RM2.25 billion with 90% decline (MACEOS, 2020). Compared to 2019, events industry had contributed RM3.9 billion expenditures and generated RM9.2 billion revenues to the country. Due to the significant decrease, a total 5610 employees had been laid off equivalent to 17% of total industry workforce. This dramatic decline of event industry also had a heavy impact on its supply chain such as destination management organizations, accommodation sector, transportation sector, tour operators, and other suppliers related to the industry. Consequently, it leaves big impact on the revenue contribution to the overall national economy generated through the business events supply chain.

In 2021, Malaysia again has gone through several phases of MCO that has ruin many local businesses in events industry due to facing cancellations and shrinkages (Naharul & Kamel, 2021). Most of the event business try to sustain by postponing the events but as MCO in unexpected and keep reoccurring, it put those businesses at risks. Therefore, to ensure the businesses' continuity, most of sectors in tourism industry including events sector have no option but to apply several relevance strategies including downsizing, firing employees, and provide an option for unpaid leaves to employees, which then leading to the increment of unemployment

(UNWTO, 2021). As an alternative, ICT has become an eminent strategic tool for most and event companies to gaining competitive advantage by exploiting the role of ICT in defining, promoting, distributing, organizing, and presenting products and services (Onan, 2020). As a strategic tool, ICT provides an infinite positive impact in the sense of efficiency of event organizations, cost reduction, differentiation, and time management (Buhalis, 2000).

Currently, Malaysia through the collaboration between MOTAC and other agencies, is working to develop Malaysia's business event industry to be top five in the world in 10 years (Bernama, 2021a). This goal is aligning with the Malaysia Business Events Strategic Marketing Plan (SMP) 2021–2030. The first initiative to achieve the goal is by making Malaysia as the first country in the ASEAN region to grasp the World Expo which it is expected to bring good impact to Malaysia's business event industry as a whole. The SMP 2021–2030 is a new long-term strategic planning to boost business event industry in Malaysia to reappear for the next decade. By highlighting business and international sporting events, this planning aim to create at least 3% of international tourist arrivals. Minister of MOTAC, Datuk Seri Nancy Shukri also state that essential for event's organizers to attract business event visitors to come to Malaysia since they normally spent three times more than the normal tourist and have a large multiplier effect, with 40% of the visitors travel with partner while 60% of them will return as tourists in the near future (Bernama, 2021a).

Additionally, the United Nation's World Tourism Organization (UNWTO) (2021) also has mentioned that business event sector is significant key drivers in generating income, employment and investment to tourism industry and the country as a whole. Thus, the SMP will focuses on developing sustainable tourism through business event sector by adapting the elements of digitalization and innovation. In line with that, the SMP 2021–2030 is expected would create about 25, 833 total employments by 2030. Previously, since 2018, Malaysia consistent shown the positive growth in business event industry, where Malaysia has become the world ranking of 33 out of 110 countries, and also being ranked number 9 within the Asia Pacific rankings (Nasir et al., 2019). Without any single doubt, business events can be recognized as part of business tourism due to the fact it provides huge effect on the economics of Malaysia and increase total of attendees among tourists.

ICT and the Revolution of Tourism

Information and communication technology (ICT) is a field that provide huge opportunities for businesses to obtaining a competitive advantage, increasing productivity, evolving with new industries and permitting new modes of management (Khan & Hossain, 2018). It is a term that widely used to explain about different electronic tools that are competent in transmission of information and data in a network. Generally, ICT is a wide-ranging of technologies of communications which includes computers, data, software, telephone, data storage and so on that allow the users to access, transmit, store and modify an information. Such an example of ICT's component that widely used in business communication are Internet, digital TV, digital radio, WIFI telecommunications, software, email, virtual reality, phones, mobile applications and social media.

The application of ICT has crucial role in the tourism industry and the management as a whole. The application of ICT in tourism industry is actually first in airline industry through the development of Computer Reservation Systems (CRSs) in the late of 1970s, subsequently to the Global Distribution Systems in 1980s such as Galileo, Sabre, Amadeus and Worldspan, and afterward the creation of the Internet in the early of 1990s (Khan & Hossain, 2018; Bockermann, 2019). This revolution has made the tourism industry amid the utmost industry in the usage of the Internet as the employment of ICT provides benefits for them to create more effective communication with their existing and potential customers. Undeniably, the growth of tourism and hospitality including all the enterprises such as rental agencies, tour operators, cruisers, travel agencies, and hotels in current world are driven by the growing effect of ICT.

Alongside the revolution of ICT, the new concept of business has emerged that referred to technology businesses. This new concept most known as e-commerce and e-business which has increase possibilities for all industries in the world, including tourism. Basically, e-commerce refers to any process of buying and selling activities, or commercial transaction through electronic platform such as Internet and mobile applications. It provides benefits to all modern business industries by permitting them to access the global market, creating new opportunities, and increasing the efficiency of operations within an organization. In the context of tourism, e-commerce has reflected to the terms of e-tourism or e-travel which refer to design, implementation and application of ICT and e-commerce solutions in the tourism industry (Khan & Hossain, 2018). It is an area which the adoption of ICT in businesses and tourist activities has transform the processes and the value chains in the tourism industry.

E-tourism concept was categorized into three specific disciplines: (1) business management; (2) information systems and management; and (3) tourism. E-tourism take an advantage of intranets to rearranging internal processes and regulates the competitiveness of the organisation. Extranets are used for increasing transactions with trusted partners of organization, while intranets are used in organization for connected with stakeholders including consumers. The e-tourism concept reflects to all business function such as e-commerce, e-marketing, e-finance, e-accounting, e-procurement, e-production, as well as e-strategy, e-planning and e-management for all sectors of the tourism industry, including tourism, travel, transport, leisure, hospitality, principals, intermediaries and public sector organisations (Bethapudi, 2013; Khan & Hossain, 2018). Thus, due to the huge concept of e-tourism and its capability to fulfil tourists' needs and demands, tourists have become more demanding, requesting high-quality services and value for their money, as well as value for their time.

Currently, the emergence of ICT has introduced mobile computing systems, or most known as smartphones, which comprise various technologies like the Internet, photography, global positioning systems and communication. It was improved the social environment completely, where the users have self-control on their online activities including travel experience at anytime and anywhere. In addition, the adoption of ICT by tourism industry has been transformed the industry from "traditional sector" to "new sector" where the interactive technique of communication offered by the components of ICT has seriously changed the nature of communication on tourism and travel sector as well as the methods of promoting and marketing. On the same note, ICT has made tourist easy to adapt with the new system, process and procedures applied by all businesses in the industry for delivering their services.

As Internet knowingly has changed the means people's communication and engagement with business, tourism industry has also been affected where Internet become tourists' main medium for information searching about a destination, tourist attraction, transportation, accommodation, and experiences shared by the previous tourist (Ngernyuang & Wu, 2020). Despite of the main role of Internet, numerous Internet channels such as social media and community forums have allow tourist to communicate with the service provider directly. These opportunities are useful for tourists' decision-making process where Internet allows them to search for the right service provider, as well as appropriate destination and attractions (Ly et al., 2020).

Besides, there are also many options provided by the Internet for the tourism service provider to interact with tourists, share an information and build a relationship with tourists. Hence, vital for the service provider to use the most effective online marketing communication platform which provide various interactive options to engage with tourists. According to Živković, Gajić and Brdar (2014), among the available Internet's platforms, social media is recognized as the most dependable channel that permits the user to disseminate content extensively. Social media and social networking sites are regarded as web-based platforms, which allow users to communicate each other, as well as to create, share and exchange stories, information, and ideas virtually in a connected community (Luttrell, 2018). Referring to Kaplan and Haenlein (2010), social media is "a group of Internet-based applications that build on the ideological and technological foundations of Web 2.0, and that allow the creation and exchange of user-generated content" (p. 161).

As social media continuously evolve, its definition and function also change and expand over time. Hence, there is no worldwide definition on social media provide by the previous scholars. Cohen (2019) had summarized the definitions of social media based on its nature and characteristics. This is including: (1) social media incorporating a plenty of formats such as image, video, text, audio and pdf; (2) social media allow cross interactions between one or more platforms through activities such as messaging, social sharing, feeds and email; (3) social media provide different level of communication such as one-to-one, one-to many and many-to-many; and (4) social media extend user meeting by creating real-time online events, expanding online interactions to offline interactions, and supplementing live events online.

Many tourism companies in the world are acknowledging social media as their marketing tools as its potential to enhance tourist engagement and collaboration.

The options of interactive communication provided in social media such as text, image and video sharing, like, comment, and live option was made social media as magnificent tool for advertising, promotion and marketing purposes. Due to the great role and contribution of media and communication technologies, many businesses have always taken advantage of the opportunities by scaling up the scope of marketing of products and services on social media with the main purpose is reaching out many prospect customers as possible (Koo et al., 2015). Moreover, social media marketing was accredited as vital for gaining competitive advantage in the digital era since it's relatively strong capacity to segment the right audience based on their social media activity.

Additionally, Internet technology and the emerging of mobile phones usage with various functions of mobile applications have dramatically enlarged the opportunities for many companies to communicate with tourists and social media has become an eminent tool for the marketing of tourism services (Law et al., 2014; Sotiriadis, 2017; Bockermann, 2019; Shankar, 2021). The social networking sites such as Facebook, Instagram, Twitter, YouTube and LinkedIn, the content uploaded, on the other hand have become main media channel for creating and spreading propaganda which usefulness for gaining tourists' attention, tourism marketing campaigns and receiving feedback from tourists.

On the other hand, online community forum like Trip Advisor, Agoda, Booking.com and Trivago has always become main references for many tourists to manage trips and make an evidence-based decision due to experiences shared by previous tourists regarding tourism product, services and destinations. This style of reaching out for information has recognized the online community forum as authentic and credible to provide dependable information and trusted services offer. Thus, with more than 200 million users around the globe, if the tourism organizations unable to get listed within the first five spots on the online community forum, they might be failed to gain the competitive advantage. Without any single doubt, the integration between ICT particularly social media and tourism industry will bring great benefits for the company, including enhance the company's image and credibility, boosting the promotion and marketing, increasing tourist loyalty, building a relationship, monitoring, and cost-effectiveness (Xiang & Gretzel, 2010; Kumar et al., 2021).

ICT and the Revolution of Tourism Festivals and Events

The main revolution caused by ICT are "everything is online" where Internet become the main supportive for doing business. Due to that, tourist nowadays greatly rely on online platform to obtain detailed information about a certain destination, tourist attractions and in planning trips (Javed et al., 2020; Ly et al., 2020). The trend of ICT adoption in tourism industry has continued by all areas of the industry such as accommodation, transportation, destination, and attraction including the tourism fairs, festivals and events. As tourism fairs, festivals and events are significant for the development of economy and socio-demographic at a specific area or destination, it should be considered as a "product" that required to satisfy tourist's need. Thus, as a product, eligible for it to be promoted thru various marketing tools including Internet marketing tools.

Over the past decades, the marketing communication for tourism industry heavily had depended on the power of traditional marketing particularly word-of-mouth (WOM). The WOM is a face-to-face communication which the information commonly delivered by family and friends. It is a powerful tool that would assist tourist's traveling planning and decision making, which also supported with an information from magazines, tourism guides and travel agencies (Kim & Chang, 2020). The emergence of ICT in the present world, however, has been extended the face-to-face WOM to a digital or electronic platform which known as electronic word-of-mouth (eWOM) (Bockermann, 2019). eWOM is defined as "all informal communications directed at consumers through Internet-based technology related to the usage or characteristics of particular goods and services or their sellers" (Litvin et al., 2008, p. 461).

eWOM from tourism perspectives can be defined as informal interactions among tourists that associated with sharing information about products or services through numerous channel such as traveller's blog, social media, and community forums. Based on characteristics of tourism like an experience good and information-intensive industry, tourists are more likely to be influenced by eWOM by obtaining reliable and credible information from various of the communication generated through eWOM (Kim & Chang, 2020). It can be expected that tourists heavily rely on eWOM as it provides an easy way to get information about tourism products and services. Furthermore, eWOM also seen as significant part of Internet marketing since it is available on mobile devices such as smartphones and tablets. This is reflecting to its characteristic which providing borderless interaction. Social media is an authoritative source of eWOM which provide both positive and negative impact (Rizky et al., 2017). Positive eWOM either via comments, photos, or videos on social media shared by previous tourists has able to enhance potential tourists' levels of motivation, on the contrary, negative eWOM from previous tourists will decrease potential tourists' level of motivation to visit the festival (Liu et al., 2018; Primack et al., 2019; Kim & Chang, 2020).

Besides, positive perception of previous tourists towards the festival will lead to the formation of positive image of the festival and organizer, which boosting the publicity for the festival (Zeng, 2013; Hanan & Karim, 2015; Ngernyuang & Wu, 2020). In this respect, Internet marketing through social media platforms become an important marketing tools for most tourism festivals and events as it is a key to reach more participants and significantly influence potential tourist's behaviour in making the right decision about attending tourism festivals and events (Garay & Morales, 2019; Kim & Chang, 2020). Moreover, tourists nowadays more trust previous tourists' opinions and recommendations more than official marketing advice from the service providers or organizer. Study by De Lira, Macdonald, Ounis, Perego, Renso and Times (2019) categorized social media usage behaviour for tourism festivals and events into three phases: before, during and after the festivals or events.

Before the Festival or Event

Generally, people will use social media in their daily lives, communicate with online friends, posting pictures and videos about the festival or event and sharing their experiences. These online activities will attract other's intention and become motivated to know more about the festival or event, reading more comments on that post, and therefore interested to attend or participate. Based on that, they will search more information through various online channel about the festival such as the accessibility, fees, facilities, benefits, eligibility and others important information that they have to know before leaving. If their decision is about to participate or attend the event, they may gladly express their feeling with the words such as "attend", "registered", "joint" and so on. Besides, the most common behaviour is sharing their feeling with family and friends and encourage them to joint too.

During the Festival or Event

During the festival or event, common for the attendees to show their participation and express their excitement, feelings, and emotion while they are at the event by posting text, photos, videos, comments, taking live recordings, with the main purpose is to share their current feelings about the event (Zhu & Kim, 2020). Activities during event will encourage their online friends or other Internet users to get to know more about the event and thus develop their perceptions about the event's image and atmosphere.

After the Festival or Event

After the festival or event, people tend to express their feeling and emotion by sharing their memories on various channels of social media. On the posting, people often to state their opinion about the positive or negative things of the festival or event, as well as advice and suggestions to the future tourists or visitors. Social media activities after the festival or event are useful for the host or organizer to access the visitors' feedback and thus develop strategies to improve their services.

Clearly, social media is a crucial tool to organizer of festivals or events for analysing tourists' attitudes and behaviours. The analysis can be confirmed based on the online activities such as likes, comments, shares and postings. As stated by Zhu and Kim (2020), tourists sharing behaviour on social media by capturing their memories about tourism experiences significantly enhance their level of satisfaction and positive behaviour like recommendation to others. Hence, social media activities are admitted will encourage potential tourists' attitude and behaviour such as intention to visit and repeat visits. It is acceptable due to the fact that tourist will trust reliable sources of information to reduce their feeling on uncertainty and increase expectation on the tourism festivals or events. As such, social media is identified as an essential tool for marketing of tourism festivals and events.

Referring to the World Travel and Tourism Council (2015), Vidal (2019), and Van Winkle and Bueddefeld (2020), there are several important aspects that need to consider by events' organizer or DMO about the integration of ICT into tourism festivals and events:

Mobile Applications

Rapid usage of smartphones and the availability of mobile applications for different purposes is a supportive trend for promoting tourism festivals and events. Tourism is seen as one of the market conquerors for certain mobile apps as tourists increasingly turn to digital information to search about interesting places, festivals and events, transport, shopping, and other information. Event mobile app were recognized as a new tool to reach audience, distribute events' news, and increase attendees. Information sharing in the apps will create excitement among audience and thus share their excitement with others in their network (Sultan et al., 2019). In a way, event mobile app helping for building and positioning the brand of tourism festivals and events. Moreover, event mobile app allows people to plan their visit especially for events that organized over several days or venues such as conference, music festivals, food festivals and cultural festivals. Despite of that, mobile app also provides an opportunity for people to create a personalized experience where people can choose and save their favourites being offered at the event they are attending. Lastly, through event mobile app, easy for host or organizer to build long-term relationship with the attendees and thus up-selling.

Online Travel Agents (OTA)

Tourist in this digital era has always looking for deals offered by service provider when booking online tourism services. Since ICT provide a systematic data management, online transactions and timely customer services, many travel agencies start to rely on online booking platforms. Booking platforms or online travel agents (OTA) refers to all-in-one platform which providing real-time information about products or services availability and price comparison to tourist such as Expedia, Agoda, Traveloka and Trivago (The ASEAN Post, 2020). OTA not only useful for tourist to find hotels rooms availability and the best deals offered but also useful to get an information about car rental companies, airlines, restaurants and other related services including festivals and events. Beside to make it easy for tourist to get limitless information, reviews provided by previous tourist on those platforms about certain product or services will help tourist to make better decisions. Due to the

COVID-19 impact, the collaboration between OTA and many service providers are now starting to restart tourism by focusing on domestic tourism and targeting local travellers. By encourage local to spend on local businesses, it helps to build the businesses and improve the entire economy of the country.

Mobile Ticketing

Mobile ticketing and cashless event methods have been applied widely in many western countries by using both Radio Frequency Identification (RFID) tags and integrated wristbands (Van Winkle & Bueddefeld, 2020). These innovations of technologies definitely have improved the efficiency of access system to the festivals and events, reduce fraud and waiting times as well as increase attendees' positive experiences (Hudson & Hudson, 2013). Basically, RFID and wristbands were first exercised at festivals and events for entrance management. Ultimately, due to its efficiency, this innovation has been fully-fledged to incorporate secure virtual payments and connecting with numerous social media network to allow the attendees to share their experiences towards festival and event virtually.

Virtual Payment Methods

One of the main benefits provided by RFID technology is cashless system which allows event attendees for faster and easier transaction without any flaws (Van Winkle & Bueddefeld, 2020). Interestingly, the integration between event and secure virtual payment methods significantly increases attendees' spending and reduce many kinds of risks such as waiting-times, missing credit cards, stolen cash and other causes. As a result, this innovation of technology will enhance attendees' experience quality and increase the festival and event revenues as a whole. As COVID-19 pandemic easy to spread among individual, virtual payment method has become important to attract people to attend festivals and events, as well as to encourage their spending during the event.

Livestreaming

Livestreaming technology, social media and artificial intelligence will enhance attendees' virtual reality experiences where they are allowed to experience the festival or event surpasses space and time (Breaker, 2018). Livestreaming events on social media such as YouTube will provide unprecedented access and creating unique experiences for people who are not able to attend an event in person. Over the years, organizers of festivals and events were avoided livestreaming events, instead encourage people to purchase tickets and attend an event in person. However, due to COVID-19 crisis where many events have been cancelled and postponed, livestreaming is extremely useful as a valuable alternative platform for organizer and attendees to experience new style of event.

Crowdfunding

Crowdfunding is a special means to raise funding from the public as a financial support of an event (Van Winkle & Bueddefeld, 2020). Crowdfunding platform is a valuable tool that has always utilized by events' organizer to enhance their brand awareness and build relationship by promoting the projects and interact with groups of audiences (Kim & Chang, 2020). Basically, the aim of crowdfunding is to fund a specific festival and event by recruiting small investments of people via an online platform. Nevertheless, it also used to generate promotional benefits for the projects simultaneously. To maximise its role, crowdfunding platform has always connected with social media channel to disseminate crowdfunding information as well as market the festival and event (Temelkov & Gulev, 2019). Spreading information on social media is one of the strategies to attract people' interest and influence their support decision. In fact, the compatibility of crowdfunding and social media eventually describes about the important of eWOM and its effect towards the successful of festivals and events.

Blockchain Technology

Blockchain is a unique database in a way of storing information that makes it difficult or impossible to change, hack, or cheat the system (Van Winkle & Bueddefeld, 2020). A notable example is music festival, which blockchain technology is useful to enhance connections between the musician and attendees. Through a platform without a gatekeeper, artists are allowed to communicate with their fans directly in a less restricted way. Based on Breaker (2018), a devolved blockchain is a platform for entertainment purposes that built by artists and for artists. Through a process of user-generated verification, blockchain technologies generates a secure audit trail of authentication (Nãsulea & Mic, 2018). Due to the great capacity, devolved blockchain technology are counted as the new revolution of ICT in the festival domain as it is not only being able to reduce administrative process and organization costs, but to increase trust. Given the integration of festival and blockchain technology, audiences can reduce cost on ticket prices and also enhance control and access to experience the festival.

Festivalscape

Generally, festivalscape refer to "the general atmosphere experienced by festival patrons" (Lee et al., 2008). Previous scholar found that visitor's satisfaction and loyalty to the festival and event significantly influenced by the atmospheric elements of the festivalscape (Mason & Paggiaro, 2012; Van Winkle & Bueddefeld, 2020). When ICT are integrated into the festivalscape, it requires the organizers to consider about the implications to the attendees' experience. The role of ICT in festivalscape include the use of video boards and social walls to project social media post in real time. Specifically, through the innovation of ICT, event attendees are able see social media contents such as photos and videos shared by other attendees at the event in real time (Breaker, 2018).

ICT and Tourism Festivals and Events in Times of Disaster

There are numerous of crises and disaster that significantly have affected human live whether it be economic, social, environmental, or political. The nature of crises and disaster are leave long-term effects on individuals, families, communities, businesses, and industry. There are various categorizations of crises and disaster, however, it was agreed by many scholars that crisis refer to internal issues while a disaster is external (Parsons, 1996; Faulkner, 2001; Stafford et al., 2002; Dillette & Ponting, 2020). Generally, crises can be ranging from natural crises to crises caused by human. Specifically, crisis is the result of something stanching from within an organization, while disasters are a result of incidences happening outside the organization (Stafford et al., 2002; Tew et al., 2008). In addition, Parsons (1996) defined a disaster as a situation "where an enterprise is confronted with sudden unpredictable catastrophic changes over which it has little control" (p. 136). Referring to that, clearly that COVID-19 pandemic meets the characteristics of a disaster, where all industries and economic sectors around the globe have been affected by the pandemic. In the case of event industry, the impact of this pandemic has been catastrophic.

COVID-19 is the newest world's disaster that has stopped tourism industry including event sector as a whole. Due to the risks of the outbreak, travel restrictions and movement control order, the only significant way to continue the events activities are by utilizing ICT. The component of ICT has become significant tool for most business including businesses in events industry where it allows the events' host and the organizer to expand their business to both local and international market. Basically, there are two main factors which will drive more and more events' host to get involve in ICT (Khan & Hossain, 2018). Firstly, technology is speedily improving, which the improvement will provide together the various of opportunities. Secondly, technology provides advantages for the host to reach groups of audiences and promote the events around the globe.

Malaysia's tourism industry was experienced facing with several crises and disaster such as SARS epidemic in 2002, tsunami on 26 December 2004, flu pandemic in 2009, Sabah earthquake on 4 June 2015, and COVID-19 pandemic in 2020. Truly, tourism industry and its sectors are vulnerable to many crises due to several external factors including economic conditions, political instability, sociodemographic, and climate change which will affect destination images and tourists' behavioral intention to visit or revisit (Jin et al., 2020; Campiranon, 2021). After a year COVID-19 pandemic has impacted the whole sectors of tourism industry, event's organizers in Malaysia are expected be more mature in predicting the need of ICT and digital platforms for organizing festivals or events. The fact that COVID-19 has created a negative impact heavily, it also has given a wake-up call for businesses around the world that their main requirement in the post-pandemic is transform current business model to the new model that inspiring by the ICT, as well as explore the innovations (Dillette & Ponting, 2020; Bernama, 2021b). Specifically, conducting an event through virtual platforms is a today's reality and it has become significant way to achieve event's aims and at the same time business can remain the continuity.

Since the COVID-19 situation are remains harmful, event industry should find ways to adapt, live with the situation and work around the challenges. According to The National Tech Association of Malaysia (PIKOM), core for businesses transformation is technology and digitalization which it will provide better spaces for social interaction, improve business communications, and evolve the way dealing with stakeholders (Bernama, 2021b). Take the CloudTheatre as an example, it is one of the Malaysia's theatre host which has affected badly by the pandemic, but now in collaborations with international host, they have developed their successful thru digital platform by streaming international arts and culture shows. Digital platforms allow them to get into a full-fledge automatic ticketing, streaming platform for online and on-ground shows, and at the same time hosting multiple shows from different countries. Currently, the CloudTheatre in progress to be hosting three upcoming cross-country virtual arts festivals, namely the George Town Festival (Malaysia), Malaysian Student Film Festival (Britain) and Let's Get Together Arts Festival (Taiwan) (Maganathan, 2021).

Conclusion

This chapter explores the role of ICT in tourism revolution during post COVID-19 pandemic, with a specific focus on festivals and events. Truly, the emergence of pandemic COVID-19 in the early of 2020 has caused tourism industry in particularly event industry appeared uncertain. In fact, all sectors in the industry also suffered with a large drop in tourist numbers either from local or international. Malaysia authorities also face with difficulties to encourage domestic tourism due to the risk of COVID-19 which easy to spread through contact people. However, because of its resilience and innovative, the event industry is already adapting completely. Part of this adapting is shifting the common physical mode of event to virtual mode by employing the component of ICT, which it is helping to bridge the distance between countries and continents. During this pandemic situation, ICT helps businesses in event industry to maintain relationships with the stakeholders until it becomes feasible again.

Thus, ICT has been acknowledged as an important tool for shifting and the continuity of event industry in the new norms as it provides new experiences, unique opportunities and flexibility for both audiences and the organizer. Specifically, by employing elements of ICT, organizers are able to organize, conduct, communicate, advertise and market their events in creative and effective way, while audiences are able to obtain information about the event, communicate their needs and demands to the organizer speedily, and experiences the event in unique way. From the marketing perspective, the revolution of ICT has highlighted social media as the most significant platform to market tourism festivals and events as it plays an important role to cater specific group of audience and developing tourist behaviour in terms of information searching and decision-making. Moreover, social media also has proven as important tool to develop potential tourists' positive behaviours towards events in particularly through the function of eWOM.

To conclude, the integration of ICT and tourism festivals and events sector would provide benefits to service providers, tourists and other stakeholders. It is worth to say that ICT is a powerful tool that brings advantage not only for promoting and marketing the festivals and events, but for strengthening the industry in particularly during the disaster of pandemic COVID-19.

References

Baldwin, R. B. W. (2020). *Economics in the time of Covid-19*. CEPR Press.

Bernama. (2021a). Malaysia aims for business events industry to be among top five in the world. *The Edge Markets*. Retrieved from: https://www.theedgemarkets.com/article. Accessed: the 30June 2021.

Bernama. (2021b). PIKOM: Covid-19 a wake-up call to transform using digital technologies. *The Edge Markets*. Retrieved from: www.theedgemarkets.com/article. Accessed: the 14 June 2021.

Bethapudi, A. (2013). The role of ICT in tourism industry. *Journal of Applied Economics and Business, 1*(4), 67–79.

Bockermann, K. J. (2019). The role of media and communication technologies for tourists, travelers and the travel industry. *Grin*. Retrieved from: https://www.grin.com/document/539645. Accessed: the 18 June 2021.

Bowdin, G. (2006). *Events management* (3rd ed.). Routledge.

Breaker. (2018). The 10 most tech-forward music festivals of 2018. *Medium*. Retrieved from: https://medium.com/. Accessed: the 18 June 2021.

Brown, A. E., Donne, K., Fallon, P., & Sharpley, R. (2020). From headliners to hangovers: Digital media communication in the British rock music festival experience. *Tourist Studies, 20*(1), 75–79.

Buhalis, D. (2000). Tourism and information technologies: Past, present and future. *Tourism Recreation Research, 25*(1), 41–58.

Campiranon, K. (2021). Redesigning events in the post COVID-19 crisis: A design thinking approach. In *Crisis management and recovery for events: Impacts and strategies*. https://doi.org/10.23912/9781911635901-4692

Cohen, H. (2019). *Social media definition: The ultimate guide that will make you smarter.* Retrieved from: http://heidicohen.com/social-media-definition/. Accessed: the 28 June 2021.

De Lira, V. M., Macdonald, C., Ounis, I., Perego, R., Renso, C., & Times, V. C. (2019). Event attendance classification in social media. *Information Processing & Management, 56*(3), 687–703.

Dillette, A., & Ponting, S. S.-A. (2020). Diffusing innovation in times of disasters: Considerations for event management professionals. *Journal of Convention & Event Management, 22*(3), 197–220.

Eck, G. V. (2018). The role of events on tourism. *Biz community*. Retrieved from: https://www.bizcommunity.com/. Accessed: the 8 June 2021.

Edwards, D., Foley, C., & Schlenker, K. (2011). Beyond tourism benefits: Measuring the social legacies of business events. *Business Events Sydney*. Retrieved from: http://www.businesseventssydney.com.au/. Accessed: the 1 July 2021.

Falassi, A. (1987). Festival: Definition and morphology. In A. Falassi (Ed.), *Time out of time*. University of New Mexico Press.

Faulkner, B. (2001). Towards a framework for tourism disaster management. *Tourism Management, 22*(2), 135–147.

Garay, L., & Morales, S. (2019). Decomposing and relating user engagement in festivals' virtual brand communities: An analysis of Sónar's Twitter and Facebook. *Tourist Studies*. https://doi.org/10.1177/1468797619873109

Getz, D. (2008). Event tourism: Definition, evolution, and research. *Tourism Management, 29*(3), 403–428.

Getz, D., & Page, S. J. (2016). *Events studies. Theory, research and policy for planned events*. Routledge.

Goeldner, C., & Ritchie, J. (2012). *Tourism* (12th ed.). Wiley.

Gössling, S., Scott, D., & Hall, C. M. (2020). Pandemics, tourism and global change: A rapid assessment of COVID-19. *Journal of Sustainable Tourism, 29*(1), 1–20.

Hanan, H., & Karim, A. S. (2015). Influence of social media in food festival destination image. In *International Conference on Tourism and Ethnicity in ASEAN and Beyond 2015.* Chiang Mai.

Hao, F., Xiao, Q., & Chon, K. (2020). COVID-19 and China's hotel industry: Impacts, a disaster management framework, and post-pandemic agenda. *International Journal of Hospitality Management, 90*, 1–11.

Hudson, S., & Hudson, R. (2013). Engaging with consumers using social media: A case study of music festivals. *International Journal of Event Festival Management, 4*, 206–223.

IATA. (2021). The impact of COVID-19 on aviation. *Airlines*. Retrieved from: https://airlines.iata.org/news/the-impact-of-covid-19-on-aviation. Accessed: the 12 June 2021.

Javed, M., Tuˇ, Z., & Jibril, A. B. (2020). The role of social media on tourists' behavior: An empirical analysis of millennials from the Czech Republic. *Sustainability, 12*(18), 7735.

Jin, X., Qu, M., & Bao, J. (2020). Impact of crisis events on Chinese outbound tourist flow: A framework for post-events growth. *Tourism Management, 74*, 334–344.

Jones, M. (2010). *Sustainable event management: A practical guide*. Earthscan.

Kaplan, A. M., & Haenlein, M. (2010). Users of the world, unite! The challenges and opportunities of social media. *Business Horizon, 53*, 59–68.

Karim, W. A. H. (2020). The Movement Control Order (MCO) for COVID-19 crisis and its impact on tourism and hospitality sector in Malaysia. *International Journal of Tourism and Hospitality.* https://doi.org/10.37227/ithj-2020-02-09

Khan, M. Y. H., & Hossain, A. (2018). The effects of ICT application on the tourism and hospitality industries in London. *SocioEconomic Challenges, 2*(4), 60–68.

Kim, H., & Chang, B. (2020). A study on the effects of crowdfunding values on the intention to visit local festivals : Focusing on mediating effects of perceived risk and e-WOM. *Sustainability, 12*(8), 3264.

Koo, C., Gretzel, U., Hunter, W. C., & Chung, N. (2015). The role of IT in tourism. *Asia Pacific Journal of Information Systems, 25*(1), 99–104.

Kumar, P., Mishra, J. M., & Rao, Y. V. (2021). Analysing tourism destination promotion through Facebook by destination marketing organizations of India. *Current Issues in Tourism.* https://doi.org/10.1080/13683500.2021.1921713

Law, R., Buhalis, D., & Cobanoglu, C. (2014). Progress on information and communication technologies in hospitality and tourism. *International Journal of Contemporary Hospitality Management, 26*(5), 727–750.

Lee, Y.-K., Lee, C.-K., Lee, S.-K., & Babin, B. J. (2008). Festivalscapes and patrons' emotions, satisfaction, and loyalty. *Journal of Business Research, 61*(1), 56–64.

Litvin, S. W., Goldsmith, R. E., & Pan, B. (2008). Electronic word-of-mouth in hospitality and tourism management. *Tourism Management, 29*, 458–468.

Liu, Y., & Chen, C. (2007). The effects of festivals and special events on city image design. *Frontiers of Architecture and Civil Engineering in China, 1*, 255–259.

Liu, H., Wu, L., & Li, X. R. (2018). Social media envy: How experience sharing on social networking sites drives millennials' aspirational tourism consumption. *Journal of Travel Research.* https://doi.org/10.1177/0047287518761615

Luttrell, R. (2018). *Social media: How to engage, share, and connect.* Roman & Littlefield.

Ly, H., Nuangjamnong, C., & Dowpiset, K. (2020). *The influencing factors in Cambodian Facebook users toward intention to use social media (Facebook) for travel decision making.* Retrieved from: https://ssrn.com/abstract=3787542. Accessed: the 14 June 2021.

Lyck, L., Long, P., & Grige, A. (2012). *Tourism, festivals and cultural events in times of crisis.* Copenhagen Business School.

MACEOS. (2020). Sektor Acara Perniagaan Catat Rugi Lebih 2.25 Bilion. *MACEOS Malaysia.* Retrieved from: https://maceos.org.my/sektor-acara-perniagaan-catat-rugi-lebih-2-25-bilion/. Accessed: the 14 June 2021.

Maganathan, D. K. (2021). Malaysian digital platform for the arts targets international festival market. *The Star.* Retrieved from: https://www.thestar.com.my/lifestyle/culture/2021/05/17/malaysian-digital-platform-for-the-arts-targets-international-festival-market. Accessed: the 13 June 2021.

Malaysia Convention and Exhibition Bureau (MyCEB). (2016). *Malaysia's business events roadmap charting Malaysia's journey to 2020 and beyond.* Retrieved from: https://www.myceb.com.my/clients/. Accessed: the 30 May 2021.

Malaysia Convention and Exhibition Bureau (MyCEB). (2019). *Malaysia's business events roadmap charting Malaysia's journey to 2020 and beyond.* Retrieved from: https://www.myceb.com.my/clients/. Accessed: the 8 July 2021.

Malaysia Convention and Exhibition Bureau (MyCEB) (n.d.). *Notice of events postponement & cancellation.* Retrieved from: https://www.myceb.com.my/notice. Accessed: the 30 May 2021.

Mapjabil, J. (2014). Penerimaan komuniti tempatan terhadap pelancongan sukan di Pulau Langkawi. In J. Mapjabil (Ed.), *Pelancongan di Pulau Langkawi - Pembangunan dan Implikasi.* Penerbit UTHM.

Mapjabil, J., Marzuki, M., Kumalah, M. J., Tangavello, L., & Zainal, M. K. (2015). Sport as a tourism attraction in Malaysia: Potential and prospects. *Geografia: Malaysian Journal of Society and Space, 11*(12), 23–31.

Mason, M. C., & Paggiaro, A. (2012). Investigating the role of festivalscape in culinary tourism: The case of food and wine events. *Tourism Management, 33*, 1329–1336.

Mazareanu, E. (2021). Coronavirus: Impact on the aviation industry worldwide-statistics & facts. *Statista.* Retrieved from: https://www.statista.com/topics/6178/coronavirus-impact-on-the-aviation-industry-worldwide/. Accessed: the 30 May 2021.

McKercher, B., Mei, W. S., & Tse, T. S. (2006). A short duration cultural festivals tourist attraction. *Journal of Sustainable Tourism, 14*, 55–66.

Naharul, M. A., & Kamel, H. (2021). Business events set for more losses on recurring MCO. *The Malaysian Reserve.* Retrieved from: https://themalaysianreserve.com/. Accessed: the 14 June 2021.

Nasir, H. M., Alagas, E. N., & Nasir, N. M. (2019). A review of business events in Malaysia: Finding the gap. *19th Kuala Lumpur International Business, Economics and Law Conference, 19*(2), 247–255.

Năsulea, C., & Mic, S. M. (2018). Using blockchain as a platform for smart cities. *Journal of E-Technology*. https://doi.org/10.6025/jet/2018/9/2/37-43

Ngernyuang, K., & Wu, P. (2020). Using social media as a tool for promoting festivals tourism. *International Journal of Computer Science & Information, 12*(3), 17–32.

Onan, G. (2020). The impact of ICT on strategic management: In the context of events in tourism organizations. In Y. A. Unvan & I. Serbestoğlu (Eds.), *Current research in economics and administrative science*. IVPE.

Parsons, W. (1996). Crisis management. *Career Development International, 1*(5), 26–28.

Primack, B. A., Karim, S. A., Shensa, A., Bowman, N., Knight, J., & Sidani, J. E. (2019). Positive and negative experiences on social media and perceived social isolation. *American Journal of Health Promotion*. https://doi.org/10.1177/0890117118824196

Rizky, R. M., Kusdi, R., & Yusri, A. (2017). The impact of e-WOM on destination image, attitude toward destination and travel intention. *RJOAS, 1*(61), 94–104.

Shankar, R. S. (2021). Social media as a credible e-source on information. A study of tourists' perception. *Journal of Brand Management, 18*(1), 48–62.

Sigala, M. (2020). Tourism and COVID-19: Impact and implications for advancing and resetting industry and research. *Journal of Business Research, 117*, 312–321.

Sigala, M., Christou, E., & Gretzel, U. (Eds.). (2012). *Social media in travel, tourism and hospitality: Theory, practice and cases*. Ashgate.

Sotiriadis, M. D. (2017). Sharing tourism experiences in social media. *International Journal of Contemporary Hospitality Management, 29*(1), 179–225.

Stafford, G., Yu, L., & Armoo, A. K. (2002). Crisis management and recovery how Washington, DC, hotels responded to terrorism. *The Cornell Hotel and Restaurant Administration Quarterly, 43*(5), 27–40.

Sultan, M. T., Sharmin, F., & Xue, K. (2019). Sharing tourism experience through social media: Consumer's behavioral intention for destination choice. *International Journal of Humanities and Social Sciences, 13*(2), 129–133.

Suryawardani, I. G. A. O., & Wiranatha, A. S. (2016). Digital marketing in promoting events and festives. A case of Sanur Village Festival. *Journal of Business on Hospitality and Tourism, 2*(1), 175–183.

Temelkov, Z., & Gulev, G. (2019). Role of crowdfunding platforms in rural tourism development. *Sociobrains International Science, 56*, 73–79.

Tew, P. J., Lu, Z., Tolomiczenko, G., & Gellatly, J. (2008). SARS: Lessons in strategic planning for hoteliers and destination marketers. *International Journal of Contemporary Hospitality Management, 20*(3), 332–346.

The ASEAN Post. (2020). *How technology is changing tourism*. Retrieved from: https://theaseanpost.com/article/how-technology-changing-tourism. Accessed: the 13 June 2021.

The Star. (2019). *Tourism contributes RM42bil to economy.* Retrieved from: thestar.com.my/. Accessed: the 30 May 2021.

Tourism Malaysia. (2020a). *Malaysia tourism statistics in brief.* Retrieved from: corporate.tourism.gov.my/statistics. Accessed: the 30 May 2021.

Tourism Malaysia. (2020b). *Tourism revenue grows +16.9%, reaches RM21.4 billion in Q1 2019.* Retrieved from: tourism.gov.my/media. Accessed: the 30 May 2021.

Tourism Malaysia. (2020c). *Cancellation of Visit Malaysia 2020 (VM2020) campaign and guest stay at tourist accommodation premises throughout the movement control order(MCO).* Retrieved from: https://www.tourism.gov.my/media/view/cancellation-of-visit-malaysia-2020-vm2020-campaign-and-guest-stay-at-tourist-accommodation-premises-throughout-the-movement-control-order-mco. Accessed: the 14 June 2021.

Tourism Malaysia. (2021). *International tourist arrivals to Malaysia Plunge 83.4% in 2020.* Retrieved from: http://mytourismdata.tourism.gov.my/. Accessed: the 13 June 2021.

UNWTO. (2021). *Tourism and COVID-19-Unprecedented economic impacts*. Retrieved on June 12, 2021, from: https://www.unwto.org/tourism-and-covid-19-unprecedented-economic-impacts. Accessed: the 13 June 2021.

Van Winkle, C., & Bueddefeld, J. (2020). Information and communication technology in event management. In *Handbook of e-Tourism* (pp. 1–22).

Vidal, B. (2019). *The new technology and travel revolution*. Retrieved from http://wearemarketing.com/. Accessed: 3 June 2021.

Wagen, L. V., & White, L. (2018). *Event management for tourism, cultural, business, and sporting events*. Retrieved from: https://books.google.com.my/. Accessed: the 22 June 2021.

World Travel and Tourism Council (*WTTC*). (2015). *Seven ways technology is changing the travel industry*. Retrieved from http://worldtraveltourismcouncil.medium.com/. Accessed: 3 June 2021.

Xiang, Z., & Gretzel, U. (2010). Role of social media in online travel information search. *Tourism Management, 31*(2), 179–188.

Yahya, S. F. S. (2020). The impact of COVID-19 on Malaysia's airline industry. *Astro AWANI*. Retrieved from: https://www.astroawani.com/berita-malaysia/impact-covid19-malaysias-airline-industry-264574. Accessed: the 22 June 2021.

Yong, Y. K. & Ghazali, K. (2018). Accelerating business tourism growth in Malaysia. *PEMANDU Associates*. Retrieved from: https://www.pemandu.org/. Accessed: the 22 June 2021.

Zeng, B. (2013). Social media in tourism. *Journal of Tourism & Hospitality, 2*(02). https://doi.org/10.4172/2167-0269.1000e125

Zhu, Z., & Kim, H. (2020). Determinants of SNS photo sharing behavior of teenagers at an outdoor music festival in China. *International Journal of Advanced Smart Convergence, 9*(2), 129–139.

Živković, R., Gajić, J., & Brdar, I. (2014). The impact of social media on tourism. In *Paper presented at Sinteza 2014 - Impact of the Internet on Business Activities in Serbia and Worldwide*. https://doi.org/10.15308/sinteza-2014-758-761.

Zubair, F., & Shamsudin, M. F. (2021). Impact of Covid-19 on tourism and hospitality industry of Malaysia. *Journal of Postgradute Current Business Research, 6*(1), 1–6.

Dr Noraihan Mohamad is a Senior Lecturer at School of Hospitality and Creative Arts, Management and Science University (MSU). She obtained Bachelor of Tourism Management (Marketing) (Hons), Master of Science in Media Management and PhD degree in Media Management from Universiti Utara Malaysia (UUM). She taught undergraduate courses on Research Methods for Service Industry, Current Issues in Hospitality and Tourism, Strategic Management for Service Industry. Apart of that, she also taught postgraduate courses on Global Trends and Issues in Hospitality and Tourism, Tourism Destination Analysis and Organizational Behavior. Her research interests are on the influence and impact of media on society, businesses, and consumers. Her scope of expertise is Malaysian mass media and new media, media and technology (including social networking), as well as tourism management. Besides presented numerous papers at national and international conference, she also contributed articles in journals and chapters in books on media. Her current contribution to Malaysia's Government is on developing a Malaysian National Unity Index.

Part VIII
Directions for Future Research

Chapter 23
Tourism Events, Festivals and Technology Application in Asia: Directions for Future Research

Ratan Ghosh and Iffat Anjum

Introduction

Event tourism can be accredited as a 'prolific game changer' to trigger the business volume of international tourism inciting sharp competition with global automobiles, information technology and petroleum industry in the contemporary world. According to United National Populations Fund (UNFPA), being the largest and most populated continent of the world, Asian sub-system accommodates 4.6 billion of total 7.8 billion inhabitants of this world demonstrating one of the most lucrative tourism destinations comprising of 59.5% demographic representation worldwide (United Nations Population Fund (UNFPA), 2021). In fact, this travel and tourism industry of Asia Pacific has sky-scraped stimulating 7.4% growth in regional GDP in the year of 2019 (World Travel & Tourism Council, 2021). Albeit pace and vogue of worldwide tourism industry has sharply dwindled during COVID-19 pandemic crisis and Asian tourism industry has also been similarly subjected to these unfortunate downfalls. Despite, owing to heterogeneous culture, heritage and ceremonials of this protracted sub-system, event tourism has band-wagoned Asian traditional rampant tourism industry since ages. Now, events can be depicted simply as ascertained periods of celebration, entertainment or other precise festive activities which invigorates tourism attraction and tourism potency of the host country alluring national and international travelers and ultimately generates marketing, profit, foreign exchange and employment for the destination. The amplitude of the

R. Ghosh
Department of Accounting and Information Systems, Bangladesh University of Professionals, Dhaka, Bangladesh

I. Anjum (✉)
Department of International Relations, Bangladesh University of Professionals, Dhaka, Bangladesh
e-mail: iffat.anjum@bup.edu.bd

A. Hassan (ed.), *Technology Application in Tourism Fairs, Festivals and Events in Asia*, https://doi.org/10.1007/978-981-16-8070-0_23

comprehensive term ‘events’ encompasses multifarious ceremonial activities of diverse objective, features and timelines. This holistic mosaic idea of ‘events’ incorporates cultural celebrations (festivals, carnivals), religious celebrations (pilgrim, congregation), business summits (conventions, themed markets), political events (royal occasions, high value individual visit), arts and entertainment (concerts, award ceremonies), sports events (games, formal tournaments), educational and scientific events (conferences, seminars) and private events (destination wedding, parties) etc. maximizing image of the host country/region as per the hierarchical tenor of these mega events (Olympic games, FIFA World Cup), special events (International Millennium Events of 2000, Royal Weddings like Wedding of Prince Harry and Meghan Markle of 2018), hallmark events or major events (Women’s Rugby World Cup, Harbin International Ice and Snow Festival of China), regional events (Dhaka Lit Fest, Holi Festival in India) and local/community events (provincial Rath Yatra of India and Jobbarer Boli Khela of Chattogram, Bangladesh) with their distinctive features (Getz & Page, 2016). However, technology and updated digital equipment have emerged as a critical addition in these diverse event tourism in contemporary Asia. Now, the prime query of this paper advances to unveil how far the incorporation of technological bestowment in events and festivals can bolster demand for tourism destination within this regional sub-system of Asia. Thereby, this paper has gradually weighed to what extent techno equipment such as digital marketing, digital opinion polls, e-ticketing and transactions, digital broadcasting and social media coverage of diverse events and festival can animate the significance of Asian sustainable tourism and destination demand in the contemporary world.

Rationale and statement of the problem of this research are straightforward. United Nations World Tourism Organization has accredited Asia to be the ‘hotbed’ of tourism growth since last several decades (United Nations World Tourism Organization (UNWTO) and Global Tourism Economy Research Centre, 2019). Apart from lucrative tourism destinations, Asian affluent culture, festivals, and attractive traditional ceremonials have metamorphosed Asia to be the fastest growing tourism region in the world. In order to experience diverse hallmark events and festivals of Asia (Chinese Ice and Snow Festival, Thai Shongkran Festival), millions of travelers tighten their shoelaces every year towards Asian destinations. Now, in parallel to fostering inherent traditions of colorful festivals, the evolution of ‘Information and Technology (IT)’ application in tourism and hospitality sector has emerged as a ‘naïve annexation’ and ‘critical tool’ in competitive Asian tourism industry today. Technology application can be considered as an inevitable requirement of twenty-first century’s tourism industry. Thereby, events and festivals of Asia are not only alluring excursionists with their unique features, but also incorporating technological bestowments such as digital marketing, social media brandling, Online Travel Agencies (OTA) and online public opinion polls to ease up the journey for them. Thai, Chinese, and Indian culture and festivals have accomplished their distinctive positions in enchanting tourists from both domestic and international domains. So, to compete and to sustain in contemporary tourism industry, Asian events and festivals have ensured technological embodiment within its capacity. China has been consecutively securing its top-ranking position in demarking maximum number of tourists in Asia since last several years. Recently, to combat with the scourge of COVID-19 pandemic based tourism crisis, China has been evidenced to ensure maximum usage of digital marketing, live streaming and online travel agencies in order to reboot its tourism reigning. Now, this paper attempts to weigh that how far overall Asian tourism events are being able to complement its festival features and tourist' demand through the application of technology in present world. In parallel, this paper also tends to explore the existing obstacles in this regard and expects to design some future directions to ensure maximum utilization of technology in event tourism to comply with the targets of sustainable tourism development in the long run.

According to United Nations World Tourism Organization (UNWTO) and Global Tourism Economy Research Centre (2019), Asia and the Pacific has secured its 'envious' global top-ranking position in triggering tourism growth since decades. The sky-scrapping motion has recorded 07% growth in international tourist arrivals in Asia demarking a total of 360 million international tourist arrivals which constitutes one-fourth of the world's total. Asia and the Pacific has evidenced to be the 'second most' visited region of the world following Europe. Asia's share of international tourist arrival has accelerated to be 25% in 2019 from that of 16% in 2000. Asia has also astounded to outperform all other regions of the world in recording highest average annual increasement of tourist arrival (). In between 2010 and 2019, Asia has secured 07% increment in average annual tourist arrival in comparison to that of 05% to be the world average. Further, Asian tourism industry has earned a grandiose of USD 443 billion in 2019 constituting 30% of the global earning. This share has almost incremented double in comparison to than that of 17% in the year of 2000. Total export revenues from tourism sector in Asian sub-system has reached the bar of USD 487 billion in the year of 2019. Now, to analyze the region wise trends, in the year of 2019 almost half of Asia's international tourist arrivals have been recorded in North-East Asia (49%) with largest tourism spending (70%) and significant increasement in annual tourism growth (+09%). Following North-East Asia, South-East Asia has perceived to be the 'emerging tourist hub' of Asian subsystem. South-East Asia represents 37% tourist arrival in Asia with a record of 17% tourism spending and average annual growth of 08% in this region. South-Asia also tends to be a potential region representing 09% tourist arrival along with a noticeable annual growth of (+10%) and 05% tourism spending. Then again, Oceania accounts for 05% of tourist arrival along with 08% tourism spending within the sub-system. Thereby, South Asia and South-East Asia are regarded to be the fastest growing sub-regions owing to their tourism potentials.

Again, according to United Nations World Tourism Organization (UNWTO) and Global Tourism Economy Research Centre (2019), from North-East Asia, China has been accredited as the most popular destination for tourists in Asian sub-system. In the year of 2019, around 66 million international tourists have been stepped in China. Not limited to Asia only, China has been recognized as the fourth most visited as well as popular tourist destination worldwide in the year of 2019. China has been followed by Thailand from South-East Asia in terms of destination popularity, around 40 million travelers have advanced towards Thailand in the same year of

2019. Thailand has also band-wagoned in the ranking chart as the eight most popular tourism destination in the world. Japan is considered as the third popular destination in Asia with 32 million tourist arrivals in 2019. After Japan, Hong Kong (China), Macao (China), Vietnam and India have been ranked respectively to be popular among tourists worldwide. However, around 79% of the Asian travelers are engendered from Asia itself which prioritizes intraregional tourism within the sub-system. As per interregional tourism, Europe (11%) and America (06%) respectively constitute largest numbers of tourists to unveil the legacy and heritage of Asia.

Now, this paper has argued 'technological application in events' to be one 'naïve' but 'important element' in raising demand of Asian tourism destinations. Asia has created one influential tourism market owing to its diverse traditions, events, and tourists spots. But to fully maximize the probable potentials of this region through the flourishment of events and festivals, there can be no alternative to the incorporation of latest information and technological applications in Asia. So, this paper attempts to establish a logical cohesion in between the increasing demands of Asian tourism destinations and the application of latest technologies in events and festivals of Asia.

To comprehend the interplay between techno-equipped tourism events and festivals and their perpetual impact on accelerating tourism destination demands in upcoming Asia, this paper has orchestrated a central research query/key research question with an attempt to inquiring into the matter more intriguingly: how far technology application in tourism events and festivals can ameliorate demand for tourism destinations in Asia?

Subsequently, this paper has also unraveled some relevant queries regarding techno-based events and acceleration of Asian tourism demand through generating some secondary research questions: first, what are the patters and features of events and festivals in Asia?; second, how do events and festivals foster Asian diverse heritage and identity?; third, how do technology and digitalization complement Asian tourism events?; fourth, hat are the prevailing obstacles in techno-equipping events and festivals in Asia?; and fifth and finally, how can persisting constrains be combatted regarding digitalization of tourism events in accelerating tourism destination demand in Asia?

To weigh the inter-relation between raising demands of Asian tourism destinations and technology application in festivals, this paper has adopted a qualitative research methodology. To measure the vigor of latest technological aids over in Asia, this paper has incorporated an explanatory research method. Further, secondary-data collection method has been embodied to compile the pedagogical understanding. Gradually, the design of the research argument has been developed and authenticity of the data has been consciously scrutinized. This paper has analyzed a cluster of diverse information related to technological bestowment and Asian event tourism from books, organizational reports, conference papers, websites of several online tourism agencies, peer reviewed journal articles, national tourism-based websites of Asia, e-book, reports compiled by tourism related Intergovernmental Organizations (IGOs) and Non-Governmental Organizations (NGOs), newspaper articles and magazines etc. This paper has derived information from both websites and organizational reports and write-ups composed by institutions like World Tourism Organization (WTO), Global Tourism Economy Research Centre (GTEC), World Travel & Tourism Council (WTTC) and Bangladesh Parjatan Corporations (BPC) etc. The authenticity, accuracy and reliability of the data inserted while developing the argument of this paper have been cross-checked and verified with major concern. To defend the argument of the paper, compiled and verified data have been analyzed using content analysis or thematic analysis technique. Finally, primary data-based survey analysis could have added some more comprehensive insights regarding the findings of the paper but due to time constrains and COVID-19 pandemic protocols these could not be implemented in real.

While venturing into the compilation of literatures related to Asian tourism events and festivals and relevance of technological application in these festivals in enhancing destination demand from the visitors, this paper has witnessed a relative vacuum of festival-based research knowledge in Asian context. Although, there persists a burgeoning collection of reports, articles, books and journals related to the comprehensive insights of event tourism and global festival tourism. But precisely literatures focusing on Asian trends, drawbacks and strengths of festival tourism have been very rare. More significantly, the contribution of technological aid in flourishing festival tourism specifically in Asia, has remained one of the very under-researched issue-area in this regard. Thereby, this paper attempts to unfold this area of knowledge while explaining the necessity of technological application, identifying the persisting gaps related to technological application and finally prescribing some future directives to fully incorporate the bestowments of technology in maximizing destination demand in Asia.

Eloquent scholar and tourism expert Donald Getz has glorified the application of technology in bolstering tourism demand (Getz & Page, 2016). Scholars possessing expertise in the similar arena have also emblazed the significance of events in harnessing tourism demand and have also advocated the complementary services of technological elements in this regard. People in all cohorts and cultures have always recognized the necessity of pleasure, work-escape, celebrations, and relaxing occasions since ages. Whether it's the arts and culture, food and wine, holidays, or religion, science and education events have always assembled people to comply with their subjective needs (Quinn, 2009). Event tourism-based research also encapsulates how these festivals have fostered heterogeneous cultural identity and traditions of people since centuries. Event tourism have been a strong demand generating factor in contemporary world. So, events and festivals have been recognized as a significant determinant of destination marketing. They not only annex critical mass of attractions available at the destination, but also they foster aged old traditional practices generating significant economic benefit for the destination (Pirnar et al., 2019).

The events industry has boosted dramatically over the last several decades, and there have been significant participation from governments, local communities, and private sectors (Uysal & Gitelson, 1994). Now, apart from these stakeholders, the significance of technological aid like digital marketing and online travel agencies have emerged as a 'game changer' in contemporary tourism events (Li et al., 2020).

Today, technologies are also adding new vigor into the aged old celebrations of festivals in different parts of the world (The ASEAN Post, 2020). Now, the application of event tourism has similarly added special significance in Asian event tourism. But still there persists a significant gap in encompassing technology in all the spheres of event tourism management (Development Asia, 2019). For example, digital branding, digital marketing, online consultation, digital literacy, e-ticket booking etc. are yet to be executed in proper manner in Asia.

However, there remains significant developments too. Social media branding, live streaming, traditional etiquette, and practice sharing allure millions of people towards Asian unique events every year. Examples are: Japanese artificial intelligence based multilingual compatible chatbot Trad-Fit is aiding tourists to avoid language barriers while travelling (Yeo, 2019). Then again, creating booking platforms through Online Travel Agencies (OTA) is becoming very popular and beneficial in South East Asian tourism growth (Thaiger, 2019). China's biggest export and import fair have incorporated digitalized elements since the inception of pandemic period (Tan et al., 2020). Online collaboration among the event based stakeholders, social media branding of Japanese and Thai food-tours (Indochina Travel, n.d.), promotion of music festivals through social media branding like WeChat and Webo in China etc. Chinadaily.com.cn. (2020) have glittered Asian event tourism in contemporary world. Now, Similar to more explored regions of Asian sub-system, South Asia also holds significant potential for technology-based event flourishment. But to maximize the elements and bestowments of technology in its fullest manner digitalization, developing ICT knowledge, website development, digital branding, uninterrupted online services etc. can pave the way towards stimulating demand and temptation for South Asian tourism destinations worldwide in future days to endeavor.

The Context

Events

To encapsulate into simpler form, events can be portrayed as a phenomenon occurring in a demarcated time and territory which might be comprised of festive celebrations commemorating any special occasions or significant matter (Westcott et al., 2015). Now, the term 'event' is an umbrella term, and it incorporates multifarious phenomenon of diverse scale, nature and kinds within its comprehensive domain. Thereby, in order to understand the features of events, it is inevitable to get acquainted with the variances of events. Although, the categorization of events does not have any concrete demarcation, so these variances might overlap, but to get enlightened regarding the interplay between events and tourism, it is prerequisite to have a generalized understanding of the variances of events. Examples are: FIFA World Cup, Indian Holy Festival.

Festivals

According to the International Dictionary of Event Management, festivals can be depicted as kind of events that incorporates kaleidoscope of conduction as well as public celebration while conveying particular meaning to both stakeholders and witnesses (Westcott et al., 2015). In alternative words, festivals can be interpreted as formal timeline, delightful and pleasurable events or happenings beholding a basic collective celebration and festive feature (Quinn, 2009). Festival also refers to the cultural resources of any particular area which make possible successful hosting of visitors over there. The of ‘festival’ is rooted in Latin linguistic. Latin word ‘festum’ is considered to be the embryonic rhizome of festival meaning ‘general fun and festivity’. But sometimes, festival is also perceived as ‘festive events’ originating from two separate Latin term ‘festum’ and ‘feria’. Here, ‘festum’ is perceived as ‘public rejoice, exhilaration and hilarity’ whereas ‘feria’; is assumed as ‘escape from scheduled work in dignifying the Gods’. Another alternative contemplation has traced the origin of festival in Latin word ‘feast’/ ‘fiesta’ indicating towards ‘collaborative celebration’. Thereby, festivals incorporate a space and time escapade from scheduled life where collaborative pleasure experiences can be created and shared (Pirnar et al., 2019). Basic feature of these festivals can be manifested as (i) collaborative celebration; (ii) public demonstration; and (iii) subjected to relevant civic rituals of festival destination. Examples are: Thai new year festival Songkran, Indian Holi festival. But festivals, too can be sub-categorized as per its diverse and versatile nature.

Event Tourism

Even tourism can be depicted as an interdisciplinary nexus in between event studies and tourism studies where tourism activities converge around the occurrences of diverse events generating ample attraction, marketing, and economic leverage for event destination. Meticulous professor and leading international tourism and event studies scholar Donald Getz has conceptualized event tourism as encompassing events and festivals and has interpreted it to be at the nexus of tourism and event studies (Quinn, 2009). Event tourism has emerged as the ‘game changer’ element in contemporary tourism industries worldwide. Event tourism affixes additional ornaments as well as attractions to the tourism destination ensuring sustainable tourism development. Events have generated triggering elements for both demand side (home/origin area) as well as for supply side (host/destination territory) (Uysal & Gitelson, 1994). In home countries, event details can be portrayed as ‘push factors’ motivating and tempting explorers towards event destinations whereas in host territory events can be assumed as ‘pull factors’ complementing development, marketing strategies and economic profiteering in tourism destination. Thus, event tourism has sky-scrapped contemporary tourism industry as image maker, marketer, catalyst. Motivator and animator of destination attractiveness.

Event Management and Tourism Management

Donald Getz advocates event tourism as a set of interrelation in between event management and tourism management. The eloquent scholar has deposited that tourism management eminently deals with tourism development scrutinizing the tourism behavior, attraction, expectation, security, and services of all kinds of tourists. Conversely, event management manifestly converges around event planning, event designing, event experiences, event marketing and managing of events. The nexus is significantly important because the more perceivable events the destination has, the more it would seem attractive to the tourists (Quinn, 2009).

Tourist Attraction

Tourist attraction can be encapsulated as persuading elements of tourist destination such as tourist spots, cultural heritage, security, accessibility, festive atmosphere, rational expenditure etc. which stimulate people's motivation to travel from home country to host country (UKEssays, 2018).

Features of Event Tourism

In order to investigate into the interplay between event tourism, technology application and sustainable tourism in Asia, it would be preeminent to highlight the basic extent and parameters of event tourism today:

First, events not only motivate excursionist towards tourism destination, but they also enliven tourism development plans and strategies engendering competitive advantages for the destination (Pirnar et al., 2019). Second, thus, tourism events can be portrayed as yardsticks of economic development in tourism destinations. Third, events are animators of destination attractiveness and promotional marketing of the destined territory. Fourth, events orchestrate accommodation, attraction, marketing, security, transport and ancillary services of perceived travelers invigorating tourism potency and capacity of the destination region. Fifth, during, less demanded off peak seasons of tourism industry, events can bridge the gap creating a significant market in tourism destinations based on the features of the events in complying with the demands of the tourists. Thus, event tourism generates substantive growth during mediocre or least tourism seasons (Dychkovskyy & Ivanov, 2020). Sixth, event tourism reduces seasonal uncertainty and maximizes influx of both international as well as national vacationists across tourism destinations. So, apart from international tourists, event tourism also incites an alternative demand stimulated by domestic travelers ultimately generating wider remit than destination based tourism (Li et al., 2020). Seventh, event tourism can booster sustainable tourism while generating economic benefit and employment of the destination territory. Eighth, vents and festivals are considered as tools to foster cultural identity and to create an optimistic image of the destination region as well. Ninth, event tourism is non-scheduled. There remains no concrete routine to harness events. Some events may occur once and for all such as royal weddings and sometimes events of divergent features might take place in the similar destination repetitively or simultaneously each having précised and distinctive socio-economic objectives (Laurell & Björner, 2018). Tenth, tourism events require the collaboration of multifarious stakeholders. Range of stakeholders might incorporate government, business, public or private industries. Proper coordination and collaboration maximize the probabilities of a successful tourism event (Getz & Page, 2016).

Sustainable Tourism Development

Sustainable tourism development has portrayed tourism to take full account of economic, environmental and societal impact for both prevailing and successor generations ensuring just and equitable requirements of travelers, tourism industry, economy, environment, identity, heritages and host communities (United Nations World Tourism Organization, n.d.). Thus, ‘sustainable tourism development’ ethics prescribe a rational balancing in between consumption and preservation of economic, environmental and socio-cultural aspects of tourism development to ascertain sustainability in the future epoch to endeavor. Sustainable Tourism Development is a continuous and contested process manifested to incorporate:

First, the optimal usages of environmental resources and global commons. While invigorating tourism industry, ecological process need to be fostered ensuring least exposure to environmental damages through carbon emission or other means. Traditional heritage, civilization and biodiversity needs to be preserved while nurturing sustainable tourism.

Second, century old traditions, ideology, ceremonials, customs, rituals and socio-cultural authenticity can be considered as intangible treasures of any particular host territory. Dignifying linguistic and cultural authenticity of tourism destination with proper empathy as well as tolerance should also be considered as inevitable component of sustainable tourism development.

Third, prevailing all the stakeholders of tourism industry should be subjected to equitable access of socio-economic conveniences from this industry. Economic development should combat unemployment triggering earning opportunities and social services in the host communities considering sustainability ethics in mind (United Nations World Tourism Organization, n.d.).

Fourth and finally, sustainable tourism development should prioritize tourist' preferences and hospitality primarily. In parallel, this process requires that travelers be aware and amicable too regarding sustainability ideas and practices while exploring the nature.

Categorization of Event Tourism

As portrayed before, within the realm of tourism studies, 'event' is used as an umbrella term. The variances of event tourism may not be concrete theoretical categorization, but it can enlighten us with the wide range of phenomenon including diverse forms of 'festivals' which are incorporated into the comprehensive definition of 'events'. To understand the domain of event tourism, this paper analyses the categorization of events based on two distinctive determinants: first, hierarchical categorization of event tourism; and second, featured categorization of event tourism. These categorizations might overlap but contains significant pedagogy regarding the nature and extent of events. For example: 'festival' is considered as an inevitable form of 'event' but not necessarily all forms of events shall incorporate festive nature. Events attracting particular cohort of national and international visitors might be arranged with pure formal and professional objectives too (Getz & Page, 2016).

Hierarchical Categorization of Event Tourism

Events trigger tourists' attraction and demand from home and abroad subjecting to their extensive occasional features, budget, uniqueness and preparedness maintaining a 'top-down' approach (Table 23.1).

Now, Donald Getz has eloquently categorized event tourism in his portfolio approach hierarchically demonstrating four specific categorizations as per the popularity and significance of the events (Getz & Page, 2016). Thereby, being persuaded by the amplitude based chronological division of the events, this paper has also classified events into five distinctive variances to understand their propagation and magnitude worldwide (Fig. 23.1).

Table 23.1 Hierarchical categorization of event tourism

Event types	Event features	Event examples
Mega Event	Colossal in extent and capacity	FIFA World Cup
	Occasional in Nature (Getz & Page, 2016)	ICC Cricket World Cup
	High tourist demand and high value	Olympic Games
	International in nature	Paralympic Games
	Both domestic and international visitants	Commonwealth Games
	Gigantic budget and vast number of attendees	Rugby World Cup
	Grandiose economic transactions	Academy Awards/The Oscar
	Substantive economic benefit	World Fairs and Expos like The Shanghai Expo
	Splendid arrangements	International Economic Summits
	International media coverage	
	High profile visits (Getz & Page, 2016)	
	Highly prestigious	
	Considered as 'must witnessed'	
Special Event	Special attribution to any significant fact	International Millennium Events of 2000
	One-timed/occasional/infrequent	Royal Weddings like Wedding of Prince Harry and Meghan Markle 2018
	Diacritical sponsoring and embodiment	Diamond Jubilees like Diamond Jubilee of Queen Elizabeth II
	High tourist demand and high value	
	International in nature	
	Both domestic and international visitants	
	Substantive number of attendees	
	Huge budget & extensive economic transactions	
	Significant economic benefit	
	Gorgeous arrangement	
	Global media coverage & high-profile visits	
	Special attention and of highly prestigious	
	In general, specific ceremonials are orchestrated	

(continued)

Table 23.1 (continued)

Event types	Event features	Event examples
Hallmark Event/Major Event	Possesses significant traditions and attractiveness	Worldwide celebration of New Year's Eve
	International in nature	Harbin International Ice and Snow Festival of China
	Periodic in nature	Women's Rugby World Cup
	High tourist demand and high value	Carnival of Rio
	Both domestic and international visitants	World Sacred Spirit Festival, India
	Large number of attendees	Octoberfest
	Significant budget	La Tomatina
	Vast economic transactions	Cannes Film Festival
	Huge economic benefit	Bishwa Ijtema/Global Congregation of Bangladesh
	Rich arrangements (Getz & Page, 2016)	Hajj of Saudi Arabia
	International media coverage	Thai new year festival Songkran
	Mostly high-profile visits (Pirnar et al., 2019)	Nozawa Onsen fire festival of Japan
	Prestigious and of precise importance	Dhaka International Trade Fair
	Generates competitive tourism advantages (New Zealand Major Events, n.d.)	
Regional Event	High tourist demand and moderated value/ medium tourist demand and medium value	Provincial Rugby Games
	Periodic/occasional/one-time	Holi Festival of South Asia
	Regional in nature sometimes might have international representation	Gion Matsuri of Japan
	Mostly domestic and sometimes international visitants	Dhaka Lit Fest
	Medium number of attendees	Pahela Falgun/First Day of Spring in Bangladesh
	Medium to high budget	Pahela Baishakh/Bangla New Year Celebration in Bangladesh
	Moderated economic transactions	Hay Festival in Dhaka
	Mediocre economic benefit	Thai/Chinese culinary Festivals
	Rational arrangements	
	Domestic and sometimes international media coverage	
	Regional significance and popularity (New Zealand Major Events, n.d.)	

(continued)

Table 23.1 (continued)

Event types	Event features	Event examples
Local Event	Low tourist demand and moderated value/low tourist demand and low value	Local fairs like provincial Rath Yatra of India and Rather Mela of Bangladesh.
	Periodic/occasional/one-time	Local festivals like Jobbarer Boli Khela of Chittagong, Bangladesh.
	Native in nature (Getz & Page, 2016)	Fundraises
	Spatial/territorial representation	Picnics
	Mostly domestic visitants	Barbecues
	Lower number of attendees	
	Low to medium budget arrangements	
	Less economic transactions and benefit (Uysal & Gitelson, 1994)	
	Mostly domestic media coverage	
	Native significance and popularity	
	Invigorate cultural plurality	
	Animates sense of belongingness and societal cohesion among different communities Staging of events foster community values (Westcott et al., 2015).	

Source: self-compilation by the authors

Fig. 23.1 Hierarchical categorization of event tourism (Getz & Page, 2016). (Source: self-developed by authors in compilation with the ideas of Donald Getz)

Featured Categorization of Event Tourism (Table 23.2)

Table 23.2 Featured categorization of event tourism

Event types	Event features and variances	Event examples	Event hierarchy
Trade and Business Events	Meetings	Global Business Summit	Mega Trade Event
	Conventions	Asian Business Summit	Special Trade Event
	Themed markets		Hallmark/Major Trade Event
	Consumer and trade shows		Regional Trade Event
	Corporate marketing events		Local Trade Event
	Trade fairs		
	Conferences		
	Summits (New Zealand Major Events, n.d.)		
Sports Events	Amateur/Professional leagues	FIFA World Cup 2019	Mega Sports Event
	Spectator/Participants	Olympic Games	Special Sports Event
	Recreational games for pleasure	Asian Games	Hallmark/Major Sports Event
	Athlete Tourism		Regional Sports Event
	Cricket/Football/Cycling tourism		Local Sports Event
Political Events	Political conferences	Rio Earth Summit 1992	Mega Political Event
	International Conventions	ASEAN Summit	Special Political Event
	Large scale congresses	SAARC Summit	Hallmark/Major Political Event
	Meetings		Regional Political Event
	High official/VIP visits		Local Political Event
	Royal occasions & Summits		

(continued)

Table 23.2 (continued)

Event types	Event features and variances	Event examples	Event hierarchy
Arts and Entertainment Events	Concerts & theatre shows	Academy/	Mega Artistic Event
	Award ceremonies	Oscar Awards	Special Artistic Event
	Crafts Exhibition	Hay Festival	Hallmark/Major Artistic Event
	Movie shows	Dhaka Lit Festival	Regional Artistic Event
	Dance shows		Local Artistic Event
	Literary event		
Education and Scientific Events	Conferences	American Geophysical Union	Mega Academic Event
	Symposiums	AAS in Asia Summits	Special Academic Event
	Seminars (Westcott et al., 2015)		Hallmark/Major Academic Event
			Regional Academic Event
			Local Academic Event
Private Event	Destination weddings	Ambani Family Weddings	Mega Private Event
	Private parties (Getz & Page, 2016)		Special Private Event
			Hallmark/Major Private Event
			Regional Private Event
			Local Private Event
Event Types	Event Features and Variances	Event Examples	Event Hierarchy
Religious Events	Religious and spiritual gatherings	Hajj of Saudi Arabia	Mega Religious Events
	Religious occasions	Bishwa Ijtema/Global Congregation of Bangladesh	Special Religious Events
	Pilgrimages		Hallmark/Major Religious Events
	Religious rites (Pirnar et al., 2019)		Regional Religious Events
			Local Religious Events

(continued)

Table 23.2 (continued)

Event types	Event features and variances	Event examples	Event hierarchy
Festivals and Cultural Events	Festive in nature (Uysal & Gitelson, 1994)	The Shanghai Expo	Mega Festivals/ Giant Festivals
	Public demonstration of rejoice (Pirnar et al., 2019)	International Millennium Events of 2000	Special Festivals
	Contains cultural/artistic elements	International Ice and Snow Festival of China	Hallmark/Major Festivals.
	Fosters customs and ceremonials (Getz & Page, 2016)	China Now Music Festival.	Regional Festivals.
		Cherry Festival, Rath Yatra	Local Festivals
	Expo	The Shanghai Expo	
	Exhibition	Thai Agri Expo	
	Nautical Festivals	Tet in Vietnam	
	Culinary Festivals	Asian Food Fest	
	Gastronomy Festivals	Istanbul Gastronomy Festival	
	Music Festivals	Sunburn Festival	
	Fairs	Thaipusam Fair	
	Parades	Victory Day Parade of Bangladesh	
	Literary Festival	Hay Festival	

Source: self-compilation by the authors

Directions for Future Research

Technology has been a blessing for the development of world tourism industry. As it benefits both the customers and service providers' tourism industry, all the stakeholders of tourism industry have grasped significance of using technology for sustainable business development. Adoption of technology in tourism industry facilitates service providers to expand their market through various kinds of digital marketing and reach to the customers of untapped market (Law et al., 2014). Moreover, application of technology in tourism sector assists the customers to make a right choice about tourism destination by providing them the opportunity to search relevant information, look into reviews of different services and compare between price of various alternatives. Technological application in tourism industry mainly deals with the use of internet-based marketing communication strategies. Some common tools of technology-based marketing approach of tourism industry are explained below:

Digital Marketing

Digital marketing is the promotional strategy adopted by business organizations to reach to its existing and potential customers by the use of internet and various kinds of digital communication channels. It includes email marketing, search engine optimization, ad click, electronic billboard, mobile marketing and electronic media marketing etc. Use of digital marketing in tourism industry helps customers to remain updated about the tourism destination, get relevant information for comparing and making purchase decision (Yasmin et al., 2015).

Social Media Content Coverage

Social media marketing is the promotional strategy by using different social media platforms including Facebook, Twitter, Instagram YouTube and so on. Social media content coverage is the way of creating valuable contents on the social media regarding the products and services which will allow the potential customers to look into it and boost their interest regarding it. When customers or service providers share the content of any event on social media, it allows other people to see the events with their eyes and grow their interest to enjoy that event in future.

Digital Opinion and Public Polls

Digital Opinion and Public Poll refers to the collection of potential customers' opinion regarding any product and services through digital platforms. Usually, a short and quick survey is used in different kinds of social media platform to collect information about customers' choice, expectation and prospective buying decision.

E-Ticketing Service

E-ticketing service is the facility to purchase tickets of bus, train and airplane online without any hassle. E-ticket is a digital ticket which enables the customer to use it without printing and customers can receive it upon paying the price online. This kind of facility allows tourist to purchase tickets at their convenient time.

Digital Broadcasting

Digital Broadcasting is the medium of showing both context and tourism information from their sources to mobile devices (Lamsfus et al., 2010). It helps the customers to get all the latest information on tourism destination or related product or services at their convenience (Table 23.3).

From the above table, it can be stated that China is very much well ahead in terms of tech-based marketing in the tourism industry. This signifies that China is prioritizing various kind of technological applications in promoting event tourism and grabbing customers' attention by providing them with the relevant information about the various events. Consequently, China seizes the top position Asia in attracting more customers in the tourism business. Moreover, new innovations in technology will help Chinese tourism industry to grow further (Wu et al., 2020). India is also following the growth track of China in fostering tourists' attention to various event tourism destinations. India is a nation of diversified culture and cultural events. To attract domestic and world tourists to the diversified cultural events, India is adopting various technology-oriented marketing strategies. However, Social media content coverage can be a misleading for potential customers of Indian tourism industry as it is burdened with many irrelevant contents. Malaysia has adopted tech-based marketing strategy for event tourism development many years ago. Malaysian government in collaboration with private entrepreneurs of the country has taken many initiatives including social media content creation, digital marketing, digital broadcasting. Moreover, Malaysian tourism service providers has recently launched virtual tourism program namely Malaysia Virtual Experience for the customers so that they can enjoy the beauty of Malaysia even in the time of COVID-19 (The Star, 2020). Japan, a nation with many seasonal festivals and events, is also adopting technology in promoting tourism sector. As most of the young Japanese are travel enthusiast, more focused should be given in social media content creation and digital broadcasting as these play the role of influencer in customers' mind. Thailand is one of the most popular tourism destinations in Asia. Before the COVID-19 pandemic, growth of Thailand tourism industry was very much impressive. However, there is still untapped market from Thailand can generate revenue. High use of technology in promotion of tourism service is the only way forward. The Maldives is not heavily engaged in technology-based marketing of event tourism business as it has very few events which can attract tourist. Bangladesh has many festival events which can bring domestic and international tourists together. However, use of technology in promoting those events are very low. Nepal is also facing similar kinds of problem in adopting technology to reach to the potential customers.

Table 23.3 Summary of technological application in event tourism

Technological application	Bangladesh	India	Nepal	Thailand	China	Japan	Malaysia	Maldives
Digital Marketing	Low	High	Low	High	High	High	Moderate	Moderate
Social Media Content Coverage	Low	Moderate	Low	Moderate	High	Moderate	High	Low
Digital Opinion and Public Polls	Low	Moderate	Low	Moderate	Moderate	Moderate	Moderate	Low
E-ticketing Service	Moderate	High	Moderate	High	High	High	High	High
Digital Broadcasting	Moderate	High	Low	Moderate	High	Moderate	High	Moderate

Source: self-compilation by the authors

Conclusion

This study aims to investigate how technological application in tourism events and festivals can increase the domestic and global demand for tourism destination in Asia. To do so, pertinent segments of events tourism across different countries of Asia has been identified. Moreover, various technological application in tourism marketing has also been pinpointed. It is found that events in Asia are heterogeneous because differences in religion, culture heritage and economic growth of different countries Asia. These differences have effect on the modality and effectiveness of proper tech-based applications in tourism industry development. However, it is found that big economies of Asia, China and India, are making commendable growth in tourism industry by adopting tech-based marketing strategies. Malaysia, Japan and Thailand remain in a decent position in making technological applications for promoting event tourism in their region. Unfortunately, Bangladesh and Nepal could not make impressive use of tech-based promotion strategy in tourism industry. Moreover, it is also identified that there are discriminations in digital literacy, internet speed, sufficient promotional channels and economic stability in many parts of the Asia. These problems affect the tech-based marketing strategy in event tourism in Asia. Resolution of these discriminations can change the entire picture of event tourism growth in Asia.

Event tourism in Asia is heterogeneous due to culture, heritage, religion, economic condition, geographical location etc. Hence, no single strategy will be applicable for all the Asian countries for tourism development. However, following recommendations can be embraced in examining the role of technological applications in event tourism promotion in Asia. Use of technology in event tourism varies among Asian countries based on economic condition. Big economies can use technology in an extensive way which results in more growth in their tourism industry. Research can be undertaken for examining the effectiveness technology adoption in tourism industries of big economies of Asia. Small and growing economies have limited resources. How they are performing by adopting tech-based marketing strategy in tourism can be researched as well. Digital literacy is not same in all the Asian Countries. Future researchers can look into the relationship of effectiveness of technological applications of tourism promotion and digital literacy of the customers. All the events may not require similar kind of attention and promotion. For example, religious events may not require more technology-based promotion but sport events or cultural events may require more use of technology-based promotions. Future studies can be undertaken on the significance of technology-based marketing strategy and types of event tourism. To communicate various tourism promotional activities with a larger group people, mobile applications can be used very frequently as almost 66% population of Asia have subscribed to mobile phone services (GSMA, n.d.). Moreover, there has been global surge in the use of mobile applications-based promotion in the tourism industry (Dorcic et al., 2019). Researchers and policy makers can examine the role mobile technology and applications on promoting festival tourism in Asia.

References

Chinadaily.com.cn. (2020). *Technology leads new future in tourism*. Retrieved from: http://www.chinadaily.com.cn/a/202012/09/WS5fd0826ba31024ad0ba9ad8b.html. Accessed: 9 Aug 2021.

Development Asia. (2019). *Harnessing technology in promoting safe tourism destinations in Central Asia*. Retrieved from: https://development.asia/policy-brief/harnessing-technology-promoting-safe-tourism-destinations-central-asia. Accessed: 9 Aug 2021.

Dorcic, J., Komsic, J., & Markovic, S. (2019). Mobile technologies and applications towards smart tourism – State of the art. *Tourism Review, 74*(1), 82–103.

Dychkovskyy, S., & Ivanov, S. (2020). Festival tourism as part of international tourism and a factor in the development of cultural tourism. *Informacijos mokslai, 89*, 73–82.

Getz, D., & Page, S. (2016). Progress and prospects for event tourism research. *Tourism Management, 52*, 593–631.

GSMA. (n.d.). The mobile economy Asia Pacific 2020. Retrieved from: https://bit.ly/2W5Kycw. Accessed: 9 Aug 2021.

Indochina Travel. (n.d.). *Culinary tours in Asia*. Retrieved from: https://www.indochinatravel.com/tours/culinary-tours-asia.html?fbclid=IwAR0OSMJS-xStPA1qVFs72Gck3zYSQwneReYwLXXmfJpo_1NAitU4-uiZKvM. Accessed: 9 Aug 2021.

Lamsfus, C., Alzua-Sorzabal, A., & Martín, D. (2010). Digital broadcasting for context-aware services in tourism. In *2010 Second Region 8 IEEE Conference on the History of Communications*. IEEE. The 3rd–5th November 2010, pp. 1–10.

Laurell, C., & Björner, E. (2018). Digital festival engagement: On the interplay between festivals, place brands, and social media. *Event Management, 22*(4), 527–540.

Law, R., Buhalis, D., & Cobanoglu, C. (2014). Progress on information and communication technologies in hospitality and tourism. *International Journal of Contemporary Hospitality Management, 26*(5), 727–750.

Li, J., Dai, G., Tang, J., & Chen, Y. (2020). Conceptualizing festival attractiveness and its impact on festival hosting destination loyalty: A mixed method approach. *Sustainability, 12*(8), 3082.

New Zealand Major Events. (n.d.). *Types of events*. Retrieved from: https://www.majorevents.govt.nz/event-security/types-of-events/?fbclid=IwAR25WoGu-nc4R6bMA9YNFUESCfwWzIrO6cOWUGXpBSNmBoNdcPKReSXYx7c. Accessed: 8 Aug 2021.

Pirnar, I., Kurtural, S., & Tutuncuoglu, M. (2019). Festivals and destination marketing: An application from Izmir City. *Ssoar.info*. Retrieved from: https://www.ssoar.info/ssoar/handle/document/66359. Accessed: 7 Aug 2021.

Quinn, B. (2009). Festivals, events, and tourism. In M. Robinson & T. Jamal (Eds.), *The SAGE handbook of tourism studies* (pp. 483–503). Sage.

Tan, CK., Hinata, Y., & Liu, C. (2020). China's largest trade fair goes online as coronavirus shuts doors. *Nikkei Asia*. Retrieved from: https://asia.nikkei.com/Business/China-tech/China-s-largest-trade-fair-goes-online-as-coronavirus-shuts-doors. Accessed: 9 Aug 2021.

Thaiger. (2019). *Asia leads the way in travel technology over next decade*. Retrieved from: https://thethaiger.com/hot-news/tourism/asia-leads-the-way-in-travel-technology-over-next-decade. Accessed: 9 Aug 2021.

The ASEAN Post. (2020). *How technology is changing tourism*. Retrieved from: https://theaseanpost.com/article/how-technology-changing-tourism. Accessed: 9 Aug 2021.

The Star. (2020). *Explore Malaysia virtually through MyVXp: Digital tourism unlimited for as low as RM15!* Retrieved from: https://www.thestar.com.my/news/nation/2020/11/30/explore-malaysia-virtually-through-myvxp-digital-tourism-unlimited-for-as-low-as-rm15. Accessed: 9 Aug 2021.

UKEssays. (2018). *Similarities and differences between tourism and event management tourism essay*. Retrieved from: https://www.ukessays.com/essays/tourism/similarities-and-differences-between-tourism-and-event-management-tourism-essay.php?vref=1. Accessed: 7 July 2021.

United Nations Population Fund (UNFPA). (2021). *World population dashboard*. Retrieved from: https://www.unfpa.org/data/world-population-dashboard. Accessed: 13 July 2021.

United Nations World Tourism Organization (UNWTO) and Global Tourism Economy Research Centre. (2019). *UNWTO/ GTERC Asia tourism trends: Executive summary 2019 edition.* Retrieved from: https://bit.ly/3fSdbRl. Accessed: 7 July 2021.

United Nations World Tourism Organization (UNWTO). (n.d.). *Sustainable tourism development.* Retrieved from: https://www.unwto.org/sustainable-development. Accessed: 7 July 2021.

Uysal, M., & Gitelson, R. (1994). Assessment of economic impacts: Festivals and special events. *Festival Management and Event Tourism, 2*(1), 3–9.

Westcott, M. Bird, G., & Briscoe, P. (2015). *Introduction to tourism and hospitality in BC.* Retrieved from: https://ecampusontario.pressbooks.pub/introtourismbc/. Accessed: 7 July 2021.

World Travel & Tourism Council (WTTC). (2021). *Travel and tourism economic impact 2021: Global economic impact and trends 2021* (p. 12). Retrieved from: https://wttc.org/Portals/0/Documents/Reports/2021/Global%20Economic%20Impact%20and%20Trends%202021.pdf?ver=2021-07-01-114957-177. Accessed: 6 July 2021.

Wu, L., Fan, A., & Shen, H. (2020). Embracing the future: New technology and mediated Chinese tourists. *Journal of China Tourism Research, 16*(4), 487–493.

Yasmin, A., Tasneem, S., & Fatem, K. (2015). Effectiveness of digital marketing in the challenging age: An empirical study. *International Journal of Management Science and Business Administration, 1*(5), 69–80.

Yeo, S. (2019). Tech in Asia - Connecting Asia's startup ecosystem. *Techinasia.com.* Retrieved from: https://www.techinasia.com/tech-southeast-asia-surging-tourist-arrivals. Accessed: 9 Aug 2021.

Ratan Ghosh has attended in Master of Business Administration in Accounting and Information Systems from University of Dhaka in 2016. Currently, Mr. Ghosh is serving as an Assistant Professor of Accounting at Bangladesh University of Professionals (BUP). Before joining at BUP, He has worked as a Lecturer in Daffodil International University, Bangladesh. Mr. Ghosh has also worked as a Research Assistant in Centre of Accounting for Capital Market Development (A project of World Bank and University Grants Commission, Bangladesh). His current research interest includes corporate governance, financial reporting, earnings quality and sustainability reporting. Ghosh is also pursuing professional accounting degree, Cost and Management Accountancy (CMA), under The Institute of Cost and Management Accountants of Bangladesh (ICMAB).

Iffat Anjum has completed her post-graduation from the Department of International Relations, University of Dhaka, Bangladesh in 2016. At present, she has been entrusted to work as an Assistant Professor in the Department of International Relations, Bangladesh University of Professionals. She fosters a key research interest in international security, gender and international relations, global environmental politics and international political economy. She has also worked with international research entities like Goethe Institute Bangladesh and Transparency International Bangladesh and has converged around publishing her research compositions on security, gender and public policies.

About the Authors

Dr Munshi Muhammad Abdul Kader Jilani has been serving as an Assistant Professor at Bangladesh Institute of Governance and Management (BIGM). He completed his Doctor of Philosophy (PhD) in Human Resource Management from the Wuhan University of Technology, China. He also completed an MBA from IISWBM, the University of Calcutta on Human Resource Management. Dr Jilani also completed his BBA and another MBA in International Management from the University of Chittagong, Bangladesh. His areas of research interest are psychological capital, knowledge management, employee resilience, and organizational behavior. He published a good number of articles in peer reviewed journals and conference proceedings, such as the Sustainability Journal, Management Research Review Journal, Journal of Innovation and Sustainability, among others.

Musanna Ahmed is a Research Assistant at IUBAT-International University of Business Agriculture and Technology, Dhaka, Bangladesh. His research interest is in Tourism and Hospitality Science and Word of Mouth marketing.

Faiza Tanaz Ahsan is a Research Assistant at North South University. Her research interests are: Women Empowerment, Edge-cutting technology in business, Artificial Intelligence, bank lending practices, credit management, cloud computing, Budget taxation and public investment, trade and globalization and so on.

Fahmida Akhter is an Assistant Professor of School of Business Administration, East Delta University, Bangladesh. Her area of concentration is Accounting & Information Systems. She started her career as an academic in 2015. Her fundamental research interest lies in the area of Corporate Governance, Corporate Social Responsibility Practices of Firms, Sustainable Environmental Practices of Corporations, Human Resource Disclosure Practices, and other voluntary disclosure practices of Companies and Sustainable Tourism. She's published articles in various national and international peer-review academic journals.

A. Hassan (ed.), *Technology Application in Tourism Fairs, Festivals and Events in Asia*, https://doi.org/10.1007/978-981-16-8070-0

Shaharin Akter is currently pursuing her MBA at the Department of Banking and Insurance under Faculty of Business Administration in University of Chittagong, Bangladesh. She is highly passionate about research with a strong desire to learn more. Her current research interests include Determinants of Profitability, Banking and Insurance Sector, Risk and Efficiency Measurement, Sustainability, and Blue Economy.

Mehtab Alam is an MPhil scholar from Riphah Institute of Public Policy, Riphah International University Islamabad, Pakistan. As a core area of Public Policy; Soft Image and Image Development of Pakistan were the primary part of research completed for degree of Master of Philosophy (MPhil). He obtained degree in Masters of Arts in English Language & Literature and also holds Bachelor of Education from Allama Iqbal Open University, Islamabad. In addition to this, he serves in various Public and Private Institutions those include the National Assembly, Parliament of Pakistan as Research Associate and Ministry of Foreign Affairs on assignments/ tasks of image development. As a writing specialist he also served in Skyscrapers Pvt. Ltd. Over the years, his interest in research is developed and allows him to keep in touch with the advancement in research on various topics and subjects. These incorporated the field of Social, Communication, & Management Sciences, and issues of international concerns like Public Diplomacy, Foreign Policy and Climate Change. During the time, Alam interacted with highly qualified and experienced individuals from various backgrounds. Alam has several articles/chapters publications.

Md Rabiul Alam is a PhD candidate at the Faculty of Education, University of Malaya, Malaysia. Prior to continue his higher study at the University of Malaya, he had been serving as a faculty in the Department of English, Asian University of Bangladesh for nearly 8 years. Besides his study at the University of Malaya, he has been working as a Graduate Research Assistant (GRA) since 2018. His research interest includes TESOL, teacher education, financial education, tourism development, digital financial inclusion and so forth. Alam has published a few of his works in reputed peer-reviewed journals.

Arslan Amzad is a Graduate Student at Xi'an Jiaotong University, China. His research interests are: tourism management, technology management in organization, job engagement in tourism and so on.

Sadia Afrin Ananya is an Assistant Professor at the Department of Tourism and Hospitality Management, Bangabandhu Sheikh Mujibur Rahman Science and Technology University (BSMRSTU), Gopalganj, Bangladesh. Ananya earned her Bachelor and Master degrees from the University of Dhaka. As the recognition of her academic feat, she earned Achieved Dean's Honor and Dean's Merit Award 2017 from FBS, DU, and Scholarship from the University of Dhaka. She has research interests in tourism, travel, hospitality, business, marketing, and statistical analysis. She has contributed to a good number of publications that are on the way to be published.

Iffat Anjum has completed her post-graduation from the Department of International Relations, University of Dhaka, Bangladesh in 2016. At present, she has been entrusted to work as an Assistant Professor in the Department of International Relations, Bangladesh University of Professionals. She fosters a key research interest in international security, gender and international relations, global environmental politics and international political economy. She has also worked with international research entities like Goethe Institute Bangladesh and Transparency International Bangladesh and has converged around publishing her research compositions on security, gender and public policies.

Dr R. S. S. W. Arachchi graduated from the Sabaragamuwa University of Sri Lanka before receiving his Master degree from Colombo University, Sri Lanka and a PhD degree from Management and Science University in Malaysia. He is currently working as the Head of, Department of Tourism Management and a Senior Lecturer of Faculty of Management Studies, Sabaragamuwa University of Sri Lanka, Belihuloya. His primary teaching and research areas are eco-tourism, community-based tourism, sustainable tourism development, responsible tourism, environment and tourism resources and homestay tourism in Asian Countries. He has published nearly 20 articles in well-recognised journals both in Sri Lanka and overseas. Dr Arachchi is an editorial board member of the Journal of Tourism and Hospitality Management, International Journal of Education Humanities and Social Science, and Global Association for Humanities and Social Science Research.

Dr Ceren Avci completed her undergraduate education at Balıkesir University School of Tourism and Hotel Management in 2008. She completed her Master's thesis and doctorate dissertetation on Tourism Management at Gazi University, Faculty of Tourism. The subject of her Master's thesis is occupational health and safety and the subject of the doctoral dissertation is on spatial relations in urban tourism. Avcı worked as a research assistant at Gazi University, Faculty of Tourism, Department of Tourism Management between 2012 and 2017. She has been working at Şırnak University since 2017. Her academic interests are tourism philosophy, tourism psychology and tourism sociology.

W. M. A. H. Bandara is a graduate of Tourism and Hospitality Management (BSc) from the Rajarata University of Sri Lanka. He is an academic member of the Department of Tourism Studies of the Uva Wellassa University of Sri Lanka. In addition, He represents several projects undergoing in the university. His research interests are Hospitality Management and Tourist Behaviors.

Assistant Professor Dr Aslı Özge Özgen Çiğdemli graduated from Ankara University, Faculty of Political Sciences, Department of International Relations in 2006. She completed a master's degree in International Relations in 2010 at

Süleyman Demirel University, with a thesis entitled "European Union Immigration Policies: Historical Background and Current Situation". She took her PhD degree in 2016 from the Hospitality and Tourism Management Department of the Akdeniz University; her dissertation was "The Comparative Analysis of Mediterranean Cruise Destinations: The Position of Turkey". Dr. Çiğdemli worked at the Republic of Turkey Ministry of Trade between 2009 and 2017 and became a faculty member at the School of Tourism and Hotel Management of Şırnak University in 2017, before moving to Osmaniye Korkut Ata University in 2020 as an Assistant Professor in the Faculty of Kadirli Applied Science, Department of Recreation Management. Her main areas of interest are tourism marketing, tourism policy, cruise tourism, technology and tourism, consumer behavior in tourism and strategic marketing management.

Ratan Ghosh has attended in Master of Business Administration in Accounting and Information Systems from University of Dhaka in 2016. Currently, Mr. Ghosh is serving as an Assistant Professor of Accounting at Bangladesh University of Professionals (BUP). Before joining at BUP, He has worked as a Lecturer in Daffodil International University, Bangladesh. Mr. Ghosh has also worked as a Research Assistant in Centre of Accounting for Capital Market Development (A project of World Bank and University Grants Commission, Bangladesh). His current research interest includes corporate governance, financial reporting, earnings quality and sustainability reporting. Ghosh is also pursuing professional accounting degree, Cost and Management Accountancy (CMA), under The Institute of Cost and Management Accountants of Bangladesh (ICMAB).

Dr Sachin Gupta has a brilliant academic Record. He has earned his Doctorate degree (PhD) from Department of Business Administration, University of Rajasthan, Jaipur. He completed his Dissertation Research during his MPhil degree on the topic "NGOs Role in developing economic condition of Rajasthan State", from Department of Business Administration, University of Rajasthan, Jaipur. He completed his Post Graduate Degrees specialties- One in Business Administration (Masters of Commerce) from Department of Business Administration, University of Rajasthan, Jaipur and another in Finance (MBA-Finance) from Rajasthan Technical University, Kota. Dr Gupta cracked UGC NET Exam in Management twice. He obtained his B.Com degree from University of Rajasthan, Jaipur. Dr Gupta completed Executive Development Program in Advanced Financial Management from XLRI-Jamshedpur, did Post Graduate Certificate Program in Finance from IIM Indore, completed UGC Sponsored Diploma Program in IFRS from Ramanujan College, University of Delhi in association with IAA (All India Accounting Association), NCR Chapter.

Dr Azizul Hassan is a member of the Tourism Consultants Network of the UK Tourism Society. Dr Hassan has been working for the tourism industry as a consultant, academic, and researcher for over 20 years. His research interest areas are technology-supported marketing for tourism and hospitality, immersive technology applications in the tourism and hospitality industry, and technology-influenced marketing suggestions for sustainable tourism and hospitality industry in developing countries. Dr Hassan has authored over 150 articles and book chapters in leading tourism outlets. He is also part of the editorial team of 25 book projects from Routledge, Springer, CAB International, and Emerald Group Publishing Limited. He is a regular reviewer of a number of international journals.

H. M. Kamrul Hassan is currently serving as Assistant Professor in the Department of Marketing, University of Chittagong. His principal interest in teaching is developing, using, and disseminating comprehensive, affordable teaching and business education platform to be maximized both by faculty members and students. He studied and lectured in sustainable marketing, entrepreneurship and tourism. He has published articles in national and international journals. Moreover, he attended many local and international conferences in Bangladesh, India, Malaysia, Thailand, Indonesia, Czech Republic and Germany. In addition, he regularly writes on concurrent business issues at the academic and non-academic journal and newspapers.

Mohammad Rokibul Hossain is currently working as an Assistant Professor of Marketing, Faculty of Business Studies at Premier University, Bangladesh. His area of concentration in Marketing. Hossain's fundamental research interests lie in Tourism Marketing, Customer Engagement, Consumer Behavior, Sustainable Marketing, and Neuromarketing. He published articles on Tourism Marketing, Customer Relationship Management and Agricultural Marketing in different national and international peer-reviewed journals

Mohammed Mainul Hossain is currently pursuing his Master of Philosophy (MPhil) at the Department of Media and Communications Studies under the Faculty of Arts and Social Sciences in University of Malaya, Malaysia. He is a dynamic and highly motivated student who has a strong passion, especially in the research area. His area of research interest is social networking applications, journalism, digital communications, information systems and information technology.

Syed Far Abid Hossain is a PhD researcher at the School of Management of Xi'an Jiaotong University, China. He graduated from the University of Sunderland, London, UK with an MBA degree. He is a Senior Lecturer of College of Business Administration, International University of Business Agriculture and Technology (IUBAT), Dhaka, Bangladesh. He has published numerous papers in indexed and reputed international journals. Also, he has achieved the 'Outstanding International Student Award', Dean's scholarship and Full bright scholarship from the Chinese government. His research interests include women entrepreneurship, innovation, mobile-phone usage, academic performance, social networking, and TPACK. He has participated numerous reputed international conferences in the USA, China, and Bangladesh. He is appointed as a reviewer by famous publishers as Elsevier, Emerald, Wiley, SAGE, Springer, Frontiers, Taylor and Francis and IGI Global.

Shazia Irshad is a Graduate Student at Xi'an Jiaotong University, China. Her research interests are: innovation, smart tourism, machine learning, AI in tourism and so on.

Synthia Islam is now working as an Adjunct Lecturer at the Department of Tourism and Hospitality Management, Bangabandhu Sheikh Mujibur Rahman Science and Technology University, Gopalganj, Bangladesh. She has received her BBA and MBA degrees from Business Administration Discipline of Khulna University with major in Marketing. She has achieved Merit Order Scholarships from Business Administration Discipline of Khulna University. Her main areas of research interest are Sustainable Tourism Marketing, Service Marketing, Entrepreneurship Development, Relationship Marketing, Strategic Marketing, Promotion and Positioning Strategies and Human Resource Management Policies. She is involved in several qualitative and quantitative research projects and has articles in both national and international journals.

Priyanka Jingar is pursuing Doctoral Degree at Department of Business Administration, Mohanlal Sukhadia University, Udaipur Rajasthan. She completed her MPhil from Mohanlal Sukhadia University, Udaipur. She has done her Post Graduation and graduation in Commerce from Maharshi Dayanand Saraswati University, Ajmer. She Qualified UGC- NET in Commerce. Her professional interest lies in the area of research. Her accolade includes Three Patent Published by Intellectual Property India, Ministry of Commerce & Industry, Government of India and a Book published by Lap Lambert Academic Publishing, Germany. She has several Research papers published with reputed journals and Chapters published with Taylor & Francis and Apple Academic Press. She has presented Research papers at National as well as International Conferences.

Dr Fatema Johara has been working as an Associate Professor at Bangladesh Army International University of Science & Technology. She did her PhD from Universiti Sains Malaysia, Malaysia. Dr Johara also did her Master of Philosophy in Management and MBA in Accounting from the University of Chittagong, Bangladesh. Her research interests lie in talent management, employee engagement, management accounting, strategic management, organizational behavior, and internal auditing. She published a good number of publications in peer reviewed journals and conference proceedings, including in the International Journal of Sustainable Strategic Management, Journal of Sustainability Science & Management, Global Business and Management Research, etc. She has been acting as a regular reviewer and editor for many international conferences and Journals.

Kaniz Kakon is an Assistant Professor, Department of Philosophy, College of Arts and Sciences, IUBAT-International University of Business Agriculture and Technology, Dhaka, Bangladesh. After receiving her Bachelor of Arts and Master of Arts degrees in Philosophy from Jahangirnagar University, Savar, Dhaka, she started her career as university faculty. As in her philosophy classes, she gets students from multi-disciplinary departments, her research interests further developed in other studies similarly. She has published research articles and conference proceedings on different issues. She became interested in the zone of Tourism to concentrate on Philosophical Issues in Tourism.

Dr B. Kalsoom is Assistant Professor at Centre for Policy Studies, COMSATS University Pakistan. Key Competent Areas in Research are evidence based opportunity mapping & Spatial distribution of Public facilities, inequality, explicit planning of public facilities at regional and local level – Big data, Urban Sustainability, Foresight & Public Policy – Public Sector Governance, Public Management & Reforms – Governance, inter-provincial & Local Government – Public Services Ethics, Social Equity, Inequality – Crowdsourcing, Social Innovation in Public Sector – behavioral insights in Public Sector, Outsourcing, Collaborative Management. She has more than 15 years of extensive experience in teaching and research. Her recent research findings appear in various International, National Journals and conference proceedings on issues of public sector reforms, public Policy, governance, sustainable Management and foreign affairs of China in higher education.

Bapon Chandra Kuri is an Assistant Professor at the Department of Tourism and Hospitality Management, Bangabandhu Sheikh Mujibur Rahman Science and Technology University (BSMRSTU), Gopalganj, Bangladesh. He received BBA and MBA degrees from the Department of Tourism and Hospitality Management, University of Dhaka. He has a good number of articles published in both national and international journals. His areas of research interest are CBT, Rebranding, and identifying employee satisfaction levels. Before joining the university, he was working in the Service Department of Radisson Blue Water Garden Hotel and in the Housekeeping Department of Hotel the Cox Today, Cox's Bazar, Bangladesh.

Hasan Mahmud is a Doctoral student at LUT School of Engineering Science, LUT University, Finland. He has 12 years of industry experience in information technology, banking, and FMCG. His research interests are in information technology and business management.

Dr Sana Maidullah received a PhD degree in Humanities and Social Science in 2019 from the Indian Institute of Technology, Jodhpur, India. She subsequently joined Lovely Professional University, Punjab, India, where she worked as an Assistant Professor. At present, she is an Assistant Professor in Tourism and Hospitality Management, Indian Institute of Management Sirmaur (IIM Sirmaur), District Sirmaur, Himachal Pradesh, India. Dr Maidullah's principal research interests are in tourism, decision making, consumer behavior, and motivation. She is also a multi-disciplinary design practitioner with experience in tourism products, communication, and environmental design.

Kiran Meena is a PhD Scholar at the Banking and Business Economic Department, Mohanlal Sukhadiya University, Udaipur. She is doing a B.Com degree in commerce from the Rajasthan University, Jaipur in 2016. She is doing an M.com degree in the Department of Commerce and Finance at the Central University of Rajasthan Bandar Sindri, Ajmer in 2018. Her research interest includes the area of Artificial Intelligence in Banking, Fintech management, the impact of technology on consumers.

Md Rashed Khan Milon has been serving as an Assistant Professor and the Chairman at the Department of English, Port City International University, Bangladesh. He mostly teaches Applied Linguistics courses, English literature and also supervises thesis students both at the undergraduate and graduate level of study. Besides, Milon is a PhD candidate at the Department of English, Jagannath University, Bangladesh. He has published a few of his papers in reputed peer-reviewed journals. Milon's research areas are Teacher Education, Second Language Acquisition, ELT Methodology, Teaching Techniques, Gender issue and Diversity in Education.

Dr Noraihan Mohamad is a Senior Lecturer at School of Hospitality and Creative Arts, Management and Science University (MSU). She obtained Bachelor of Tourism Management (Marketing) (Hons), Master of Science in Media Management and PhD degree in Media Management from Universiti Utara Malaysia (UUM). She taught undergraduate courses on Research Methods for Service Industry, Current Issues in Hospitality and Tourism, Strategic Management for Service Industry. Apart of that, she also taught postgraduate courses on Global Trends and Issues in Hospitality and Tourism, Tourism Destination Analysis and Organizational Behavior. Her research interests are on the influence and impact of media on society, businesses, and consumers. Her scope of expertise is Malaysian mass media and new media, media and technology (including social networking), as well as tourism management. Besides presented numerous papers at national and international conference, she also contributed articles in journals and chapters in books on media. Her current contribution to Malaysia's Government is on developing a Malaysian National Unity Index.

Kazi Mohiuddin is a PhD Researcher at Shanghai Maritime University, China. He has published numerous papers in well-indexed reputed international journals. His research interests are: tourism management, maritime management, Artificial Intelligence in tourism and so on.

Md. Ruhul Amin Mollah is an Assistant Professor at Jagannath University, Bangladesh. He obtained Master's degree in Tourism Management from Hebei University at China and MBA & BBA from the University of Chittagong at Bangladesh. He is currently a PhD Candidate in the Department of Tourism, Sport and Hotel Management at Griffith University, Australia. He is also a Teaching and Research Assistant at Griffith University. His research interest includes tourism technology, sport tourism, religious tourism, contemporary management issues and sustainable tourism. He has presented papers in peer-reviewed conferences such as The Council for Australasian Tourism and Hospitality Education (CAUTHE) and The Greater Western Chapter of the Travel & Tourism Research Association (GWTTRA) and also published 10 peer-reviewed articles in reputed journals. His publications appeared in the Journal of Sport & Tourism, Man in India, ABAC journal and several others.

Dr Rakiba Nabi is Professor in the Department of Law at the University of Chittagong, Chattogram, Bangladesh. Dr Nabi graduated with an LLB (Honors), and LLM from the University of Chittagong in 2004, did a second LL.M in Human Rights from the University of Hong Kong in 2012-13, and completed her SJD in Indigenous Law from the University of Kansas, the United States in 2020. Her present areas of interest are Human Rights, Refugee Law, Indigenous Law and Policy, Law of International Organization, Tourism and Sustainable Development. She published 10 scholarly articles in national and international journals. She also works in several projects funded by Research Cell, CU, and Law Ministry, Bangladesh.

Armana Hakim Nadi is a Research Assistant at IUBAT-International University of Business Agriculture and Technology, Dhaka, Bangladesh. She enjoys writing and relishes to play with words. As she pursues her career in teaching, it helped her to engage with the topic involving education and have a superior perspective over it.

A. M. D. B. Nawarathna is a graduate in Hospitality, Tourism and Events Management (BBM) from the Uva Wellassa University of Sri Lanka and currently reading for an MBA in Tourism from the Sabaragamuwa University of Sri Lanka. He is a full-time faculty member of the Department of Tourism Studies. He also is a member of different committees and projects in the university and has coordinated several events organized in the university. His research interests are in Event Management, MICE tourism, Travel Agency and Tour Operations Management, Tourism Planning and Management, Destination Management and Marketing etc.

Hafsa Neamah is a Research Assistant at IUBAT-International University of Business Agriculture and Technology, Dhaka, Bangladesh. Her research interests are tourism and economic development.

Dr Sharina Osman is a Senior Lecturer and the Head of Tourism Section at Universiti Kuala Lumpur Business School. She earned her PhD degree in Management from the University of Exeter, United Kingdom; Master of Human Resource Management and BBA (Marketing) Honors from Universiti Utara Malaysia (UUM), Malaysia. Her research interest is around Identity, Image and Reputation, Organizational Culture, Gastronomic Tourism, Tourism Destination Identity & Branding, Participative Tourism Development and e-Tourism. Dr Osman carries with her an array of experience which ranges of Human Resource to Marketing (Image & Branding) and Tourism. She is also the Editorial Advisory

Board for several journals and was also invited as Keynote or Guest Speakers in seminars. Being in an academic and in the research environment, coupled with vast hands-on experience in the industry, she helps organizations improve the competency and capacity building through trainings and consultations. She believes in interacting and experiential learning and has conducted many workshops to different sectors, including government and private at both local and international.

Putu Diah Sastri Pitanatri is a Lecturer in the Hospitality Department, Bali Tourism Polytechnic, Indonesia. She is currently a PhD candidate of Tourism Studies at Gadjah Mada University. She is active in various national and international associations. She edited the book "Homestay: A Mosaics of Community Based Tourism" with I Wayan Mertha in 2018, and "Culinary Tourism: A New Attributes of Ubud" with Nyoman Darma Putra in 2016. She also wrote a book chapter titled "Big Data Research for Tourism Marketing in Indonesia" in 2021. She has received 1st place in Win Way Award (2018) as the best Lecturer under the Ministry of Tourism Republic of Indonesia and best paper presenter in OBOROT-Palembang 1st International Conference in 2018. Till date, she is also an active reviewer in national journals and speaker in many international conferences. Her research interests ar on big data, tourist studies and sustainable tourism management.

Mohammed Shahedul Quader studied and lectured International Business, Marketing, and Shipping Management, working as Associate Professor in the Department of Marketing, University of Chittagong. He has completed his MBA in Shipping Management, MA in Marketing Management from University of Leicester and Middlesex University, UK respectively. He has more than 15 years of teaching experience in the field of international business. Besides teaching, he has published more than 30 articles in reputed local and international journals and joined international conferences local and abroad.

Dr Muhammad Khalilur Rahman currently works in the Faculty of Entrepreneurship and Business at Universiti Malaysia Kelantan (UMK), Malaysia. He has been actively involved in research activities and published over 55 articles and book chapters in leading service management outlets. He has a wide interest in tourism and service management research which includes medical tourism, eco-tourism, halal tourism, service quality, brand equity, supply chain management, operation management, green and sustainable development. Rahman is a reviewer of Kybernetes, Journal of Hospitality and Tourism Insights, International Journal of Contemporary Hospitality Management, Journal of Islamic Marketing, and BMC Public Health.

Md. Sohel Rana is a PhD researcher in the Faculty of Business and Accountancy, University of Malaya, Malaysia. He holds a bachelor's degree in business administration and an MBA in Management Studies from the University of Rajshahi, Bangladesh. His area of expertise in research includes Microfinance & Financial Inclusion, Islamic Microfinance, Poverty and Development Economics, Productivity

& Efficiency, Tourism, Halal Tourism, Insurance, Environmental Economics, etc. He has already published a number of book chapters, journal articles in the Web of Science and Scopus indexed journals and several editorial opinions in newspaper. He has also reviewed several journal articles. He has the experience of working as a full time Researcher at the University of Malaya, Malaysia. He also holds six years of industrial experience in financial institutions and export-oriented companies in Bangladesh.

Dr A H M Salimullah is Professor in the Department of Economics at the University of Chittagong, Chattogram, Bangladesh. Dr Salimullah graduated with a BSS (Honors) and MSS degree in Economics from the University of Chittagong in 2003, did a second MA degree in Financial Economics from the Middle Tennessee State University (MTSU) in 2014, and completed his PhD in Econometrics and Quantitative Economics from the same university in 2018. His present areas of interest are Labor Economics, Dynamic Macroeconomics, and Economics of Tourism. He published 10 scholarly articles in national and international journals. Besides, he presented several papers at different national and international conferences. The notable conferences that he attended and presented his research were the Academy of Economics and Finance (AEF) Conference in 2018 at Houston, Texas, and Dr Gregory P. Domin Graduate Conference in 2017 at Columbus State University, Columbus, Georgia. He worked as a reviewer for a journal named *Scientia et Humanitas*, sponsored by the University Honors College at MTSU. He is a proud winner of the Deans' Distinguished Essay Award for his scholarly publication in the Honors College journal of MTSU. Besides, he won important prizes during scholars' week university-wide poster presentation at MTSU. He is chosen as a proud member of the MTSU Chapter of Beta Gamma Sigma due to his scholarly achievement.

Saroop Roy B. R. BTech, MTA is currently Assistant Professor in Travel and Tourism as well Co-ordinator of Centre for Responsible Tourism at Kerala Institute of Tourism and Travel Studies (KITTS). Before that, he had worked with Thenmala Ecotourism Promotion Society and was also the Kerala Co-ordinator of EQUATIONS, a Bangalore based research, campaign and advocacy organization on tourism policies and issues for 8 years. His expertise areas include tourism policy formulation, ecotourism, responsible tourism, rural tourism and community linked tourism projects. He participated in the SBSTTA 13 Meeting of the Convention on Biological Diversity held at Rome in February 2008. He also led the Indian delegation of Ecotourism practitioners to attend the South African and Indian Practitioners' Workshop to exchange lessons and best practices on using Nature Tourism as a force for Poverty Alleviation at Johannesburg in May 2008. He has played a significant role in the formulation and implementation of Responsible Tourism Initiative in the state of Kerala. On the policy front, he was the drafting committee member of the Kerala Tourism Policy 2012, Revamping of Homestay Classification Scheme for Kerala 2015 and preparation of Charter for Green Carpet initiative for Kerala Tourism 2016. He co-authored the book "An Introduction to business of tourism" published by SAGE India in May 2017 bringing out various dimensions of tourism with an academic perspective.

Emmanuel Sebata is currently a PhD Candidate at the Sport Business School, Beijing Sport University Beijing, China. He is a Teaching and Research Graduate Fellow in the Department of Sport science, Faculty of Science, Kyambogo University, Kampala, Uganda. He was also selected as an IOC Young Leader 2021-2024. His areas of research interest are: sport tourism, sport economy, Olympics, leisure, event heritage, festivals, and sport and tourism development in developing economies. He completed Master's Degree in Tourism Management biased in sport tourism at Hebei University, China and a Bachelor of Science in Sport and Leisure Management at Kyambogo University, Uganda.

Dr Md. Aftab Uddin has been teaching as an Associate Professor in the Department of Human Resource Management, University of Chittagong, Bangladesh 4331. He completed his PhD and MBA majoring in Human Resource Management from the Wuhan University of Technology in China. Previously, Dr Uddin also completed his first MBA (International Management) and BBA (Management) from the University of Chittagong, Bangladesh. Dr Uddin jointly published a book entitled "HRM Cases". He also conducts research on industrial psychology, cognitive psychology, environmental psychology, AI adoption in workplace, which were published in peer-reviewed journals and conference proceedings at home and abroad with different indexations (SSCI/SCIE/CPCI/ESCI/Scopus/ABDC/Scimago/ABI/ABS, etc.).

Luh Yusni Wiarti is a Lecturer and the Head of Tourism Department, Bali Tourism Polytechnic, Indonesia. She is currently a PhD Candidate of Tourism Studies at Udayana University. Previously, she obtained her Masters in Double Degree Program at Udayana University (Tourism Studies) and University of Angers (Tourism and Hospitality Management), France. She actively participated in various national and international conferences such as ITSA Conference in Greenwich University London, INTACT Conference in Gajahmada University Indonesia, ISOT and IHTC Conference in UPI Bandung Indonesia, the ASEAN Tourism Research Conference at James Cooks University in Singapore, the International Conference on Tourism and Business (ICTB) in Lucerne University of Applied Science and Art in Switzerland, etc. She won the best paper awarded at INTACT conference (2016). She has written book chapters titled "Homestay's opportunity and challenge: A study case of local participation in KEK Mandalika Lombok, Indonesia" in 2016, and "Gastronomy in Ubud: Opportunity and challenge of the legendary cultural destination model in Bali" in 2021. Her current research focuses on events and festivals, as well as topics in destination management.

The manufacturer's authorised representative in the EU is Springer Nature Customer Service Centre GmbH, Europaplatz 3, 69115 Heidelberg, Germany. If you have any concerns regarding our products, please contact ProductSafety@springernature.com

Printed and bound by CPI Group (UK) Ltd, Croydon, CR0 4YY

30/09/2026

02225233-0011